World Soils Book Series

Series editor

Prof. Alfred E. Hartemink
Department of Soil Science, FD Hole Soils Laboratory
University of Wisconsin–Madison
Madison,
USA

Aims and scope

The *World Soils Book Series* brings together soil information and soil knowledge of a particular country in a concise and reader-friendly way. The books include sections on soil research history, geomorphology, major soil types, soil maps, soil properties, soil classification, soil fertility, land use and vegetation, soil management, and soils and humans.

International Union of Soil Sciences

More information about this series at http://www.springer.com/series/8915

Rachel Creamer · Lilian O'Sullivan
Editors

The Soils of Ireland

Springer

Editors
Rachel Creamer
Wageningen University and Research
Atlas—Wageningen Campus
Wageningen
The Netherlands

Lilian O'Sullivan
Teagasc Agriculture and Food Development Authority
Johnstown Castle Environment Research Centre
Wexford
Ireland

ISSN 2211-1255 ISSN 2211-1263 (electronic)
World Soils Book Series
ISBN 978-3-030-10024-7 ISBN 978-3-319-71189-8 (eBook)
https://doi.org/10.1007/978-3-319-71189-8

Printed on acid-free paper

This Springer imprint is published by the registered company Springer International Publishing AG
part of Springer Nature
The registered company address is: Gewerbestrasse 11, 6330 Cham, Switzerland

Foreword, Undersigned and Dated

Our soil is an irreplaceable natural resource that has a critical role in food and feed production, as well as in services such as water purification, the provision of a home for biodiversity and as a major store of carbon. However, our soils are threatened by: soil degradation from agriculture and forestry; sealing from urban sprawl and contamination from industrial activities. As pressure to intensify soil use continues, we must balance the variety of demands on our soils in a sustainable manner. This relies, in the first instance, on a comprehensive knowledge of soil and its properties.

There is a proud history of mapping soils in Ireland. Indeed, An Foras Talúntais undertook the first soil survey in the 1970s, which mapped the soils of approximately half of the country. This book builds upon and completes this historic data by including the Irish Soil Information System, funded by the EPA Research Programme and co-funded by Teagasc.

This book will help us understand how this natural and national resource can be managed sustainably, how it is impacted by our decisions and how we can respond collectively to the most significant challenges of our time. It is intended that this book will be a rich resource for students, educators and researchers and the Environmental Protection Agency and Teagasc are delighted to support its authors in its completion.

Director
April 2017

epa Research
Climate - Water - Sustainability
Identifying pressures • Informing policy • Developing solutions

Frank Ó Mara
Director of Research
April 2017

Teagasc
Agriculture and Food Development Authority

Preface

The range of soil related themes covered in this volume reminds us of the connective nature of the soil, between the lithosphere, biosphere, atmosphere and geosphere. Although soil only accounts for a small portion of the Earth's surface, it has enormous importance in terms of the functions that it performs particularly in overall ecosystem functioning that is critical to sustain life on earth.

In the process of developing this book, the collaboration of many researchers and organisations across Ireland has been an important aspect of the work and one which we believe adds value to the manuscript. It is our aspiration that this book will represent an important educational resource and to that end, the book has been written and developed with that goal in mind. Equally, the collaborative nature of this book can therefore be considered to support and add value with respect to addressing the needs of Irish students to those with an interest in Irish soils alike. The demand for soils knowledge is driven by the global challenges of the day, such as food security, climate change, energy and water security as soil fulfils a direct role in responding to these challenges. We recognise the role of soils in implementing the Global Sustainable Developments Goals which include goals that specifically cite soil, reflective of the growing recognition of the importance of the soil resource. We hope that the work contained in this book will help to stimulate interest in Irish soils and that it will help to broaden understanding and interest in this vital natural resource.

In an effort to understand where we are now, we open this volume of the World Soils Book Series, with a visit to the past. We provide a brief overview of soil related research in Ireland with an emphasis on soil pedology. This is followed with an introduction to soil formation, followed by soil classification and soil mapping in Ireland. In our chapter on soil formation, expertise from Geological Survey Ireland and Met Éireann allow the geological origins of Irish soils and the climatology of the country to be described. The completion of the 3rd Edition National Soil Map of Ireland, i.e. the development of a harmonised legend for Irish soils for the first time, was an important advancement in relation to classification and mapping in Ireland. Considerable attention has been paid to this part of the book as it largely informs subsequent chapters and forms the basis for the descriptions of the soils across the landscapes presented. In relation to this, it is important to acknowledge the work and efforts of those who have gone before. The content of this book relies in no small part on their efforts. Development of the Irish Soil Information Systems, while embracing the opportunities of new technologies and affording new possibilities through open online access, respectfully builds upon the legacy work such as that conducted by An Foras Talúntais and the National Soil Survey.

Unlike other volumes in this series, Ireland being a relatively small island, coupled with an emphatic influence in relation to glaciation means that soils here do not so readily divide into regional distributions. However, we have endeavoured to describe landscapes in natural groupings that best represent our landscapes largely based on elevation, and by parent material in the lowlands. While peatlands have a dedicated chapter, they appear repeatedly throughout the book reflective of their importance, not only in relation to their prevalence but also for the important societal benefits that they provide. Mountain landscapes represent only a small portion of the Irish landscape, much lower than landscapes of similar geological origin in

Great Britain or Scandinavia. The lowland landscapes are described separately based on their parent materials. The drumlin belt of the north midlands of Ireland that stretches across parts of Northern Ireland into the Republic of Ireland is one of the most extensive drumlin belts in Europe (described in Chap. 11). Finally, the landscape chapters conclude with a description of urban soils (Chap. 12). These soils tend to be underrepresented in soil surveys generally, leading to information gaps. Urbanisation has been rapid in Ireland in the period 2000–2006 and so attention must be paid to the impact that this represents both directly and indirectly for the soil resource.

Research in Ireland is responding to the failure to ratify the Soil Framework Directive and nowadays soil research, whilst acknowledging threats to soil quality adopts a more optimistic approach to soils, focusing on the functionality of soils and how to optimise soil quality. The latter chapters delve more into the capacity of Irish soils to function in the delivery of important soil based ecosystem services. Ireland is traditionally an agrarian society and this remains the major use of the land base today. Excess soil moisture is a major constraint in Irish soils from a productivity and environmental perspective. Here, we describe the current land based agricultural activities and provide a history of land drainage schemes for Ireland. The fertility and nutrient cycling of Irish soils is described as are some best management practices to support the nutrient cycling process. This naturally leads on to the role of soil in water quality followed by the potential of Irish soils to store carbon. Thereafter, the living soil is described with an overview of the role and extent of biodiversity in our soils. The final chapters in this section explore our archaeological soils and the options for sustainable spatial planning. Finally, the book concludes with an overview of the current policy context and the potential future role of soils in policy.

The Soils of Ireland book was generously co-funded by the Irish Environmental Protection Agency and Teagasc. The project was led by Teagasc and included the participation of researchers from other research and academic institutes as well as other stakeholders across Ireland. We are very grateful to have had this opportunity to contribute, work and collaborate on such an important topic.

Wageningen, The Netherlands Rachel Creamer
Wexford, Ireland Lilian O'Sullivan

Contents

Editors and Contributors

About the Editors

Prof. Rachel Creamer is chair of Soil Biology and Biological Soil Quality at Wageningen University and Research. She is an author and co-author of over 50 scientific papers and the lead scientist on a number of national and European projects. Prof. Creamer has a background in pedology and soil quality, with research experience in the UK, Ireland and now the Netherlands. She was the lead scientist of the Teagasc Soil Quality and Classification sub-programme in Johnstown Castle from 2008 to 2016, where she led the development of the 3rd edition soil map of Ireland and associated soil information system.

Lilian O'Sullivan was awarded a Research Fellowship from the Irish Environmental Protection Agency to coordinate the Irish contribution to the World Soils Book Series contained here. She is based at Teagasc, Environment Research Centre in Co. Wexford where she has worked formerly as part of the Soil Quality Assessment Research project. Her research interests and work include soils and sustainability, agricultural development, food security and food policy.

Contributors

Tom Bolger UCD School of Biology and Environmental Science, University College Dublin, Belfield, Dublin 4, Ireland

Fiona Brennan Teagasc Environment, Soils and Land-Use Department, Johnstown Castle, County Wexford, Ireland

Kenneth A. Byrne Department of Biological Sciences, School of Natural Sciences, University of Limerick, Limerick, Ireland

James F. Collins University College Dublin, Dublin 4, Ireland

Rachel Creamer Wageningen University and ResearchAtlas—Wageningen Campus, Wageningen, The Netherlands; Teagasc Agriculture and Food Development Authority, Johnstown Castle Environment Research Centre, Wexford, Ireland

Thomas Cummins UCD School of Agriculture and Food Science, University College Dublin, Dublin 4, Ireland

Karen Daly Teagasc Environment Research Centre, Johnstown Castle, County Wexford, Ireland

Stephen Davis UCD School of Archaeology, University College Dublin, Belfield, Dublin 4, Ireland

Alan D. W. Dobson School of Microbiology and Environmental Research Institute, University College Cork, Cork, Ireland

Reamonn Fealy Teagasc, Rural Economy and Development Programme, Ashtown, Dublin, Ireland

Owen Fenton Teagasc Environment Research Centre, Johnstown Castle, Wexford, County Wexford, Ireland

Rochelle Fritch UCD School of Agriculture and Food Science, University College Dublin, Dublin 4, Ireland

Vincent Gallagher Department of Communications, Climate Action and Environment, Geological Survey Ireland, Dublin 4, Ireland

Mairéad Glennon Department of Communications, Climate Action and Environment, Geological Survey Ireland, Dublin 4, Ireland

Maryann Harris UCD School of Biology and Environmental Science, University College Dublin, Dublin 4, Ireland; Parks and Landscape Services Division, Dublin City Council, Dublin, Ireland

Phil Jordan Geography and Environmental Sciences, Ulster University, Coleraine, Northern Ireland, UK

Hendrik Van Der Kamp School of Transport Engineering, Environment and Planning, Dublin Institute of Technology, Dublin, Ireland

Gary Lanigan Teagasc Environment Research Centre, Johnstown Castle, Wexford, County Wexford, Ireland

Helen Lewis UCD School of Archaeology, University College Dublin, Belfield, Dublin 4, Ireland

Brian McConnell Department of Communications, Climate Action and Environment, Geological Survey of Ireland, Dublin 4, Ireland

Michael J. McLaughlin Soil Science, School of Agriculture, Food and Wine, The University of Adelaide, CSIRO Land and Water, Glen Osmond, Australia

Per-Erik Mellander Teagasc Environment Research Centre, Johnstown Castle, Wexford, County Wexford, Ireland

Franc Myles Archaeology and Built Heritage, Dublin, Ireland

Clíodhna Ní Lionáin Devenish Nutrition, Belfast, Ireland

Daire Ó hUallacháin Teagasc Environment Research Centre, Johnstown Castle, Wexford, County Wexford, Ireland

Lilian O'Sullivan Teagasc Agriculture and Food Development Authority, Johnstown Castle Environment Research Centre, Wexford, Ireland; Irish Environmental Protection Agency, Johnstown Castle, County Wexford, Ireland

Brian Reidy Teagasc Moorepark, Animal and Grassland Research and Innovation Centre, Fermoy, County Cork, Ireland

Florence Renou-Wilson School of Biology and Environmental Science, University College Dublin, Dublin, Ireland

Karl Richards Teagasc Environment Research Centre, Johnstown Castle, Wexford, County Wexford, Ireland

Ray Scanlon Department of Communications, Climate Action and Environment, Geological Survey of Ireland, Dublin 4, Ireland

Olaf Schmidt UCD School of Agriculture and Food Science, University College Dublin, Belfield, Dublin 4, Ireland

Rogier Schulte Teagasc, Environment Research Centre, Johnstown Castle, County Wexford, Ireland; Wageningen University and Research, Wageningen, The Netherlands

Sophie Sheriff School of the Environment, University of Dundee, Dundee, Scotland

Iolanda Simo Smart Soil & Land, Lleida, Spain

Brian Tobin UCD Forestry, School of Agriculture and Food Science, University College Dublin, Dublin 4, Ireland

Pat Tuohy Moorepark, Animal and Grassland Research and Innovation Centre, Teagasc, Fermoy, County Cork, Ireland

Sara E. Vero Agricultural Catchments Programme, Teagasc, Wexford, County Wexford, Ireland

David P. Wall Teagasc, Environment Research Centre, Johnstown Castle, County Wexford, Ireland

Séamus Walsh Met Éireann, the Irish Meteorological Service, Dublin 9, Ireland

Author Biographies

Tom Bolger is Professor and Chair of Zoology at University College Dublin. His primary research interests are in the effects of global change on biodiversity and nutrient dynamics in soil systems and in the evaluation of the status of and changes in that biodiversity. He is subject editor of several journals including Soil Biology and Biochemistry.

Fiona Brennan, Ph.D. is a Research Officer in Teagasc, Johnstown Castle and an Adjunct Lecturer in NUI Galway. She leads a research group on Soil and Environmental Microbiology whose main research focus is on the soil microbiome, functional microbial communities involved in nutrient cycling, plant–soil–microbial interactions, microbial adaptation to the environment and the fate of enteropathogens in soil.

Kenneth A. Byrne teaches soil science and agricultural science at the Department of Biological Sciences, School of Natural Sciences, University of Limerick. His main research interests are soil carbon stocks and greenhouse gas emissions, carbon storage in wood products and peatland management.

James F. Collins, M Agr Sc, Ph.D. is a retired soil scientist whose main research interests are in pedology and in relationships between Irish soils and climate. He has authored/coauthored over 100 publications (including 'Quickening the earth: soil minding and mending in Ireland', 2008) and currently writes on aspects of local history. He was Secretary/Treasurer of the Soil Science Society of Ireland from 1976 to 2000.

Thomas Cummins teaches soil science at University College Dublin. His research is in anthropogenic soils and long-term forest monitoring.

Karen Daly, M.Sc., Ph.D. is a Research Officer with Teagasc, the Irish Agriculture and Food Authority. Her main research interests include nutrient cycling in soil, soil biogeochemical processes and water quality modelling at catchment scale.

Stephen Davis, Ph.D. is a Lecturer in archaeology at University College Dublin who specialises in environmental archaeology (insect remains, pollen, and testate amoebae) and remote sensing. He has led projects in the Boyne Valley in Ireland, specifically within the Brú na Bóinne World Heritage Site, Co. Meath.

Alan D. W. Dobson studied biochemistry at the National University of Ireland, Galway and following his Ph.D., worked as a Postdoc at Baylor College of Medicine in Houston, in the area of eukaryotic gene regulation. He then moved to the School of Microbiology at University College Cork, where he is Professor of Environmental Microbiology. His group employs both genomic- and metagenomic-based approaches to exploit the microbiota associated with marine and terrestrial environments with the goal of identifying novel bioactive compounds with potential biopharmaceutical applications.

Reamonn Fealy is a Senior Researcher at the Spatial Analysis Unit in Teagasc. While his research portfolio spans a broad range of agri-environmental issues, he has spent a large part of his career working on the development and improvement of soil mapping and informatics in Ireland.

Prof. Dr. Owen Fenton is a researcher with Teagasc in Ireland. Research focuses on soil hydrology and the source, fate and transformation/ remediation of nutrients along with all surface and subsurface pathways in agricultural catchments.

Rochelle Fritch, Ph.D. is an Ecologist in UCD School of Agriculture and Food Science.Her main research topics are grassland ecology and ecosystem functioning, specifically plant and invertebrate ecology, and their interactions with the environment in agricultural systems.

Vincent Gallagher, Ph.D. has since 1987 worked mainly for Geological Survey Ireland as a geochemist on various topics including ore deposit research, radon, mineral exploration and contaminated land. He is currently Project Geochemist for the Tellus project, Geological Survey Ireland's national geochemical and geophysical mapping project.

Mairéad Glennon manages the Tellus programme at Geological Survey Ireland which undertakes nationwide airborne geophysical and ground-based geochemical mapping. Her background is in geography and hydrogeology and she has previously worked in the private sector on contaminated land projects. She is a Non-executive Director of the Institute of Geologists of Ireland.

Maryann Harris is a landscape architect and ecologist with experience in managing biodiversity of urban ecosystems at Dublin City Council in Ireland. She is currently a Doctoral Research Fellow in environmental science at University College Dublin researching the structure of urban green spaces to support biodiversity in Dublin Bay UNESCO Biosphere.

Phil Jordan is Professor of Catchment Science, School of Geography and Environmental Sciences, Ulster University, Coleraine (N. Ireland). His research areas relate to the dynamics and fate of nutrients and sediments in river and lake catchments, high-resolution spatio-temporal water quality observation/analysis and agri-environmental policy evaluation.

Hendrik van der Kamp is the Head of the School of Transport Engineering, Environment and Planning in Dublin Institute of Technology. As well as providing planning advice to community groups, architects, consultants and private developers, he has given advice and acted as an expert evaluator to a range of public sector bodies including Government Departments and the European Commission. He is a member of the Executive Committee of the European Council of Spatial Planners (www.ectp-ceu.eu) and is currently Vice-President.

Dr. Gary Lanigan is a Principal Research Officer in Teagasc. He heads up the Greenhouse Gas and Climate Change sub-programme at Johnstown Castle Environmental Research Centre, having previously held positions at Trinity College Dublin and Cambridge University. His main interests are carbon and nitrogen biogeochemical cycling.

Helen Lewis, Ph.D. is an Associate Professor at University College Dublin School of Archaeology. Her research areas include geoarchaeology, landscape archaeology, and cave archaeology.

Dr. Brian McConnell is a Senior Geologist at Geological Survey Ireland, where he manages the onshore bedrock and sub-soils mapping programme.

Michael J. McLaughlin is a Professor in Soil Science at the University of Adelaide, Australia and is a Fellow of the Australian Commonwealth Scientific and Industrial Research Organisation (CSIRO). His main research interests are soil and fertiliser chemistry, crop nutrition and soil contamination.

Per-Erik Mellander, Ph.D. is a Research Officer in Catchment science, Department of Environment, Soils and Landuse, Teagasc. He has a Ph.D. in Environmental Assessment from the Swedish University of Agricultural Sciences in Upssala, Sweden. His research focuses on the understanding of influences of climate and human impact on ecosystem processes and he is the author or co-author of 40 published scientific papers.

Franc Myles is Principal of a Dublin-based consultancy, Archaeology and Built Heritage. His research interests include the post-mediaeval expansion of Dublin and early island monasticism on the western seaboard.

Dr. Clíodhna Ní Lionáin is a licenced archaeologist with considerable commercial and research field experience in Ireland and Iberia and is employed by Devenish as Project Archaeologist for the Lands at Dowth. Her research interests include prehistoric archaeology, rock art, photogrammetry, and Community Archaeology.

Dr. Daire Ó hUallacháin is a Functional Ecologist with the Environment, Soils and Land-Use Department in Teagasc (Ireland). He has over 20 years of research experience in agri-ecology, terrestrial & aquatic Ecology, Water Quality and Sediment Dynamics.

Brian Reidy, Ph.D. is a soil ecology consultant. His research has focussed primarily on soil carbon, sequestration and land use change. In recent years, soil taxonomy and classification, coupled with extensive soil survey and correlation to WRB, have become areas of importance. Current research is in hydrology, drainage and soil/river catchment interaction.

Florence Renou-Wilson, Ph.D. teaches at University College Dublin (Ireland) where she is also the Director of the Masters of Science in Global Change: Ecosystem Science and Policy. She investigates peatlands in an integrative way at the crossroads of ecology, forestry, agriculture, climate change and sustainable management.

Karl Richards is a Principle Research Officer in Teagasc Johnstown Castle. He is author and co-author of over 85 peer-reviewed papers. His research focuses on understanding the fate, transport and mitigation of pollutants through soils to water and the atmosphere from agricultural sources.

Ray Scanlon is a Principal Geologist in the Geological Survey of Ireland. He holds a B.Sc. in Geology and an M.Sc. in Geochemistry, both from University College Dublin. His research interests include applications of soil geochemical and geophysical data to environmental, mineralogical and for radon prediction and the development of policy in these areas.

Olaf Schmidt, Ph.D. is Associate Professor at the School of Agriculture and Food Science, and a member of the UCD Earth Institute, University College Dublin, Ireland. His main research interests are in the biology, ecology and biochemistry of soils, the management of invertebrate crop pests and the application of stable isotope ratio analysis in food authentication.

Prof. Rogier Schulte worked at Teagasc as leader in Translational Research on Sustainable Food Production, which meant that he worked closely with policymakers in Ireland and the EU. He developed the concept of Functional Land Management and is currently extending and applying this framework in projects across five continents. More recently, he has moved to Wageningen University and Research in the Netherlands, to take up the position of Chair of the Farming Systems Ecology group.

Sophie Sheriff, Ph.D. is a Postdoctoral Research Associate at the University of Dundee, Scotland. Her primary research interests are catchment-scale soil erosion, sediment and hydrology dynamics, food and water security, water–food–energy nexus and future water resources projections to support water quality and quantity management.

Iolanda Simo, Ph.D. is an independent Research Associate at Smart Soil & Land (Spain). She has expertise in soil science, environmental science, agronomy, drainage and GIS. Her main research topics include soil quality, carbon in soils, desertification, soil cartography and soil classification.

Brian Tobin, Ph.D. works for UCD Forestry in the School of Agriculture and Food Science in University College Dublin (Ireland). His research interests include forest ecosystem carbon sequestration, silviculture and tree species productivity and biomass allocation as well as above- and belowground woody decomposition.

Pat Tuohy, Ph.D. is a Research Officer on the Heavy Soils Programme at the Teagasc, Animal and Grassland Research and Innovation Centre. His main research interests are the design, installation and characterisation of land drainage systems, soil physical and hydrological characterisation and the optimisation and management of poorly drained grassland soils.

Sara E. Vero, Ph.D. is a researcher in the Teagasc Agricultural Catchments Programme, based in the Environmental Research Centre, Johnstown Castle, Co. Wexford. Her main research topics include hydrochemistry, nutrient transport, soil physics, numerical modelling and vadose zone instrumentation. She is also part of an on-going collaboration which is developing a conceptual model for the hydrology of American prairie landscapes.

David P. Wall is a Biogeochemist and leader of the Nutrient Efficiency Research Programme at Teagasc—the Irish Agriculture and Food Development Authority. He is Editor of the national soil fertility advice 'Major and Minor Nutrients for Productive Agricultural Crops' in Ireland, and his research interests relate to better understanding of nutrient cycling in agricultural landscapes.

Séamus Walsh is Head of the Climatology and Observations Division of Met Éireann. He previously worked as a weather forecaster and in the provision of climate services to the agricultural and environmental sectors.His main areas of interest are past data analysis, observation networks and generation of long term datasets for Ireland. He has contributed to various publications on the climate of Ireland.

A History of Soil Research with Emphasis on Pedology

1

Lilian O'Sullivan, James F. Collins, Thomas Cummins, Reamonn Fealy, and Rachel Creamer

Abstract

While commentary on the productivity of Irish pastures can be dated as far back as 55 BC, direct references to the different kinds of soil only begin in the eighth century AD. Early scientific information on Irish soils typically came from the endeavours of geologists or other professionals who studied rock and outcropping patterns. The Geological Survey of Ireland, a constituent of the Geological Survey of Great Britain and Ireland, developed small-scale county soil maps, largely for the southern counties in the mid-1800s. In relation to pedology, John Hodges, an Irish soil analyst described soil profiles in the 1850s in terms identical with those dated to the Dokuchaev School in 1880s Russia. Soil science in Ireland has typically been delivered with agricultural and environmental science at University level. Agricultural extension dates back to 1731 through the Dublin Society employment of itinerant instructors. The soil test, pioneered by P H Gallagher in the 1930s became the single most important tool for the advisory service. In their essays on the characteristics of Irish soil types he, together with Dr. Tom Walsh, laid the groundwork for future classification and mapping. In 1945, the Department of Agriculture established a research centre at Johnstown Castle, Co. Wexford. The soil division of An Foras Talúntais was headquartered there and the National Soil Survey of Ireland established. In 1980 the second edition, General Soil Map of Ireland, was published. This survey was discontinued in 1988. In 1998, the Spatial Analysis Unit was established in Teagasc a key output of which was the first nationally complete digital subsoil and indicative soil map in 2009. Not long after, the 3rd Edition National Soil Map of Ireland was launched in 2014, through the collaboration of the Spatial Analysis Unit and Teagasc Johnstown Castle.

Keywords

Geology • Soil testing • Pedology • Johnstown Castle Soil survey

L. O'Sullivan (✉) · R. Creamer
Teagasc Agriculture and Food Development Authority, Johnstown Castle Environment Research Centre, Wexford, Ireland
e-mail: Lilian.OSullivan@teagasc.ie

L. O'Sullivan
Irish Environmental Protection Agency, Johnstown Castle, County Wexford, Ireland

J. F. Collins
University College Dublin, Belfield, Dublin 4, Ireland

T. Cummins
UCD School of Agriculture and Food Science, University College Dublin, Belfield, Dublin 4, Ireland

R. Fealy
Teagasc, Rural Economy and Development Programme, Ashtown, Dublin 15, Ireland

R. Creamer
Wageningen University and Research, Atlas—Wageningen Campus, Droevendaalsesteeg 4, Building 104, 6708 PB Wageningen, The Netherlands

1.1 In the Past…

Favourable comments on the productivity of Irish pastures can be dated as far back as 55 BC (Murphy 2015) but direct comments or narratives on the different kinds of soil begin only in the eighth century AD. Using early Irish Law tracts, Kelly (1997) deduced that three classes of arable land and three classes of rough land were recognised. The main distinguishing criteria being vegetation (trees, shrubs..), access to water/wetness, and stoniness/shallowness. Stock-carrying capacity was rated in numbers of *milch* (24, 20, 16) or dry (16, 12, 8) cows per *cumal.* Some centuries later, Gerald of Wales (*ca.* 1183) noted soil diversity: '… *soil soft and watery… rather sandy than rocky… fruitful and rich in its soils'. 'In some places a blackish earth, in others clay and in many parts a mix of both together, as likewise there are sandy places where the ground is mixed earth and sand,*

R. Creamer and L. O'Sullivan (eds.), *The Soils of Ireland*, World Soils Book Series,
https://doi.org/10.1007/978-3-319-71189-8_1

sand and clay, gravel and clay or earth... The plains are well clothed with grass, and the haggards are bursting with straw'. Dealing with manorial centres in the 1250/1350 period, Murphy (2015) listed some of the details in manorial extents (a description and valuation of all the items on the manor). This included the principal types of land-use: arable, meadow, pasture and wood. Many English sources wrote effusively about Irish soil in the late sixteenth century including Stanihurst who described *…so rich a kingdom, the Pastures are greene [in winter]…*; Payne *…soil for the most part fertil, and apte for wheat, rye, barley…* and Dymmock who found that *…the land lyeth very low and therefore very waterish and full of marishes…* A comprehensive account of agriculture was written by Boate (1652) ranging over many aspects including soils, crops, animals manures, burning lime and procuring sea-sand and shells. Subsequently, William Petty in the 1660s who confined himself to what was visible over-ground assigned Irish land to two groups: 'profitable' and 'unprofitable' (Simington 1931–61). The former group included arable, pasture, meadow and timberland, while the latter consisted of shrubby wood, rocky land, red bog and mountain. For some counties, Petty's commentary contained many references to agricultural practices such as liming, sanding and gravelling, aimed at increasing productivity (Simington 1931–61).

1.2 Geology and Geological Origins of Soil Research

Much of the early scientific information on Irish soils came from the endeavours of geologists or those other professionals who studied local rocks and their outcropping patterns. Boate (1652) may have pointed the way, having referenced the qualities of the Clare/Galway limestone-based mould, soils of various textures and 'in some places a blackish earth'. In his treatises, Smith noted the contrasts between the soils and agriculture of the Devonian and Carboniferous sandstones in Co. Waterford (1746); the potential value of a soil map in Co Cork (1750) and soils in Co. Kerry (1756). Later that century, Young (1780) commented on the rockiness of Irish soilscapes: '*…but the rocks here are clothed in verdure...the softest and most beautiful mould imaginable…*' In Co. Longford, Young noticed that even a two-inch thick layer of blue clay was sufficient to induce water retention. In a recent assessment of the roles of geology and soil in supporting cattle rearing, the contributions of Kirwan 1793, Meade 1806, Newenham 1809, Radcliffe 1814; and Shaw-Mason 1814–18, were noted (Collins 2016). Rock types and their associated soils were common themes in many of the (Royal) Dublin Society statistical surveys of the early 1800s, those of Fraser (1801, Wicklow), Sampson (1802, Londonderry), Tighe (1802, Kilkenny) and Townsend, (1810, Cork,) being particularly informative. The last-mentioned, Townsend, should be credited with introducing the pedologists' term, 'parent material', into the literature. Referring to the east/west valleys in Co. Cork, where sandstone debris from the mountains partially covers the limestone, he wrote: '*The limestone is certainly not the parent of the soil which is found over it... the earth incumbent on the limestone owes its birth to another parent*' (Townsend 1810). While the primary objective was bog reclamation, the reports of the Bog Commissioners, 1814–18, (which included R Griffith, geologist) alerted many to a range of calcareous materials suitable for neutralising the growing problem of soil acidity.

The Geological Survey of Great Britain and Ireland, instituted in 1845, conducted geological surveys and research across the UK and Ireland. The Geological Survey of Ireland (GSI) was a constituent part and survey work in Ireland was conducted under its local director, Henry James (Herries Davies 1983, 1995). The early results, largely from southern counties, were presented in small-scale county maps. These were superseded by one-inch maps based upon the field survey work conducted between 1857 and 1890 (Kilroe 1907). Early geological mapping was motivated by the search for economic minerals with the examination of soils and subsoils being an integral part of the mission. Du Noyer was the chief collector of samples taken across Cork, Waterford and Kilkenny. The GSI Memoir Series accompanying map sheets for a greater part of the country contain references to the use and availability of a range of soil amendments, mostly calcareous sands, gravels and marls, as well as locations of limestone rock suitable for burning to lime. As the survey proceeded through the 1870s/80s the emphasis on pedological matters lessened and in parts of the sandstone areas of South Munster the drift (till) was so thin that commentary on the soil was not worthwhile. However, George Kinahan highlighted the case of soils in his many publications (Kinahan 1878, 1885/9, 1908a, 1908b) especially in his 1908 booklet: 'Soils'.

1.3 Pedology, Agriculture, Education and Advisory

The recognition of the soil profile as the basic unit in soil survey has been dated to the 1880s in Russia (Dokuchaev 1883), but some decades previously, John Hodges, an Irish academic, soil analyst and author, described two soils in terms identical with those of the Dokuchaev School. In 1855/6, Hodges published compositional and descriptive data for the soils and subsoils of the Albert (Glasnevin) and Munster (Cork) Model Farms, together with advice on draining, trenching, crop selection and fertilisation reflecting the interwoven nature of the above-named disciplines in the 1850s.

Hodges' site, descriptive, compositional and commentary data were assessed and his work placed in context with the agricultural and educational developments of the period (Collins 2008). His interest and writings captured legacies and student progression that ascended from the Model Agricultural School system and transitioned to the Albert College and on to the Queen's universities. Other eminent persons of the period include Robert Kane, Edmund Murphy and Wm Sullivan. Writing on pedological themes, Antisell (1845, 46) reviewed the works of Griffith, Hamilton, Oldham, Mallet, Wilkinson and Kane, and having discussed weathering (1845, Chap. 3), addressed the contribution of various Irish rock types to soil formation. The pioneering work and writings of James Murray on soil phosphorus during these decades deserves special mention (Cooper and Davis 2004).

Irish agriculture was also well-served during the 1800s by a wide range of agricultural societies, many publishing journals, pamphlets and gazettes. The nature and properties of soils featured regularly in most of them, the subsoil getting the most notice in those articles dealing with drainage, trenching and ridge-and-furrow construction. The most impressive tome was the 1280-page first issue of the *Irish Farmer's Journal* in 1845; it included articles and notes on virtually every agricultural topic of the time including one entitled '*On the application of geology to agriculture*' (p. 213). One of the most influential and prolific agricultural educators of the 1800s was Martin Doyle, co-editor with Edmund Murphy of *Irish Farmers' and Gardeners' Magazine and Register of Rural Affairs* in the 1830s. Between 1830 and 1834, no fewer than a dozen books and pamphlets, including a two-volume collected works, were published under the Martin Doyle pseudonym (Cadogan 1989). Cadogan concluded '*…at a time when organic farming and self-sufficiency are enjoying a new vogue, his essays are as topical as when they first saw the light of day over 150 years ago…*'

As a discipline, soil science in Ireland has typically been delivered with agricultural and environmental science at University level. Agricultural extension in Ireland dates back to the establishment of the Dublin Society (later Royal Dublin Society) in 1731 and its employment of 'itinerant instructors'. State employment of instructors commenced with the establishment of the Department of Agriculture and Technical Instruction (DATI) in 1900. It is acknowledged that the soil testing, pioneered by P H Gallagher of UCD in the 1930s, became the single most important tool for the advisory service, leading to the transformation of Irish agriculture (Kirley 2008). Although the uptake of advisory services declined somewhat in the late 1980s, policy changes such as the Common Agricultural Policy (CAP) Reforms, the Rural Environment Protection Scheme (REPS) and the Green, Low-Carbon, Agri-Environment Scheme (GLAS) have seen an increased demand for advisory services. As well as Teagasc Agriculture and Food Development Authority, many private sector entities currently compete in the area of advisory service provision in Ireland.

1.4 Official Pedology, Pedogenic Investigations and Surveys

The establishment of DATI brought together diverse educational, scientific and administrative institutions including the Royal College of Science (a successor to the Museum of Irish Industry) and the County Committees of Agriculture (subsets of County Councils, established in 1898). One of its first commissioned publications, '*Ireland, industrial and agricultural*' (Coyne 1902), included an eight-page article '*The soils of Ireland*' by J R Kilroe; this was a forerunner for a much larger tome, '*A description of the Soil-Geology of Ireland*' by Kilroe in 1907. The Geological Survey, under Grenvile Cole, joined in 1905 (Herries Davies 1995), turning its attention to soils and employing Timothy Hallissy as a soil physicist. The soils of the new Agricultural College farms at Athenry, Ballyhaise and Clonakilty and a Forestry School at Avondale, Co. Wicklow, were studied. Unacquainted with the new thinking from Russia or from USA, Hallissy and co-authors drew heavily on their geological experiences. Their map of the 800-acre Ballyhaise Farm, showing the spatial distribution of seven soil types (scale 1:5280) attracted widespread approval. Surveyed by Hallissy and Kilroe, their map was colour-printed in 1909 and was awarded a Grand Prix Diploma. A few years later the GSI undertook a soil survey of Clare Island, Co Mayo, the findings of which were published by the Royal Irish Academy (Hallissy 1914) and in a special Survey Memoir (Cole et al. 1914). Hallissy may have been echoing the thoughts of many when he asserted that 'the soils occur in such a variety that a complete description of them is impossible, and were it possible, the labour necessary for full investigation of their properties would not be justified by the results'.

The study of both edaphology and pedology and, indeed, education/advisory to farmers, progressed significantly following the appointment of P H Gallagher at UCD, in 1928 (See Collins 2008, pp 462–7 and bibliography). While soil fertility and mineral nutritional considerations were addressed during the 1930s, it was Gallagher's research into the characteristics of Irish soil types that gained most prominence. Co-authored with T Walsh, their essay laid the groundwork on which future mapping and classification was to progress. In a 1944 paper ('The Soil'), Gallagher criticised those who would be satisfied with classifications such as 'Arable No.1', 'Arable No. 2', etc. and he outlined his scheme of using colour indicator dyes to aid soil advisory work (Gallagher 1944). On joining the State Department of Agriculture (Soils division), Walsh's position on the staff of the Faculty of Agriculture, UCD, was taken by W D (Bill) Brickley. Most of those who followed a career in

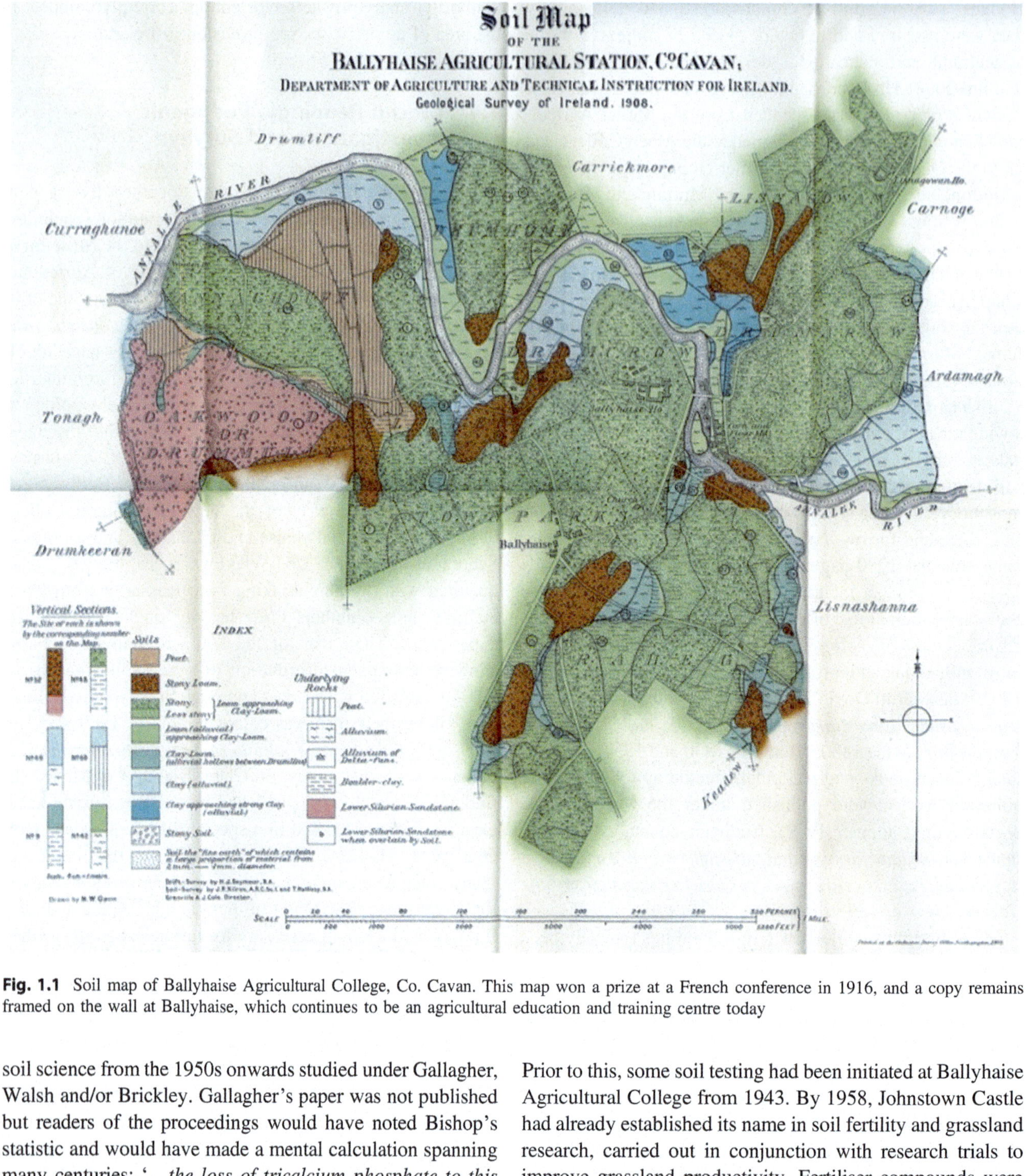

Fig. 1.1 Soil map of Ballyhaise Agricultural College, Co. Cavan. This map won a prize at a French conference in 1916, and a copy remains framed on the wall at Ballyhaise, which continues to be an agricultural education and training centre today

soil science from the 1950s onwards studied under Gallagher, Walsh and/or Brickley. Gallagher's paper was not published but readers of the proceedings would have noted Bishop's statistic and would have made a mental calculation spanning many centuries: '…*the loss of tricalcium phosphate to this country mainly through the export of cattle during the past few years has been somewhere between 20,000 and 30,000 tons per year*' (Bishop 1944). Much of the pedogenic research emanating from the Faculty of Agriculture, UCD, has been carried out on projects associated with the National Soil Survey (Fig. 1.1).

In 1945, the Department of Agriculture established a research centre at Johnstown Castle, Co. Wexford (Fig. 1.2). Prior to this, some soil testing had been initiated at Ballyhaise Agricultural College from 1943. By 1958, Johnstown Castle had already established its name in soil fertility and grassland research, carried out in conjunction with research trials to improve grassland productivity. Fertiliser compounds were developed based on the research of that time (Miley 2008b). The post-war European Recovery Plan or Marshall Plan was established in 1947 to rebuild Europe after World War II with money from the United States of America. Ireland was neutral during the war but was eligible for funding due to its role as a food exporter to the UK (Miley 2008a). Joesph Carrigan, head of the Economic Co-operation Administration in Dublin, first proposed the development of a research institute in 1949. In

Fig. 1.2 Johnstown Castle near Wexford town, where the Department of Agriculture established a research centre in 1945. The later An Foras Talúntais, the Agricultural Institute, and its descendant Teagasc were headquartered on this campus. Photo: Brian Murphy

late 1958 an institute for agricultural research, *An Foras Talúntais* (AFT), was established. In 1988, Teagasc Agriculture and Food Development Authority succeeded both AFT and *An Chomhairle Oiliúna Talmhaíochta,* the Agricultural Training Council (ACOT) which was the division responsible for education and advisory services. For over 50 years, AFT, ACOT and Teagasc have occupied a central role in research, advisory and training related to soil research in Ireland.

In the post-war period, soils in Ireland were significantly deficient in fertility. According to G A Holmes, a New Zealand government's agricultural attaché in London, it was a miracle that grass was found to grow, given the economic and political instability coupled with the limited use of fertiliser inputs (Miley 2008b). In 1950, research by Walsh et al. (1950) found that over 90% of soil samples were either 'very' or 'moderately' deficient in phosphorus (P), over 50% were deficient in potassium (K) and fewer than 50% were satisfactory for lime. Setting up its Soil Division headquarters at Johnstown Castle, AFT established the National Soil Survey of Ireland, under the guidance of Pierce Ryan. Amongst its other programmes, most urgent was the correction of soil acidity, mainly with ground limestone, and the avoidance of mineral nutrient deficiencies.

On joining the European Economic Community (EEC) in 1973, the role of agriculture as a major growth area for the Irish economy was widely recognised. Adequate knowledge of the land resource was considered essential for agricultural development and was a major driver of the soil survey (Gardiner and Radford 1980). In 1980, the second edition, General Soil Map of Ireland (Scale 1:575,000) and an accompanying bulletin by Gardiner and Radford (1980) were published (Fig. 1.3). The explanatory bulletin, No 36, included a description of the major soil types and assessed their land-use potential. Covering approximately 44% of the country, the soils of ten full counties were mapped and published: Wexford, Carlow, Limerick, Clare, Westmeath, Meath, Laois, Kildare, Leitrim and Offaly. Partial county surveys are published for west Mayo, west Donegal, Tipperary North Riding and West Cork. The National Soil Survey was discontinued in 1988, even though 'Soils of Co. Waterford' was published as Soil Survey Bulletin No 44, in 2011 (Diamond and Sills 2011). In the 1990s, environmental concerns, such as potential loss of nutrients from land, risking eutrophication of freshwaters, changed the emphasis of soil studies, moving away from agricultural production toward ecosystem science.

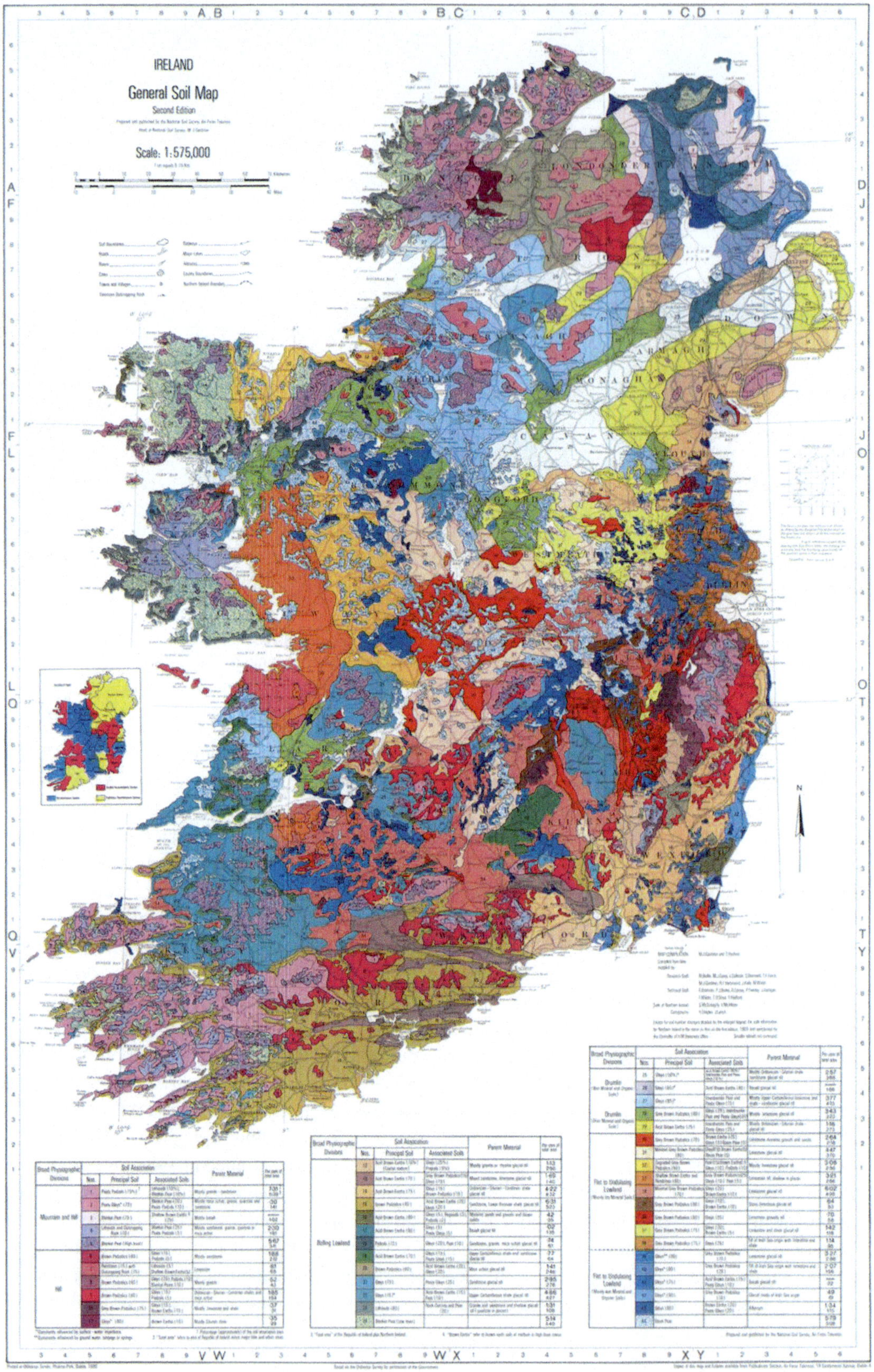

Fig. 1.3 The second edition national soil map of Ireland published by an Foras Talúnais in 1980, to accompany Gardiner and Radford's *Soil Associations of Ireland and their Land Use Potential*

Fig. 1.4 Dr. Tom Walsh, Director of An Foras Talúntais for its first 22 years (1958–1980) during which time he firmly established AFT as a leading centre in soil science research

Dr. Thomas Walsh (1914–1988) MAgrSc, DSc, PhD, LLD, ScD, MRIA (Fig. 1.4).

Born in Piercestown, Co. Wexford, Dr. Walsh graduated from University College Dublin (UCD) in 1937 with an honours degree in Agricultural Science. The following year he completed his master's degree in the same subject, his Ph.D. in 1941, and he was awarded a DSc for his published work in 1947. During his time at UCD (1938–1945), Dr. Walsh lectured in soil science, after which he worked as Soil Advisory Officer in the Department of Agriculture (1945–1954) and it was he who led the campaign on tacking the issue of soil fertility (O'Dwyer 2008). Dr. Walsh is considered the single most influential figure in the development of the internationally renowned agricultural and soil research infrastructure in Ireland which shaped the agriculture and food industry of today (O'Dwyer 2008).
During his lifetime, Dr. Walsh was very active in a large number of scientific organisations. He was elected a member of the Royal Irish Academy (MRIA) in 1956 where he served on the Council, as Senior Vice-President, as Science Secretary and as Secretary. He was a member of the Science Committee of the Royal Dublin Society and in 1969 was awarded the Society's Boyle Medal, for outstanding contribution to science in Ireland. He served on the Commission for Higher Education and was Chairman of the National Council for Educational Awards (NCEA) during the 1970s. He was appointed Director of An Comhairle Oiliúna Talmhaiochta (ACOT, the national farm advisory body) in 1980. He retired from public life in 1983.

In 1998, the Spatial Analysis Unit (SAU) was established in Teagasc to conduct work on the Irish Forest Soils project (FIPS-IFS). At the outset of the FIPS-IFS project, less than half of the country's soils had been surveyed and published at detailed reconnaissance (1:126,720) scale. The FIPS-IFS project resulted in a methodology to create an indicative soil classification and map for those areas not previously surveyed by the National Soil Survey, using satellite and aerial remote sensing and Geographical Information Systems (GIS). Inherent in this was the production of a soil parent material map. The SAU extended these efforts through the Environmental Protection Agency (EPA)/Teagasc Soil and Subsoil Mapping project leading to the first nationally complete digital subsoil and indicative soil maps created by a unified methodology (Fealy et al. 2009).

Notwithstanding the range of soil research and mapping undertaken since the 1950s, a comprehensive inventory by Daly and Fealy (2007) found that soil data coverage of Ireland was incomplete in both detail and extent. Shortly afterwards, the Irish Soil Information Systems (SIS) project was established in 2008, jointly funded by the EPA and Teagasc. A third edition National Soil Map of Ireland (Creamer et al. 2014) was developed using a combination of traditional soil survey methods and new GIS predictive mapping methods (see Chaps. 3 and 4). The 3rd Edition map also incorporated the legacy data generated in the earlier soil surveys and developed a harmonised legend for the soils of Ireland. This map and its soil database were launched in September 2014, coinciding with the centenary year of Dr. Walsh's birth.

The Joint Working Group on Applied Agricultural Meteorology (AGMET) has published soil relevant works including *Climate, Weather and Irish Agriculture* (Keane and Collins (1986), *Agro-Meteorological Modelling—Principles, Data and Applications* (Holden 2001, Chap. 4: Terestrial Data) and the *Agroclimatic Atlas of Ireland* (Collins and Cummins 1996). In '*Quickening the Earth—Soil Minding and Mending in Ireland*' Collins (2008) provides an account of agriculturalists' interaction with, and management of, land and soil over many centuries. The Soil Science Society of Ireland, formed in 1977, promotes the knowledge and use of, Irish soils. The members are professionally active in research, teaching, administration, consultancy, advisory services, state agencies, civil service and industry (SSSI 2016). The Irish Quaternary Association, Irish Peatland Conservation Council, and member organisations of the Irish Geoscience Network all continue to work to improve understanding of soils and the integration of soil knowledge into other disciplines.

References

Antisell T (1845) A manual of agricultural chemistry, with its application to the soils of Ireland. Hodges and Smith, Dublin

Bishop DW (1944) Mineral resources. In: The natural resources of Ireland: a series of discourses… RDS, Dublin, pp 55–61

Boate G (1652) Ireland's natural history, being a true description… Hartlib, London

Cadogan T (1989) Rev. William Hickey/'Martin Doyle' (1778–1875): agricultural educator and essayist. Bandon Hist J 5:36–42

Cole GAJ, Kilroe JR, Hallissy T, Newell Arber EA (1914) The geology of Clare Island, Co. Mayo. Memoirs of the geological survey of Ireland, DATI, Dublin

Collins JF (2008) Quickening the earth: soil minding and mending in Ireland. UCD School of Biology and Environmental Science, Dublin

Collins JF (2016) Geology soils and cattle production chapter 1. In: O'Connell M, Kelly F, McAdam JA (eds) Cattle in ancient and modern Ireland Cambridge Scholars publishing, Newcastle upon Tyne, UK

Collins JF, Cummins T, (1996) (eds) Agroclimatic Atlas of Ireland. AGMET, Dublin

Cooper M, Davis J (2004) The Irish fertiliser industry: a history. Irish Academic Press, Dublin

Creamer R, Simo I, Fealy R, Hallett S, Hannam J, Holden A, Jones B, Massey P, Mayr T, McDonald E, Reidy B, Sills P, Truckell I, Zawadzka J, Schulte R (2014) National soil map of Ireland. Teagasc Agriculture and Food Development Authority, Cranfield University, Environmental Protection Agency, Washington, D.C.

Daly K, Fealy R (2007) Digital soil information system for Ireland (2005-S-DS-22), Final Report. Environmental RTDI programme 2000–2006, Environmental Protection Agency, Johnstown Castle, Wexford

Diamond J, Sills P (2011) Soils of Co. Waterford, Soil Survey Bulletin No. 44. National Soil Survey of Ireland, Teagasc, Co, Carlow

Dokuchaev VV (1883) Russian chernozem (trans: Kaner N). Jerusalem, Israel Programme for Scientific Translations

Fealy RM, Green S, Loftus M, Meehan R, Radford T, Cronin C, Bulfin M (2009) Teagasc EPA soil and subsoils mapping project-final report. vol I. Teagasc. Dublin

Fraser R (1801) General view of the agriculture and mineralogy, present state and circumstances of the County Wicklow: with observations on the means of their improvement, drawn up for the consideration of the Dublin society. Fraisberry & Campbell, Dublin

Gallagher PH (1944) In: 'The soil', a lecture by Dr P.H. Gallagher at the Sir Robert Kane Centenary symposium held by the Royal Dublin Society, 14 Apr 1944. Typescript Courtesy of Soil Science Group, UCD

Gardiner MJ, Radford T (1980) Soil associations of Ireland and their land use potential. Explanatory bulletin to the soil map of Ireland 1980. Soil survey bulletin No. 36. An Foras Talúntais. Dublin

Hallissy T (1914) Clare Island survey: geology. Proceedings of the Royal Irish Academy, 31 (1911–15), section 1, part 7

Herries Davies GL (1983) Sheets of many colours: the mapping of Ireland's rocks, 1750–1890. Royal Dublin Society, Dublin

Herries Davies GL (1995). North from the Hook, 150 years of the geological survey of Ireland. Geological Survey of Ireland

Holden NM (2001) (ed) Agri-meteorological modelling:-principles, data and applications. AGMET, Dublin

Keane T, Collins JF (1986) (eds). Climate, weather and Irish agriculture, joint working group on applied agricultural meteorology (AGMET), c/o Met Éireann, Dublin

Kelly F (1997) Early Irish farming. Dublin Institute for Advanced Studies, Dublin

Kilroe JR (1902) The soils of Ireland. In: Coyne WP (ed) Ireland industrial and agricultural. Brown & Nolan Ltd., Dublin, pp 28–35

Kilroe JR (1907) A description of the soil-geology of Ireland based upon geological survey social inquiry society of Ireland, vol XVIX, part V, 1956/1957, pp 104–136

Kinahan GH (1878) Manual of the geology of Ireland. Keagan Paul, London

Kinahan GH (1885/9) Economic geology of Ireland. Royal Geological Society of Ireland, Dublin

Kinahan GH (1908a) Superficial and agricultural geology—Ireland. No. 1 Lime. Sealy, Bryers and Walker, Dublin

Kinahan GH (1908b) Superficial and agricultural geology—Ireland. No. 2 Soils. Sealy, Bryers and Walker, Dublin

Kirley T (2008) A century of advisory services for farmers. In: Miley M (ed) An Foras Taluntais—50 years of agriculture and food research, Teagasc Dublin

Miley M (2008a) Establishment and evolution of an Foras Talúntais. In: Miley M (ed) An Foras Taluntais—50 years of agriculture and food research, Teagasc Dublin

Miley M (2008b) Soils, grass and environmental research at Johnstown Castle. In: Miley M (ed) An Foras Taluntais—50 years of agriculture and food research, Teagasc Dublin

Murphy M (2015) Manor centres, settlement and agricultural systems in Ireland, 1250–1350. In: Murphy M, Stout M (eds) (2015) Agriculture and settlement in Ireland. Four Courts Press, Dublin, pp 69–100

O'Dwyer T (2008) Destined for high office—a profile of the 'Doc'. In: Miley M (ed) An Foras Taluntais—50 years of agriculture and food research, Teagasc Dublin

Sampson GV (1802) Statistical survey of the county of Londonderry: with observations on the means of improvement. Graisberry and Campbell, Dublin

Soil Science Society of Ireland (SSSI) (2016) Soil Science Society of Ireland (website) http://www.ucd.ie/sssi/Home.html. Accessed 05/07/2016

Simington RC (1931–61) The civil survey A.D, 1654–1656. Irish Manuscripts Commission, Dublin

Smith C (1746) The antient and present state of the county and city of Waterford, Reilly, Dublin

Smith C (1750) The antient and present state of the county and city of Cork. Reilly, Dublin

Smith C (1756) The antient and present state of the county of Kerry. Ewing et al, Dublin

Tighe W (1802) Statistical observations relative to the county of Kilkenny: made in the years 1800 & 1801. Graisberry and Campbell, Dublin

Townsend H (1810) Statistical survey of the county of Cork. Graisberry and Campbell, Dublin

Walsh T, Mannion L, Ryan PF (1950) Some aspects of the fertility status of Irish soils. Trans Int Cong Soil Sci, vol 3

Young A (1780) A tour in Ireland, with general observations on the present state of that Kingdom in 1776–78, 2 vols. Messrs. Whitstone et alia, Dublin

2 Soil Formation

Lilian O'Sullivan, Brian McConnell, Ray Scanlon, Séamus Walsh, and Rachel Creamer

Abstract

Parent material, topography, organisms and climate represent the key soil forming factors that interact over a timespan of millennia to produce soil. These soil forming factors produce soil through physical, chemical and biological weathering occurring overtime. The relative influence of the individual factors is responsible for differences found in soils. Soils in Ireland are relatively young, having formed since the retreat of the last ice age approximately 15,000 years ago. Parent materials in Ireland are broadly categorised into solid bedrock geology and bedrock derived glacial geology, with the latter accounting for the majority of parent materials across the Irish landscape. The Irish climate is strongly influenced by its position on the eastern side of the Atlantic Ocean and on the western fringes of continental Europe resulting in a mild maritime climate with prevailing south-westerly winds. The high rainfall rates in Ireland are a dominant driver of soil genesis with leaching and gleying as the two main processes driving soil development.

Keywords

Parent material • Weathering • Topography
Time • Climate • Organisms • Soil profile
Soil horizon

2.1 Soil Forming Factors

Soils are formed as a result of the interplay of several factors that underpin soil formation across a timeline spanning millennia, dating back to the oldest rocks found on earth from 4.03 billion years ago. To understand the soils of Ireland requires an examination of the soil forming factors and the key events that occurred in Ireland over time, to arrive at the soil and landscape, as we know it today. In 1941, Hans Jenny published 'Factors of Soil Formation a System of Quantitative Pedology', which is a useful resource for assessing how soils are formed. In it, he proposes the key soil forming factors (Fig. 2.1) as parent material, topography, time, climate and organisms (Jenny 1941) and today, these remain the cornerstones of investigation when seeking to assess soil formation anywhere. These soil forming factors produce soil through physical, chemical and biological weathering occurring over time, and in response to climatic factors and landscape position.

Parent material represents an important factor of soil formation, the majority of which is of geological origin. There are many different soil types and some of the differences in soil are due to the wide variety of rocks from which the soils are formed. Rocks contain minerals and it is the weathering of rocks that determines the mineralogy of soil. Weathering refers to the process by which rocks get broken down into smaller pieces. Three main types of weathering occur; physical, chemical and biological:

- **Physical weathering** refers to the mechanical breakdown of rocks into smaller pieces. Physical weathering occurs due to changes in temperature, freeze/thaw action,

L. O'Sullivan (✉) · R. Creamer
Teagasc Agriculture and Food Development Authority, Johnstown Castle Environment Research Centre, Wexford, Ireland
e-mail: Lilian.OSullivan@teagasc.ie

L. O'Sullivan
Irish Environmental Protection Agency, Johnstown Castle, County Wexford, Ireland

B. McConnell · R. Scanlon
Department of Communications, Climate Action and Environment, Geological Survey of Ireland, Beggars Bush, Haddington Road, Dublin 4, Ireland

S. Walsh
Met Éireann, the Irish Meteorological Service, Glasnevin Hill, Dublin 9, Ireland

R. Creamer
Wageningen University and Research Atlas—Wageningen Campus, Wageningen, The Netherlands

R. Creamer and L. O'Sullivan (eds.), *The Soils of Ireland*, World Soils Book Series,
https://doi.org/10.1007/978-3-319-71189-8_2

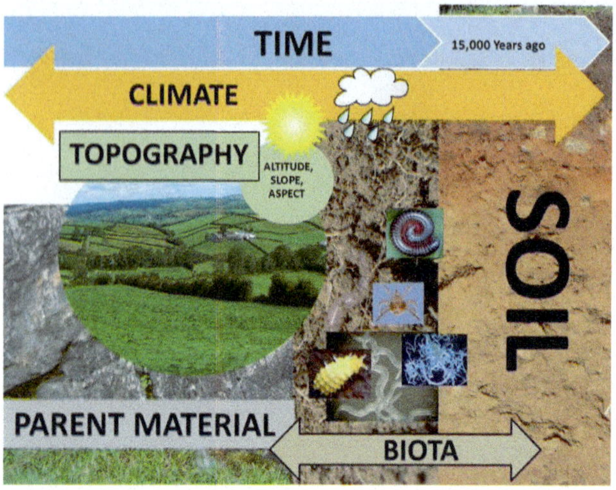

Fig. 2.1 Soil forming factors presented with background images taken during the Irish Soil Information System field campaign

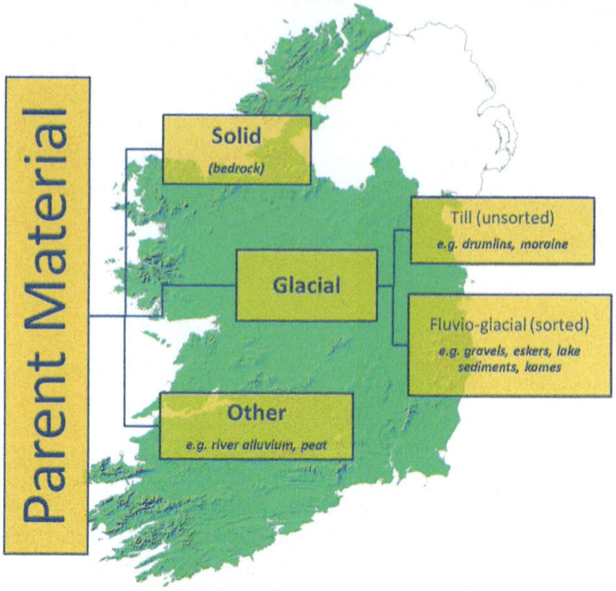

Fig. 2.2 Parent material types in Ireland

wetting, drying and erosion by wind, water and ice. Physical weathering does not significantly alter the mineral composition of the rocks.

- **Chemical weathering** refers to the breakdown of materials through chemical reactions. Carbon dioxide (CO_2) in the air dissolves in rainwater to form a weak carbonic acid (H_2CO_3) solution. When this acid solution comes into contact with rocks, it wears them down over time. Rainwater is, therefore, a primary weathering agent as it contains ions from the atmosphere and from the soil through which it moves. Unlike physical weathering, chemical weathering alters the mineral composition of rocks into secondary minerals such as clays. Physical and chemical weathering processes are complementary. Physical weathering increases the surface area of exposed rock, which in turn increases the areas susceptible to chemical weathering.
- **Biological weathering** occurs when organisms break down materials both physically and chemically. Physically, plant roots penetrate cracks in rocks thereby extending the exposed cracked surface area. Chemically, plants produce CO_2 that combines with rainwater to produce H_2CO_3. Organisms catalyse chemical reactions and also biologically mix (bioturbation) the structure of the parent material. This can increase the pathways for water flow, facilitating weathering to occur through the soil.

The nature and properties of parent material will influence important soil characteristics including the texture and the mineral composition, which in turn, defines other aspects of the soil such as the movement of water through the soil profile and soil fertility. Parent material also determines the content and type of clay present in the soil. Clay content is important as soils with higher clay content are more fertile. This is because the negatively charged clay surfaces adsorb and hold on to positively charged cations such as magnesium (Mg), potassium (K) and calcium (Ca) which are important for plant productivity.

In Ireland, parent materials are broadly grouped into two main types, solid geology and drift geology. Solid geology refers to weathered materials overlying the consolidated bedrock. Drift geology refers to unconsolidated deposits or loose sediments transported by ice, water, wind or gravity largely derived from the underlying solid geology (Fig. 2.2). Drift parent material accounts for the majority of parent materials across the Irish landscape. Geology is described in more detail in Sect. 2.2.

Topography is essentially created from geological activity and Ireland records a very dynamic geology associated with lithospheric plate tectonic movements. Topography refers to the shape of the landscape and includes macro- and microrelief dimensions. Macrorelief includes: altitude which is the height above sea level; the aspect, which refers to the direction of the site and whether it is south or north facing, and slope which refers to steepness. Microrelief refers to smaller localised variations, such as the height of the water table (Fig. 2.3). These attributes will determine the interaction between the site and climatic factors, such as solar radiation or precipitation. South-facing slopes in Ireland will receive more solar radiation and are consequently warmer and dryer with higher levels of evapotranspiration than slopes of a north-facing aspect. In Ireland, it is estimated that a 10° north-facing slope receives 40% less solar radiation than a similar south-facing slope (Collins et al. 2004). Relief determines the extent to which gravity influences the movement of water, soil and parent material to different parts of the landscape (Graham 2006). Undulating topography and slope gradient will greatly influence the movement of parent material and soil downslope, with steeper slopes encouraging natural erosion of the surface layers. Soils found at higher elevations will differ significantly compared to soils downslope. In this sense, Brady (1974) indicates that topography is not only a modifier of climatic and vegetative effects but often acts as a major control in local areas. The speed of water movement across the landscape is also determined by relief, with flatter landscapes more likely to produce uniform weathering. Topography can therefore, speed up or delay the force with which these factors interact, affecting soil formation.

In Ireland, the majority of the mountain or hill areas (>500 m) are found along the western seaboard but mountain areas can also be found in Wicklow, Waterford and Tipperary. Elsewhere, hill (150–500 m) and drumlin landscapes, while not reaching a significant altitude, may sustain slopes in excess of 12° (Gardiner and Radford 1980). These slopes are likely to be subjected to gravitational forces sufficient to influence the movement of soil and water across the landscape. Topography also influences land use and management, as slopes with a gradient greater than 12° make tractor operations difficult (Mockler et al. 2013). This can determine anthropogenic factors that influence soil development, such as cultivation or fertilisation.

Climate represents a major influence in soil formation, in particular temperature and precipitation. Climate is a major determinant of vegetation which further influences soil formation. Temperature is important in chemical reactions and according to Van't Hoff's temperature rule, every 10 °C rise in temperature increases the velocity of chemical reactions by a factor of two to three (Jenny 1941). Soil temperatures in Ireland are higher in winter compared to most other parts of the world at the same latitude due to the moderating effect of the Gulf Stream (Gardiner and Radford 1980). Overall, soil moisture related to precipitation represents the dominant climatic factor affecting soil formation in Ireland. Water impacts soil formation in a number of ways, it can contribute to mechanical weathering through freeze/thaw action increasing the surface area exposed to weathering. Also, water is essential for chemical weathering (Duchaufour 1982; Shoji et al. 2006). Water performs a vital role in dissolution and transportation, acting as a medium for acids produced by living organisms that breakdown minerals into secondary minerals including clay, along with facilitating the growth of vegetation that acts on soil formation (Buol et al. 1997). In Ireland, precipitation exceeds evaporation and as a result, water movement down the soil profile is one of the dominant soil forming processes (Gardiner and Radford 1980). This results in leaching, where clay particles and nutrients are dissolved and moved down the profile, a process that accelerates with rainfall, however it is dependent on the texture of the soil. Most free-draining soils in Ireland are likely to require a continuous renewal of soil nutrients and lime above that which is removed via herbage and crop production alone (Gardiner and Radford 1980). A more detailed description of the Irish climate is presented in Sect. 2.3.

Fig. 2.3 Topographical features that affect soil formation shown across an Irish landscape. *Photo* Rogier Schulte

Soil biota includes a wide diversity of living organisms from microorganisms such as bacteria, fungi and algae to soil animals, including protozoa, nematodes, arthropods, insects and earthworms to higher animals and plants. Living organisms play a major role in profile development as they are necessary for organic matter accumulation, profile mixing, nutrient cycling and structural stability (Brady 1974). Soil biology alters the physical conditions of the soil through bioturbation and creates a range of chemical reactions. For example, soil biota may accelerate the rate of mineral weathering through the production of organic acids facilitating mineral solubilisation (Killham 1994), increasing nutrients in the soil solution. Living organisms play a vital role in the decomposition of organic matter with noticeable differences related to vegetation.

Soil profile development differs also in relation to vegetation. Exposed mineral material is gradually colonised by vegetation creating stable aggregates and soil structure (Duchaufour 1982). Gradually, the soil thickens and stabilises with the succession of plant communities (Duchaufour 1982). Soils under grassland systems have a large root biomass which contributes soil organic matter into the soil developing dark surface horizons (Brady 1974). In contrast, organic matter accumulation under forestry creates a shallow surface organic layer made up of leaf and woody litter. Vegetation present can define the cycling of nutrients, for instance, coniferous forestry is likely to remove a lot of base nutrients from the soil while the litter returned tends to be very acidic, low in cations, thereby increasing the potential for soil acidity. This affects soil organisms in relation to species diversity, as the conditions will favour more acid tolerant species, such as enchytraeid worms (Killham 1994).

As part of the CréBeo project (2005-S-LS-8) a baseline of keystone species important for soil functioning and ecosystem services in Irish soils were identified. The study was carried out on 61 sites, selected across a range of land uses (Schmidt et al. 2010). Mycorrhizas are formed by an association between a fungus and a plant root with a wide variety of vascular plants involved in these associations (Killham 1994). This association can enhance plant uptake of nutrients and can also enable the plant host to access nutrients in inorganic form that would otherwise be unavailable (Stribley and Read 1980; Killham 1994). The mycorrhizal fungi recovered during the CréBeo project included 446 arbuscular mycorrhizal fungi with significant differences found in relation to species richness and fungal community composition associated with different sampling methods and plant species (Schmidt et al. 2010). Arbuscular mycorrhizal fungi are fungi that can penetrate the surface cells of the vascular plant roots. The research also recorded a total of 19 out of 27 recorded species of earthworm (three anecic, seven epigeic and nine endogeic taxa). Endogeics, the most prevalent species found, exist usually in the upper 25 cm of the soil, but some species can burrow very deeply. They create hortizontal burrows which they back-fill as they move through the soil, eating the soil and excreting as they move through. They tend to be relatively pale in colour. A total of 92 nematode taxa were recorded to at least genus level from the 61 sites, with 13 new nematode species recorded for Ireland, two of which were new for both Ireland and the British Isles (Schmidt et al. 2010). One hundred and eight mite taxa were recorded in 48 of the sites, with 65 oribatid and 43 mesostigmatid taxa (Schmidt et al. 2010). In general, living organisms are a critical factor in determining soil character with rich fertile soil conditions containing a high number of active soil organisms.

Time is an important consideration as soil formation may take hundreds and thousands of years and it is on this basis that soil is considered a non-renewable resource. Soil formation is a dynamic process, representing a transition from one state of equilibrium to another, from parent materials to mature soils (Jenny 1941) and time will determine the duration for which a set of factors is active. In Ireland, the majority of soils have been formed in the Holocene time period, which is the period since the end of the last ice age, approximately 15,000 years ago. This means that soils in Ireland are considered young in a global context as parent material is still young and also because soil processes are comparatively slow (see Stages of soil evolution, Orgiazzi et al. 2016, p 110). In contrast, many believe that the soils in the tropical regions of Africa and South America may be millions of years old (Orgiazzi et al. 2016).

Anthropogenic factors are also responsible for changes or differences in soil character and may alter the natural development of soils through land use and management practices. Cultivation, forest clearance, land drainage, applications of lime and fertilisers are amongst the anthropogenic practices that will influence soil development. The interactions of land use and land cover shape our environment and our soils with farming, forestry, landscape, biodiversity and the built environment, all influencing soil development. Land cover describes what is visible on the land surface whereas land use refers to the use of the land from a human perspective (Wall et al. 2016). An assessment of Ireland in 2012 (Fig. 2.4) indicates that agriculture is the primary land use and land cover type accounting for two-thirds of the land base, followed by wetlands (15.56%) and forestry (9.37%) (Wall et al. 2016). Chapter 14 describes Irish soils in relation to agriculture and forestry, with soil biodiversity described in Chap. 18.

Study Box 2.1: Anthropogenic Impact on Soil Formation

Brown Podzolic soils represent an example whereby the podzolisation process can be actively delayed or inhibited through management. Unlike Podzols, this means that these represent good agricultural soils with a wide-use range. These soils are found further downslope from Podzols so that as well as experiencing lower rainfall, they are more accessible from a management perspective.

Brown Podzolic soils still display the reddish subsoil indicating acidic conditions at some point during soil formation (Fig. 2.5). Although they may highlight some leaching, they have not developed the poor acidic mineral topsoil or the bleached sandy layer characteristic of Podzols. Often, this may be the result of management where the acidification process has been offset through appropriate nutrient and liming amendments and changes in vegetation from ericaceous vegetation to grassland (Fig. 2.5).

Collectively, all of these factors control the rate of weathering of rocks, the constitution and composition of soils and their alterations through the profile (Gardiner and Radford 1980). The relative influence of the individual factors is responsible for the differences found in soils. None of the individual factors are uniform. Parent material, climate, topography and organisms are all highly variable, as is the age of the land surface. This means that there are countless combinations of soil formation factors, giving rise to a multitude of different soils. Despite this variability, soil distribution is not altogether random. Each soil reflects the

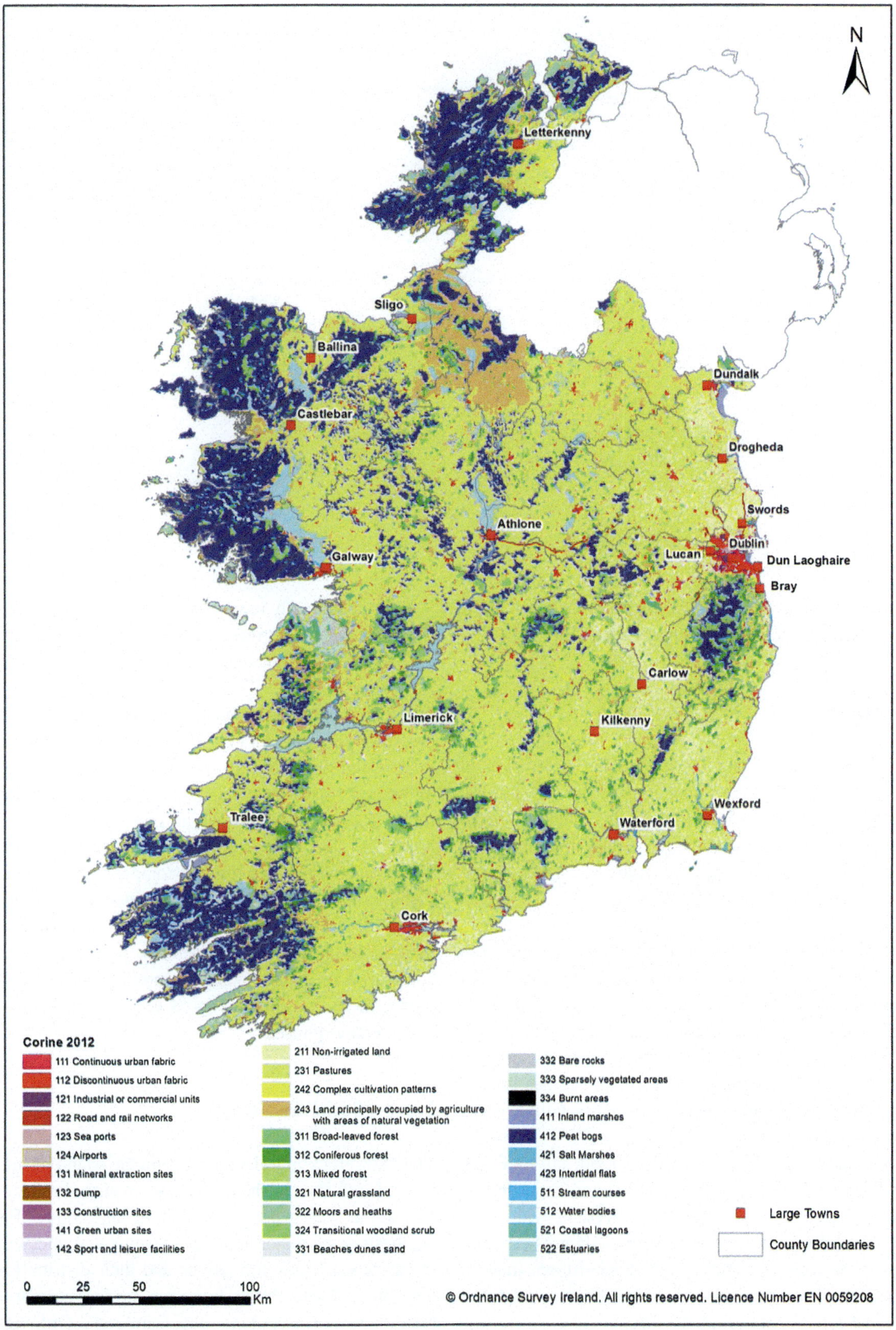

Fig. 2.4 CORINE 2012 land use and land cover map. *Source* EPA GIS stock

Fig. 2.5 Podzol (left) with acid mineral topsoil overlying a sandy bleached layer. Brown Podzolic (right) with red subsoil indicating acidic conditions during soil formation with a topsoil that is rich in nutrients, potentially due to additions of lime, manure or fertiliser. *Source* Irish Soil Information System

environment in which it has formed, occupying a definite geographic area and occurs in patterns with other soils.

2.2 Geology

Rocks contain minerals and the weathering or erosion of rocks largely determines the mineralogy of the soil. Indeed, weathered granite will result in acidic soils while weathered limestone produces alkaline soils. In Ireland, there is a wide range of underlying rock types, generally classified as igenous, sedimentary and metamorphic rocks.

- Igenous rocks are formed from hot molten magma from which all rocks are essentially derived. Igenous rocks occur when the hot molten magma is injected into the earth's crust. If the magma cools within the earth's crust, it forms coarsely crystalline rocks such as granite. Alternatively, the magma may reach the surface as volcanic eruptions forming fine-grained crystalline rocks such as basalt.
- Sedimentary rocks are formed from sand, silt and clay rock fragments that are eroded, carried and deposited by wind and water. As sediments are overlaid with other materials, the sediments become compressed and cemented to form new rocks, such as sandstone, shale or limestone. Cementation arises from fluids containing dissolved salts flowing through the sediment.
- Metamorphic rocks are igenous or sedimentary rocks that have been altered due to pressure or temperature changes associated with movement or fracturing of the earth's crust. Slate is formed from shales, quartzite from quartz (silica) rich sandstone and marble from limestone.

Table 2.1 builds on the early work by Kilroe (1907) who presented the soils of Ireland in relation to their geological origins and on more recent work by the Geological Survey of Ireland (Sleeman et al. 2004). The classification of the rocks in Table 2.1 provides an overview of the main rock types and the typical soil characteristics associated with weathering and altered mineral composition. Two groups are defined; first, rocks that are stratified and second, those where no true bedding is observable. The first group is mainly composed of sedimentary rocks while the latter consists of igenous rocks. Metamorphic rocks are classified within these two groups, based upon the original rock that was metamorphosed.

2.2.1 Solid Geology

The oldest rock in Ireland occurs on the island of Inishtrahull located off the Inishowen coast, dating back to almost 1.8 billion years ago (Sleeman et al. 2004). Figure 2.6 synthesises the major geological events that affected Ireland, the associated rock formations and their distribution in Ireland from Precambrian times up to the current time. In relation to solid geology; the Cambrian, Ordovician, Silurian, Devonian, Carboniferous and Permian Periods collectively form the *Palaeozoic Era,* and most of the

Table 2.1 Classification of rock groups and potential for soil development

Rock group class[a]	Basis	Soil formation
Group I—stratified rocks: aqueous—deposited from and lie in strata		
Mechanical	**Clay (marl), shale, *slate, mica-schist*** Hydrated aluminium silicate is base for these rocks with variable proportions of sand (silica) resulting in sandy loams, or strong clay soils. Neither constituent provides soil fertility but if potassium feldspar, or calcium carbonate, other sulphates, chlorides and phosphates are present, these provide fertility.	**Marl**—typically calcareous clay but often contains high calcium carbonate content **Shale**—moderately hardened clay ***Slate***—more hardened clay but very fissile, splitting easily along planes of weakness but requires prolonged exposure to become friable mould or clay ***Mica-schist***—crystalline rock forming clay and fine sand on ultimate weathering
	Sand, sandstone, grit, *quatrzite*, with associated conglomerates silica (sand) is the basis of this group, highly variable depending upon the content of other minerals. Many may be argillaceous (clayey), or calcareous graduating into arenaceous (sandy) limestone or feldspathic. Texture will depend on the size of the contributing particles which includes conglomerates.	The disintegrated materials of these rocks are mixed debris, being sandy or loamy depending on the original rocks.
Chemical	**Rock salt, gypsum** Deposits of rock salt and gypsum occur when waters evaporate from lakes.	
Animal	**Lime, Limestone, dolomite, *marble*** these occur in layers of biogenic sediments that form on sea-floors when calcareous shells of invertebrate animals disintegrate. Dolomite and marble are produced as a result of subsequent changes. All can contain other matter, such as clay or sand and the colours can be grey, yellow, blue or black: dolomite is usually yellow and marble is often white. All of these rocks weather freely, especially the more earthy and sandy varieties. The constituents of these rocks contribute to soil fertility.	**LimeLimestone**—base is calcium carbonate but may graduate to shaly or sandy limestone, or calcareous shale or grit when carbonate content is low **Dolomite**—usually limestone where a portion of the calcium carbonate is replaced by magnesium carbonate ***Marble***—is metamorphosed limestone and crystalline
Vegetable	**Peat lignite, coal, *anthracite and graphite*** The accumulation of organic matter called peat which derives from dead and decaying plant material occurs under high water saturation conditions. Vegetable matter is transformed and where excess water fills pores, results in low oxygen diffusion and anaerobic conditions. Acidic conditions result from anaerobic respiration producing organic acids and a low pH.	**Peat** soils are predominantly organic material with a minimum thickness of 40 cm or more of peat in a reference section from 40 to 80 cm depth. Soils accumulated either as autochthonous (formed in present position) or as constituents of sub-aquatic sediments
Group II—crystalline or non-stratified rocks: igenous—formed of molten magma, solidified in various forms. No true bedding observable		
Volcanic: lava ejected at surface		
Acid (>60% silica)	**Felsite, trachyte, obsidian, ash**	**Felsites** including rhyolite and obsidian on weathering deliver soil that contains potash
Basic	**Andesitic and basaltic lavas, ash**	**Andesite** is intermediary between the acid and basic rock groups. Composed of sodium–calcium feldspar, hornblende or augite
		Basalt is dark in colour. Clays derived from basalt tend to weather rapidly and supply several nutrients but also toxins depending on minor minerals present
Plutonic: molten material failed to reach surface at time of injection		
Acid (>60% silica)	**Granite, felsite, gneiss**: Granite, Gneiss: crystalline aggregates of feldspar, quartz and mica of variable proportions Feldspar is mainly orthoclase Quartz usually colourless and limpid Mica is typically white, black, dark green or weathering brown	**Granite and gneiss**—yield relatively unfavourable soils **Felsites** including rhyolite and obsidian on weathering deliver soil that contains potash
Basic	**Diorite, gabbro, basalt, dolerite, hornblende-schist**: Igenous rocks rich in ferromagnesian silicate minerals	**Diorite**—similar to andesite and can contain some quartz (acid). They provide more lime to soil than acid rocks, and may contain phosphates **Gabbro, dolerite**—these are similar to basalt, but coarse-grained. • Hornblende-schist—metamorphic equivalent the last two

[a]Metamorphic rocks shown here in italics

Geological Period	Million years ago	Environmemt and events in Ireland	Associated rock deposits	Occurrence
Quaternary	2.6-0	Ice Age glaciations and warm periods	Tills, glaciofluvial sands and gravels, peat	Widespread
Tertiary	66-2.6	Crustal stretching led to period of volcanism in Ireland 62 -58 million years ago. Tropical weathering resulted in aluminium-rich clays.	Ash, basalt flows, acid igneous rock, gabbro, granophyre	Carrick-a-rede (ash). Giant's Causeway, Antrim; Ring of Gullian and the Cooley Mountains at Carlingford.
Cretaceous	145-66	Upper Cretaceous marine deposition occurred as warm shallow sea covered Ireland.	White chalk, limstone.	Cliffs around Antrim and Causeway Coast and Limavady, Co. Derry.
Jurassic	201-145	Ireland - 40°N of equator.	Marine sediments.	Larne (early Jurassic mudstones). Offshore basins
Triassic	252-201	Terrestrial dessert conditions with deposition of sandstone and siltstone. Later lacustrinal conditions resulted in thick sediments and evaporite accumulation.	Sandstones, halite (rock salt).	Belfast, Co. Antrim, the Porcupine and Irish Sea basins in Celtic Sea basins.
Permian	299-252	Ireland in tropical latitude north of the equator, New Red Sandstone Desert.	Cyclical deposition of limestone, shale and salt layers. Dolomites occur in north but gypsum deposits occur in the south.	Permo-Triassic rocks mainly preserved in northeast. Kingscourt outlier straddling counties Cavan, Monaghan and Meath.
		Vaisian Orogeny - Late Carboniferous/Early Permian, resulting in folding -anticlines and synclines.	Sandstones, mudstones of Devonian Old Red Sandstone in anticlines and Carboniferous valley floors.	Munster ridge and valley province.
Carboniferous	359-299	Equitorial latitude, wet climate, coal deposition; variable sea level.	Deep water shales, detaic sandstones containing coal measures.	Shale- Shannon Estuary in Counties Limerick and Clare. Sandstones in West Clare at Cliffs of Moher and Bridges of Ross. Coal deposits- Arigna, Co. Roscommon, Castlecomer, Co. Kilkenny, Bridgetown, Co. Wexford and Kish Basin in Dublin Bay.
		Central plain covered in broad shallow waters. Crustal stretching leads to block and basin, with shallow shelf seas and deep basins.	Limestones from biogenic sediment - shallow areas with relatively pure limestones, calcareous mudstones in deeper basins.	The Burren, Co. Clare, The Marble Arch Caves of the Cuilcagh Mountains of Cavan-Fermanagh. Extensive poorly exposed areas in Midlands.
		Tropical latitudes, sea encroached from the southwest of Ireland due to global rising sea level. Early Carboniferous - estuarine, beach and shallow marine deposits.	Sand and mud in marine area, Old Red Sandstone further north.	Marine sandstones in south Cork, terrestrial and later marine sandstones in Roscommon
Devonian	419-359	Ireland on edge of Old Red Sandstone Continent by mid-Devonian moved from 35°S to equator. River processes dominated depostion. Semi-arid climate and low water table cause oxidation giving sandstone red colour.	Old red sandstone unconformably on eroded sequences of silurian rocks, with conglomerates, sandstones, siltstones and mudstones.	Munster Basin - thickest non-marine in Europe at 6 km on Iveragh Peninsula. Northern margin extending eastwards from Dingle Bay to Galtee Mountains and Comeragh Mountains to Waterford coast.
			Intermontane Basins - old red sandstone with conglomerates and alluvial fan deposits.	Limited to northern part around Clew Bay, Curlew Mountains, Fintona and Cushendall.
		Crust subduction during closure of Iapetus Ocean; Ireland amalgamates in Acadian orogenic phase, forms mountain chain, part of Caledonian orogeny. Granites intruded and crystallised. Mountains worn down through aggressive erosion.	Sedimentary - hugh volumes of mud, sand, gravel, pebbles	
			Granite plutons.	Donegal, Mayo - Sligo, Galway, Carlow - Wicklow - Dublin, Armagh - Down.
Silurian	443-419	Iapetus Ocean narrowed creating smaller distinct sedimentary basins. Deep marine, shallow water or subaerial river deposition. Silurian volcanism in Dingle.	Basalt lavas and rhyolitic tuffs	Clogher Head, Co. Kerry.
			Shallow deposition sedimentary - marine mudstones and sandstones rich in fossils such as shellfish and corals.	Dingle area of Co. Kerry; Croagh Patrick-Louisburgh , C. Mayo
			Sedimentary - deep water	Central and SE Ireland including the uplands of the south midlands - Slieve Bloom to Galtee mountains and east Kildare.
Ordovician	485-443	Northward subduction creates accretionary margin; Ordovician to Siluran oceanic sediments added to northern continental margin in slivers	Bands of sandstone and shale	Longford - Down Inlier: Down, Monaghan, Cavan and Longford.
		Collision between northern volcanic arcs and Laurentia resulting in Grampian Orogeny, intrusion of magma and metamorphism of Dalradian sedimentary rocks.	Gabbro to granite intrusions; Dalradian metamorphism rocks.	Connemara, N Mayo, Ox Mountains, Donegal
		Deep water environment	Oceanic sedimentation, turbidites	Wicklow-Wexford slate belt; south Mayo
		Subduction of seafloor as Iapetus Ocean gradually closed resulting in volcanic arcs	Lavas, tuffs	South Mayo, Wicklow to Waterford
Cambrian	541-485	Iapetus Ocean reaches maximum width (5,000 km) Fossils appear due to evolution of multi-cellular marine organisms with hard parts (shells etc.)	Marine sedimentary rocks	East and southeast - Howth Head, Bray Head, Sugar Loaf Wicklow and between Cahore Point and Waterford Harbour
Precambrian	1780-541	Thick marine sedimentary sequence deposited on stretched Laurentian continental margin	Dalradian Supergroup, metamorphosed marine sedimentary	Donegal, north Mayo, Connemara
		Ancient continental rocks: Inistrahull and Annagh Head- oldest rocks in Ireland are part of Laurentia, Rosslare Complex part of Gondwana, on opposite sides of Iapetus Ocean	Coarsely crystalline banded gneisses	Inishtrahull, Co. Donegal & Mullet Peninsula, C. Mayo; Rosslare, Co. Wexford

Fig. 2.6 Geological events timeline for Ireland from Precambrian period to date

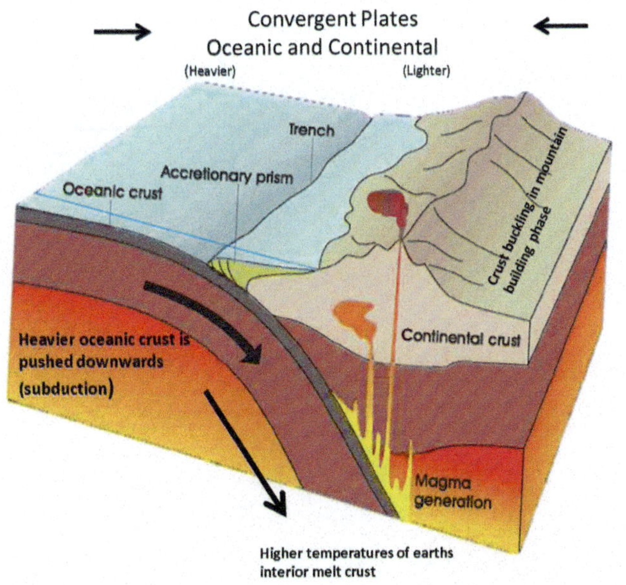

Fig. 2.7 Model of a subduction and volcanic zone adapted from Sleeman et al. (2004). When two lithospheric plates meet, one plate is pushed down to the Earth's interior in a process of subduction. Continental plates cannot be subducted due to a relatively low density and when both plates are continental a huge crumple zone pushes up mountains

underlying geology of Ireland dates from that timeframe (Fig. 2.6). As representing the most widespread periods, they are described in further detail below.

Ordovician Period 495–440 million years ago: From the Cambrian Period on, the presence of marine organisms with hard parts made the preservation of fossils and specific geological dating much easier. Fossil evidence from the Ordovician Period indicates distinct communities on the then, two halves of Ireland, on either margin of a wide ocean called Iapetus. Volcanic and marine sedimentary rocks occurring in the south-east, from Wicklow to Waterford, and in the north-west, in Mayo and Tyrone, were formed in subduction zones as Iapetus gradually closed (Fig. 2.7) (Graham 2001). A wide band of sandstone and shale running from Down, through Monaghan and Cavan to Longford, called the Longford—Down Inlier, formed as a result of scraping off of Ordovician and Silurian marine Accretionary prism deposits on the continental margin during subduction.

Silurian Period 440–420 million years ago: The Silurian Period was a time of sedimentation in the closing Iapetus Ocean and in several small, distinct sedimentary basins on its margins. Silurian rocks of the south midlands, such as the Slieve Blooms, the Galtee Mountains and East Kildare, were formed by deep-water sedimentation by turbidity currents (Fig. 2.8), which are rapidly moving underwater sediment-laden currents. Other Silurian sedimentary rocks can be found between Killary Harbour and Lough Mask.

A mountain-building phase called the Caledonian Orogeny occurred during the late Silurian and early Devonian Periods when Iapetus finally closed and the continents on either side collided. The mountains of the Caledonian Orogeny were eroded and provided the huge volume of sediment for the sedimentary rocks of the subsequent Devonian and Carboniferous Periods. During the orogeny, large volumes of magma were injected into the earth's crust and cooled below the surface to produce granite. This granite can be found in the counties of the north-west and the east of the country.

Fig. 2.8 The internal structure of a typical turbidite bed from Sleeman et al. (2004). Turbidity currents are sediment-laden currents that travel downslope under gravity in subaqueous environments, such as oceans. The rocks deposited by turbidity currents are called turbidites and can form extensive beds (2 cm–2 m thick), occurring in sequence over a long time period spanning hundreds and thousands of years (Sleeman et al. 2004). Much of Ireland's Ordovician and Silurian sedimentary rock sequence is composed of turbidites

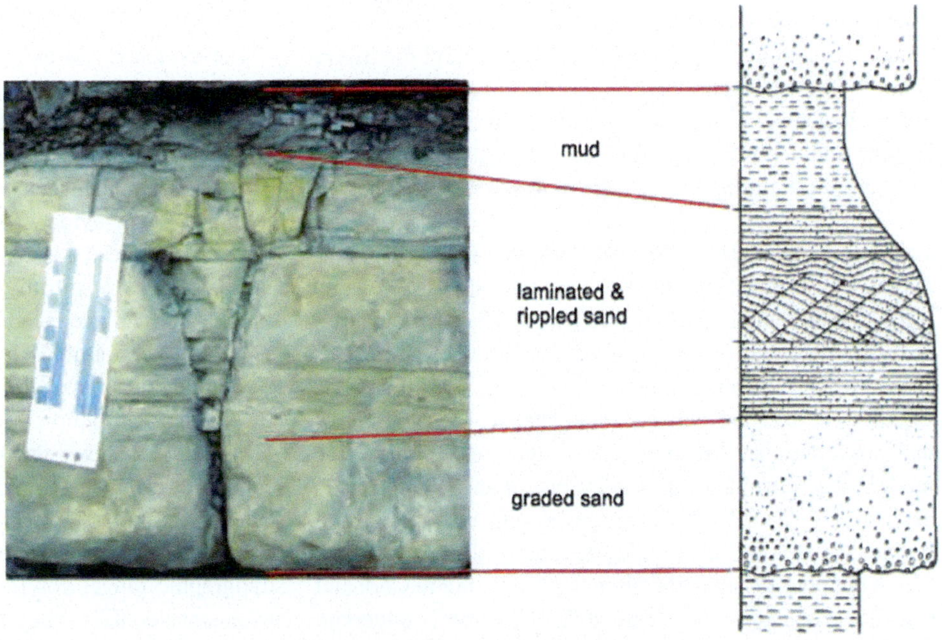

Fig. 2.9 Image from West Cork where old red sandstone represents the underlying geology. *Photo* Rogier Schulte

Devonian Period 420–335 million years ago: During the Devonian period, Ireland was situated on the 'Old Red Sandstone' continent. Ireland experienced aggressive erosion and deposition of the sediment from the Caledonian Mountains during this time. The Munster Basin of this time is the thickest non-marine Devonian sequence in Europe, stretching from Dingle Bay, through the Galtee and Comeragh Mountains to the Waterford coast (Sleeman et al. 2004). In places, the Old Red Sandstone is found to reach 6 km in thickness along the peninsulas in West Cork (Fig. 2.9) and Kerry (Pracht and Sleeman 2002). Late Devonian rocks can be found surrounding upland areas including around the Slieve Blooms, Slieve Aughty and Slieve Phelim (Sleeman et al. 2004). Volcanic activity also occurred during this period. Erosion and deposition continued until inundation of the sea from the south during the Late Devonian Period.

Carboniferous Period 355–298 million years ago: This Period was a time of limestone formation due to biogenic $CaCO_3$ sedimentation from invertebrate shellfish. Invertebrate faunas, contributed enormous quantities of calcareous grains, which consolidated overtime to create solid limestone (Sleeman et al. 2004). Physical and chemical weathering of the limestone has created one of the most remarkable modern-day karst landscapes in Europe, found in the Burren in Co. Clare (Chap. 9). The Cuilcagh Mountains of Cavan and Fermanagh are other areas of karst, home to the Marble Arch Caves but pockets are found elsewhere including in Cork and Galway. Altogether, more than 40% of Ireland is underlain with Carboniferous rocks, and the entire Central Plain is extensively Carboniferous strata with the exception of some scattered inliers (Gardiner and Radford 1980), which are older rocks that are completely surrounded by younger rocks. Soil pH has an important role in soil fertility and soils that include limestone in their genesis are likely to have a pH that favours microbiological activity for better nutrient recycling and release. Within the Late Permian, the Palaeozoic Era was coming to a close and formation of the rocks with greatest influence on soil formation in Ireland had already taken place.

The evidence and extent of deposits associated with the Late Permian, Triassic, Jurassic and Cretaceous Periods are mostly limited to the north-east of the island, in Northern Ireland. The main phase of volcanism during the Tertiary period occurred between 62 and 58 million years ago, and again is mostly evidenced in Northern Ireland. Volcanic eruptions produced basalts including the spectacular Giant's Causeway in Co. Antrim. During the Tertiary Period, tropical weathering produced a land that was covered in a residual of aluminium rich clays. By now, Ireland was land and the continents and the oceans had almost reached their current day distribution.

Figure 2.10 shows the Bedrock Geology of Ireland map reproduced from the Geological Survey of Ireland original scale of 1:1,000,000 map and includes the offshore and onshore geology. The bedrock geological legend shows the stratigraphic ranges for Ireland as a whole, but individual successions contain many more breaks.

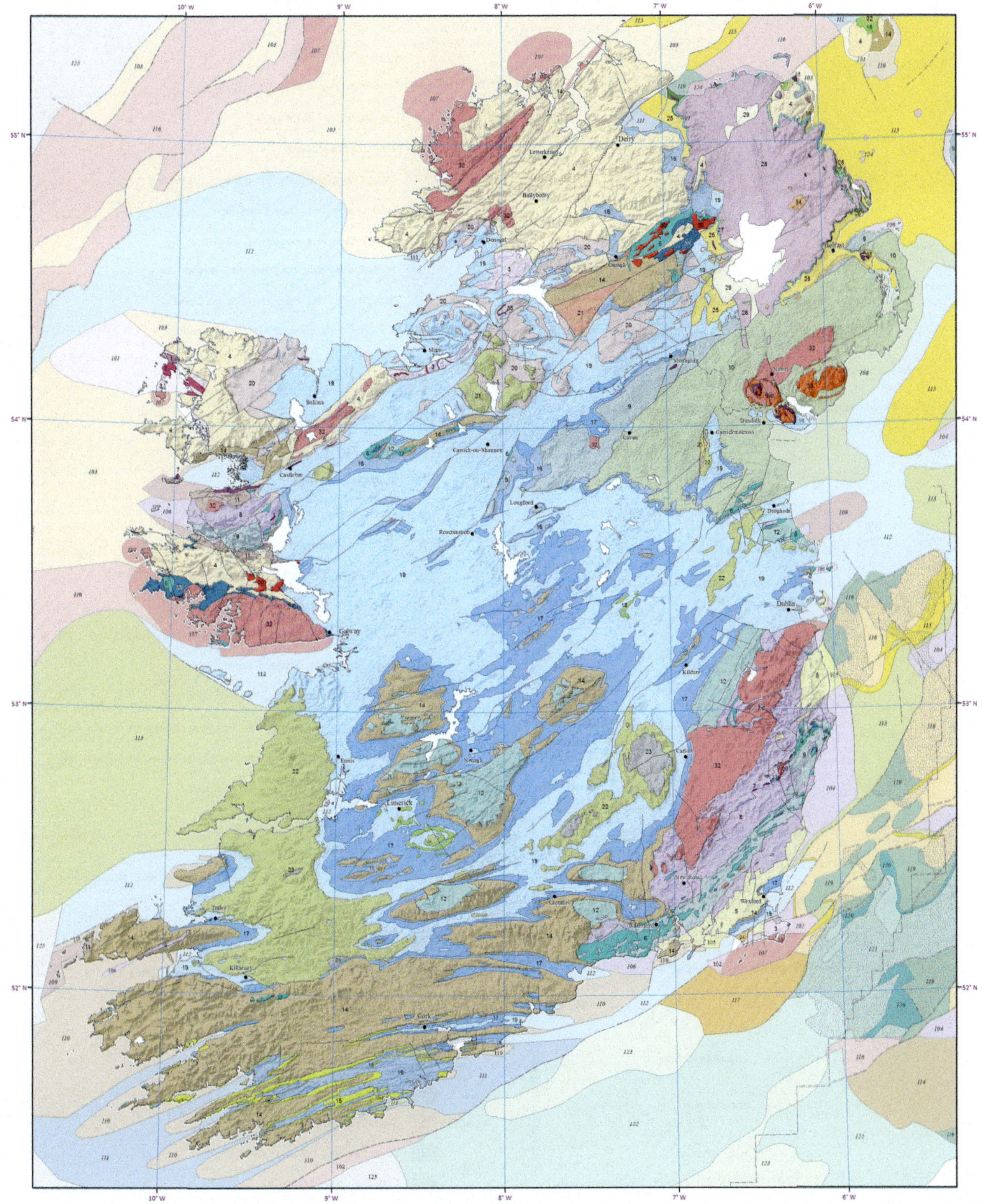

Fig. 2.10 Bedrock geology of Ireland map from Geological Survey of Ireland (2014)

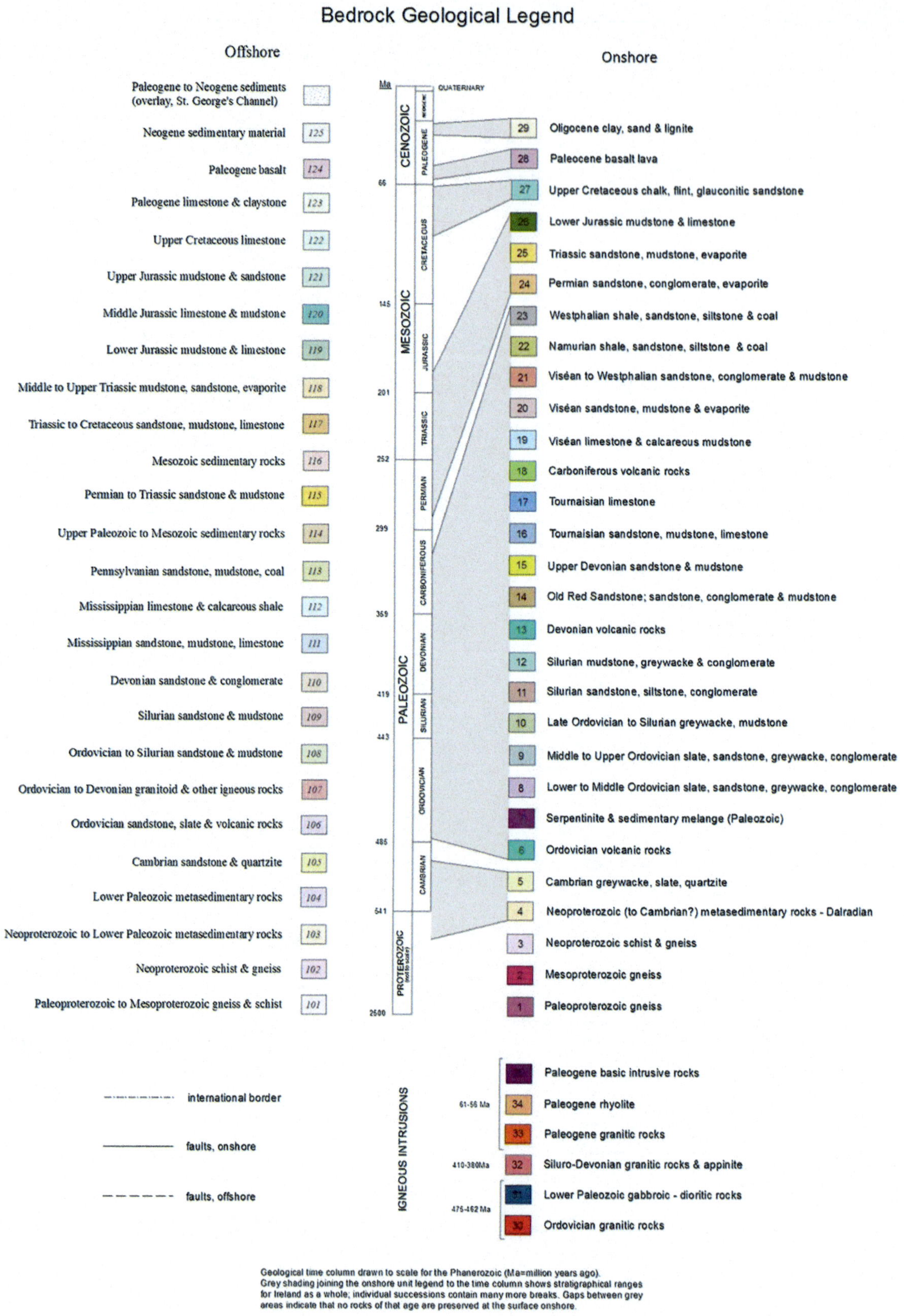

Fig. 2.10 (continued)

2.2.2 Quaternary Geology

In geological timelines, the last 2.6 million years are referred to as the Quaternary Period which can be divided into two series or epochs, the Pleistocene from 2.6 million years ago until 11,700 years before the present day, and the Holocene, occurring from that time until the present day (GSI 2016). Glacial deposition refers to the laying down of sediment that had once been transported by a glacier. The deposited material is referred to as glacial drift which is considered to cover about 85% of Ireland. Glacial drift therefore represents the majority of parent material in Ireland. Glacial geology can be further subdivided into till and fluvio-glacial deposits. Glacial till or boulder clay is unsorted material that has been directly deposited by the ice, such as drumlins. Landform features from glacial till include moraines (Fig. 2.11) and drumlins. Sorted glacial deposits are referred to as fluvio-glacial as these have been sorted by glacial meltwaters. Common fluvio-glacial features include gravels, eskers, lake sediments and kames (Fig. 2.12). Overall, moraines, eskers and drumlins are amongst the most commonly found glacial deposits in Ireland.

Study Box 2.2: Moraines—Sorted Till

Moraine glacial features are commonly found across the Irish landscape and are comprised of glacial till that has been deposited. Several types of moraine features exist including; ground moraines, lateral moraines and terminal moraines. Ground moraines occur where glacial till has been deposited on the valley floor typically found where the glacier meets the rock underneath (Fig. 2.11). Lateral moraines occur along the valley walls where the valley walls are broken up by frost thereby adding material to the glacial ice surface. The material is carried along until the ice melts leaving a ridge of material along the valley side. The moraine that occurs at the furthest extent of the ice is referred to as the terminal moraine.

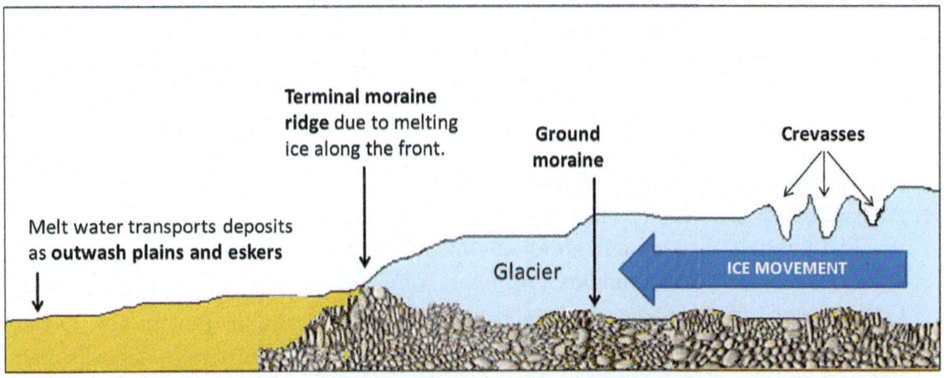

Fig. 2.11 Visual representation of moraine landforms associated with glacial deposition

Fig. 2.12 Hummocky topography of a kame and kettle landscape in Co. Wexford. *Photo* Rogier Schulte

Eskers

Eskers are fluvio-glacial deposits formed of washed sand and gravel and are aligned parallel to the ice flow in which they were carried. Eskers often take the form of long sinuous sand/gravel ridges and are free-draining in nature usually found with steep sides. Where they occur on valley floors they occur within the former ice margins defined by the moraine systems and show to form beneath glaciers. The eskers of the midlands of Ireland are considered among the finest in the world. Historically, eskers have been excavated for sand and gravels for the construction industry as well as providing a stable foundation for many roads in the midlands which is often a boggy or difficult terrain (Hammond and Brennan 2003).

Kame and Kettle Topography

Kames are formed in depressions on the glacier surface that contain fluvio-glacial sands and gravel. When the ices melt the sand and gravel that was previously trapped in cavities fall to the ground surface to form irregular hills and mounds. These are often found with kettle holes which are depressions that remain when blocks of ice from the ice sheet melted. Together, these create a 'kame and kettle' hummocky topography which can be found in the midlands of Ireland and in the south-east in Co. Wexford (Fig. 2.12) around Curracloe.

Study Box 2.3: Drumlins—Unsorted Till

As glaciers surge across the landscape they accumulate material through plucking and abrasion which is later deposited, sometimes as drumlins. Drumlins, from the Gaelic word '*dromnín*' meaning small hill, are oval shaped hills that consist of boulder clay that was deposited as the glacier retreated. Drumlins can indicate the direction of the glacier movement as they are oblong, being steeper on the 'stoss', upstream end of a glacier receding motion, with a more gentle slope on the 'lee' side (Fig. 2.13). Drumlins typically occur in swarms and Ireland is home to one of the most extensive drumlin fields in Europe stretching from Strangford Lough to South Donegal (Gardiner and Radford 1980). The friability and nutrient status of drumlin soils is generally determined by the parent material of the Drumlin. Drumlin soils can range in terms of drainage status from reasonably dry soils to very heavy clay wet soils. Where the parent material is entirely composed of drift core will give rise to wet drumlins, whereas those with a rock core may be dry drumlins. Drumlins are considered difficult from an agricultural perspective due to the slope gradient which often limits the use of agricultural machinery.

Glaciation in Ireland

Ireland was affected by three successive glaciations, the Elster, the Munsterian (Saale) and the most recent glacial event, the Midlandian (Weichsel). During glaciation, ice flowed westwards from Scotland, England and Wales filling the Irish Sea basin and even depositing drift as far as the west of Ireland (Flint 1957). Few glacial deposits from the Elster glaciation can be found, so the latter are the main glacial events of the Irish Pleistocene.

During the Munsterian glaciation, a powerful ice cap developed in the Scottish Highlands, sending a massive ice sheet along the bed of the Irish Sea resulting in shelly calcareous till depositions across Ireland, predominantly along the east coast, but also west towards Cork, and north-west to Malin Beg in Co. Donegal (Gardiner and Radford 1980). Munsterian deposits can be distinguished from younger glacial deposits on the basis of their subdued relief, strong decalcification and soil development and the depth of cryoturbation (churning of material with frost) (Finch and

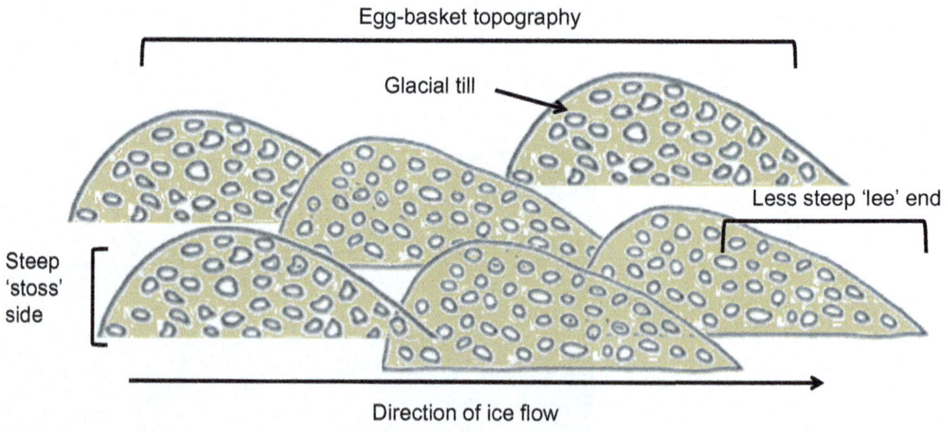

Fig. 2.13 Visual representation of a swarm drumlin landscape

Synge 1966). Landforms in the west of Ireland include hummocks and ridges that overlie glacially stripped and striated bedrock surfaces, comprised of undifferentiated till (Knight et al. 2004). In the east, the Wicklow Mountains and the surrounding areas are characterised by Irish Sea deposits that form hummocky sand and gravel landforms, and flat till plains (Culleton 1978). Glaciers formed over land in the east contained limestone and granite erratics within till (Knight et al. 2004) while the Irish Sea ice deposited till containing sand and gravel, with a mixture of marine shell fragments and local erratics, such as Leinster granite (Culleton 1978).

The cold-climate period of the Midlandian was associated with less extensive ice masses and comprised of a number of substages; glacial and non-glacial. During the main stadial (cold) phase, Scottish ice advanced once more, but unable to surmount the Antrim coast, the ice was forced to the north and to the south. As a result, the Irish Sea ice mainly affected a narrow belt along the east coast of Ireland (Gardiner and Radford 1980), with clay-rich deposits occurring in Wicklow and Wexford (Mitchell and Ryan 1986), sometimes 60 km from the coastline. These heavy clay deposits can today be associated with heavy, silty clay soils, known as the Macamores in Co. Wexford. The resultant soils are heavy from a farming perspective as they are considered difficult to work with. While the ice mass of the Midlandian did not reach the southern parts of the country, its impact is widespread across the landscapes of central and Northern Ireland evidenced by topographic features such as kames, eskers, moraines and drumlins. Where independent ice caps formed in upland valleys, corries were scooped out, noticeable in the Macgillycuddy's Reeks in County Kerry. In upland areas, the soil was generally stripped, providing one source of drift material. The subsequent deposition of drift material in the lowland areas, often gave rise to fertile agricultural soils.

Altogether, glaciation in Ireland represented a grand scale reworking of Irish parent materials, such that the landscape became one draped in several metres of surface glacial till, or fluvio-glacial till of very mixed origin. Soil formation in Ireland is mostly initiated from this glacial till which forms the majority of parent material in Ireland. The retreat of the ice at the end of the glaciation of the Holocene Epoch means that currently, Ireland is experiencing an interglacial period. The majority of soils in Ireland have formed since that time, less than 15,000 years ago which means that they are young soils.

2.3 Climate of Ireland

Ireland's climate is strongly influenced by its position on the eastern side of the Atlantic Ocean and on the western fringes of the continent of Europe. This results in a mild maritime climate characterised by prevailing south-westerly winds. This ensures cool winters and relatively mild summers with rainfall well distributed throughout the year. Climate represents a major influence on soil formation, in particular temperature and precipitation, which combined are responsible for weathering. Moisture is involved in most physical, chemical and biological processes that occur in the soil (Collins et al. 2004). Temperature influences the rate at which these processes operate (Jenny 1941). Together, these parameters are key controls that define soil properties, soil development and land use (Collins et al. 2004).

2.3.1 Rainfall

Rainfall is one of the most variable meteorological parameters exhibiting high variation on an hourly, daily, weekly, monthly, seasonal and annual basis. Rainfall also highlights considerable localised effects due to the influence of topography on rain-bearing winds (Keane and Sheridan 2004).

Mean annual rainfall ranges from 750 to 1000 mm in midland and eastern areas to between 1000 and 1400 mm elsewhere, with the higher totals in western coastal areas. The average annual rainfall over all of Ireland is approximately 1230 mm (Fig. 2.14). Rainfall increases significantly with height, with amounts over 3000 mm experienced on the highest mountains in the southwest, west and north-west. The difference between western and eastern parts represents a variation increase of 75% in the west (Keane and Sheridan 2004). The wettest areas of Ireland coincide with the Ombrotrophic (rain-fed) Peats of the western seaboard. Peatlands are characterised by an accumulation of organic matter derived from dead and decaying plant material under high water saturation conditions. The presence of water limits decay such that plant production exceeds decay, allowing the accumulation of peat. The wet mineral Surface water Gley soils of Ireland can similarly attribute their development in large part to wetness associated with the volume of rainfall received (Collins et al. 2004). Also, the development of Podzol and Brown Podzolic soils are directly related to colder, wetter conditions that result in the development of lower horizons rich in Fe, Al and humus due to leaching. Luvisols are associated with clay being washed down the profile and redeposited.

Seasonally, spring and summer are the driest seasons with approximately 260 mm of rain while winter and autumn are wettest, with approximately 350 mm (Fig. 2.15). The driest months are April, May, June and July, with an all Ireland average of approximately 80 mm each month; February, March, August and September have average rainfall totals of approximately 100 mm, while October, November, December and January have all Ireland averages of approximately 130 mm (Fig. 2.16).

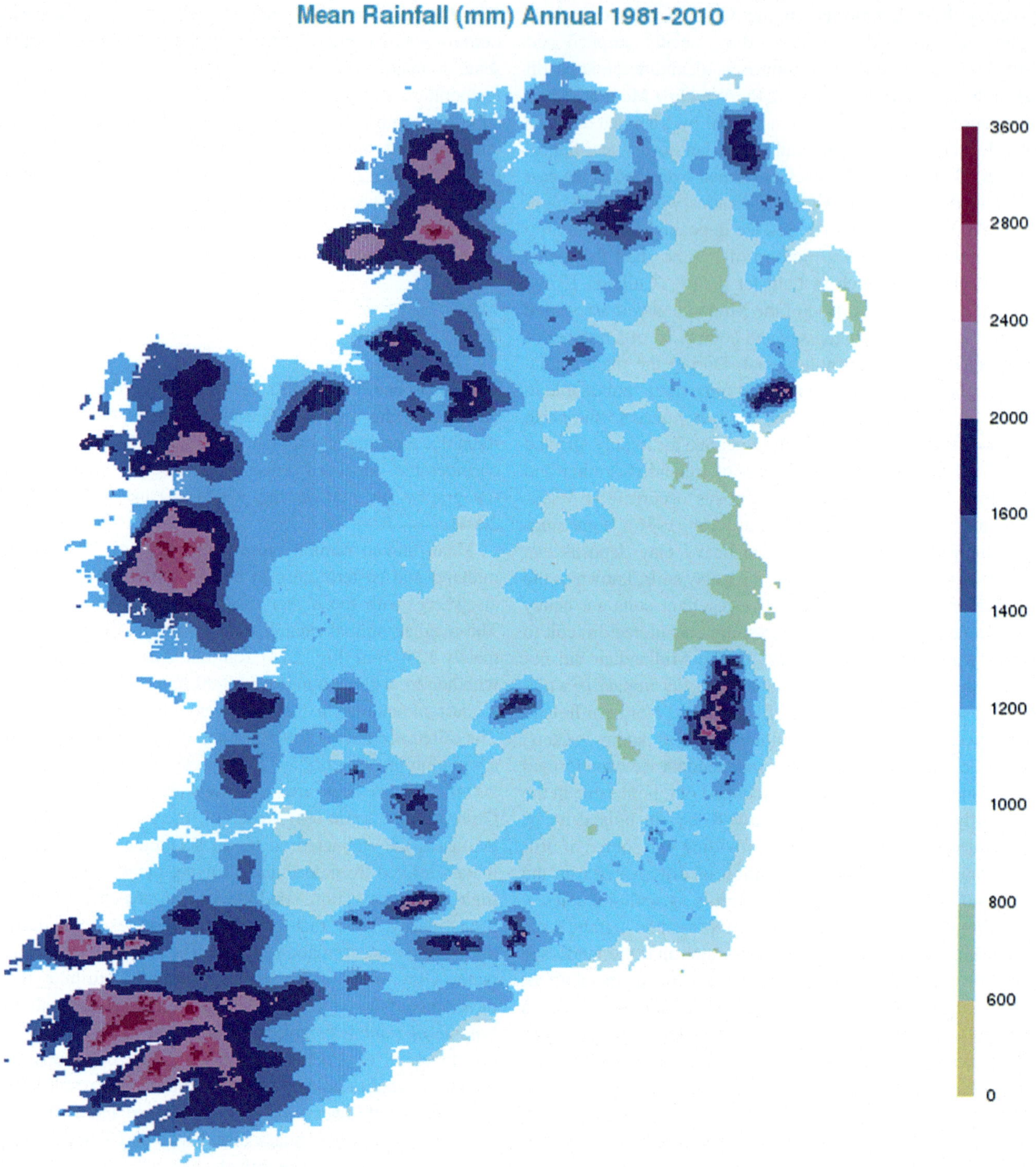

Fig. 2.14 Mean annual rainfall ranges for Ireland from 1981 to 2010 by Walsh (2012) based on data from Met Éireann, The Irish Meteorological Service. The average annual rainfall averaged over all of Ireland is approximately 1230 mm. Rainfall increases significantly with height, with amounts over 3000 mm experienced on the highest mountains in the southwest, west and north-west

Rainfall shows considerable variation on an annual, seasonal and monthly basis. Average annual and seasonal totals since 1942 are shown in Fig. 2.17.

The average number of days with rainfall exceeding 0.2 mm, 1 mm and 10 mm show a considerate bias towards more rain days in western areas and over mountains (Fig. 2.18).

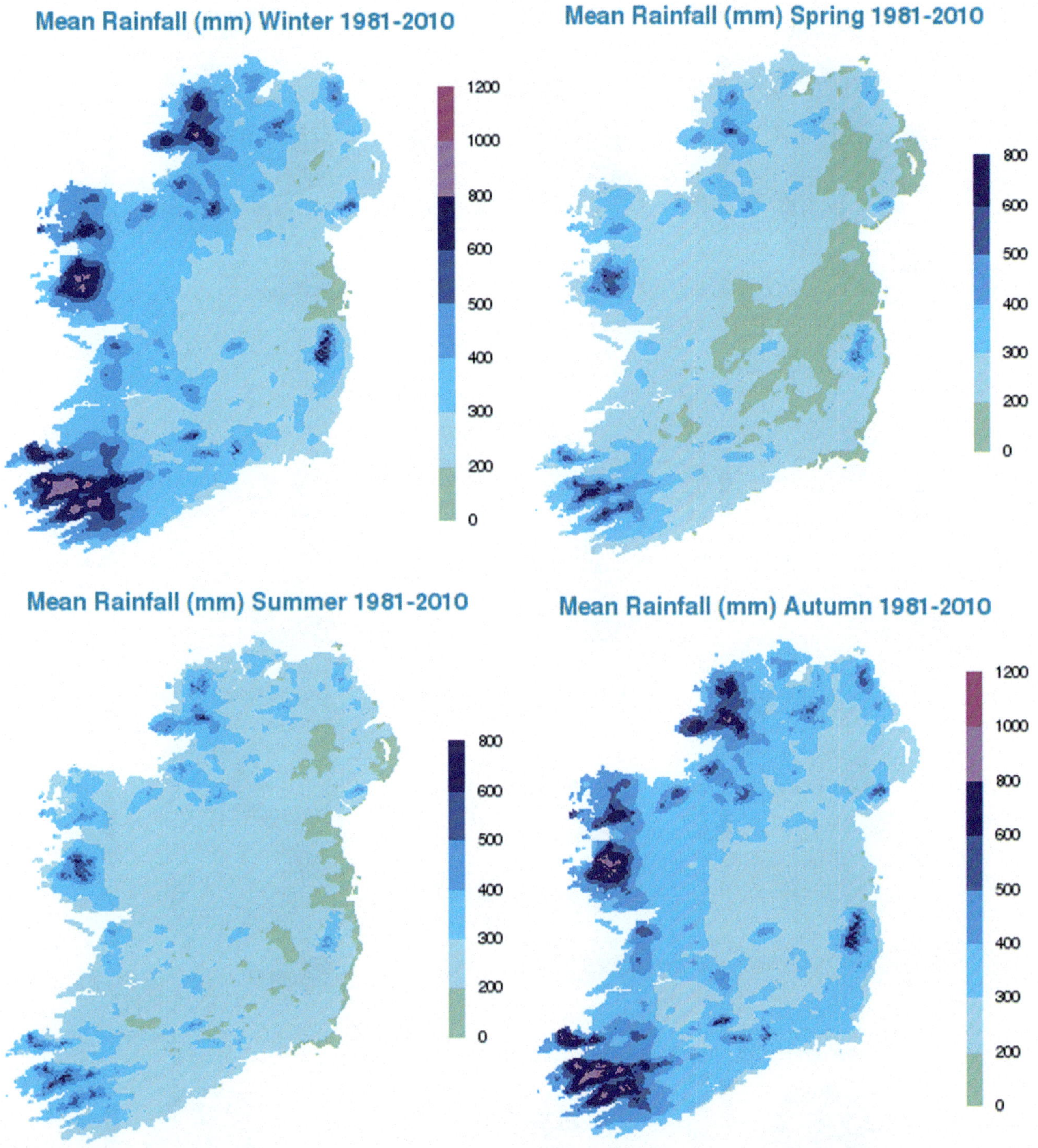

Fig. 2.15 Mean seasonal rainfall ranges for Ireland from 1981 to 2010 by Walsh (2012) based on data from Met Éireann, The Irish Meteorological Service

2.3.2 Air Temperature

Mean annual air temperatures generally range from 9 °C in the north-east to 11 °C in the southwest, upland areas are significantly cooler as temperature decreases at a rate of approximately 6.5 °C per 1000 m (Fig. 2.19).

Summer is the warmest season with mean temperatures ranging between 14 and 16 °C, followed by autumn and spring, winter is the coldest season with mean temperatures between 4 and 7 °C (Fig. 2.20). Highest maximum temperatures occur inland during the summer. Monthly values show great variation (Fig. 2.21). July is the warmest month

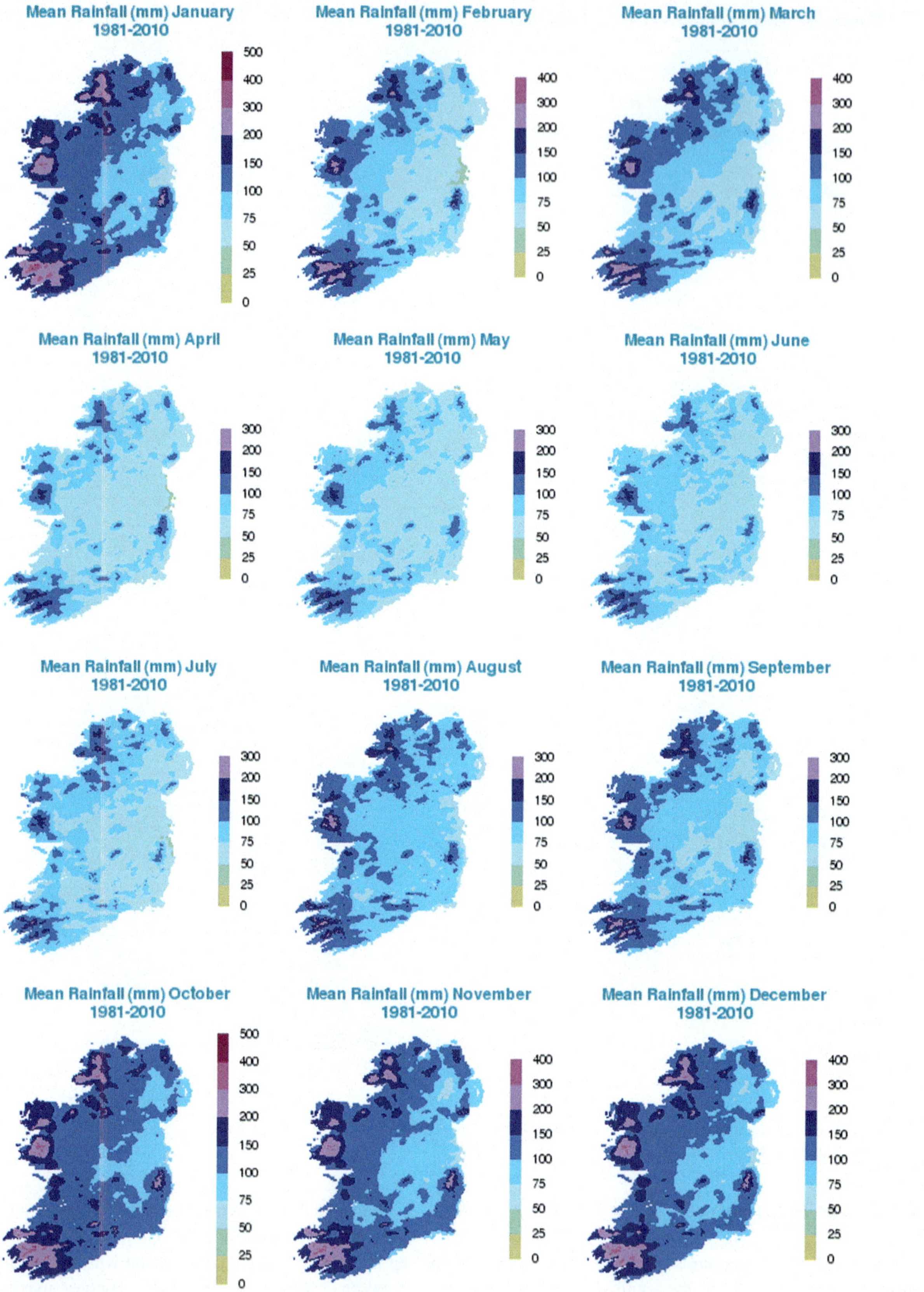

Fig. 2.16 Mean monthly rainfall ranges for Ireland from 1981 to 2010 by Walsh (2012) based on data from Met Éireann, The Irish Meteorological Service

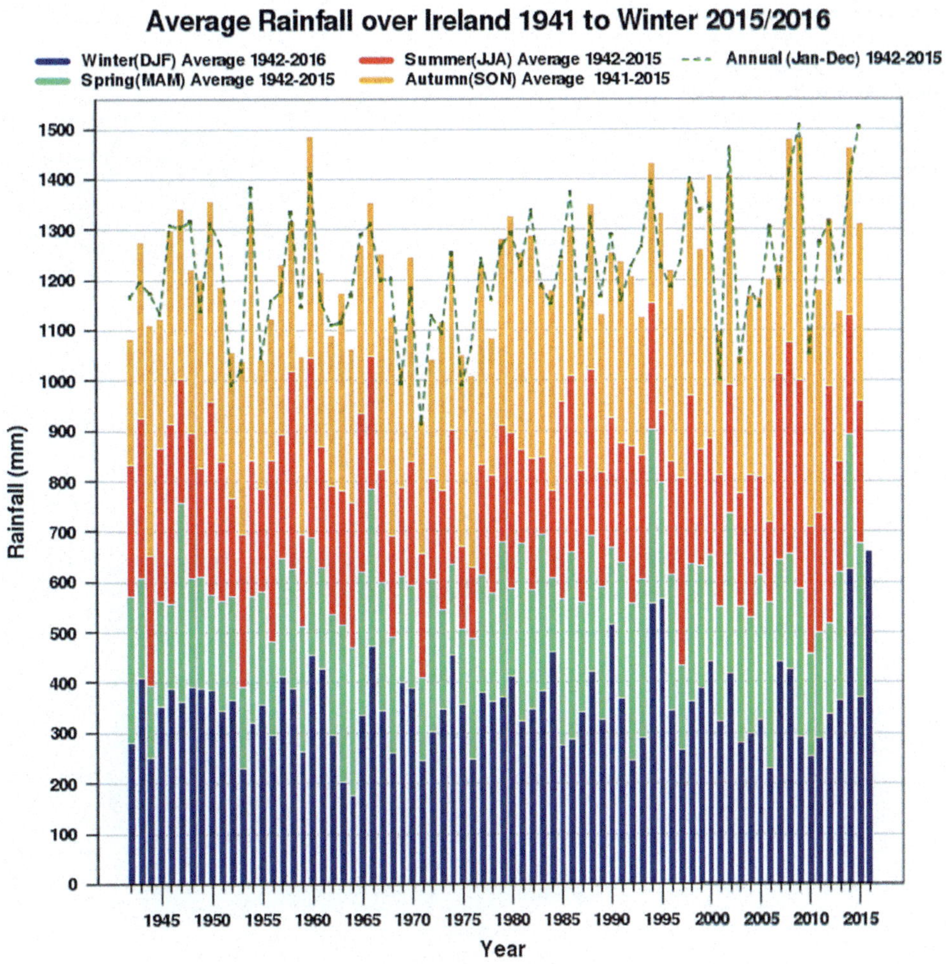

Fig. 2.17 Average annual and seasonal rainfall totals since 1942 based on data from Met Éireann, The Irish Meteorological Service

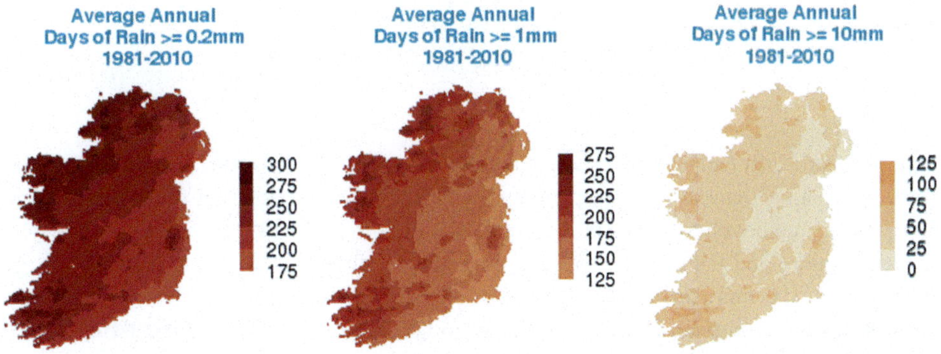

Fig. 2.18 Average numbers of days with rainfall ≥ 0.2, ≥ 1 and ≥ 10 mm by based on data from the Met Éireann, The Irish Meteorological Service

on average, with mean monthly maxima between 18 and 20 °C. The coldest month is January with mean monthly maxima ranging between 6 and 9 °C. During winter, coastal areas are mildest due to the moderating influence of the relatively warm surrounding seas. Extremes of heat and cold are relatively rare.

2.3.3 Soil and Earth Temperature

Soil temperature has a major influence on the growth and development of plants. Low temperatures inhibit the growth of most agricultural crops. The temperature of the soil is primarily due to the amount of radiation received from the

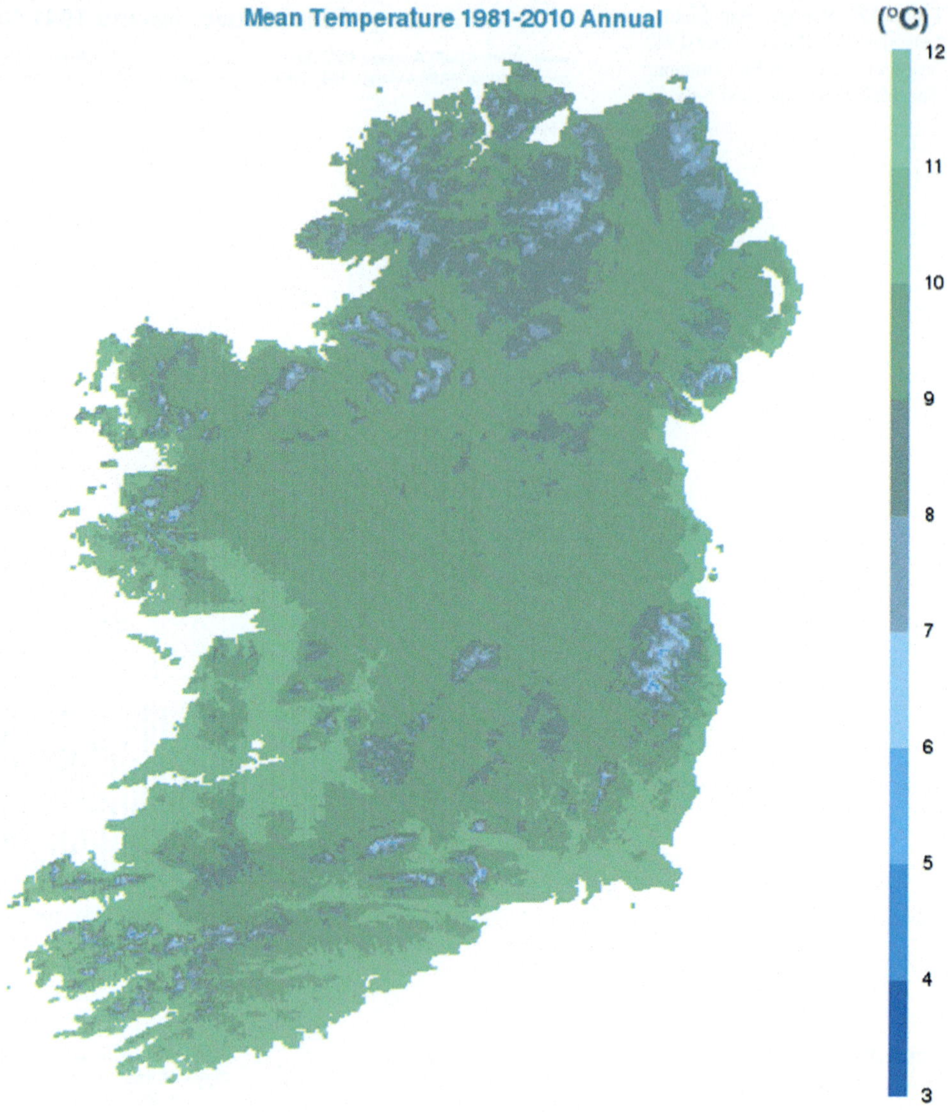

Fig. 2.19 Mean annual air temperatures from 1981 to 2010 from Walsh (2012) based on data from Met Éireann, The Irish Meteorological Service. Mean annual air temperatures range from 9 to 11 °C with upland areas is significantly cooler

sun. Soil and earth temperatures show much less diurnal and annual variation than surface temperatures which are measured at 1.2 m above ground level. Figure 2.22 shows mean air temperature and the average soil and earth temperatures at depths of 100 mm (soil), 300, 600 and 1200 mm (earth) for Casement Aerodrome, Co. Dublin. Closer to the surface, temperatures show greatest variation and follow air temperature patterns with a small time lag. The time it takes for the soil to transmit heat downwards, means that seasonal maximum values occur later, the greater the depth. Soil type and soil moisture regime also alter the time it takes for soils to heat up. The higher springtime temperatures in the southern counties allow earlier grass growth on free-draining soils that warm up more quickly (Collins et al. 2004). In contrast, the wet drumlin soils of the north-central counties take longer to warm up as more energy is required to heat them due to their higher moisture content. Overall, moisture content, air circulation, plant cover, cloud cover, aspect and slope all impact mean daily soil temperatures (Collins et al. 2004)

2.3.4 Potential Evaoptranspiration

Potential Evaoptranspiration (ET) is the flux of water vapour to the atmosphere through evaporation and transpiration under unrestricted moisture availability. It depends on meteorological factors such as radiation, temperature, wind speed and relative humidity. ET can be measured with lysimeters, but it is more usual to calculate a reference evapotranspiration (ET_o) for a standard grass surface by application of a mathematical formula (e.g. FAO56), using meteorological input data. Daily rates of ET_o vary from less than 1 mm/day in winter to 4 or 5 mm/day in summer. Average annual ET_o generally ranges

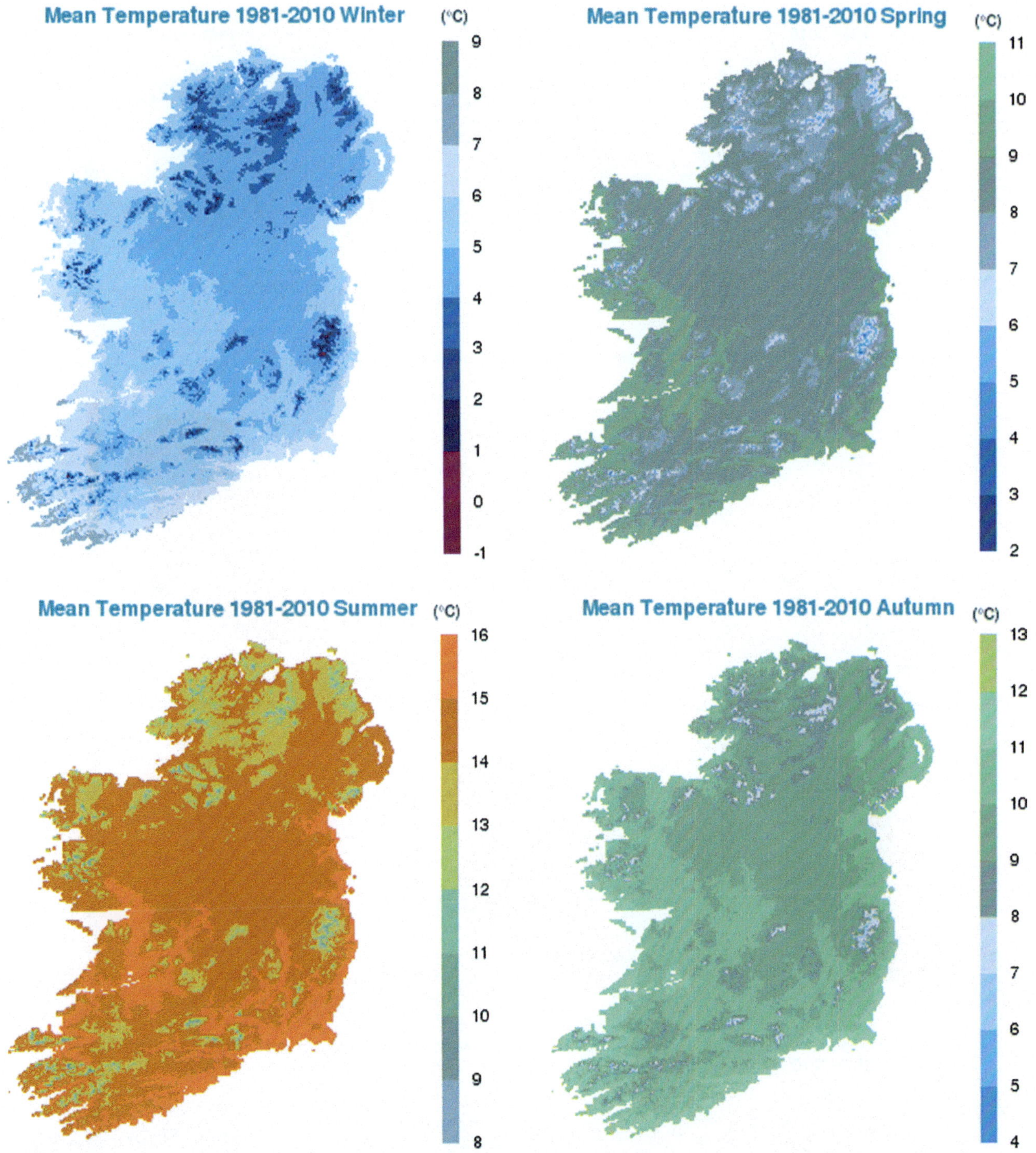

Fig. 2.20 Mean annual air temperatures from 1981 to 2010 by season from Walsh (2012) based on data from Met Éireann, The Irish Meteorological Service. Summer is the warmest season followed by autumn and spring. Winter is the coldest season with mean temperatures between 4 and 7 °C

from 470 to 500 mm in midland regions to 550 to 600 mm in coastal regions, highest values in the south and southwest. The actual evapotranspiration which occurs (ET_a) depends on moisture availability within the soil; it is always less than or the same as ET_o, if soil is saturated ET_a and ET_o are the same. Average ET_a and ET_o for moderately drained soils at Valentia Island Observatory and Casement Aerodrome are shown in Fig. 2.23.

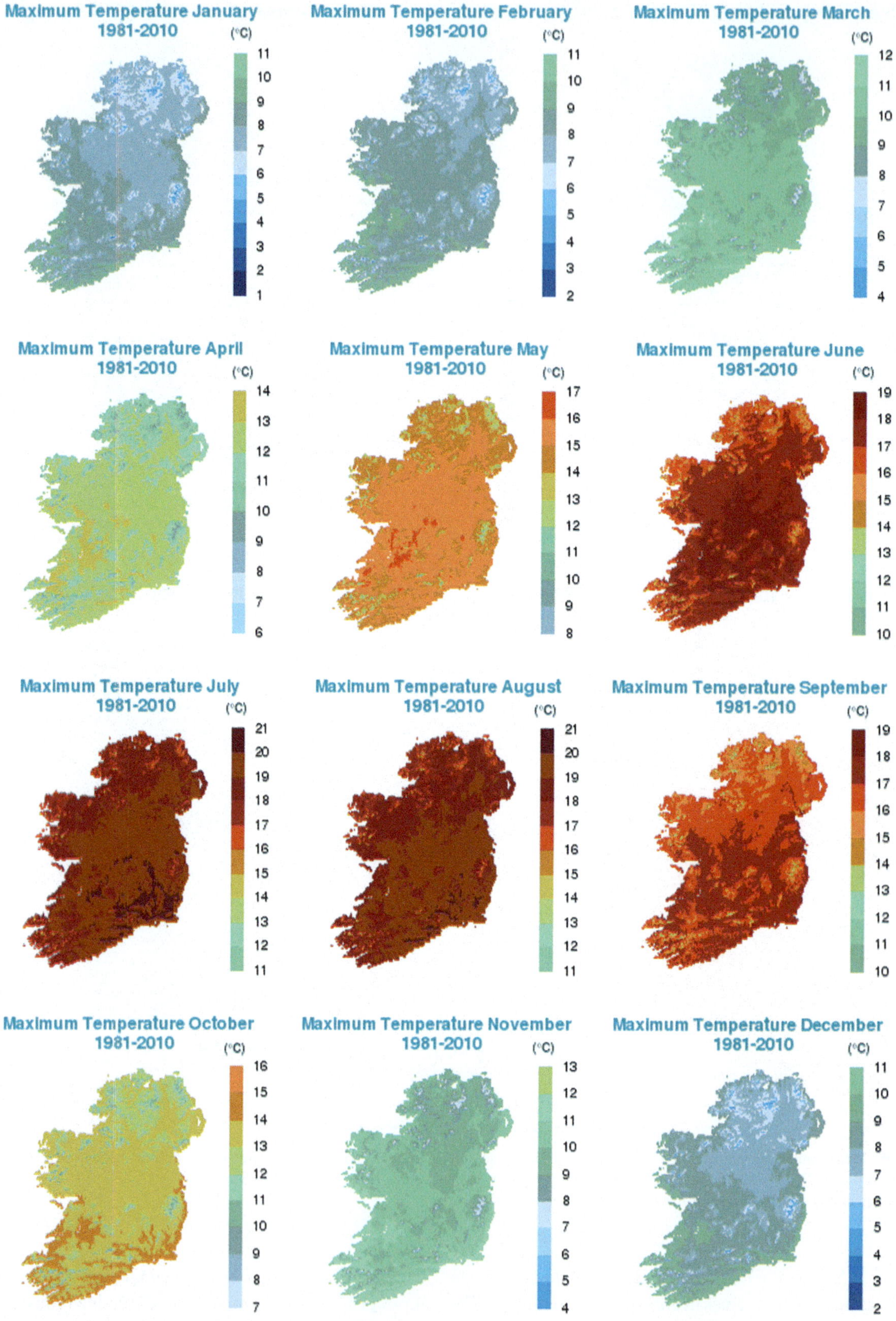

Fig. 2.21 Mean maximum temperatures by month from 1981 to 2010 by Walsh (2012) based on data from Met Éireann, The Irish Meteorological Service. January is typically the coldest month and July the hottest

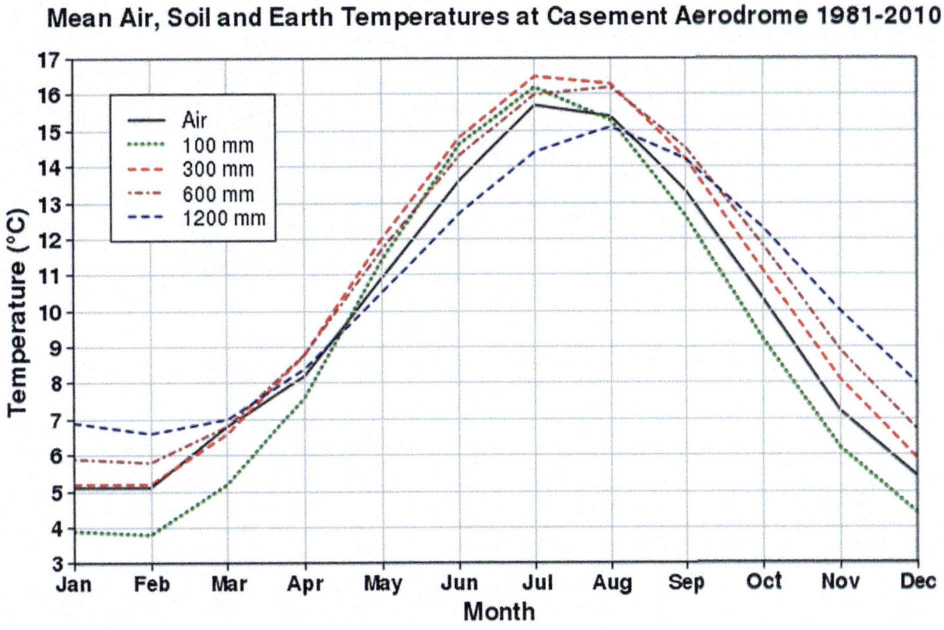

Fig. 2.22 Mean air, soil and earth temperatures from 1981 to 2010 based on data from the Met Éireann, The Irish Meteorological Service

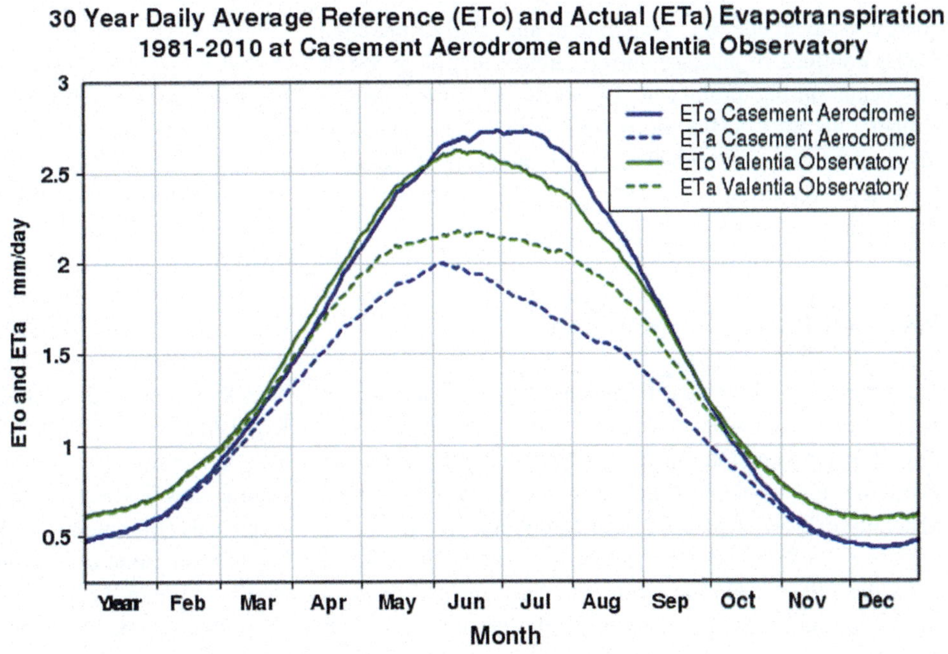

Fig. 2.23 Average reference and actual evapotranspiration from 1981 to 2010 at Casement Aerodrome and Valentia Observatory based on data from the Met Éireann, The Irish Meteorological Service

Annual precipitation exceeds actual evapotranspiration in all regions of the country, but the excess can be considerably greater in the west. On a seasonal basis, even in summer, precipitation for the most part exceeds evapotranspiration in the west.

2.3.5 Soil Moisture Deficits

Soil moisture deficit (SMD) is the amount of water in mm of precipitation required to bring soil water content back to field capacity. The SMD is effectively a balance between rainfall, actual evapotranspiration (ET_a) and drainage; it is dependent on the seasonal patterns of precipitation and ET_a. Negative SMDs mean that the soil is waterlogged, while small positive SMDs indicate that soil is close to field capacity and may not be trafficable. As SMDs increase significantly above 20 mm, moisture availability for growth becomes increasingly restricted. During winter, soils remain for the most part, near or above field capacity with SMD

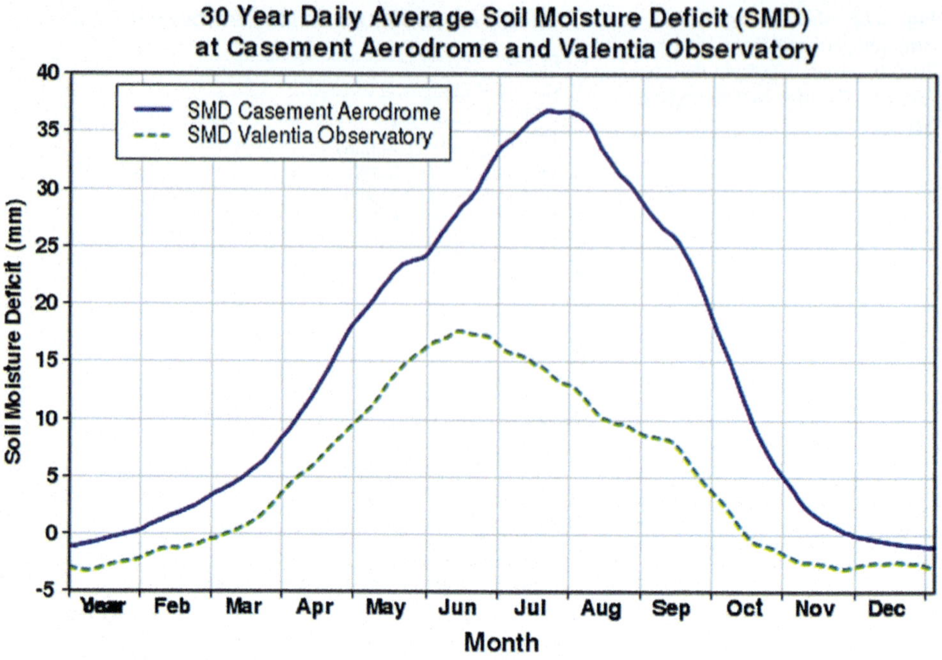

Fig. 2.24 Mean daily average soil moisture deficit from 1981 to 2010 at Casement Aerodrome and Valentia Observatory based on data from the Met Éireann, The Irish Meteorological Service

close to zero; in spring as temperatures increase, more energy is available for evaporation and SMDs increase; SMDs continue to increase during summer and large SMDs can occur during prolonged dry spells; in autumn as the temperature falls SMDs decrease close to field capacity. Figure 2.24 shows average daily soil moisture deficits, at Casement Aerodrome and Valentia Observatory, for moderately drained soils.

2.4 Soil Profile Development

The interaction of soil forming factors results in the development of a soil profile. A soil profile is a vertical section of the soil beginning at the surface and extending down to, and including the geological parent material. A soil profile generally exhibits a succession of layers that are called soil horizons. The horizons above the parent material are referred to as the *solum*, from the Latin legal term for soil, land or parcel of land (Brady 1974) and refer to the A and B horizons.

The nature of the soil profile is important in relation to many aspects of soil ecosystem functioning including aspects of plant growth, such as root development and moisture holding capacity. Ecosystem services are the benefits people obtain from ecosystems and include provisioning services such as food; regulating services such as regulation of drought; supporting services such as nutrient cycling; and cultural services such as recreational benefits (Millennium Ecosystem Assessment (MEA) 2005).

2.4.1 Soil Horizons

A soil horizon is a layer of soil that reflects distinct characteristics as a result of soil forming processes. Soil Horizons occur roughly parallel to the soil surface. The characteristics of the soil profile in relation to physical and chemical properties form the basis for differentiating one soil from another. Typically, three main soil horizons are identified by the letters, A, B and C (Fig. 2.25).

The A horizon is the uppermost layer in mineral soils and being closest to the surface is first to be reached by rainfall. As a result, this layer is more leached than lower horizons. The A horizon is the part of the soil with the most abundant living matter and in which organic matter is most plentiful. The A horizons of most Irish soils have been depleted of soluble chemical substances and of the fine clay particles in some soils (Hammond and Brennan 2003). In contrast, the B horizon is a zone of accumulation and is often high in iron (Fe), aluminium (Al), humus or clay that has been leached from the overlying horizons. The structure of the B horizon tends to be more blocky or prismatic particularly where it has been clay enriched (Hammond and Brennan 2003). The C horizon refers to the material beneath the solum consisting of geological material, including glacial drift. This horizon is less weathered and has less organic matter than overlying horizons. Surface horizons with partly decomposed organic matter occurring usually in poorly drained mineral soils or organic peat soils are designated O horizons. The organic matter content of the O horizon is much greater than the underlying mineral horizons.

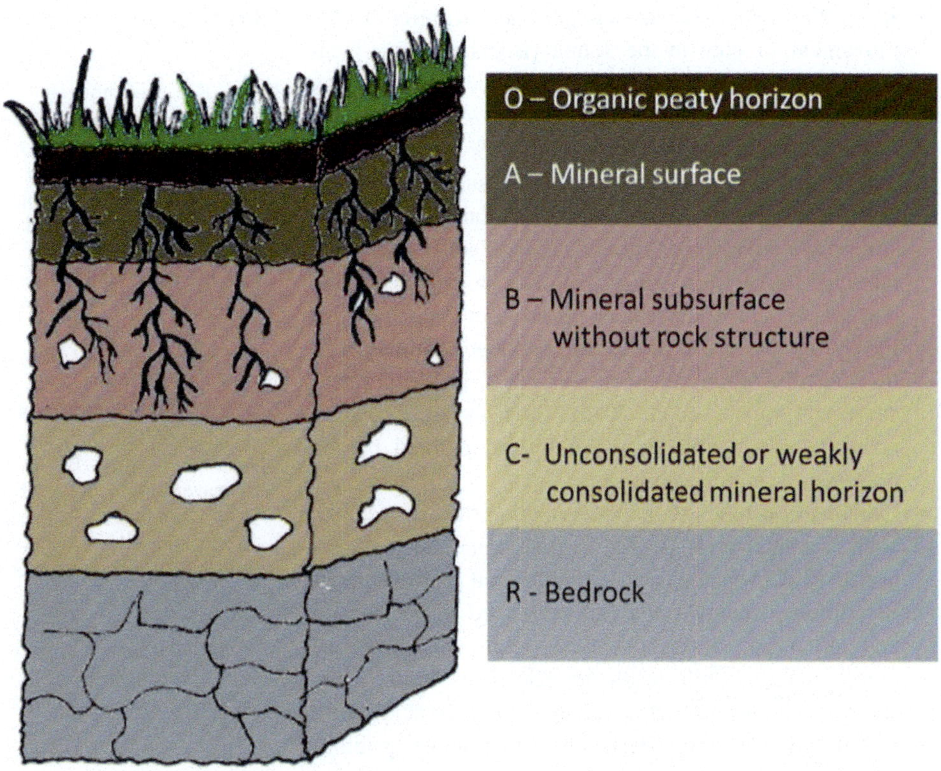

Fig. 2.25 Soil profile with O, A, B, C and R horizons shown

Subhorizon suffix' and definitions are described in Chap. 3 on soil classification.

2.4.2 Principal Processes of Soil Profile Development in Ireland

Bockheim and Gennadiyev (2000) identify 17 generalised soil forming processes that can be linked to soil taxa and diagnostic horizons, properties and materials in major classification systems including the World Reference Base. The main processes of soil profile development in Ireland are climate-driven, in particular by rainfall. High rainfall rates in Ireland result in leaching and gleying of the soil profile. Here, we present the main processes of profile development in Irish soils.

- **Leaching** is the removal of materials in solution from the soil. In Ireland, precipitation exceeds evapotranspiration in all parts of the country in the wetter parts of the year resulting in the constant downward and outward movement of base nutrients by leaching (Collins et al. 2004). Soluble constituents are percolated downwards (eluviated) with gravitational water to be redistributed further down the profile. Soil texture affects nutrient leaching losses, with sandy soils leading to a higher percolation rate compared with finer textured clay soils. The latter also have a higher nutrient adsorbing power (Brady 1974). As insoluble constituents such as Fe, Al and humus are eluviated the soil becomes more acid overtime (Gardiner and Radford 1980). This can result in **Podzolisation** which is a collection of processes that includes eluviation of base cations, weathering transformations of Fe and Al compounds that are mobilised in surface horizons and translocated (Bockheim and Gennadiyev 2000). Podzolisation is characterised by the accumulation of organic matter at the surface resulting in an elluviated E horizon and an iron-rich subsurface horizon.
- **Gleying** is another principal soil profile development process in Ireland and results from aquic or waterlogged conditions, evidenced by features of gleying or mottling. Waterlogging occurs due to the slowly permeable nature of the subsoil itself or due to the presence of a high water table. In these conditions, leaching is limited due to restricted water movement. Waterlogging results in anaerobic conditions which means that Fe and manganese (Mn) constituents of the soil are reduced to Fe^{2+} forms. The reduction of Fe and Mn results in a blue-grey colour indicative of gleying. Reoxidation results in these are referred to as reductimorphic and redoximorphic features (Bouma 1983; Blume and Schlichting 1985). Gleying in Irish soils is closely related to the volume of rain received, with a westward prominence of gleys in alignment with precipitation patterns (Collins et al. 2004).

- **Illuviation (also known as argilluviation)** refers to the movement of clay in the solum (Bockheim and Gennadiyev 2000). This process is evidenced by clay movement resulting in an argic (Bt) horizon, which must contain a minimum clay increase of 20% relative to the eluviated horizon above (Bockheim and Gennadiyev 2000). This process of clay translocation is a common soil profile development process in Irish soils where clay particles move from one soil horizon to another. Clay particles are moved mechanically from the upper part of the soil by percolating water and are deposited deeper in the soil (illuviation) onto soil surfaces or in soil pores or cavities (Orgiazzi et al. 2016). Luvisols are a widely occurring Great Soil Group in Ireland defined by the presence of an argic horizon.
- **Calcification** refers to the redistribution of carbonates in soils (Schaetzl et al. 1996). In Irish soils, calcium carbonate ($CaCO_3$) is redistributed in the soil profile without its complete removal and can be related to weathering and the nature of the chemicals released. Calcareous parent material will result in calcareous soils while siliceous parent materials weather to produce acidic soil.
- **Paludization or paludification** refers to deep accumulation of organic matter on the landscape usually in marshy or peat areas (Bockheim and Gennadiyev 2000). Prolonged saturation can be associated with anoxic conditions resulting in an absence of oxygen necessary to oxidise organic material so that over time organic matter accumulates.

2.5 Conclusion

Soils in Ireland are relatively young, having formed since the retreat of the last ice age approximately 15,000 years ago. Parent materials in Ireland are broadly categorised into solid bedrock geology and subsequently reworked glacial geology, with the latter accounting for the majority of parent materials across the Irish landscape. The Irish climate is strongly influenced by its position on the eastern side of the Atlantic Ocean and on the western fringes of continental Europe resulting in a mild maritime climate with prevailing south-westerly winds. The high rainfall rates in Ireland are a dominant driver of soil profile development with leaching and gleying as the two main processes driving soil development.

References

Blume HP, Schlichting E (1985) Morphology of wetland soils. In: Wetland soils: characterization, classification and utilization. International Rice Research Institute, Los Banõs, Philippines, pp 161–176

Bockheim JG, Gennadiyev AN (2000) The role of soil-forming processes in the definition of taxa in soil taxonomy and the world soil reference base. Geoderma 95:53–72

Bouma J (1983) Hydology and soil genesis of soils with aquic moisture regimes. In: Wilding LP, Smeck NE, Halls GF (eds) Pedogenesis and soil taxonomy; I. Concepts and interactions. Elsevier, Amsterdam, pp 253–281

Brady N (1974) The nature and properties of soils, 8th edn. Macmillan Publishing Co. Inc, New York

Buol SW, Hole FD, McCracken RJ, Southard RJ (1997) Soil genesis and classification, 4th edn. Iowa State University Press, Ames

Collins JF, Larney FT, Morgan MA (2004) Climate and soil management. In: Keane T, Collins JF (eds) Climate, Weather and Irish agriculture, 2nd ed. Joint Working Group on Applied Agricultural Meterology (AGMET) Met Éireann, Dublin

Culleton EB (1978) Characterisation of glacial deposits in south Wexford. Proc R Irish Acad 78B:293–308

Duchaufour P (1982) Pedology pedogenesis and classification (trans: Paton TR). George Allen and Unwin Ltd, London

Finch T, Synge FM (1966) The drift and soils of west Clare and the adjoining parts of Counties Kerry and Limerick. Irish Geogr 5: 161–172

Flint RF (1957) Glacial and pleistocene geology. Wiley, New York

Gardiner MJ, Radford T (1980) Soil associations of Ireland and their land use potential explanatory bulletin to soil map of Ireland 1980. National Soil Survey of Ireland, An Foras Talúntais, Dublin

Geological Survey of Ireland (GSI) (2014) Bedrock geology of Ireland, 1:1,000,000 Scale (map). Available at: https://www.gsi.ie/NR/rdonlyres/B7F1DBC1-C306-4730-90AB-87123139C04B/0/GSI_Bedrock_1million.pdf. Accessed: 13 Dec 2016

GSI (2016) Quaternary and geotechnical. Geological Survey of Ireland (website). Available at: https://www.gsi.ie/Programmes/Quaternary+Geotechnical/. Accessed: 27 Jan 2016

Graham JR (2001) Ordovician. In: Holland CH (ed) Geology of Ireland. Dunedin Academic Press, Edinburgh

Graham RC (2006) Factors of soil formation: topography. In: Giacomo Certini (ed) Soils: basic concepts and future challenges. Cambridge University Press, Cambridge, pp 151–164. Available from: Cambridge Books Online: http://dx.doi.org/10.1017/CBO9780511535802.012. Accessed: 07 Apr 2016

Hammond RF, Brennan LE (2003) Soils of Co. Offaly soil survey Bulletin No. 43. National Soil Survey of Ireland, Teagasc

Jenny H (1941) Factors of soil formation a system of quantitative pedology. Dover Publications Inc., New York

Keane T, Sheridan T (2004) Climate and soil management. In: Keane T, Collins JF (eds) Climate, weather and Irish agriculture, 2nd edn. Joint Working Group on Applied Agricultural Meterology (AGMET) Met Éireann, Dublin

Killham K (1994) Soil ecology. University Press Cambridge, United Kingdom

Kilroe JR (1907) A description of the soil-geology of Ireland based upon geological survey maps and records with notes on climate.

Department of Agriculture and Technical Instruction for Ireland, Dublin
Knight J, Coxon P, Marshall McCabe A, McCarron SG (2004) Pleistocene glaciations in Ireland. In: Ehlers J and Gibbard PL (eds) Quaternary glaciations—extent and chronology part 1: Europe. Elsevier
Millennium Ecosystem Assessment (MEA) (2005) Ecosystems and human well-being: a framework for assessment. Island Press, Washington, DC
Mitchell F, Ryan M (1986) Reading the Irish landscape. Town House, Dublin
Mockler E, Bruen M, Desta M, Misstear B (2013) Pathways project final report volume 4: catchment modelling tool. STRIVE Report No. 56. Irish Environmental Protection Agency, Johnstown Castle, Ireland
Orgiazzi A, Bardgett RD, Barrios E, Behan-Pelletier V, Briones MJI, Chotte JL, De Deyn GB, Eggleton P, Fierer N, Fraser T, Hedlund K, Jeffrey S, Johnson NC, Jones A, Kandeler E, Kaneko N, Lavelle P, Lemanceau P, Miko L, Montanarella L, Moreira FMS, Ramirez KS, Scheu S, Singh BD, Six J, van der Putten WH, Wall DH (eds) (2016) Global soil biodiversity Atlas. European Commission, Publications Office of the European Union, Luxembourg, p 176
Pracht M, Sleeman AG (2002) A geological description of West Cork and adjacent parts of Kerry to accompany the bedrock geology 1:100,000 scale map series, sheet 24, West Cork, with contributions by G. Wright (groundwater) and W. Cox (minerals). Geological Survey of Ireland, 79 pp
Schaetzl RJ, Frederick WE, Tornes L (1996) Secondary carbonates in three fine and fine-loamy Alfisols in Michigan. Soil Sci Am J 60:1862–1870
Schmidt O, Keith AM, Arroyo J, Bolger T, Boots B, Breen J, Clipson N, Doohan FM, Griffin CT, Hazard C and Niechoj R (2010) CréBeo—baseline data, response to pressures, functions and conservation of keystone micro- and macro-organisms in Irish soils (2005-S-LS-8). STRIVE Report prepared for the Environmental Protection Agency by University College Dublin
Shoji S, Nanzyo M, Takahashi T (2006) Factors of soil formation: climate. As exemplified by volcanic ash soils. In: Giacomo C (ed) Soils: basic concepts and future challenges. Cambridge University Press, Cambridge, pp 131–150. Available from: Cambridge Books Online http://dx.doi.org/10.1017/CBO9780511535802.011. Accessed 07 Apr 2016
Sleeman A, McConnell B, Gatley S (2004) Understanding Earth processes, rocks and the geological history of Ireland a companion to the 1:1,000,000 scale bedrock geological map of Ireland. Geological Survey of Ireland
Stribley DP, Read DJ (1980) New Phytol 86:368–371
Wall B, Derham J, O'Mahony T (eds) (2016) Ireland's environment 2016—An assessment. Environmental Protection Agency, Co. Wexford, Ireland
Walsh S (2012) A summary of climate averages for Ireland. Climatological Note No. 14. Met Éireann

Soil Classification

3

Rachel Creamer, Iolanda Simo, Brian Reidy, Thomas Cummins, and Lilian O'Sullivan

Abstract

In 2014 the 3rd Edition National Soil Map of Ireland was published at a resolution of 1:250,000, accompanied by an associated online soil information system (SIS) database. The 3rd Edition National Soil Map was developed using a combination of novel digital mapping techniques, traditional field survey methods, and historic soil survey detail. This produced for the first time, a consistent national level legend for Irish soils. Irish soil classification 2014 consists of a three-tiered taxonomy of 'Great Soil Group', 'Subgroup' and 'Series', ordered from the most general to the most specific. This chapter explains how soils are classified including the reference section, how to describe a soil profile, horizon definitions and soil diagnostic criteria. Finally, the correlation of the Irish SIS with World Reference Base (WRB) classification is described and the classification of urban soils using the WRB system.

Keywords

Soil classification • Diagnostic criteria • Great Soil Groups • Subgroups • Soil series • World Reference Base

3.1 Introduction

The launch of the Irish Soil Information System (SIS) in 2014 represented a major advancement in the availability of soil information at a national level in Ireland. A comprehensive inventory of soil data in Ireland compiled in 2007 highlighted that soil data coverage was incomplete in detail and extent (Daly and Fealy 2007), and that an updated soil map was necessary to satisfy a growing demand to meet sustainability targets (Creamer et al. 2014). The Irish SIS project was designed to satisfy this growing need, a key output of which was the development of the 3rd Edition National Soil Map of Ireland, a digital map at a resolution of 1:250,000 (Chap. 4). The Irish SIS combined novel digital mapping techniques with traditional field survey methods and historic soil survey detail, to establish a consistent national legend for Irish soils. A brief overview of the World Reference Base for Soil Resources (WRB) with specific attention to the correlation of the Irish SIS with the WRB classification system is also presented.

As described in Chap. 2, soil formation and development depends upon the interactions of several factors. Soil classification within the Irish SIS recognises these soil forming factors and processes, morphology and soil geographic approaches which have been complemented with laboratory data for verification (Simo et al. 2015). The heterogeneity of soil insists that similar members are grouped together so that the great number of soils, which vary from one another in different degrees of contrast, can be clustered into progressively higher categories (USDA 1960). Thus, soil classification is the process of grouping soils together where a similar range of chemical, physical and biological properties

R. Creamer (✉) · L. O'Sullivan
Teagasc Agriculture and Food Development Authority, Johnstown Castle Environment Research Centre, Wexford, Ireland
e-mail: Rachel.creamer@wur.nl

R. Creamer
Wageningen University and Research
Atlas—Wageningen Campus, Wageningen, The Netherlands

B. Reidy
Teagasc Moorepark, Animal and Grassland Research and Innovation Centre, Fermoy, County Cork, Ireland

T. Cummins
University College Dublin Soil Science, UCD School of Agriculture and Food Science, University College Dublin, Belfield, Dublin 4, Ireland

L. O'Sullivan
Irish Environmental Protection Agency, Johnstown Castle, County Wexford, Ireland

I. Simo
Smart Soil & Land, Jaume II St., 11, Lleida, Spain

R. Creamer and L. O'Sullivan (eds.), *The Soils of Ireland*, World Soils Book Series,
https://doi.org/10.1007/978-3-319-71189-8_3

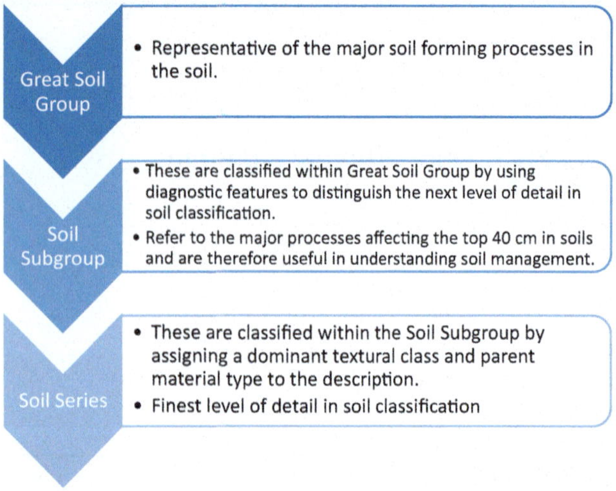

Fig. 3.1 Three-tiers of soil taxonomy in the 2014 Irish soil classification

have been observed, into units that can be geo-referenced and mapped. The Irish SIS consists of a three-tiered taxonomy of 'Great Soil Group', 'Subgroup' and 'Series', ordered from the most general to the most specific (Fig. 3.1).

3.2 Reference Section

The reference section refers to the vertical section of the soil profile which is described and classified. Within this section, diagnostic horizons are identified following set principles outlined in the Irish SIS handbook. The depth or thickness of this section varies depending upon the soil parent material. In Irish soil classification, four main reference section depths are recognised:

1. Soils in peat have a reference section that extends from 40 to 80 cm with organic material deeper than 80 cm (Fig. 3.2). Utilising a thickness of 40 cm or more of peat to classify peat soils is consistent with the WRB definition of Histosol (IUSS Working Group WRB 2006). These soils are composed of organic material largely derived of partially decomposed plant remains that accumulated under waterlogged conditions, either formed in situ or as constituents of sub-aquatic sediments such as organic lake muds (Avery 1980).

   In peat soil, the mineral fraction has little textural significance as peats are mostly composed of organic materials. Soils with more than 50% organic matter are defined as peats. Soils with less than 50% but greater than 20% organic matter and with 50% or more sand content in the mineral fraction are defined as Sandy Peat, or with less than 50% sand content defined as Loamy Peat. Other criteria for differentiation of soils in peat of less than 50% organic matter: Soils with organic carbon (OC) <20% (calculated by organic carbon analysis), two categories are distinguished; soils with more than 12% OC and more than 50% Sand, they are defined as Peaty sand; and soil with OC >14.5% and amount of clay $\geq$50%, defined as Peaty loam.
2. **Histic Organic soils over bedrock** mainly consist of organic material over a layer of coherent bedrock or skeletal products of weathering. These soils extend from the organic soil surface to the upper surface of this bedrock/skeletal layer, whichever is at the shallowest depth (30 cm). The peaty (> 20% OC) O horizon must have a thickness of 7.5 cm or more.
3. **Mineral soils over bedrock or lithoskeletal substrate** are comprised of mineral material overlying bedrock or skeletal weathered bedrock fragments. This includes two subtypes (Clayden and Hollis 1984): Firstly, where the surface layer is no greater than 30 cm thick, typically occurring as Rendzina or Lithosol soils. Secondly, where the profile depth is greater than 30 cm but bedrock occurs within 80 cm (Simo et al. 2014).
4. **All other substrates**, which includes any mineral soils comprised in thick drift. In general, soil material must be at least 80 cm thick. These substrates are mostly glacial till or fluvio-glacial deposits and recent alluvium.

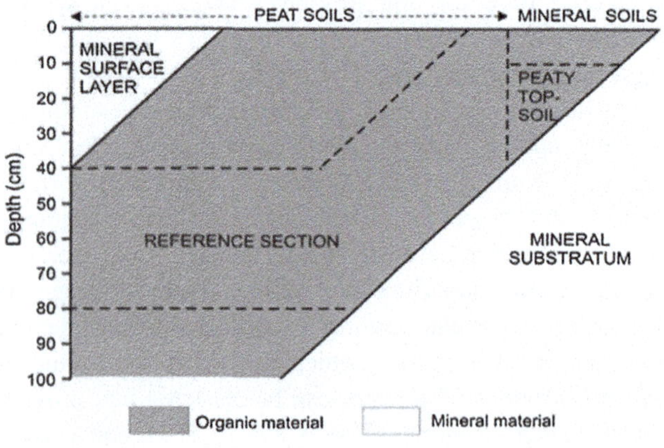

Fig. 3.2 Reference section for peat soils (from Simo et al. 2014, based upon Clayden and Hollis 1984 p. 13)

3.3 Describing a Soil Profile

Part of the process of describing the soil profile includes a site description. Local landscape description is an important part of the soil classification process. On arrival GPS locational data and the principal landscape features should be recorded, such as relief, topography, aspect and land use (Fig. 3.3). Once the profile pit is dug to a depth of approximately 1.2 m, the profile face must be cleared and photographed, along with recording the parent material. Thereafter, each horizon should be described in detail with characteristics of field estimated

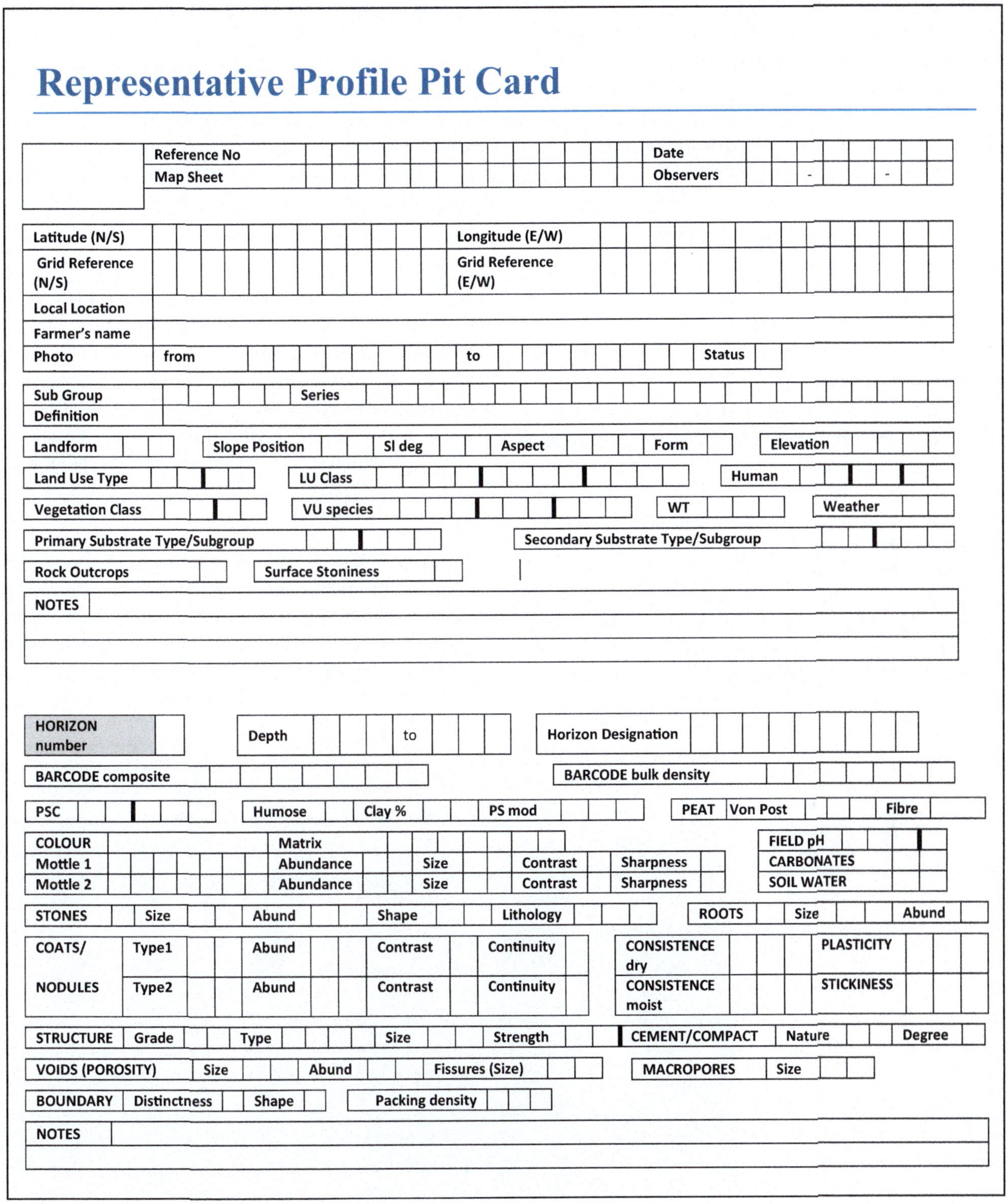

Representative Profile Pit Card

Reference No | Date
Map Sheet | Observers - -

Latitude (N/S) | Longitude (E/W)
Grid Reference (N/S) | Grid Reference (E/W)
Local Location
Farmer's name
Photo from to Status

Sub Group | Series
Definition

Landform | Slope Position | Sl deg | Aspect | Form | Elevation
Land Use Type | LU Class | Human
Vegetation Class | VU species | WT | Weather
Primary Substrate Type/Subgroup | Secondary Substrate Type/Subgroup
Rock Outcrops | Surface Stoniness
NOTES

HORIZON number | Depth to | Horizon Designation
BARCODE composite | BARCODE bulk density
PSC | Humose | Clay % | PS mod | PEAT | Von Post | Fibre
COLOUR | Matrix | FIELD pH
Mottle 1 | Abundance | Size | Contrast | Sharpness | CARBONATES
Mottle 2 | Abundance | Size | Contrast | Sharpness | SOIL WATER
STONES | Size | Abund | Shape | Lithology | ROOTS | Size | Abund
COATS/ NODULES | Type1 | Abund | Contrast | Continuity | CONSISTENCE dry | PLASTICITY
Type2 | Abund | Contrast | Continuity | CONSISTENCE moist | STICKINESS
STRUCTURE | Grade | Type | Size | Strength | CEMENT/COMPACT | Nature | Degree
VOIDS (POROSITY) | Size | Abund | Fissures (Size) | MACROPORES | Size
BOUNDARY | Distinctness | Shape | Packing density
NOTES

Fig. 3.3 Profile description card to record the site description and horizon detail for each horizon. During the SIS campaign a field tablet with a supporting database and interface was used to record data directly in the field

Table 3.1 Sampling of horizons in Irish SIS (Creamer et al. 2014)

Chemical properties (2 kg composite fresh soil from each horizon)	pH (in H_20 and $CaCl_2$) Total: C, N Organic carbon Organic matter content (Loss on ignition) Cation exchange capacity (CEC) Base saturation (BS) Fe/Al (oxalate extraction)
Physical properties	Bulk density: three per horizon reported as average value (cm^3) with standard deviation (σ) (Fig. 3.4). Texture

texture, structure, colour, carbonates, stone content and type, root content and depth and the presence of biota recorded (Creamer et al. 2014).

In relation to analysing soil profile pits, along with the classification criteria described, a suite of measurements were analysed at horizon level (Table 3.1) to provide a more complete description of the soil profile.

Figure 3.4 (left) shows an open profile pit with three bulk density rings taken per horizon. Final bulk density values are recorded on a horizon basis and reported as an average value (cm^3) with standard deviation (σ).

3.4 Horizon Definitions

When characterising or classifying soil type, it is essential to define the differing layers (horizons) found within the soil profile. This characterisation is done by defining certain characteristics such as structure, colour, porosity, mottling, textural changes, pH, etc. These characteristics are then used to classify genetic horizons. The classification of soils at Subgroup level requires that the various horizons of the soil profile are described and defined. Horizon definitions within the Irish SIS are fully described in the field handbook by Simo et al. (2014). Overall, there are three main horizons, A, B and C that occur in a majority of soil profiles as shown in Fig. 3.5. The A and B horizons constitute the 'true soil' whilst the C horizon refers to parent material. Some soils lack a B horizon and are said to have AC profiles. On other soils, organic layers (O horizons) may overlie the mineral horizons. Underlying rock is indicated by the symbol R.

Organic horizons

- Peaty O horizons accumulate under wet conditions. They are saturated with water for at least 30 consecutive days in most years, or have been artificially drained, and include fibrous, semi-fibrous and amorphous peat. Sub-horizons include: Of, Omf, Oh, Op.

Fig. 3.4 Three bulk density rings per horizon as a physical soil measurement shown on left, method for bulk density measurements shown on right (Photo: Jeremy Emmet-Booth)

Fig. 3.5 Sample soil profiles annotated with soil horizon designations. *Source* Irish Soil Information System

- L horizon represents fresh litter deposited during the previous annual cycle. It is normally loose and the original plant structures are little altered.
- F, H are organic horizons originating as litter deposited or accumulated at the surface and seldom saturated with water for more than a month at a time. The F has partly

Table 3.2 Suffix and definition

Suffix	Property[a]	Horizon identifier
f		The presence of an Ironpan
g	Gleying	Intense reduction of iron or reduction due to seasonal or continuous waterlogging. A greyish colour due to wetness can be found and oxidised Fe and Mn mottles and/or coatings are concentrated on ped surfaces
h	Humic	Considerably more organic matter than mineral matter, with thickness over 15 or 10–15 cm if over bedrock
k	Calcareous	Calcareous in nature within 40 cm of surface having formed in calcareous parent material, with >2% $CaCO_3$
p	Ploughing	Evidence of disturbance by cultivation
r	Regolith	Loose superficial material covering solid rock
s	Spodic	Enrichment of organic matter complexed with Fe and Al due to illuviation and/or biochemical weathering in situ
t	Argillic	Accumulation of translocated clay.
u	Unconsolidated	Unconsolidated, normally apedal but may show stratification
w	Weathering	Evidence of alteration
x	Fragic	Properties include firmness, brittleness with a high bulk density reflecting a compact but uncemented nature

[a]Where suffix letter appears capitalised indicates a dominance of that characteristic

decomposed or comminuted litter, remaining from earlier years, in which some of the original plant structures are visible to the naked eye. The H has well decomposed litter, often mixed with mineral matter, in which the original plant structures cannot be seen.

Mineral horizons

- The A horizon refers to the mineral horizon formed at or near the surface. An A horizon is characterised by incorporated humified organic matter, or by evidence of cultivation, or both. The incorporation of organic matter is presumed to be due to biological activity or artificial mixing. Horizon designations include: Ah, Ap, Ahg, Apg, AB.
- The E horizon is a subsurface mineral horizon that contains less organic matter and/or dithionite-extractable iron and/or silicate clay than the immediately underlying horizon, presumably as a result of loss or washing out of one or more of these constituents.
- The B horizon is the mineral subsurface horizon, without rock structure characterised either by illuviation or by an alteration of original material such as the solution and removal of carbonates. Horizon designations that occur include: Bf, Bg, Bh, Bs, Bt, Btg, Bw, Bx, Bk, BC.
- The C horizon is usually unconsolidated or a weakly consolidated mineral horizon that retains soil structure or otherwise lacks the properties of the overlying horizons. C horizons may have been modified by gleying or by an accumulation of carbonates or soluble salts or they may have developed brittleness and associated properties of fragipans. Horizon designations include: Cu, Cr, Ck, Cg.

Many soils are subdivided on the basis of a significant difference within the main horizons. These differences are identified by the horizon designation accompanied by a suffix identifier (Table 3.2). Where lithological discontinuity occurs and two different geologic parent materials are present, the number '2' is placed in front of the horizon symbol for horizons developed in the second parent material.

3.4.1 *Diagnostic Criteria*

Table 3.3 describes the criteria for the diagnostic features included in the Irish SIS for the classification of soils at Subgroup level.

3.5 Classification of Great Soil Groups

Great Soil Groups represent the most general soil classification unit in the Irish SIS. Great Soil Groups are organised based upon the dominant soil forming factors. Soils within Great Soil Groups have common characteristics and soil forming conditions; they are developed under the influence of environmental factors (vegetation and climate) active over a considerable geographic range and have one or more Subgroups of soil. Soils occurring in a Great Soil Groups have the same arrangement and degree of expression of horizons in the soil profile. In Ireland, 11 Great Soil Groups have been identified and are coded 01–11: (01) Ombrotrophic Peat Soils; (02) Mineratrophic Peat Soils; (03) Rendzina; (04) Lithosols; (05) Alluvial Soils;

Table 3.3 Diagnostic features to classify soil subgroup

0 None	There are no distinguishing characteristics associated with this feature
1 Histic	This is a peaty (>20% OC) (O) horizon of ≥ 7.5 cm thickness or more, overlying mineral soil or rock material. It generally occurs at the surface or beneath a thin layer of more or less decomposed litter (L, F). This diagnostic horizon is conceptually equivalent to the histic horizon in WRB (IUSS Working Group WRB 2006) but is allowed to be thinner than the ≥ 10 cm specified by the WRB
2 Gleyic	This feature occurs in soils affected by seasonal or continuous waterlogging. The presence of Fe and manganese (Mn) concretions and/or nodules of grey colouring due to the continuous presence of a water table, this can be indicative of wetness. A gleyed subsurface horizon has greyish colours caused by the reduction of Fe and Mn in the lower part of the profile and/or inside the peds, with oxidised Fe and Mn mottles and/or coatings being concentrated on ped surfaces Requirements: 1. Qualifies as Eag, Eg, Bg, Btg, BCg, Cg or CG within 80 cm depth, and lacks a slowly permeable sub-surface layer; and 2. At least 5 cm thick and extends below 30 cm depth; and 3. In general, the lower subsoil horizons will have grey matrix colours: a. Hue 7.5YR chroma = 1; b. Hue 10YR or yellower chroma $\leq$ 2; or c. Greyish (Hue N, GY), greenish (Hue GY,G) and bluish (Hue BG, B) hues
3 Stagnic	This is a subsurface horizon (within 40 cm) that is at least 15 cm thick, which displays evidence of reduced Fe and Mn as a result of a perched water table, forming a significant barrier to water movement for a period long enough to allow reducing conditions to occur. This soil reflects mottling where the surfaces of the peds (or parts of the soil matrix) are lighter (at least one Munsell value unit more) and paler (at least one chroma unit less), and the interiors of the peds (or parts of the soil matrix) are more reddish (at least one hue unit) and brighter (at least one chroma unit more) than the non-redoximorphic parts of the layer (IUSS Working Group WRB 2006). These soils tend to have slowly permeable subsurface horizons, i.e. as a result of fragipan characteristics
4 Fe–Ac	This is a B horizon in which amorphous materials containing Al and/or Fe that have accumulated, either by illuviation or by biochemical weathering in situ. It meets the following requirements (An Foras Talúntais unpublished): 1. Qualifies as an Iron-pan horizon or 2. The moist chroma is greater than 3, or the value is ≤ 3; and 3. Starts within 120 cm depth; underlies an E, A, H or O horizon, and extends below 15 cm depth, excluding fresh or partially decomposed litter (L, F); or 4. The horizon must be at least 10 cm thick if there is no overlying albic E horizon and it consists only of a Bs horizon; at least 2.5 cm thick if it consists only of a Bh horizon below an albic E horizon. This diagnostic horizon is similarly defined to the spodic horizon (FAO 2006)
5 Calcareous	Soils in which a calcareous horizon is present within 40 cm of the surface, have a calcareous B horizon and are formed in calcareous parent materials. Calcareous nature is defined as > 2% $CaCO_3$, where a visible effervescence is displayed when applying a few drops of 10% Hydrochloric Acid
6 Humose	This is a soil which contains an A horizon with significantly more organic matter, than mineral matter. It can occur as a sequence of H or Oh and Ah horizons that meet the following requirements over a thickness of more than 15 cm or 10–15 cm if directly over bedrock (R or Cr): 1. Moist rubbed colour with value and chroma of 3 or less 2. Humose or partly humose and partly organic (<7.5 cm thick) As a guideline in the field, humose soil materials generally have moist rubbed colour value/chroma of 3/2 but in addition the soil should feel soapy and slippery due to the organic matter, the presence of which is more difficult to detect in silty soils because organic and silty materials have a similar feel. The presence of a humose topsoil usually qualifies as a mollic horizon if it has high base saturation (FAO 2006) or umbric horizon in WRB if it is base depleted (IUSS Working Group WRB 2006). In podzols, humic refers to the presence of a Bh horizon
7 Drained	Drained soils are primarily where human activity has altered the drainage status of the soil from poorly drained to imperfectly to moderately drained status. In organic soils this activity results in an earthy soil in peaty surface layers (Op or Oh). These need to be at least 20 cm thick and contain less than 15% w/w visible plant remains (fibres). It sometimes overlies an organic subsurface horizon and usually has a distinct granular or subangular blocky structure
8 Cut-over	Cut-over Ombrotrophic Peat Soils have had the surface peat removed by hand cutting methods but the peat thickness still meets the criteria for peat soils
9 Anthropic	This diagnostic horizon corresponds to the thick A1 horizon of De Bakker and Schelling (1966), and includes Hortic, Plaggic and Terric horizons in WRB (IUSS Working Group WRB 2006). Most commonly Anthropic features are used to describe a terric horizon (from Latin terra, earth) which is a human-induced surface horizon >40 cm thick. These horizons develop through the addition of earthy manures, compost, beach sands, calcareous sands and shells or mud over a long period time. It builds up gradually and may contain stones, randomly sorted and distributed. It commonly contains artefacts such as pottery fragments, cultural debris and refuse, which are typically very small (<1 cm in diameter) much abraded and occupy <20% volume of the horizon. This horizon may also include soil material that has been imported to a site causing a significant alteration of the surface horizon, such as soil inversion

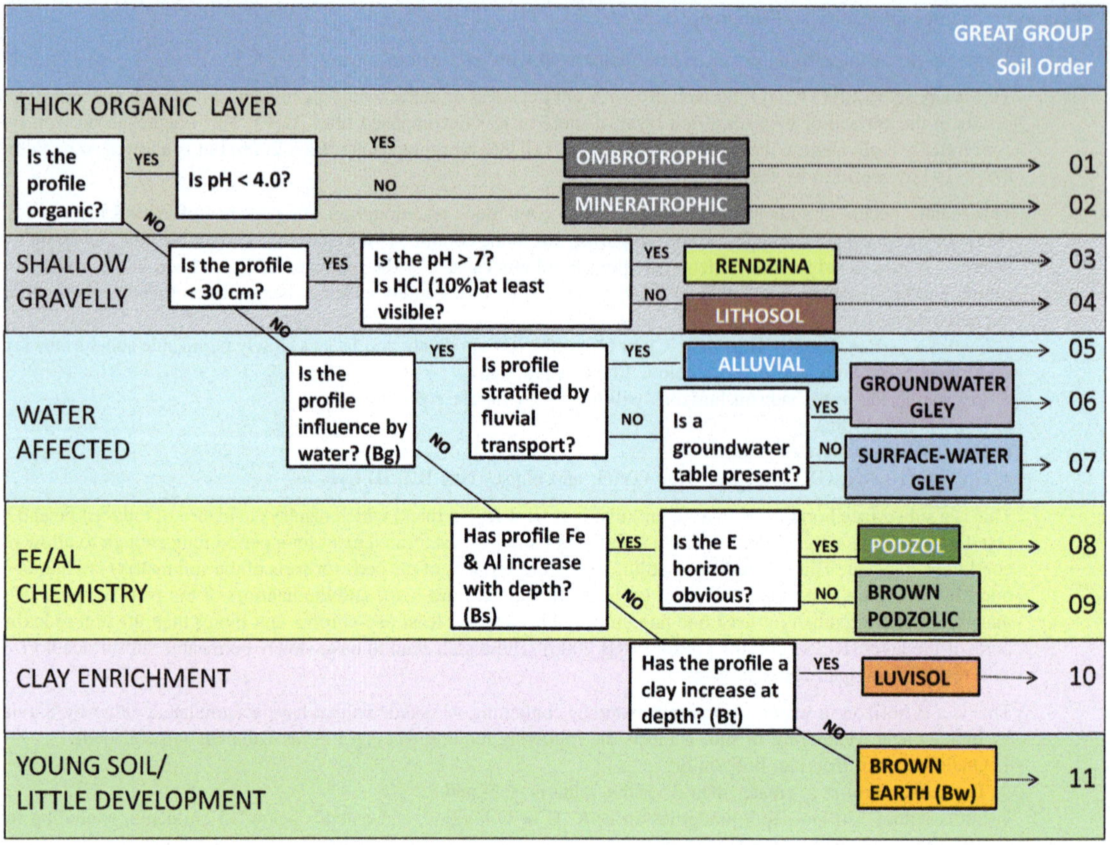

Fig. 3.6 Key to determine Great Soil Group classification based upon Irish SIS Classification 2014

(06) Groundwater Gleys; (07) Surface-water Gleys; (08) Podzols; (09) Brown Podzolic; (10) Luvisols and (11) Brown Earths. The following key allows a simplified determination of Great Soil Group within the Irish SIS by following a quick yes/no system that is applied from the top to the bottom. The most distinguishing features are at the top of the key, and Brown Earths are at the bottom representing the group with the fewest distinguishing features (Fig. 3.6).

This key is based upon the following principles, with criteria synthesised in Table 3.4:

- Organic soils are first differentiated from mineral soils and are further defined on the basis of whether they are rain-fed (Ombrotrophic peat) or groundwater-fed systems (Minerotrophic peat). If pH is <4 they are Ombrotrophic. If pH >4 they are classified as Mineratrophic peat soils.
- Next shallow soils (<30 cm) with a severe limitation to rooting are identified as Lithosol or Rendzina soils. If the soil pH is >7, or if the presence of calcium carbonate ($CaCO_3$) is induced by a hydrochloric acid (HCl) (10%) reaction these soils are classified as Rendzina soils.

Table 3.4 Sequencing of Great Soil Groups (Simo et al. 2014)

Criteria	Code	Great Soil Group
Soils with thick organic layers	1	Ombrotrophic Peat
	2	Minerotrophic Peat
Shallow or extremely gravelly soils (<30 cm)	3	Rendzina
	4	Lithosol
Soils influenced by water	5	Alluvial
Within 40 cm	6	Groundwater Gley
	7	Surface-water Gley
Soils affected by Fe/Al chemistry increase	8	Podzol
	9	Brown podzolic
Soils with clay enriched subsoil	10	Luvisols
Relatively young or soils with little profile development	11	Brown Earth

- Thirdly soils that are influenced by water are identified due to the presence of mottling in the soil profile. If the profile shows stratification associated with fluvial transport and deposition, these soils are classified as Alluvial soils. If there is no evidence of Alluvial bands, the profile is checked for a groundwater influence. This should be clearly observed within the surface 40 cm. If present, the profile will exhibit a grey uniform colour throughout the profile and these soils are classified as Groundwater Gleys. These are typically influenced by location and are often found near rivers or inter-drumlin hollows where the water table is high. If no groundwater influence is found, these soils are classified as Surface-water Gleys which are poorly drained due to presence of a slowly permeable subsurface layer.

- If the profile has not been influenced by water, the profile is next checked for iron (Fe) and/or aluminium (Al) chemistry. Where Fe and Al have played a major role in the soil formation, soils are classified as Podzols and Brown Podzolics. Where an obvious E horizon exists, indicating a zone of eluviation is present, these soils are classified as Podzols. This is evidenced by the presence of a bleached white sandy layer. Soils where no E horizon is evident, but where clear Fe/Al accumulation has occurred in the Bs horizon are defined as Brown Podzolics.
- Step five, soils with clay illuviation from surface and/or sub-surface horizons to lower down the profile are classified as Luvisols. These soils have an increase of clay in the lower horizon by 20% (presence of a Bt horizon).
- Finally, Brown Earths which are relatively young soils or soils with limited profile development are identified and represent the group with no clear distinguishing features.

Soils vary due to a suite of soil forming factors and processes, including topography (see Chap. 2). As a result, the catena sequence of soils varies from hilltop to valley floor. As the Irish SIS considers these soil forming factors and processes, a generalised description of the soil catena sequence can be presented (Fig. 3.7) with the 11 Great Soil Groups of Ireland described across a landscape position.

3.6 Classification of Soil Subgroups

The second level of classification in Irish SIS is the Soil Subgroup. Altogether, the Irish SIS contains 67 Subgroups, based upon the diagnostic features found within the 11 Great Soil Groups. The character of the various horizons in the soil profile, reflect the influential processes of soil formation that represent the true characteristics of a soil that are important for its use and management. Diagnostic features are soil properties that describe soil characteristics and are used to classify the Soil Subgroup.

Altogether nine main diagnostic features are included in the Irish SIS, coded 1–9 (Fig. 3.8). 0 is also included in the classification key and indicates that there are no additional distinguishing characteristics present. The Irish SIS has developed a key based upon the following principles:

1. The first number in the code refers to the Great Group (01–11) following the key system (Fig. 3.6).
2. The second number is the dominant diagnostic feature (Fig. 3.8) code (0–9).
3. The third number is the secondary diagnostic feature code (0–9) (Fig. 3.8).

The nomenclature at Subgroup level provides the most information on the soil processes, as Subgroups are named on the basis of the most important features that occur in the reference section to a maximum of two, described as

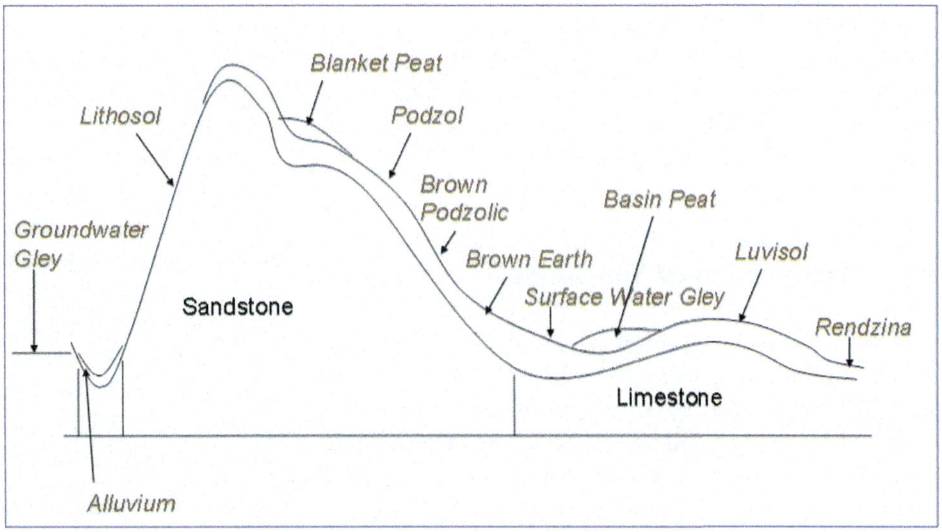

Fig. 3.7 Great Soil Groups described across a landscape position (Simo et al. 2014)

Fig. 3.8 Key codes applied according to the subgroups and diagnostic features

	Diagnostic Feature	None	Histic	Gleyic	Stagnic	Iron-pan	Calcareous	Humose	Drain	Cut	Anthropic
Subgroup	Code	0	1	2	3	4	5	6	7	8	9
Ombrotrophic Peat	01		01.1.0						01.7.0	01.8.0	01.9.0
Mineratrophic Peat	02		02.1.0						02.7.0	02.8.0	
Rendzina	03	03.0.0	03.1.0					03.6.0			
Lithosol	04	04.0.0	04.1.0					04.6.0			
Alluvial	05	05.0.0	05.1.0				05.5.0 05.5.1 05.5.6	05.6.0	05.7.0 05.7.2 05.7.6		
Groundwater Gley	06	06.0.0	06.1.0		06.3.0 06.3.6		06.5.0 06.5.1 06.5.6	06.6.0			06.9.0
Surface-water Gley	07	07.0.0						07.6.0			07.9.0
Podzol (Fe-Ac)	08	08.0.0		08.2.0	08.3.0	08.4.3		08.6.0			08.9.0
Brown Podzolic	09	09.0.0		09.2.0	09.3.0 09.3.6			09.6.0			09.9.0
Luvisol	10	10.0.0		10.2.0 10.2.6	10.3.0 10.3.6			10.6.0			10.9.0
Brown Earth	11	11.0.0		11.2.0 11.2.6	11.3.0 11.3.6		11.5.0 11.5.3 11.5.6 11.5.9	11.6.0			11.9.0 11.9.2 11.9.3 11.9.6

additional features to the Great Soil Group. For example, a Humic Calcareous Brown Earth is coded as 11.5.6 (Fig. 3.8) where 11 indicates the Great Group; the soil is predominantly calcareous (within 40 cm) indicated by a five, 11.5 and finally, the soil also has a humose surface horizon represented by a six, therefore: 11.5.6. Figure 3.8 shows the Subgroups that exist within Irish SIS.

3.7 Classification of Soil Series

Soil Series is the final level of classification and represents the most specific category in the taxonomy. In Ireland, of the final 213 Soil Series classified, 73 originated and were described by the AFT soil survey. Series are defined on the basis of the following hierarchy; Great Soil Group, Soil Subgroup, texture and parent material. Series names are derived from the place names, such as town lands, from where the soil was first defined or where it was best expressed. Soil Series are identified by a unique code made by concatenating the Subgroup code with a code based upon the series name; for example, 'Ballylanders' is a Typical Brown Earth (1100) and has a map code 'BY', hence the unique full series code is '1100BY' (Simo et al. 2013).

Texture refers to the relative proportions of the various sized particles in the mineral fraction of the soil, specifically the sand, silt and clay less than 2 mm in diameter (Gardiner and Radford 1980). Texture is described at horizon level. Soil texture is one of the most important physical characteristics of the soil due to its influence on soil properties, such as moisture retention. The Irish SIS applied the particle-size distributions as adopted by USDA (Fig. 3.9) to maintain consistency with earlier soil survey work carried out by AFT. Particle sizes are described as follows: sand (2 mm–50 picometre (pm)), silt (50–2 pm), and clay (<2 pm) sized particles.

Overall soil profile texture: Four broad particle-size groupings are defined for generalising the overall texture of the reference section; sandy (Sy), loamy (Ly), silty

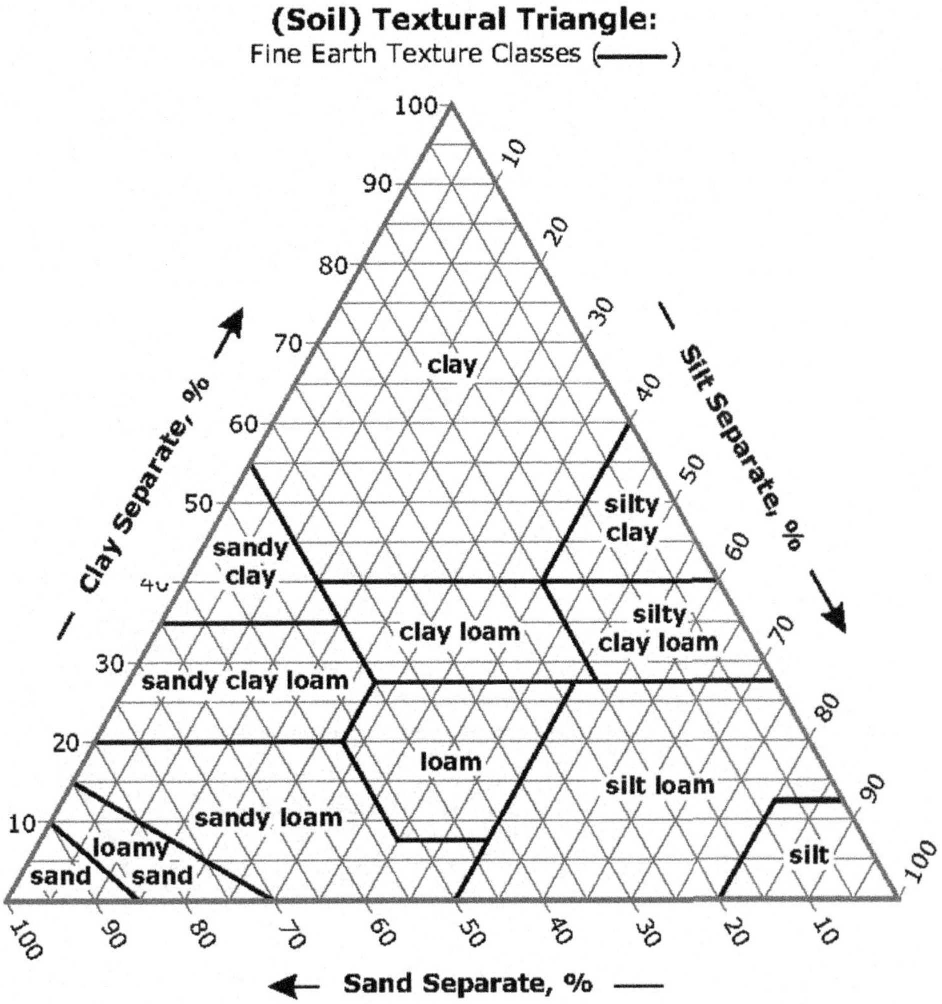

Fig. 3.9 Chart with percentages of clay (<0.002 mm) silt (0.002–0.05 mm) and sand (0.05–2.0 mm) in the basic soil texutre classes (Soil Survey Staff 1993)

(Zy) and clayey (Cey). When describing the total profile, as opposed to individual horizons, texture is described on the basis of particle size classes (Fig. 3.10). For agriculturally important soils, the loamy and silty groups are further subdivided into coarse and fine categories. To define the soil as either coarse (cLy) or fine loamy (fLy) the dominant texture within 80 cm should be used to define the broad textural class. The following procedures are applied (Jones et al. 2014):

1. if topsoil ~40 cm thick cLy over fLy, classify soil as cLy
2. if topsoil ~40 cm thick fLy over cLy, classify soil as fLy
3. if topsoil <40 cm thick fLy over cLy, classify soil as cLy
4. if topsoil <40 cm thick cLy over fLy, classify soil as fLy

In the Podzol and Lithosol Great Soil Groups, no distinction is made between coarse and fine loamy and Soil Series are described as Sandy, Loamy or Clayey (in Lithosols only).

Parent material is also used to define the Soil Series unit. Substrates in Ireland occur in five different states; peat, bedrock, drift, alluvium and anthropogenic (Table 3.5).

Peat is a biogenic deposit having developed in the post-glacial period of the last 10,000 years. Three main peat formations are recognised in Ireland (1) raised bogs of the Central Plain; (2) blanket bogs of the western seaboard and upland regions and (3) fen peat. Their genesis has been influenced by drainage, climate, hydrology, geomorphology, nutrient status and glacial geology but over time, these deposits have also been altered by man's activities.

Bedrock is the hard rock beneath surface materials such as soil and gravels. These soils comprise mineral material overlying a layer of coherent bedrock, or skeletal weathered bedrock fragments (that are at least 15 cm thick and begins above and extends below 80 cm depth). Where possible, effort to distinguish between bedrock and skeletal substrates should be made.

Drifts, the third substrate group, are composed mostly of Quaternary deposits including glacially-derived tills and fluvioglacial deposits and recent alluvium. The main classes

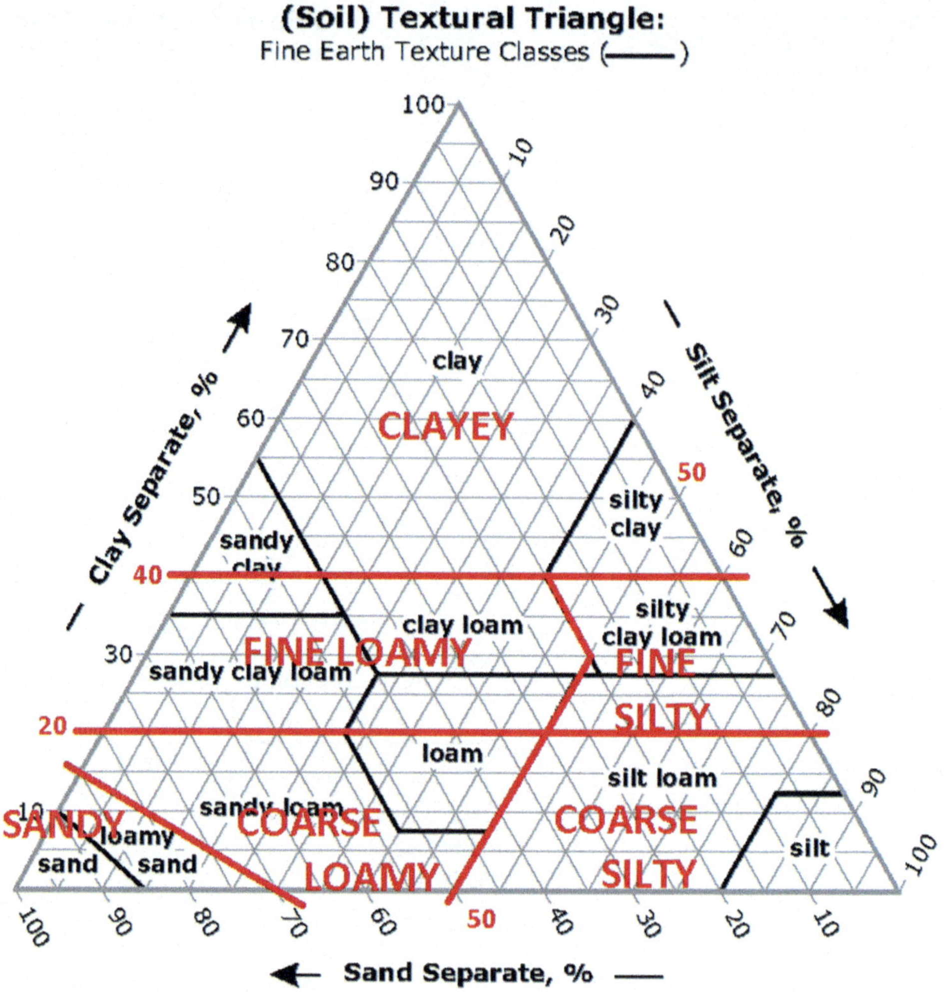

Fig. 3.10 Soil particle-size classes and broad groups for Soil Series definition (Jones et al. 2014) used in Irish soil classification 2014

of drift in Ireland found are: (1) drift with siliceous stones with deposits dominated by sandstone, slate, shale or chert stones. Soils in this substrate are usually non-calcareous to at least 120 cm; (2) drift with limestones that have calcareous material within 80 cm depth, or a non-calcareous B horizon within 80 cm but that passes conformable within 120 cm into calcareous material; (3) drift with igenous and metamorphic stones where these constitute the dominant local lithology and, (4) stoneless drift soils.

Alluvium substrate are soils that include all thick drifts in which loamy or clayey marine, river, or lacustrine sediments of recent age extend below 30 cm depth (Avery 1980) of which three types are used to define Soil Series; river alluvium, marine/estuarine alluvium and lake marl. However, the distinction between each type is not always clear because intrinsic properties such as structure, porosity, pH, $CaCO_3$ content, exchangeable cations and colour do not provide a consistent means of identification.

3.8 Correlation of Irish SIS to World Reference Base

Global population growth, land degradation and climate change, along with the interdependence of countries for the supply of food and agricultural produce, prompted the need for a global harmonised soil information system. On this basis the Food and Agriculture Organization of the United Nations (FAO) called on the active participation of the soil community to come together to develop a framework for the harmonisation of existing soil classification systems (IUSS Working Group WRB 2006). The first meetings were held in 1980 and 1981, hosted by the Poushkarov Institute of Soil Science and Yield Programming in Sofia, Bulgaria. At the first meeting it was decided to launch a programme to develop an International Reference Base for Soil Classification (IRB). At the second meeting, the general principles

Table 3.5 Substrate types used to define soil series in Ireland (Simo et al. 2014)

State	Substrate type	Materials
Peat	Blanket and Raised bog, fen cutover, industrial	
Bedrock	Basic igenous	Basalt, gabbro
	Acid igenous	Granite, rhyolite
	Limestone	Limestone
	Sandstone	Old Red Sandstone, millstone grit, quartzite, chert
	Shale/slate	Shale, slate, calcareous shale
	Sandstone and shales	Interbedded sandstone and shales Mica schist
	Gneiss and schist	
Drift (Till or fluvioglacial sediments)	With siliceous stones	Sandstones, shale, slate, quartzite, chert
	With limestones	Limestone
	With igenous and metamorphic stones	Basalt, gabbro, granite, gneiss, schist, aeolian, glaciolacustrine
	Stoneless Non-calcareous gravels Calcareous gravels	
Alluvium	River Estuarine Marine Lake marl	
Anthropogenic	Made ground	

of the joint programme towards the development of the IRB were defined (IUSS Working Group WRB 2006). In 1992, the IRB was renamed the World Reference Base (WRB), with the working group established at the 15th Congress of ISSS. The work culminated in the release of the first edition of the WRB for soil resources at the 16th World Congress of Soil Science at Montpellier in 1998. A second edition was subsequently released at the 18th World Congress at Philadelphia in 2006. The first edition comprised of 30 Reference Soil Groups (RSGs) with the second edition including 32 RSGs. Following a period of eight years, and an intensive worldwide testing and data collection effort, the third edition was launched in 2015 to reflect new updates and the earlier drafts and editions of the WRB (IUSS Working Group WRB 2015).

The development and revision of the publication of the WRB for soil resources has greatly contributed to a better understanding of soil science and has been used by the European Commission as the basis for harmonised assessment for the European Union and beyond. Many countries have adopted the WRB as a higher level to their national soil classification system (Reidy et al. 2014). The WRB was not designed as a national classification system but rather to serve as a mechanism for harmonisation to facilitate communication at an international level.

In the Irish SIS Final Technical Report 8, Reidy et al. (2014) describe the correlation of the Irish soil classification system to the WRB 2006 system. During the SIS field campaign 225 soil pits were described in detail with samples analysed to 1 m depth. The description and the associated analytical data were completed with consideration of correlation to the WRB system (Reidy et al. 2014). Within WRB, two categorical levels are defined: the 1st level contains the 32 RSGs with groups of soils based upon dominant identifiers such as soil forming factors or processes that indicate the soil condition. The RSG represent all major regions of the world. The 2nd level includes prefix and suffix qualifiers that are added to the RSG to allow precise characterisation and classification of individual soil profiles (Reidy et al. 2014). Within WRB the classification of soils is based on soil properties defined in terms of diagnostic horizons, properties and materials that take into account their relationship with soil forming processes. Diagnostic features are then included that are of significance to soil management

Table 3.6 Soil great group code and soil group in Irish classification and correlation to the WRB (Reference Soil Group) from Reidy et al. (2014) with dominant WRB reference group in bold

Soil group code	Soil group Irish classification	Reference soil group WRB
01	Ombrotrophic Peat	**Histosol (Ombric)**
02	Minerotrophic Peat	**Histosol (Rheic)**
03	Rendzina	**Regosol (Calcaric), Phaeozem (Calcaric)**
04	Lithosols	**Leptosol**, Histosol, Regosol, Cambisol
05	Alluvial	**Fluvisol**, Cambisol, Gleysol, Regosol, Stagnosol, Planosol
06	Groundwater Gleys	**Gleysol**, Phaeozem, Stagnosol, Umbrisol
07	Surface-water Gleys	**Stagnosol**, Phaeozem, Cambisol
08	Podzols	**Podzol,** Umbrisol, Arenosol
09	Brown podzolics	**Cambisol, Podzol,** Phaeozem, Umbrisol, Acrisol
10	Luvisols	**Luvisol**, Cambisol, Lixisol, Gleysol, Albeluvisol, Planosol, Phaeozem
11	Brown Earths	**Cambisol, Regosol**, Luvisol, Phaeozem, Umbrisol, Stagnosol, Lixisol

Table 3.7 Garbic Ekranic Technosol (Densic, Toxic, Transportic) based upon definitions of formative elements for second-level units of WRB from IUSS (2015)

Classification	Meaning	Prefix/suffix intermediate (Abbreviation)	Definition of formative elements for second-level units of WRB
Technosol	Soil sealed with "technic" hard material within 5 cm of the soil surface covering 95% or more of the horizontal extent of the soil. Soil contains many artefacts	Ekranic (ek)	Rock-like sealing of concrete, tarmacadam, asphalt having technic hard rock within 5 cm of surface. Only associated with Technosols
		Garbic (ga)	Soil contains significant amount of material 20 cm or more thick within the first 100 cm. The artefacts contain >35% or more organic wastes in Technosols
		Densic (dn)	Natural or artificial compaction within 50 cm of the soil surface prohibiting root penetration
		Toxic (tx)	Has a layer with toxic concentrations of organic/inorganic substances other than ions of Al, Fe, Na, Ca and Mg, within 50 cm of the soil surface
		Transportic (tn)	A layer >=30 cm with liquid or solid material that has been moved from source by intentional human activity

and should be selected, to the extent possible, at a high level of generalisation. Similarly, within the Irish SIS, classification at Great Soil Group level considers the main soil forming processes with classification at Soil Subgroup level taking diagnostic criteria into account. Table 3.6 by Reidy et al. (2014) shows the RSG that Irish Great Soil Groups are correlated to. For example, Irish Great Soil Group 01, Ombrotrophic Peat correlates with RSG Histosol, with the Ombric suffix qualifier. A full table of the current WRB correlation for each representative soil profile at Soil Series level is also available from Reidy et al. (2014).

While some WRB prefix and suffix qualifiers may not so readily correlate and might require laboratory data not measured in the SIS, currently most Great Soil Groups readily correlate with the RSG of the WRB. The WRB is an on-going evolving process and correlation of the Irish SIS with WRB designations is important for collaboration and the presentation of work at an international level.

Fig. 3.11 Urban soils. **a** Ekranic Technosol, a pavement with constructed deposits below, formerly a tramline, road base, and embankment. **b** Regosol, formed during soil relocation as cut-and-fill for highway construction; this will eventually be given a topsoil cover and grass seed sown. **c** Garbic Technosol (Thapto-Fluvisolic), reclamation deposits of commercial garbage, with many artefacts such as bottles; beach sediments lie a further 80 cm below the base of this hand-dug pit. **d** Regosol (Relocatic), a reclamation deposit over a worked-out sand pit, showing remnants of former site soils, roughly mixed

3.9 WRB Classification for Urban Soils

Concepts for discussing urban soils have been developed by the SUITMA series of conferences (Soils of Urban, Industrial, Transport, Military and Mining Areas), and by ICOMANTH (International Committee on Anthropogenic Soils of the United States Department of Agriculture). In relation to classification, many informative names for urban soils come from the WRB for Soil Resources (IUSS Working Group WRB 2015). The WRB for Soil Resources includes many concepts that can help with describing urban soils, and how to recognise human action in soils everywhere. "Transportic", "Relocatic" and "Aric" qualifiers in soil names refer to soil moved long distances (such as in trucks), short distances (using diggers), and mixed without moving, so soil layers are disrupted. These transformations occur during construction, but also in many historical agricultural operations such as burying dead animals, bringing up limestone to spread on the land surface, installing and removing drains and earthbanks. Soils that have been mixed do not have horizons, and are called Regosols. The soil name "Technosol" was coined in 2006, and represents soils with sealing (including by buildings and roads, carparks and other "pavements") OR soils with a high proportion of artefacts (including all products of manufacture, wastes, ash, oil, and mine spoil). A narrower concept is Ekranic Technosol, with the "Ekranic" qualifier specifically referring to soils sealed by constructed rock-like material such as concrete, tarmacadam or asphalt. This qualifier is only associated with the Technosol group and goes before the name, as do qualifiers for intermediates between groups. Further qualifiers are listed in brackets after the name, giving very informative soil names. Table 3.7 provides an example of an urban soil type, using the WRB approach to naming:

The Garbic Ekranic Technosol (Densic, Toxic, Transportic) soil is primarily a Technosol. It has two characteristics typically associated with Technosols, the more-important one being that it is sealed by a "technic hard" material such as concrete or asphalt (Ekranic), and secondly that it includes a significant amount of material derived from organic wastes (Garbic). The order of Irish SIS nomenclature is consistent with the order described here for the WRB. Further characteristics identifiable under field conditions are listed, in alphabetical order. The soil is denser than will allow growth of roots (Densic); it contains substances toxic to organisms (Toxic); and it has been moved from another site (Transportic).

A few more examples show some variation among urban soils:

Arenosol (Relocatic)

Regosol (Densic, Technic)

Spolic Technosol (Thapto-Fluvisolic)

Arenosol is a sandy soil, while Relocatic identifies it as having been reconstructed without being removed from the site. A Regosol is a soil with no distinct horizons from soil-formation processes; Technic shows it as having significant artefact content (but not enough to qualify as a Technosol). Spolic Technosols represent industrial waste or mine spoil, while Thapto-Fluvisolic identifies underlying buried (Thapto-) sediment derived from flowing water (Fluvisol[ic]).

The full system, for naming all soils in the world, is published by IUSS Working Group WRB (2015). Examples of urban soils with their WRB names are given in Fig. 3.11.

Acknowledgements For the most part, this chapter explains the classification system for Irish soils as developed within the Irish Soil Information Systems project (2007-S-CD-1-S1). The authors here acknowledge the "Soil Information System" project funded as part of the Environmental Protection Agency (EPA) Research Programme, administered on behalf of the Department of the Environment, Heritage and Local Government by the EPA which has the statutory function of co-ordinating and promoting Irish environmental research. The project included four main partners: Teagasc—Environment, Soils and Land-use Centre, Teagasc—Spatial Analysis Group, Rural Economy and Development Programme, Cranfield University—National Soil Resources Institute and University College Dublin. It is important too, to recognise the many staff and PhD students, placement students, FÁS placements and the Steering Committee who all contributed so positively to this project. The authors here too acknowledge An Foras Talúntais and express gratitude to all of the technical, field survey and research teams who worked over a time period of decades to compile a wealth of legacy soil data that formed much of the basis for the Irish Soil Information Systems.

References

Avery BW (1980) Soil classification for England and Wales [Higher categories]. In: Soil survey technical monograph no. 14, Harpenden, p 67

Clayden B, Hollis JM (1984) Criteria for differentiating soil series. In: Soil survey technical monograph no. 17, Harpenden, p 159

Creamer R, Simo I, Reidy B, Carvalho J, Fealy R, Hallet S, Jones R, Holden A, Holden N, Hannam J, Massey P, Mayr T, McDonald E, O'Rourke S, Sills P, Truckell I, Zawadzka J, Schulte R, (2014) Irish soil information synthesis report (2007-S-CD-1-S1) EPA STRIVE Programme, Wexford. Available for download from http://erc.epa.ie/safer/reports

Daly K, Fealy R (2007) Digital soil information system for Ireland (2005-S-DS-22), Final report. Environmental RTDI Programme 2000–2006, Environmental Protection Agency, Johnstown Castle, Wexford

De Bakker H, Schelling J (1966) Systeem van bodemclassificatie voor Nederland. De Hogere Niveaus (with English summary). Pudoc, Wageningen

Food and Agriculture Organization (FAO) (2006) Guidelines for soil description, 4th edn. FAO, Rome, p 97

Gardiner MJ, Radford T (1980) Soil associations of ireland and their land use potential explanatory bulletin to soil map of Ireland 1980. National Soil Survey of Ireland, An Foras Talúntais, Dublin

IUSS Working Group WRB (2006). World reference base for soil resources—a framework for international classification, correlation and communication. World soil resources reports 103. ISSN 0532-0188, FAO, Rome, p 128

IUSS Working Group WRB (2015). World reference base for soil resources 2014, update 2015 international soil classification system for naming soils and creating legends for soil maps. World soil resources reports no. 106. FAO, Rome

Jones R J A, Hannam J A, Palmer R C, Truckell I G, Creamer R E, McDonald E (2014) Irish SIS final technical report 1: interim classification, Harmonisation and generalisation of county soil maps of Ireland, EPA STRIVE Programme, Wexford. Available for download from http://erc.epa.ie/safer/reports

NRA (undated) A guide to landscape treatments for national road schemes in Ireland. National Roads Authority, Dublin

Reidy B, Simo I, Spaargaren O, Creamer RE, (2014). Irish soil information systems correlation of the irish soil classification system to world reference base—2006 system (2007-S-CD-1-S1) final technical report 8. EPA STRIVE Programme, Wexford. Available for download from http://erc.epa.ie/safer/reports

Simo I, Creamer RE, Hannam JA, Jones RJA (2013) Irish soil information systems legend (2007-S-CD-1-S1) final technical report 13. EPA STRIVE Programme, Wexford. Available for download from http://erc.epa.ie/safer/reports

Simo I, Creamer RE, Reidy B, Hannam JA, Fealy R, Hamilton B, Jahns G, Massey P, McDonald E, Schulte RPO, Sills P, Spaargaren O (2014) Irish SIS final technical report 10: soil profile handbook. EPA STRIVE Programme, Wexford. Available for download from http://erc.epa.ie/safer/reports

Simo I, Schulte RPO, Corstanje R, Hannam JA, Creamer R (2015) Validating digital soil maps using soil taxonomic distance: a case study of Ireland. Geoderma Reg 5:188–197

Soil Survey Staff (1993) Soil survey manual. United States Department of Agriculture handbook no. 18, Washington DC, p 437

United States Department of Agriculture (USDA) (1960) Soil classification a comprehensive system 7th approximation, United States Department of Agriculture

Soil Geography and Development of the Third Edition National Soil Map

4

Lilian O'Sullivan, Reamonn Fealy, Iolanda Simo, and Rachel Creamer

Abstract

The geography of the soils of Ireland in relation to the distribution of Great Soil Groups, the distinct Soil Subgroups that occur within each Great Soil Group and a reference profile description at series level from the online SIS database are presented in this chapter. This follows with a description of the development of the 3rd Edition National Soil Map of Ireland. Prior to the Irish Soil Information Systems (SIS) project, systematic pedology in Ireland had largely been related to mapping exercises, such as the previous National Soil Survey (NSS) conducted by An Foras Talúntais (AFT) who mapped 44% of the country at a scale of 1:126,720; developed a General Soil Map of Ireland and a National Peatland map, both at 1:575,000 scale. In 2009 the Irish SIS project was established to bring together all existing pedological data and map the remaining land area. A scale of 1:250,000 was determined as appropriate for the development of the 3rd Edition National Soil Map of Ireland based on recommendations of the European Soil Bureau Network (ESBN) Technical Working Group responsible for soil monitoring and harmonisation. The 1:250,000 national soil map of Ireland is a soil map composed of soil associations of the 213 soil series, and altogether a total of 58 soil associations, excluding peat, alluvium, urban, rock and marsh, are defined. A harmonisation of legacy data (*Terra Cognita*) was completed and served as the basis for the generation of soil-landscape models which were then used to model areas where detailed soil surveys had not previously been carried out. Validation of this methodology was carried out over a 2.5-year field survey, in which 11,000 locations were assessed for soil type using an auger survey approach. Where previously unidentified combinations of soilscape units were found, soil profile pits were excavated, sampled and described at representative locations across the country to define this new soil information. These 225 pits were described and sampled in detail and were used to support the development of a new soil classification (described in Chap. 3).

Keywords

Soil geography • National soil map • Terra Cognita
Legacy data • Mapping

L. O'Sullivan (✉) · R. Creamer
Teagasc Agriculture and Food Development Authority, Johnstown Castle Environment Research Centre, Wexford, Ireland
e-mail: Lilian.OSullivan@teagasc.ie

L. O'Sullivan
Irish Environmental Protection Agency, Johnstown Castle, County Wexford, Ireland

R. Fealy
Teagasc, Rural Economy and Development Programme, Ashtown, Dublin, Ireland

R. Creamer
Wageningen University and Research Atlas—Wageningen Campus, Wageningen, The Netherlands

I. Simo
Smart Soil & Land, Jaume II St., 11, Lleida, Spain

4.1 Soil Geography of Ireland

The following provides an overview of the soil geography of Ireland. Each Great Soil Group is presented with a description of the associated Subgroups that occur within that Great Soil Group. Thereafter, the description includes a sample Soil Series data sheet that can be accessed through the online Irish Soil Information System (SIS) database. This provides an indication of the detail that was recorded for each of the reference profile pits as part of the field campaign for the SIS project 2014.

R. Creamer and L. O'Sullivan (eds.), *The Soils of Ireland*, World Soils Book Series,
https://doi.org/10.1007/978-3-319-71189-8_4

4.1.1 Ombrotrophic Peat Soils (01)

Ombrotrophic Peat soils are rain-fed soils found in lowland (raised bog) and upland (blanket bog) positions that are rich in organic matter. They are organic rich, with an O horizon >40 cm, found within the upper 80 cm (Fig. 4.1). Ombrotrophic soils are oligotrophic with pH < 4.0 (in $CaCl_2$ 1:2.5 undried, equivalent to pH 4.5 in 1:2.5 H_2O) throughout the Reference Section (Jones et al. 2011). These soils correlate with the Histosol Reference Soil Group of the WRB (IUSS Working Group WRB 2006) with an Ombric qualifier. In Counties Cork, Clare, Donegal, Kerry and Sligo upland peats have developed under extreme influence from the ocean, with a distribution pattern demarcated along the lines of earlier orogenies (Hammond 1981). Also, the distribution of peat soils in Ireland aligns with the higher precipitation patterns of the western seaboard reflected in a higher occurrence of rain-fed peat soils found in this region. As a result, the majority of peats in Ireland are blanket bogs, however, lower level Atlantic type peats also occur and are found in Counties Galway, Mayo, Cork, and Donegal and to a lesser extent in Kerry (Gardiner and Radford 1980). The development of blanket bog in the eastern half of the country is limited to the wetter higher locations, such as the Wicklow Mountains (Fig. 4.1). This Great Group comprises four distinct Subgroups, which differ by key characteristics:

01.1.0 ***Natural Ombrotrophic Peat Soils*** have fibrous (Of), semi-fibrous (Om) or humified peat (Oh) horizons and they have not been disturbed by human influence, other than grazing by animals. Representative profile description sheet Fig. 4.2.

01.7.0 ***Drained Ombrotrophic Peat Soils*** have as earthy topsoil due to management practices, such as draining and cultivation/fertilisation.

01.8.0 ***Cut Ombrotrophic Peat Soils*** have had the surface peat removed by hand cutting methods but the peat thickness still meets the criteria for peat soils.

01.9.0 ***Anthropic Ombrotrophic Peat Soils*** have had the surface peat removed by industrial methods such as machine sod removal and milling but the peat thickness still meets the criteria for peat soils.

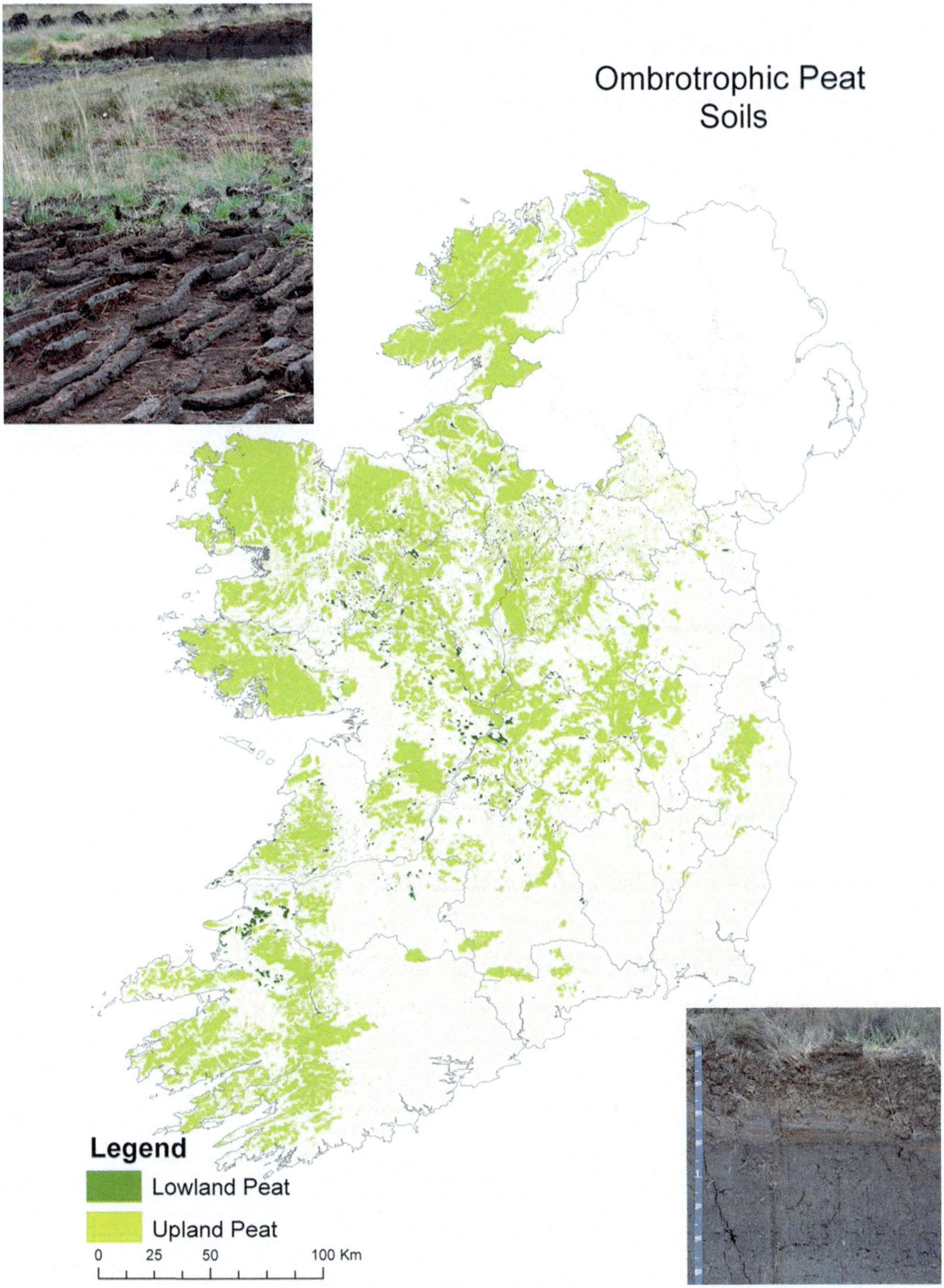

Fig. 4.1 Ombrotrophic Peat Soil: Landscape image (left), geographic distribution (centre) and soil profile (right)

REPRESENTATIVE PROFILE DESCRIPTION

SERIES: AUGHTY

Reference profile: 184
County: Laois
Weather:
Eleavation:

TOPOGRAPHY
Position:
Slope Degree:
Slope Form:
Aspect:

PARENT MATERIAL
Substrate Type: Peat
Substrate Subgroup: Ombrotrophic

TEXTURAL CRITERIA
Textural Class: Peat
Texturally contrasting:

LAND USE
Land use: Heath / Scrub
Human technologies:
Vegetation: Eriophorum vaginatum (Bog cotton), Calluna vulgaris (Heather; Erica tetralix (Cross-leaved heath), Narthecium ossifragum (Bog asphodel) locally abundant

WATER TABLE

ROCK OUTCROPS

SURFACE STONE

IRISH CLASSIFICATION (2013)
Soil subgroup: 01.1.0 Natural Ombrotrophic Peat Soils

Series: Aughty
Definition: Peat (Blanket Bog)

WRB CLASSIFICATION (2006)
Ombric Dystric Sapric Histosol

DESCRIPTION

0 - 25 cm **Oh1**
MATRIX COLOUR: 5YR2.5/1.**BOUNDARY:** Clear, Smooth.

25 - 80 cm **Oh2**
MATRIX COLOUR: 5YR3/2.

80 - 150 cm **Oh3**
MATRIX COLOUR: 5YR3/2. **BOUNDARY:** Abrupt, Smooth.

150 - 158 cm **BC1**
MATRIX COLOUR: 10YR5/2. **TEXTURE:** Loam. **STRUCTURE:** Massive **STICKINESS:** Sticky. **BOUNDARY:** Clear. **NOTES:** Fine fossil roots.

158 cm + **BC2**
MATRIX COLOUR: 10YR7/2. **TEXTURE:** Loam. **STRUCTURE:** Massive. **NOTES:** Fossil roots.

Fig. 4.2 Reference profile pit description of 01.1.0, a natural Ombrotrophic Peat, the Aughty Soil Series

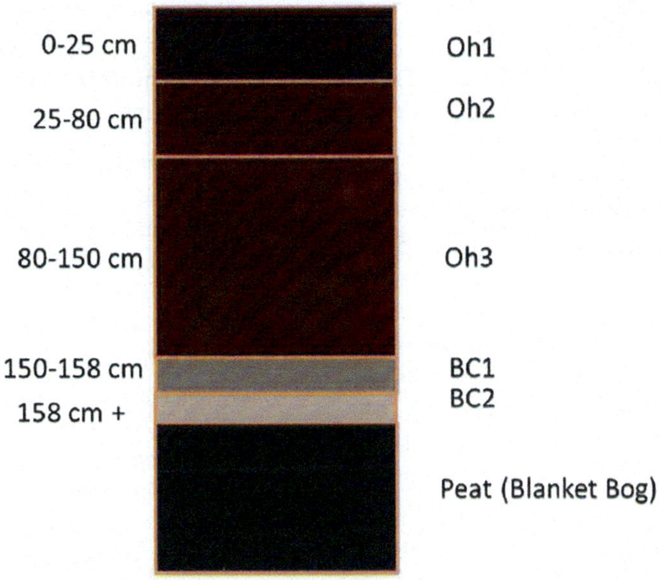

LABORATORY ANALYSES

Horizon	pH	Total (%)		C/N
		Nitrogen	Carbon	
Oh1	3.6			
Oh2	3.5			
Oh3	3.8			
BC	3.6			

PARTICLE SIZE (%)			Textural Class USDA	Bulk Density g/cm^3	Standard Deviation
Sand 2000-50 μm	Silt 50-2 μm	Clay <2 μm			
				0.11	
				0.09	
				0.1	

PEAT ANALYSIS							
Moisture (%)	Ash (%)	Fibre	PP Colour	WHC	PPI	EXC Ca/Mg	Spec Index
556	3.5	6					7
554	1.8	6					6
598	1.6	6					7
411	1.7	6					6

Fig. 4.2 (continued)

4.1.2 Mineratrophic Peat Soils (02)

Mineratrophic Peat soils are organic-rich soils with an O horizon of more than 40 cm (Fig. 4.3) within the upper 80 cm and are groundwater fed peat systems occurring in river valleys, inter-drumlin hollows and on the periphery of raised bogs (Fig. 4.3). They are eutrophic and have a pH > 4.0 (in $CaCl_2$ 1:2.5 undried, equivalent to pH 4.5 in 1:2.5 H_2O) in at least some part of the Reference Section (Jones et al. 2011). The high groundwater table creates anaerobic conditions for most of the year resulting in poor decomposition rates. These peats are often referred to as 'fen' soils and compared to 'bogs' they tend to have a higher pH. Historically, fens in Ireland have been drained for agricultural production. As a result, undisturbed fens are rare and the Pollardstown Fen in County Kildare is the largest remaining calcareous spring-fed fen in Ireland (Fig. 4.3).

Mineratrophic Peat soils correlate with the Histosol Reference Soil Group of WRB (IUSS Working Group WRB 2006), with a Rheic qualifier. This Great Group comprises of three distinct subgroups:

02.1.0 ***Natural Mineratrophic Peat Soils*** have well-humified, amorphous Oh horizons, fibrous (Of) or semi-fibrous (Om) upper horizons. These soils have not been disturbed by human influence, other than grazing by animals (Fig. 4.4).

02.7.0 ***Drained Mineratrophic Peat Soils*** are drained Ombrotrophic Peat Soils but have earthy topsoil due to management practices, such as draining, cultivation and/or fertilisation.

02.8.0 ***Cut Mineratrophic Peat Soils***, cutover Ombrotrophic Peat Soils have had the surface peat removed by hand cutting methods but the peat thickness still meets the criteria for peat soils.

Fig. 4.3 Mineratrophic Peat Soil: profile pit (left), geographic distribution (centre) and landscape image (right)

REPRESENTATIVE PROFILE DESCRIPTION

SERIES: POLLARDSTOWN

Reference Profile: 679
County: Offaly
Weather:
Elevation: 60 m

TOPOGRAPHY
Position:
Slope degree:
Slope form: 0
Aspect:

PARENT MATERIAL
Substrate Type: Peat
Substrate Subgroup: Minerotropic

TEXTURAL CRITERIA
Textural Class:
Texturally contrasting:

LAND USE
Land use: Fen
Human technologies
Vegetation: Fen species, Potamogeton natans (Pondweed), Menyanthes trifoliate (Bogbean), Mentha aquatic (mint)

WATER TABLE 0 cm

ROCK OURCROPS

SURFACE STONE

IRISH CLASSIFICATION (2013)
Soil subgroup: 02.1.0 Natural Minerotrophic Peat Soil

Series: Pollardstown
Definition: Peat (Fen, Unchained)

WRB CLASSIFICATION (2006)
Limnic Rheic Histosol

DESCRIPTION

0 - 15 cm **Oh1**

MATRIX COLOR (moist): 5YR2.5/1. **STRUCTURE:** Structureless. **ROOTS:** Common. **BOUNDARY:** Clear, Smooth. **Notes:** No recognisable fossil plant remains.

15 - 45 cm **Ohg**

MATRIX COLOR (moist): 5YR2.5/1. **ROOTS:** Many. **BOUNDARY:** Clear, Smooth. **Notes:** Amorphous, greasy, with no vsible recognisable plant structures.

45 - 100 cm **Cg**

BOUNDARY: Abrupt. **Notes:** Shell marl.

100 - 150 cm **Oh2**

Amorphous peat

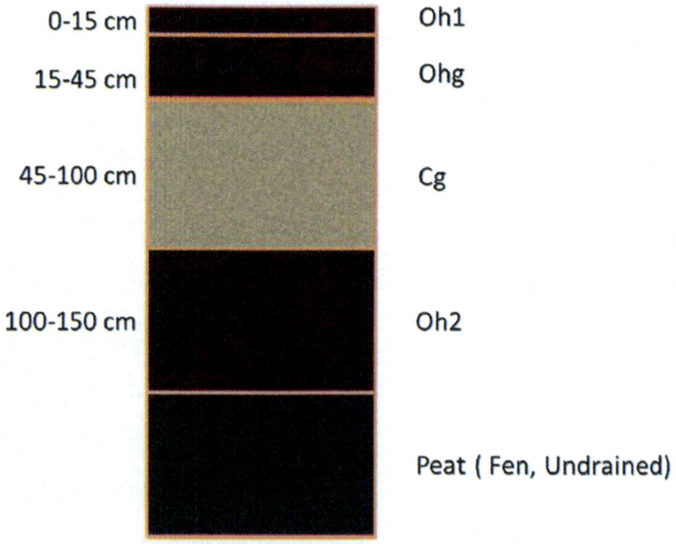

Fig. 4.4 Reference profile pit description of 02.1.0, a natural Mineratrophic Peat Soil, Pollardstown Soil Series

LABORATORY ANALYSES

Horizon	pH	Total (%)		C/N
		Nitrogen	Carbon	
Oa1	6.2	2	47.7	23.9
Oa2g		2.19	46.9	21.4
Cg		0.95	10.4	10.9

Fig. 4.4 (continued)

4.1.3 Rendzina Soils (03)

Rendzinas are shallow soils developed over calcareous parent rock material, such as limestone bedrock. They are <30 cm deep, calcareous (>2% $CaCO_3$) soils with hard rock (R) or skeletal material comprising >80% coarse fragments (>2 mm esd) at or above 30 cm (Jones et al. 2011). Their calcareous nature is reflected in a high soil pH (pH > 7). These soils can be very stony due to the weathering of rock material close to the soil surface (Fig. 4.5). A majority of Rendzina soils in Ireland are found in the west of Ireland and they represent the dominant soil type in the Burren region of County Clare (Fig. 4.5). The shallow depth of these soils may represent an agricultural limitation and accordingly, these soils are mostly suited to extensive grazing (Fig. 4.5) but where they are sufficiently deep, these can make suitable tillage soils (Gardiner and Radford 1980). Rendzina soils primarily correlate with the Leptosol Reference Soils Group of WRB (IUSS Working Group WRB 2006) having developed in or on calcareous bedrock with a calcareous qualifier. There are three distinct Subgroups within this Great Group:

- ***03.0.0*** ***Typical Rendzina*** are soils with a distinct topsoil, without any further distinguishing features.
- ***03.1.0*** ***Histic Rendzina*** have a peaty surface horizon (>20% organic carbon) less than 40 cm thick.
- ***03.6.0*** ***Humic Rendzina*** have a humose surface horizon (Fig. 4.6).

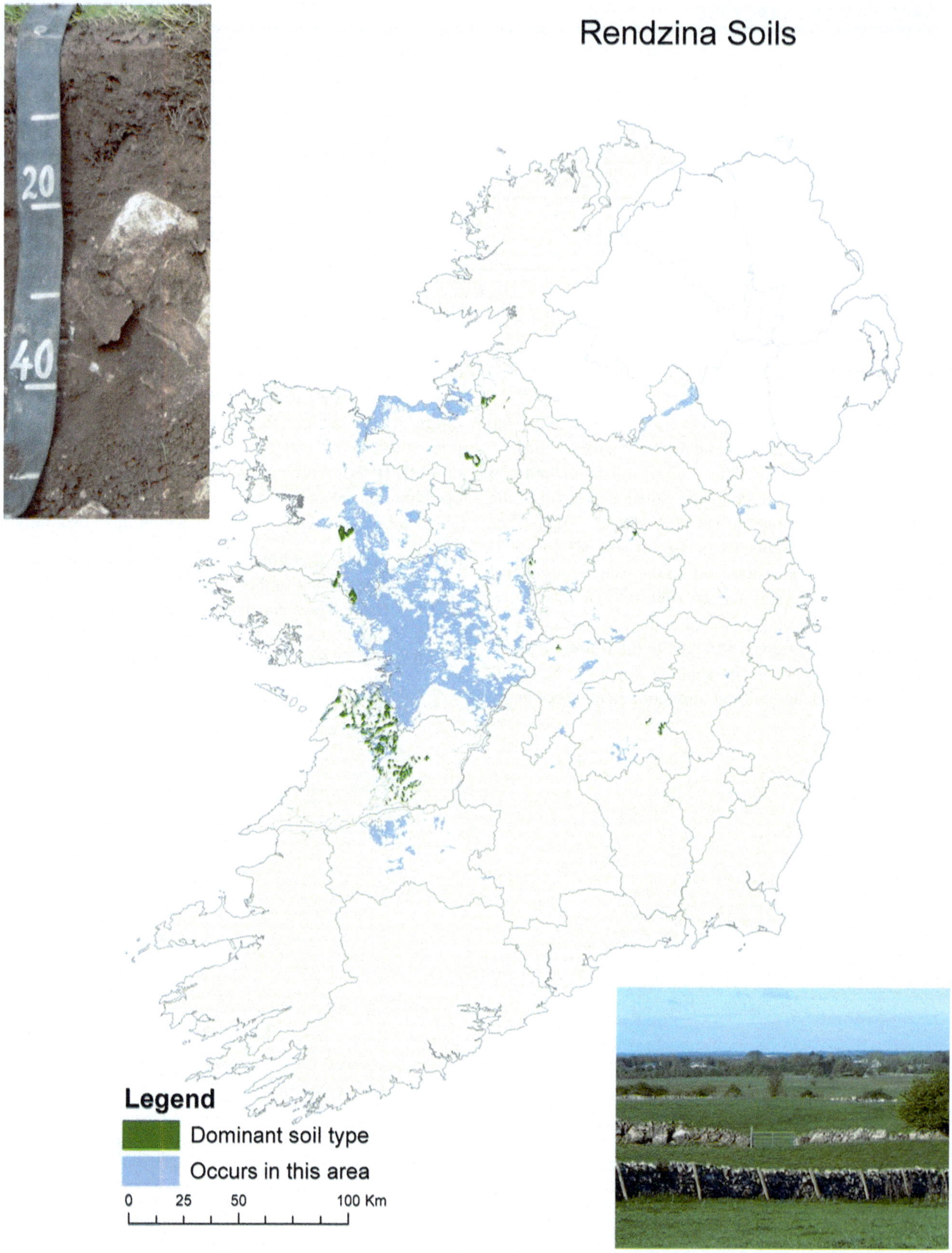

Fig. 4.5 Rendzina: profile pit (left), geographic distribution (centre) and landscape image (right)

REPRESENTATIVE PROFILE DESCRIPTION

SERIES: BURREN

Reference profile: 389
County: Leitrim
Weather:
Eleavation:

TOPOGRAPHY
Position:
Slope Degree:
Slope Form:
Aspect:

PARENT MATERIAL
Substrate Type: Bedrock
Substrate Subgroup: Limestone

TEXTURAL CRITERIA
Textural Class: Fine loamy
Texturally contrasting:

LAND USE
Land use:
Human technologies:
Vegetation:

WATER TABLE

ROCK OUTCROPS

SURFACE STONE

IRISH CLASSIFICATION (2013)
Soil subgroup: 03.6.0 Humic Rendzina

Series: Burren
Definition: Loamy over limestone bedrock

WRB CLASSIFICATION (2006)
Eutric Leptic Regosol (Humic)

DESCRIPTION

0 - 8 cm **A1**

MATRIX COLOUR: 10YR4/3. **TEXTURE:** Clay loam. **STRUCTURE:** Weak to moderate, Crumbly, Fine, Parting to subangular blocky. **CONSISTENCY:** Friable. **ROOTS:** Many, Fine and medium . **BOUNDARY:** Clear, Wavy.

8 - 17 cm **A2**

MATRIX COLOUR: 10YR5/4. **TEXTURE:** Clay loam. **STRUCTURE:** Weak, Subangular blocky, Fine. **CONSISTENCY:** Friable. **ROOTS:** Many, Very fine and fine . **BOUNDARY:** Clear, Wavy.

17 - 35 cm **Cr**

MATRIX COLOUR: 10YR5/8. **STONES:** Abundant, Stones and boulders. **TEXTURE:** Clay loam. **STRUCTURE:** Massive. **PLASTICITY:** Non-plastic. **STICKINESS:** Slightly sticky. **ROOTS:** Many, very fine and fine . **BOUNDARY:** Abrupt, Wavy.

35 cm + **R**

Fig. 4.6 Reference profile pit description of 03.6.0 Humic Rendzina, Burren Soil Series

LABORATORY ANALYSES

Horizon	pH	Total (%)		C/N
		Nitrogen	Carbon	
A1	5.6		7.2	
A2	7.2		3.6	
Cr	7.5		1.4	

EXCHANGEABLE COMPLEX				
CEC (cmol kg^{-1})	TEB (cmol kg^{-1})	Base Sat. (%)	Free Iron (%)	TNV (%)
43.6	19.03	44	1.4	
35.39	21.43	61	1.5	
22.8	16.8	74	1.4	

PARTICLE SIZE (%)			Textural Class USDA	Bulk Density g/cm^3	Standard Deviation
Sand 2000-50 µm	Silt 50-2 µm	Clay <2 µm			
37	34	29	Clay loam		
40	31	29	Clay loam		
38	28	34	Clay loam		

Fig. 4.6 (continued)

4.1.4 Lithosol Soils (04)

Lithosols are shallow (<30 cm depth) non-calcareous acidic soils, frequently overlying hard rock or skeletal and gravelly material made up of >80% coarse material (>2 mm). Lithosols are commonly associated with mountain landscapes and may represent the dominant soils in some higher locations, such as in the Wicklow Mountains in the east, or in the higher coastal locations of the south-west and north-west (Fig. 4.7). Unlike Rendzinas, these soils are found on acidic parent rock, such as sandstone, shale, granite, gneiss and schist. In general, these are very stony soils due to the weathering of rock material close to the surface and are often associated with rock outcrops (Jones et al. 2011) (Fig. 4.7). The pH of the soil is usually less than pH 6.5. These soils are often found with exposed rock outcrops and their use-range is typically limited to rough grazing (Gardiner and Radford 1980). This Great Group correlates primarily with the Leptosol Reference Soil Group of WRB (IUSS Working Group WRB 2006), but may also correlate with the Histosol and Regosol Reference Soil Groups.

04.0.0 ***Typical Lithosols*** have distinct topsoil, without any distinguishing features.

04.1.0 ***Histic Lithosols*** have a peaty surface horizon (>20% organic carbon) less than 40 cm thick (Fig. 4.8).

04.6.0 ***Humic Lithosols*** have a humose surface horizon.

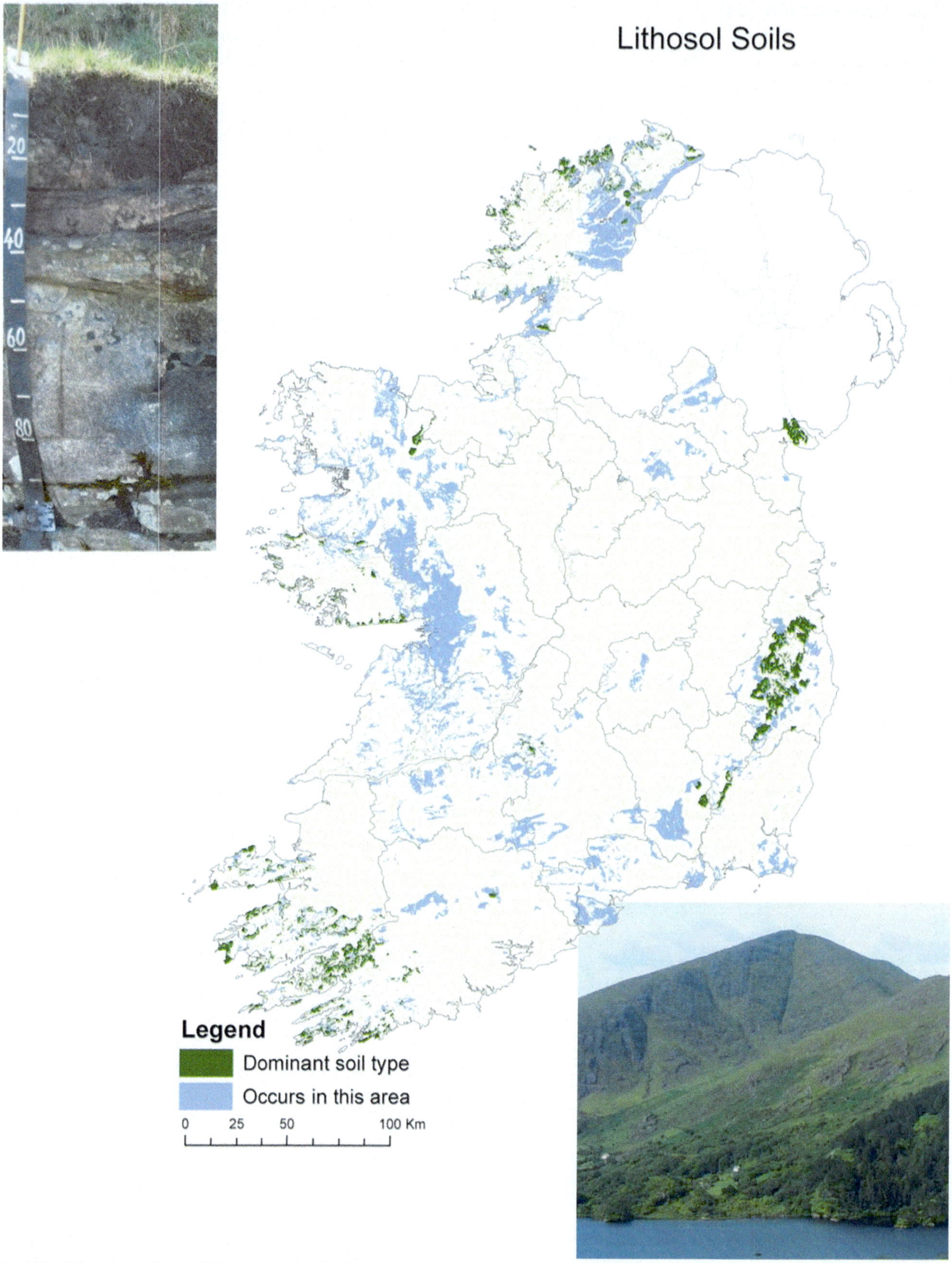

Fig. 4.7 Lithosol: profile pit (left), geographic distribution (centre) and landscape image (right)

REPRESENTATIVE PROFILE DESCRIPTION

SERIES: CARRIGVAHANAGH

Reference profile: RPO21BR01
County: Wicklow
Weather: Partly cloudy
Elevation: 119

TOPOGRAPHY
Position: Middle slope
Slope degree: 1
Slope Form: Straight
Aspect: N

PARENT MATERIAL
Substrate Type: Bedrock
Substrate Subgroup: Igneous acid

TEXTURAL CRITERIA
Textural Class:
Texturally contrasting: Coarse loamy

LAND USE
Land use: Scrub, Blanket bog
Human technologies: Clearing, Pollution,
Vegetation: Heather, Deciduous
Species: Hawthorn, Furze, Birch

WATER TABLE None

ROCK OUTCROPS None

SURFACE STONE Few

IRISH CLASSIFICATION (2013)
Soil subgroup: 04.1.0 Histic Lithosol

National Soil Series: Carrigvahanagh
Peat over lithoskeletal acid igneous bedrock

DESCRIPTION

0 - 4 cm **L**

MATRIX COLOR: 75YR32. **STONES (%):** None. **TEXTURE:** Fibrous peat. **CONSISTENCY:** Loose. **ROOTS:** Many, Fine. **PACKING DENSITY:** Low. **POROSITY:** High. **MACROPORES:** Common. **BOUNDARY:** Abrupt, Smooth.

4 - 20 cm **Oh**

MATRIX COLOR: 10YR21. **STONES (%):** Common, 6mm -2 cm, Angular, Quartzite. **TEXTURE:** Amorphous. **CONSISTENCY:** Very friable. **ROOTS:** Many, Fine and Coarse. **PACKING DENSITY:** Low. **POROSITY:** High. **MACROPORES:** Medium. **BOUNDARY:** Abrupt, Smooth.

20 cm + **Cr**

MATRIX COLOR: 10YR43. **STONES (%):** Abundant, 6mm -2 cm, Angular, Quartzite; Abundant, 2-6 cm, Angular, Quartzite. **TEXTURE:** Sandy Loam. Structure: Massive. **CONSISTENCY:** Very firm. **ROOTS:** Many, Fine and Coarse. **PACKING DENSITY:** High. **POROSITY:** Low.

Fig. 4.8 Reference profile pit description of 04.1.0 Histic Lithosol, Carrigvahanagh Soil Series

LABORATORY ANALYSIS

Horizon	pH	Total (%)		Organic Carbon (%)	Loss-on-ignition (%)
		Nitrogen	Carbon		
1(L)	4.2	2.09	40.67	45.60	85.7
2(Oh)	3.7	1.40	28.77	29.63	51.5
3(Cr)	3.8	0.07	1.39	1.12	

EXCHANGEABLE COMPLEX					
CEC (cmol kg^{-1})	Exchangeable Bases (cmol kg^{-1})				Base Saturation (%)
	Na^+	K^+	Mg^{2+}	Ca^{2+}	
28.99	0.70	3.29	6.70	16.60	94
22.73	0.50	1.65	3.78	4.43	46
2.95	0.08	0.06	0.08	0.09	11

PARTICLE SIZE (%)			Textural Class USDA
Sand 2000-50 µm	Silt 50-2 µm	Clay <2 µm	
68	26	6	Sandy Loam

Fig. 4.8 (continued)

4.1.5 Alluvial Soils (05)

Alluvial soils are formed in deposits of river, lake, estuarine or marine alluvium (Fig. 4.9). The majority of series described are associated with recent rivers and streams and can therefore be found throughout the country (Fig. 4.9). The lake alluviums found in Ireland are mostly associated with depressions at the sites of glacial or postglacial lakes. These soils are often flooded and can be waterlogged for prolonged periods of the year. The soils are similar to Groundwater Gleys in their characteristics; however, alluvial soils display distinct layers of soil material associated with each flooding or deposition event (Fig. 4.9). Alluvial soils reflect predominantly gleyic properties, but stagnic properties may also be observed, where permeability is problematic due to a heavy texture, compacted subsurface horizon. They can have loamy or clayey B horizons that may overlie coarse substratum.

There are two main groups of Alluvial soils found in Ireland:

1. Those formed in alluvial deposits and remain poorly drained due to the influence of shallow groundwater.
2. Drained Alluvial soils formed on alluvial deposits which have been drained.

Within this Great Group, eight distinct Subgroups are found:

05.0.0 *Typical Alluvial Soils* show evidence of reduced conditions (gleying) within 40 cm, without any other distinguishing features (Fig. 4.10).

05.1.0 *Histic Alluvial Soils* have evidence of reduced conditions (gleying) within 40 cm, with a peaty surface horizon (>20% organic carbon) less than 40 cm thick.

05.5.0 *Typical Calcareous Alluvial Soils* These soils have evidence of reduced conditions (gleying) within 40 cm and a calcareous subsurface horizon starting at or within 40 cm.

05.5.1 *Histic Calcareous Alluvial Soils* have evidence of reduced conditions (gleying) within 40 cm and a calcareous subsurface horizon starting at or within 40 cm, with a peaty surface horizon (>20% organic carbon) less than 40 cm thick.

05.5.6 *Humic Calcareous Alluvial Soils* have evidence of reduced conditions (gleying) within 40 cm, a calcareous subsurface horizon starting at or within 40 cm and a humose topsoil.

05.6.0 *Humic Alluvial Soils* have evidence of reduced conditions (gleying) within 40 cm, with a humose topsoil.

05.7.0 *Typical Drained Alluvial Soils* have no distinguishing features in the topsoil and are well-to-imperfectly drained.

05.7.2 *Gleyic Drained Alluvial Soils* have been drained, but indication of groundwater resulting in reduced conditions present, with gleying observable between 40 and 80 cm.

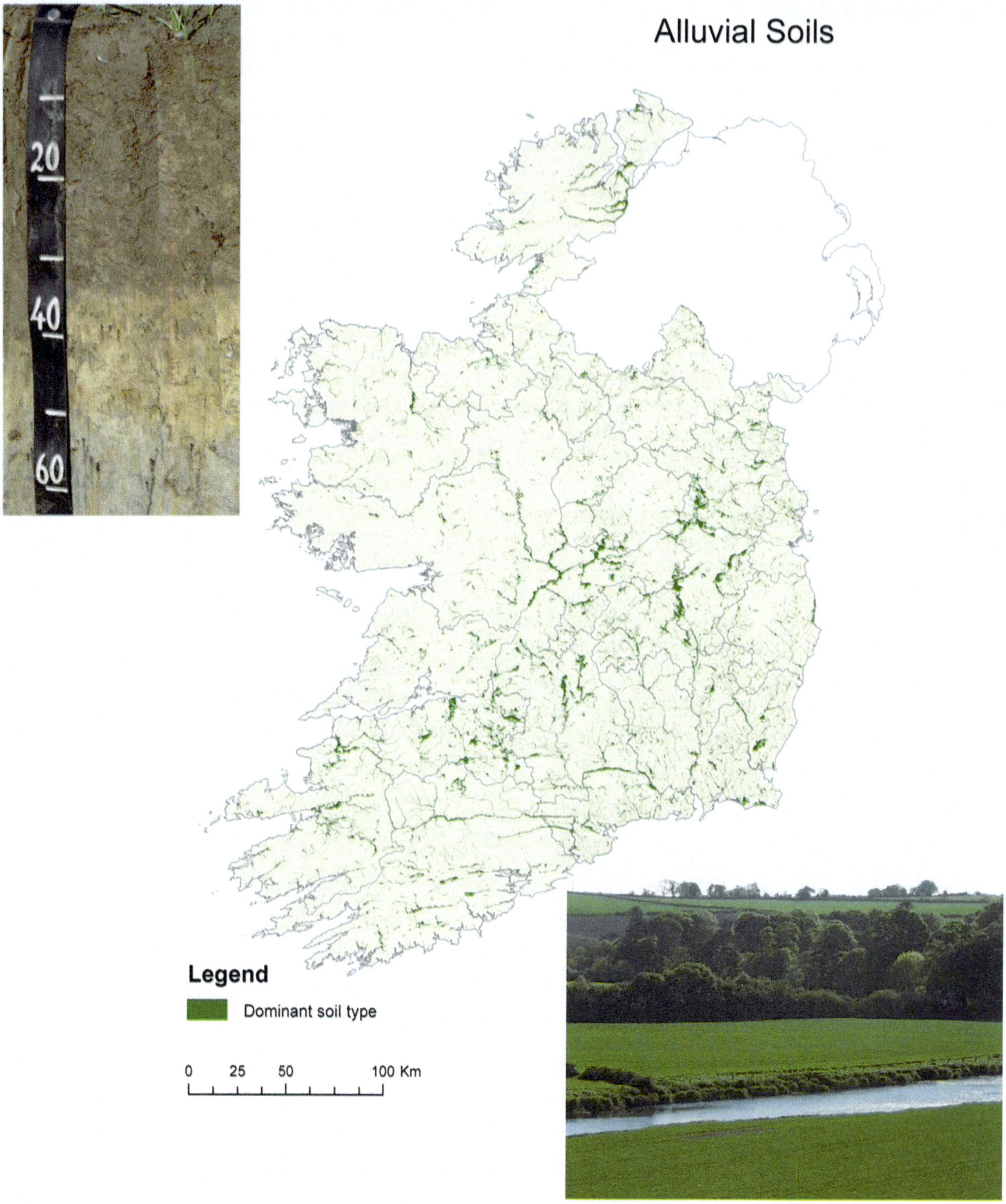

Fig. 4.9 Alluvial: profile pit (left), geographic distribution (centre) and landscape image (right)

REPRESENTATIVE PROFILE DESCRIPTION

SERIES: KILMANNOCK

Reference profile: 491
County: Wexford
Weather:
Eleavation:

TOPOGRAPHY
Position:
Slope Degree:
Slope Form:
Aspect:

PARENT MATERIAL
Substrate Type: Alluvium
Substrate Subgroup: Estuarine

TEXTURAL CRITERIA
Textural Class: Silty
Texturally contrasting:

LAND USE
Land use: Pasture
Human technologies:
Vegetation: Rush (Juncus) dominated

WATER TABLE

ROCK OUTCROPS

SURFACE STONE

IRISH CLASSIFICATION (2013)
Soil subgroup: 05.0.0 Typical Alluvial Gley

Series: Kilmannock
Definition: Silty estuarine alluvium

WRB CLASSIFICATION (2006)
Calcaric Mollic Gleyic Fluvisol (Humic)

DESCRIPTION

0 - 25 cm **Apg**

MATRIX COLOUR: 10YR3/2. **MOTTLE:** 10YR5/4, Few, Fine, Distinct. **TEXTURE:** Silt loam. **STRUCTURE:** Weak, Crumbly, Medium. **CONSISTENCY:** Firm. **STICKINESS:** Sticky. **ROOTS:** Many. **BOUNDARY:** Clear, Smooth.

25 - 53 cm **Bg**

MATRIX COLOUR: 2.5Y 5/2. **TEXTURE:** Silt loam. **STRUCTURE:** Massive. **STICKINESS:** Sticky. **ROOTS:** Few. **INCLUSIONS:** Soft, Ferruginous, Cylindrical, Segregations. **BOUNDARY:** Clear, Smooth.

53 - 74 cm **BCG**

MATRIX COLOUR: 5Y 4/1. **MOTTLE:** 5YR3/4, Common, Coarse diffuse. **TEXTURE:** Sandy loam. **STRUCTURE:** Massive. **STICKINESS:** Sticky. **BOUNDARY:** Clear, Smooth.

74 - 91 cm **CG**

MATRIX COLOUR: 5Y 4/1. **TEXTURE:** Silty clay loam. **STICKINESS:** Sticky.

Fig. 4.10 Reference profile pit description of 05.0.0 Typical Alluvial Gley, Kilmannock Series

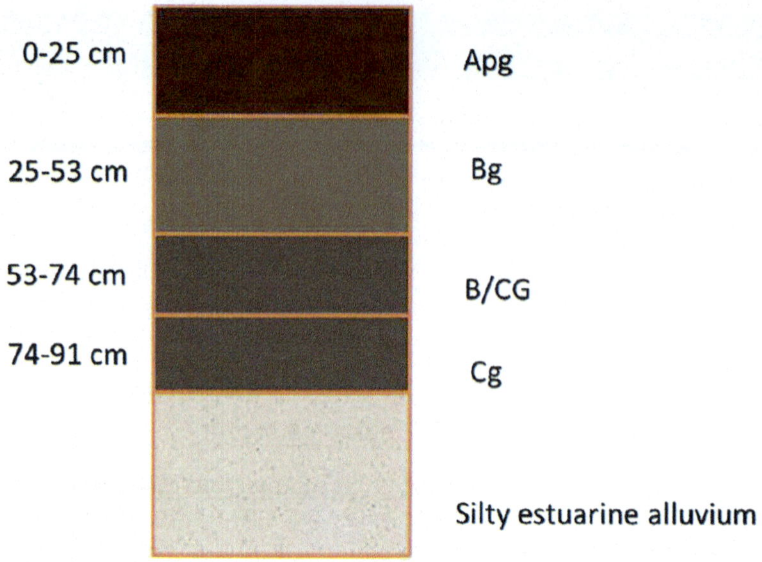

LABORATORY ANALYSES

Horizon	pH	Total (%)		C/N
		Nitrogen	Carbon	
Apg	7.3	0.26	3.85	14.8
Bg	7.8	0.15	1.67	11.2
BCG	8	0.07	1.08	15.4
CG	8.3	0.17	1.56	9.21

EXCHANGEABLE COMPLEX				
CEC (cmol kg^{-1})	TEB (cmol kg^{-1})	Base Sat. (%)	Free Iron (%)	TNV (%)
12.8			1.12	16.2
10.8			1.33	15.6
3.6			1.08	12.3
9.8			1.34	84.90

PARTICLE SIZE (%)			Textural Class USDA	Bulk Density g/cm^3	Standard Deviation
Sand 2000-50 µm	Silt 50-2 µm	Clay <2 µm			
1.5	74.6	23.9	Silt loam		
3.2	69.9	26.9	Silt loam		
73.6	17.1	9.3	Sandy loam		
2.7	68.8	28.5	Silty clay loam		

Fig. 4.10 (continued)

4.1.6 Groundwater Gleys (06)

Groundwater Gley soils occur due to a high groundwater table. These soils reflect 'gleyed' subsurface horizons, evidenced either by prominent mottling or by a uniformly grey subsoil to within 40 cm depth (Fig. 4.11). Gleying occurs due to periodic waterlogging for prolonged periods, resulting from a shallow, fluctuating groundwater table. The gleyed subsurface horizons and the substratum are permeable. Due to the water table influence, these soils can typically be found near rivers, or in the inter-drumlin hollows where the water table is high (Fig. 4.11). Topographically, they are found towards the base of slopes and the highest prevalence of these soils can be found at the lower elevations in the midlands of Ireland (Fig. 4.11). The colour matrix is important in identifying Groundwater Gleys. In these soils, the lower subsoil horizons generally have grey matrix colours:

1. Hue 7.5YR chroma = 1; or
2. Hue 10YR or yellower chroma $\leq$ 2; or
3. greyish (Hue N, GY), greenish (Hue GY, G) and bluish (Hue BG, B) hues

Groundwater Gleys correlate primarily with the Gleysol WRB Reference Soil Group (IUSS Working Group WRB 2006). These soils tend to have coarse textures (coarse-loamy or sandy) subsoil horizons. Within this Great Group, nine distinct soil Subgroups are included:

06.0.0 ***Typical Groundwater Gleys*** have evidence of gleying within 40 cm due to the presence of a groundwater table within the profile and distinct topsoil, without any other distinguishing features (Fig. 4.12).

06.1.0 ***Histic Groundwater Gleys*** have evidence of gleying within 40 cm due to the presence of a groundwater table within the profile and a peaty surface horizon (>20% OC) less than 40 cm thick.

06.3.0 ***Stagnic Groundwater Gleys*** show evidence of gleying within 40 cm due to the presence of a groundwater table within the profile and stagnation due to a slowly permeable horizon.

06.3.6 ***Humic Stagnic Groundwater Gley*** shows evidence of gleying within 40 cm due to the presence of a groundwater table within the profile and stagnation due to a slowly permeable horizon. The surface horizon is humose.

06.5.0 ***Calcareous Groundwater Gleys*** have evidence of gleying within 40 cm due to the presence of a groundwater table within the profile and a distinct topsoil with a calcareous subsurface horizon starting within 40 cm.

06.5.1 ***Histic Calcareous Groundwater Gleys*** have evidence of a calcareous gleyed subsurface horizon starting within 40 cm and a peaty surface horizon (>20% organic carbon) less than 40 cm thick.

06.5.6 ***Humic Calcareous Groundwater Gleys*** have evidence of a calcareous gleyed subsurface horizon starting within 40 cm and a humose topsoil.

06.6.0 ***Humic Groundwater Gleys*** have evidence of gleying within 40 cm and humose topsoil.

06.9.0 ***Anthropic Groundwater Gleys*** have evidence of gleying within 40 cm, due to the presence of a groundwater table within the profile and the surface horizons have been altered by significant human activity which has altered the nature of the surface horizon. This may include; terric applications such as calcareous sea sands and shells, artefacts, importation of soil material, inversion of soils in situ.

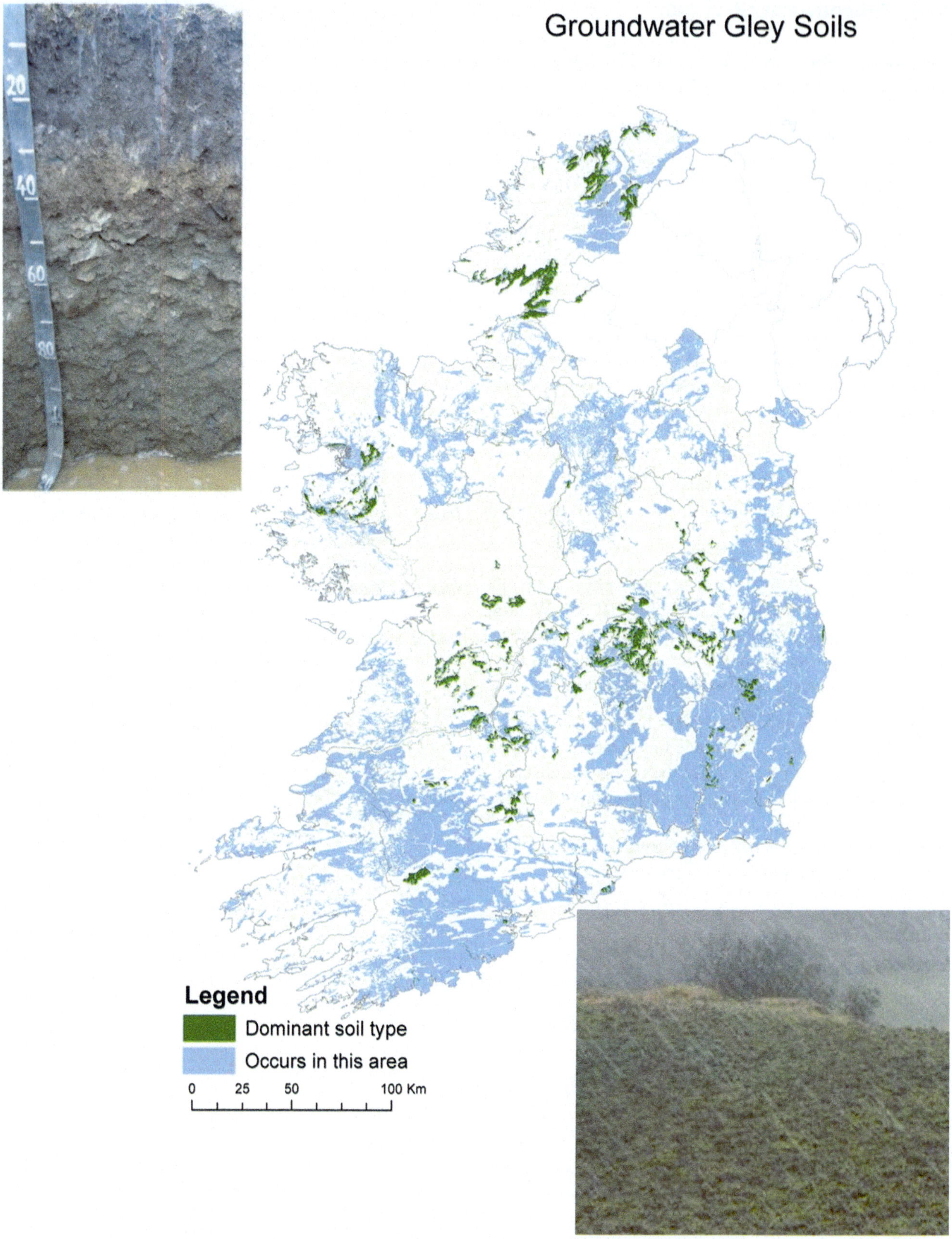

Fig. 4.11 Groundwater Gley: profile pit (left), geographic distribution (centre) and landscape image (right)

REPRESENTATIVE PROFILE DESCRIPTION

SERIES: KILPIERCE

Reference profile: 470
County: Wexford
Weather:
Eleavation:

TOPOGRAPHY
Position:
Slope Degree:
Slope Form:
Aspect:

PARENT MATERIAL
Substrate Type: Glacial drift
Substrate Subgroup: Ordovician shale

TEXTURAL CRITERIA
Textural Class: Fine loamy
Texturally contrasting:

LAND USE
Land use: Unimproved pasture
Human technologies:
Vegetation: Old pasture dominated by rush (Juncus)

WATER TABLE

ROCK OUTCROPS

SURFACE STONE

IRISH CLASSIFICATION (2013)
Soil subgroup: 06.0.0 Typical Groundwater Gley

Series: Kilpierce
Definition: Fine loamy drift with siliceous stones

WRB CLASSIFICATION (2006)
Eutric Gleysol (Humic)

DESCRIPTION

0 - 15 cm **Ap1**
MATRIX COLOUR: 10YR4/2. **TEXTURE:** Clay loam. **STRUCTURE:** Weak, Crumbly, Fine. **ROOTS:** Many. **BOUNDARY:** Gradual, Smooth.

15 - 25 cm **Ap2**
MATRIX COLOUR: 10YR5/2. **TEXTURE:** Clay Loam. **STRUCTURE:** Weak, Crumbly, Fine. **CONSISTENCY:** Friable. **ROOTS:** Many. **BOUNDARY:** Clear, Smooth.

25 - 41 cm **Bg**
MATRIX COLOUR: 10YR5/1. **MOTTLE:** 5YR4/6, Many, prominent. **TEXTURE:** Loam. **STRUCTURE:** Weak, Columnar, Fine to massive. **CONSISTENCY:** Friable. **ROOTS:** Few. **BOUNDARY:** Clear, Smooth.

41 - 61 cm **BCG**
MATRIX COLOUR: 10YR5/1. **MOTTLE:** 5YR4/6, Many, Medium, Distinct. **TEXTURE:** Loam. **STRUCTURE:** Massive. **ROOTS:** No.

Fig. 4.12 Reference profile pit description of 06.0.0 Groundwater Gley, Kilpierce Soil Series

LABORATORY ANALYSES

Horizon	pH	Total (%)		C/N
		Nitrogen	Carbon	
Ap1	5.6	0.39	4.9	12.6
Ap2	5.6	0.23	2.7	11.3
Bg	5.9	0.08	0.8	8.9
BCG	6	0.06	0.2	3.3

EXCHANGEABLE COMPLEX				
CEC ($cmol\ kg^{-1}$)	TEB ($cmol\ kg^{-1}$)	Base Sat. (%)	Free Iron (%)	TNV (%)
15.6	5.3	34	2.0	
13.1	4.8	37	1.78	
6.6	3.8	58	1.78	
6.8	4.5	66	1.91	

PARTICLE SIZE (%)			Textural Class USDA	Bulk Density g/cm^3	Standard Deviation
Sand 2000-50 µm	Silt 50-2 µm	Clay <2 µm			
32.5	37.2	30.3	Clay loam		
36.3	35.3	28.3	Clay loam		
37.3	38.5	24.2	Loam		
34.2	38.8	27	Clay loam		

Fig. 4.12 (continued)

4.1.7 Surface-Water Gleys (07)

Surface-water Gleys are poorly drained soils with a stagnic horizon and a slowly permeable subsurface horizon occurring within 40 cm of the surface. The slowly permeable subsurface horizon impedes downward water movement from in situ precipitation and/or lateral run off from upslope positions resulting in seasonal waterlogging of the top 40 cm for prolonged periods of the year. In general, the slowly permeable subsurface horizon is due to higher clay content, or due to very dense material in the subsoil. These soils tend to have fine texture (fine-loamy or clayey) in the lower subsoil horizons and typically show prominent mottling above 40 cm (Jones et al. 2011). It is identified in the field by the following properties:

1. Greyish or mottled colours occur in parts of the soil matrix immediately above the horizon in which water movement is restricted, causing a reduction of iron and manganese. Peds usually have greyish faces and ochreous mottles internally (Bg, Btg or BCg horizon), though these features may be absent in slowly permeable subsurface horizons derived from reddish parent materials; OR
2. High packing density (Hodgson 1976, p. 39), very small (<5% w/w) air capacity (Thomasson 1975), or both; OR
3. Compacted subsurface horizon induced by inappropriate cultivations or other practices.

Surface-water Gleys differ from groundwater gleys, as the gleying is due to the slow movement of water through the soil profile rather than the presence of a high water table (Fig. 4.13). Surface-water Gleys are the dominant soils found in North Cork, North Kerry and West Limerick (Fig. 4.13). The drumlin belt of the border counties also reflects a high prevalence of Surface-water Gleys (Fig. 4.13). In County Leitrim, the drumlin topography is characterised with Surface-water Gleys associated with heavy wet soils. Surface-water Gleys correlate primarily with the Stagnosol WRB Reference Soil Group (IUSS Working Group WRB 2006). Three main Subgroups occur (Fig. 4.13):

07.0.0 ***Typical Surface-water Gleys*** have evidence of stagnic features within 40 cm and distinct topsoil, without any other distinguishing features (Fig. 4.14).

07.6.0 ***Humic Surface-water Gleys*** have evidence of stagnic features within 40 cm and humose topsoil.

07.9.0 ***Anthropic Surface-water Gleys*** have evidence of stagnic features within 40 cm as a result of a slowly permeable horizon and the surface horizons have been altered by significant human activity which has altered the nature of the surface horizon. This may include; terric applications such as calcareous sea sands and shells, artefacts, importation of soil material, inversion of soils in situ.

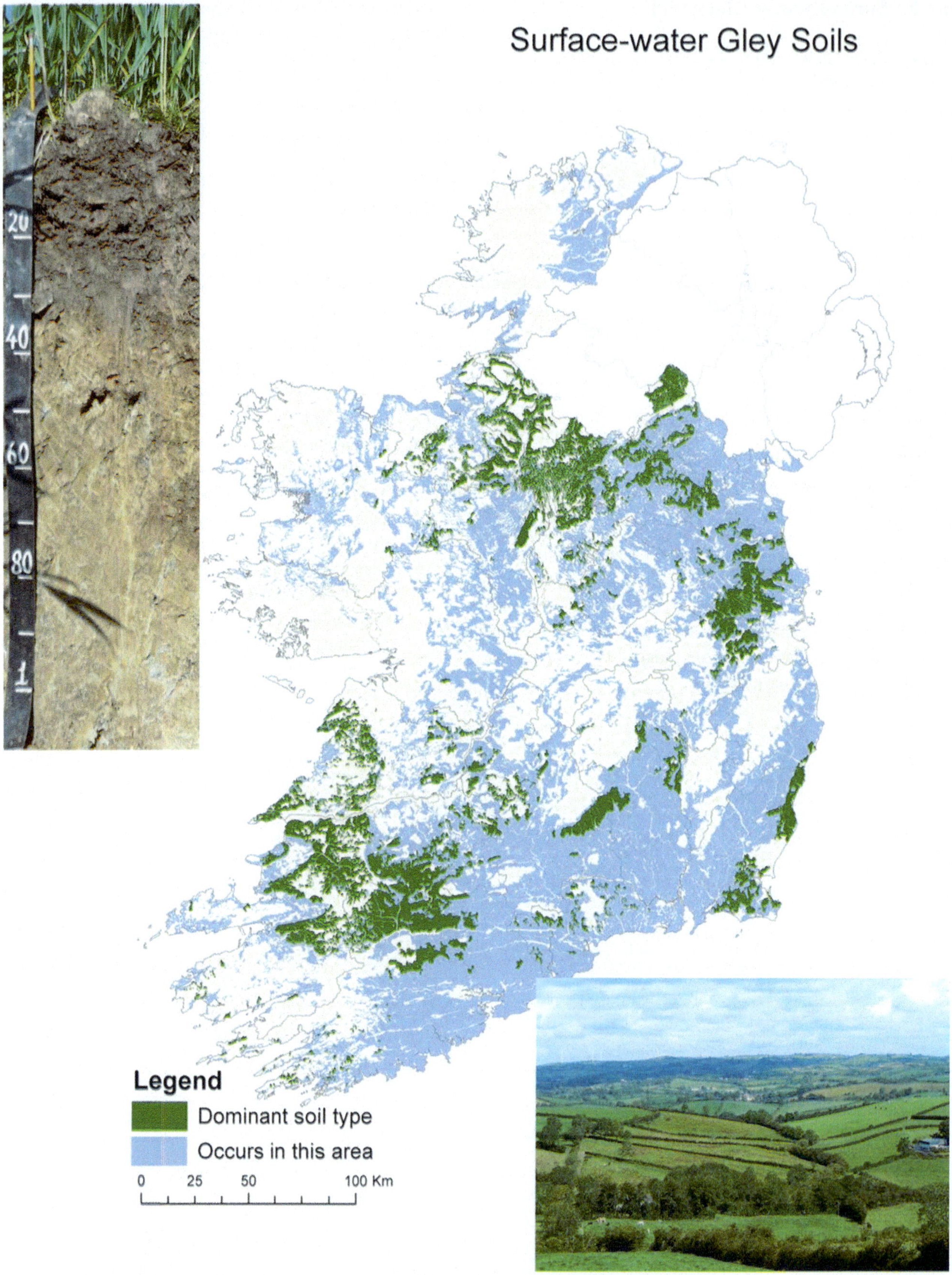

Fig. 4.13 Surface-water Gley: profile pit (left), geographic distribution (centre) and landscape image (right)

REPRESENTATIVE PROFILE DESCRIPTION

SERIES: KILRUSH

Reference profile:	RPS24RC21
County:	Tipperary South
Weather:	Sunny/clear
Elevation:	191

TOPOGRAPHY

Position:	Middle slope
Slope degree:	2
Slope Form:	Straight
Aspect:	S

PARENT MATERIAL

Substrate Type:	Drift
Substrate Subgroup:	Siliceous stones

TEXTURAL CRITERIA

Textural Class:	Fine loamy
Texturally contrasting:	

LAND USE

Land use:	Grassland Improved
Human technologies:	Ploughing, Fertilizer applications

WATER TABLE None

ROCK OUTCROPS None

SURFACE STONE None

IRISH CLASSIFICATION (2013)
Soil subgroup: 07.0.0 Typical Surface-water Gley

National Soil Series: Kilrush
Fine loamy drift with siliceous stones

DESCRIPTION

0 - 25 cm **Apg**

MATRIX COLOR: 10YR42. **MOTTLE:** 5YR46, Common, Fine, Distinct, Sharp. **STONES (%):** Common, 2-6 cm, Sub angular, Sandstone. **TEXTURE:** Clay loam. **STRUCTURE:** Moderate, Sub-angular blocky, Fine. **PLASTICITY:** Slightly plastic. **STICKINESS:** Slightly sticky. **ROOTS:** Many, Fine. **PACKING DENSITY:** Medium. **BOUNDARY:** Abrupt, Smooth.

25 - 55 cm **Bg**

MATRIX COLOR: 10YR53. **MOTTLE:** 5YR44, Many, Medium, Prominent, Sharp. **STONES (%):** Common, 2-6 cm, Sub rounded, Sandstone. **TEXTURE:** Clay loam. **STRUCTURE:** Moderate, Sub-angular blocky, Fine to medium. **CONSISTENCY:** Very firm. **PLASTICITY:** Slightly plastic. **STICKINESS:** Slightly sticky. **ROOTS:** Common, Fine. **COATS:** Manganese, Many, Prominent. **PACKING DENSITY:** High. **BOUNDARY:** Clear, Smooth.

55 - 80 cm **BCg**

MATRIX COLOR: 10YR53. **MOTTLE:** 10YR56, Many, Fine, Prominent, Sharp; 10YR42, Common, Medium, Distinct, Clear. **STONES (%):** Many, 6mm -2 cm, Sub rounded, Sandstone; 6-20 cm, Angular, Shale. **TEXTURE:** Sandy clay loam. **STRUCTURE:** Moderate, Prismatic, Fine. **PLASTICITY:** Very plastic. **STICKINESS:** Very sticky. **COATS:** Clay, Common, Faint, Continuous, Manganese, Few, Distinct. **ACCUMULATIONS (Fe/Mn):** 2-5 %. **PACKING DENSITY:** High. **BOUNDARY:** Gradual, Wavy.

80 - 110 cm **Cg1**

MATRIX COLOR: 10YR42. **MOTTLE:** 25Y43, Common, Medium, Distinct; 10YR46, Common, Fine, Prominent. **STONES (%):** Many, 6mm -2 cm, Sub rounded, Sandstone. **TEXTURE:** Clay loam. **STRUCTURE:** Massive. **PLASTICITY:** Very plastic. **STICKINESS:** Very sticky. **COATS:** Manganese, Common, Distinct. **PACKING DENSITY:** High. **BOUNDARY:** Gradual, Smooth.

110 - 140 cm **Cg2**

MATRIX COLOR: 25Y43. **MOTTLE:** 25Y51, Common, Medium, Prominent, Sharp; 10YR56, Common, Fine, Prominent, Sharp. **STONES (%):** Many, 6-20 cm, Sub angular, Shale; Old Red sandstone. **TEXTURE:** Clay loam. **STRUCTURE:** Massive. **PLASTICITY:** Very plastic. **STICKINESS:** Very sticky. **COATS:** Manganese, Few. **PACKING DENSITY:** High.

Fig. 4.14 Reference profile description of 07.0.0 Typical Surface-water Gley, Kilrush Soil Series

LABORATORY ANALYSIS

Horizon	pH	Total (%)		Organic Carbon (%)	Loss-on-ignition (%)
		Nitrogen	Carbon		
1(Apg)	4.9	0.49	4.15	3.43	
2(Bg)	5.6	0.09	0.58	0.37	
3(BCg)	5.7	0.08	0.53	0.35	
4(Cg1)	6.1	0.08	0.29	0.28	
5(Cg2)	6.2	0.07	0.37	0.25	

EXCHANGEABLE COMPLEX					
CEC (cmol kg^{-1})	Exchangeable Bases (cmol kg^{-1})				Base Saturation (%)
	Na^+	K^+	Mg^{2+}	Ca^{2+}	
8.72	0.08	0.17	0.57	5.72	75
5.66	0.08	0.08	0.44	4.45	89
7.83	0.08	0.10	0.70	5.88	86
11.46	0.08	0.14	1.05	8.93	89
10.81	0.08	0.13	1.34	8.86	96

PARTICLE SIZE (%)			Textural Class USDA	Bulk Density g/cm^3	Standard Deviation
Sand 2000-50 µm	Silt 50-2 µm	Clay <2 µm			
34	35	31	Clay Loam	1.04	0.05
39	34	27	Clay Loam/Loam	1.50	0.11
45	24	31	Sandy Clay Loam/Clay Loam	1.48	0.02
33	37	30	Clay Loam	1.58	0.04
34	37	29	Clay Loam		

Fig. 4.14 (continued)

4.1.8 Podzols (08)

Podzols are soils that consist of a peaty layer over poor acidic mineral topsoil. They have been subjected to the podzolisation process, where a depletion of nutrients has occurred due to leaching, in particular, iron and aluminium from the surface horizons (Gardiner and Radford 1980). Podzols have a dark, humose or peaty surface horizon and an albic horizon (E) overlying a podzolic B horizon (Bs). If the topsoil is peaty it should be <40 cm thick (Fig. 4.15). The albic horizon has a moist colour value >4. An ironpan may be present and if the topsoil is peaty a Bh horizon may also develop. Gleying can also occur within the profile (Jones et al. 2011). This definition correlates with the Podzol Reference Soil Group of WRB (IUSS Working Group WRB 2006). In general, these soils have developed in cold, wet acidic conditions typical at higher elevations (Fig. 4.15), and are typically covered with a heather moor or coniferous forest stand in higher elevations (Fig. 4.15). Organic acids derived from ericaceous vegetation complex (chemically bond) with nutrients in the upper horizon to make sesquioxides. The sesquioxides are mobile and are easily washed down the profile leaving the distinctive banding of dark organic peaty topsoil over a bleached white sandy (albic) layer from which nutrients have leached to a richer reddish subsoil where the nutrients have been deposited (Fig. 4.15). These soils have a high lime and fertiliser requirement but the difficulty of mechanical reclamation along with the limited cultivation potential means that these soils are usually used for forestry (Gardiner and Radford 1980).

There are five soil Subgroups recorded in the Podzol Great Group:

08.0.0 ***Typical Podzols*** have a Bs horizon. A Bh horizon is absent or incipient (<2.5 cm thick).

08.2.0 ***Gleyic Podzols*** have a gleyed subsurface horizon (Bg) below the Bs derived from a shallow fluctuating groundwater table in low lying sites.

08.3.0 ***Stagnic Podzols*** occur on raised sites, and a slowly permeable subsurface layer at depth impedes drainage below the Bs that commonly qualifies as a Btg.

08.4.3 ***Stagnic Ironpan Podzols*** have peaty topsoil above a gleyed albic horizon (Eg) that rests directly on a cemented, continuous thin ironpan (Bf) that impedes vertical water movement. A Bh horizon is commonly absent or very thin (<2.5 cm thick) (Fig. 4.16).

08.6.0 ***Humic Podzols*** have a Bh horizon (at least 0.6% organic carbon and it should be more than 2.5 cm thick) that overlies a Bs horizon.

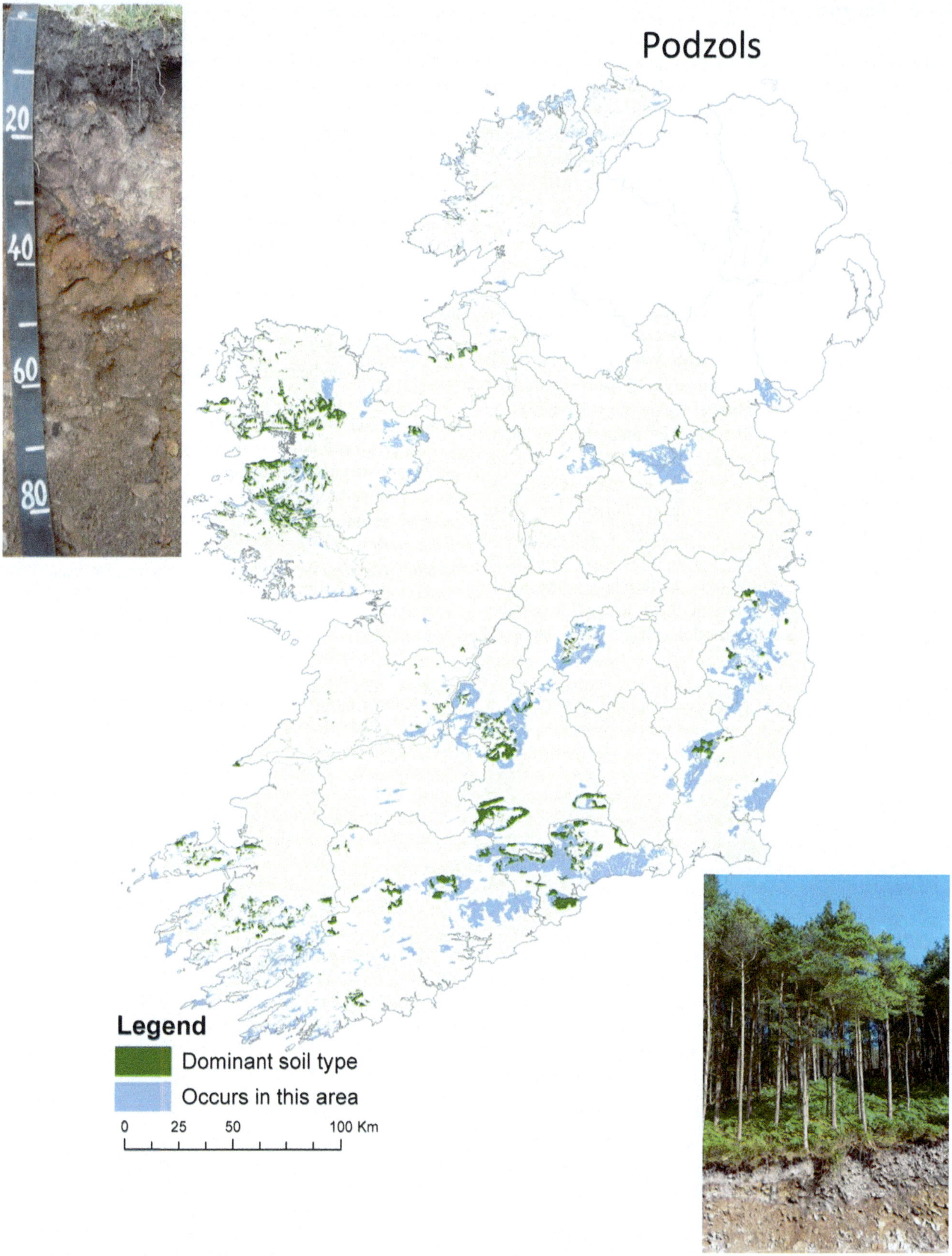

Fig. 4.15 Podzol: profile pit (left), geographic distribution (centre) and landscape image (right)

REPRESENTATIVE PROFILE DESCRIPTION

SERIES: KNOCKASTANNA

Reference profile: RPW69BR01
County: Cork
Weather: Snow
Elevation: 290

TOPOGRAPHY
Position: Upper slope
Slope degree: 3
Slope Form: Straight
Aspect: WSW

PARENT MATERIAL
Substrate Type: Bedrock
Substrate Subgroup: Shale

TEXTURAL CRITERIA
Textural Class: Coarse loamy
Texturally contrasting:

LAND USE
Land use: Grassland unimproved,
Human technologies: Clearing, Ditching, Surface compaction
Vegetation: Grassland
Species: Rushes

WATER TABLE None

ROCK OUTCROPS None

SURFACE STONE None

IRISH CLASSIFICATION (2013)
Soil subgroup: 08.4.3 Stagnic Iron-pan Podzol

National Soil Series: Knockastanna
Loamy over shale bedrock

WRB CLASSIFICATION (2006)
Albic Histic Podzol (Endoskeletic)

DESCRIPTION

0 - 15 cm **Ofm**

MATRIX COLOR: 10YR21. **STONES (%)**: Few, 2-6 cm, Angular, Shale. **TEXTURE:** Semi-fibrous peat, Peat, H5 Moderately decomposed, Mesic (semi-fibrous). **STRUCTURE:** Weak, Sub-angular blocky, Medium. **CONSISTENCY:** Very friable. **PLASTICITY:** Slightly plastic. **STICKINESS:** Non-sticky. **ROOTS:** Many, Fine. **PACKING DENSITY:** Very Low. **POROSITY:** Very High. **MACROPORES:** Fine. **BOUNDARY:** Abrupt, Wavy.

15 - 35 cm **Eg**

MATRIX COLOR: 10YR73. **MOTTLE:** 10YR34, Common, Medium, Prominent, Sharp. **STONES (%):** Common, 2-6 mm, Angular, Shale; Common, 2-6 cm, Angular, Shale. **TEXTURE:** Sandy loam. **STRUCTURE:** Weak, Angular blocky, Medium. **CONSISTENCY:** Firm. **PLASTICITY:** Non-plastic. **STICKINESS:** Non-sticky. **ROOTS:** Few , Fine. **PACKING DENSITY:** High. **POROSITY:** Low. **BOUNDARY:** Clear, Wavy.

35 - 38 cm **Bh**

MATRIX COLOR: 10YR32. **STONES (%):** Common, 2-6 mm, Angular, Shale; 2-6 cm, Flat/platy, Shale. **STRUCTURE:** Massive. **CONSISTENCY:** Very friable. **PLASTICITY:** Slightly plastic. **STICKINESS:** Sticky. **ROOTS:** Very few, Fine. **COATS:** Humus, Common, Distinct, Continuous. **PACKING DENSITY:** Medium. **POROSITY:** Medium. **BOUNDARY:** Abrupt, Wavy.

38 - 50 cm **Bs**

MATRIX COLOR: 75YR46. **MOTTLE:** 5YR56, Few, Medium, Distinct, Clear. **STONES (%):** Many, 6 mm - 2 cm, Angular, Shale; Few, 20-60 cm, Sub-angular, Quartzite. **TEXTURE:** Loam. **STRUCTURE:** Weak, Sub-angular blocky, Fine. **CONSISTENCY:** Very friable. **PLASTICITY:** Slightly plastic. **STICKINESS:** Slightly sticky. **ROOTS:** Very few, Fine. **PACKING DENSITY:** Medium. **POROSITY:** Medium. **MACROPORES:** Fine. **BOUNDARY:** Gradual, Wavy.

50 - 70 cm **BC**

MATRIX COLOR: 10YR46. **STONES (%):** Abundant, 6 mm - 2 cm, Sub-angular, Shale; 20-60 cm, Sub-angular, Sandstone. **TEXTURE:** Sandy loam. **STRUCTURE:** Massive. **CONSISTENCY:** Friable. **PLASTICITY:** Non-plastic. **STICKINESS:** Slightly sticky. **PACKING DENSITY:** Medium. **POROSITY:** Medium. **BOUNDARY:** Gradual, Wavy.

70 cm + **Cr**

Fig. 4.16 Reference profile description of 08.4.3 Stagnic Iron-pan Podzol, Knockastanna Soil Series

LABORATORY ANALYSIS

Horizon	pH	Total (%)		Organic Carbon (%)	Loss-on-ignition (%)
		Nitrogen	Carbon		
1 (Ofm)	5.1	1.33	24.80	26.80	42.0
2 (Eg)	4.6	0.06	1.04	1.00	
3 (Bh)					
4 (Bs)	4.8	0.08	1.72	0.60	
5 (BC)	4.9	0.03	0.56	0.26	

Horizon	OXALATE EXTRACTABLE		EXCHANGEABLE COMPLEX					
	Fe (g kg^{-1})	Al (g kg^{-1})	CEC (cmol kg^{-1})	Exchangeable Bases (cmol kg^{-1})				Base Saturation (%)
				Na^+	K^+	Mg^{2+}	Ca^{2+}	
1 (Ofm)	2.68	0.69	36.9	0.30	0.45	9.21	21.9	86.2
2 (Eg)	1.16	0.55	5.01	0.08	0.04	0.57	0.45	23.0
3 (Bh)								
4 (Bs)	22.92	4.02	5.35	0.08	0.03	0.26	0.24	11.7
5 (BC)	4.94	1.64	0.81	0.08	0.03	0.08	0.09	36.6

Horizon	PARTICLE SIZE (%)			Textural Class USDA
	Sand 2000-50 μm	Silt 50-2 μm	Clay <2 μm	
1 (Ofm)				
2 (Eg)	59	28	13	Sandy Loam
3 (Bh)				
4 (Bs)	50	34	16	Loam
5 (BC)	53	40	7	Sandy Loam

Fig. 4.16 (continued)

4.1.9 Brown Podzolic Soils (09)

Brown Podzolic soils show features of incipient podzolic processes but these are insufficiently expressed for these soils to be classified as Podzols. They have a podzolic B horizon (Bs) but lack an albic (E) horizon (Fig. 4.17). The podzolic B horizon has a moist chroma >3. In some cases, there may also be a thin discontinuous ironpan present. Although these soils tend to have lost their nutrients, in particular, Fe and Al due to acidic conditions and high rainfall over time, appropriate management may help to maintain a nutrient-rich topsoil. These soils respond well to management, including the addition of manure and fertiliser, and as a result, can make good agricultural soils. They tend to be free draining naturally and do not require additional drainage. Brown Podzolic soils still have a reddish subsoil, indicating acid conditions that were present at some time during their formation. For example, the slightly acidic Brown Podzolic soils dominant in the south of County Cork may be attributed to the sandstone bedrock (see Chap. 3) prevalent in this area (Fig. 4.17).

This Great Group includes five distinct soil Subgroups:

09.0.0 ***Typical Brown Podzolic Soils*** have distinct topsoil, without any distinguishing features (Fig. 4.18).

09.2.0 ***Gleyic Brown Podzolic Soils*** display gleyic features due to the presence of a shallow fluctuating groundwater table.

09.3.0 ***Stagnic Brown Podzolic Soils*** display stagnic properties as a result of the presence of a slowly permeable subsurface horizon.

09.3.6 ***Humic Stagnic Brown Podzolic Soils*** have humose topsoil and display stagnic properties as a result of the presence of a slowly permeable subsurface horizon.

09.6.0 ***Humic Brown Podzolic Soils*** have humose topsoil.

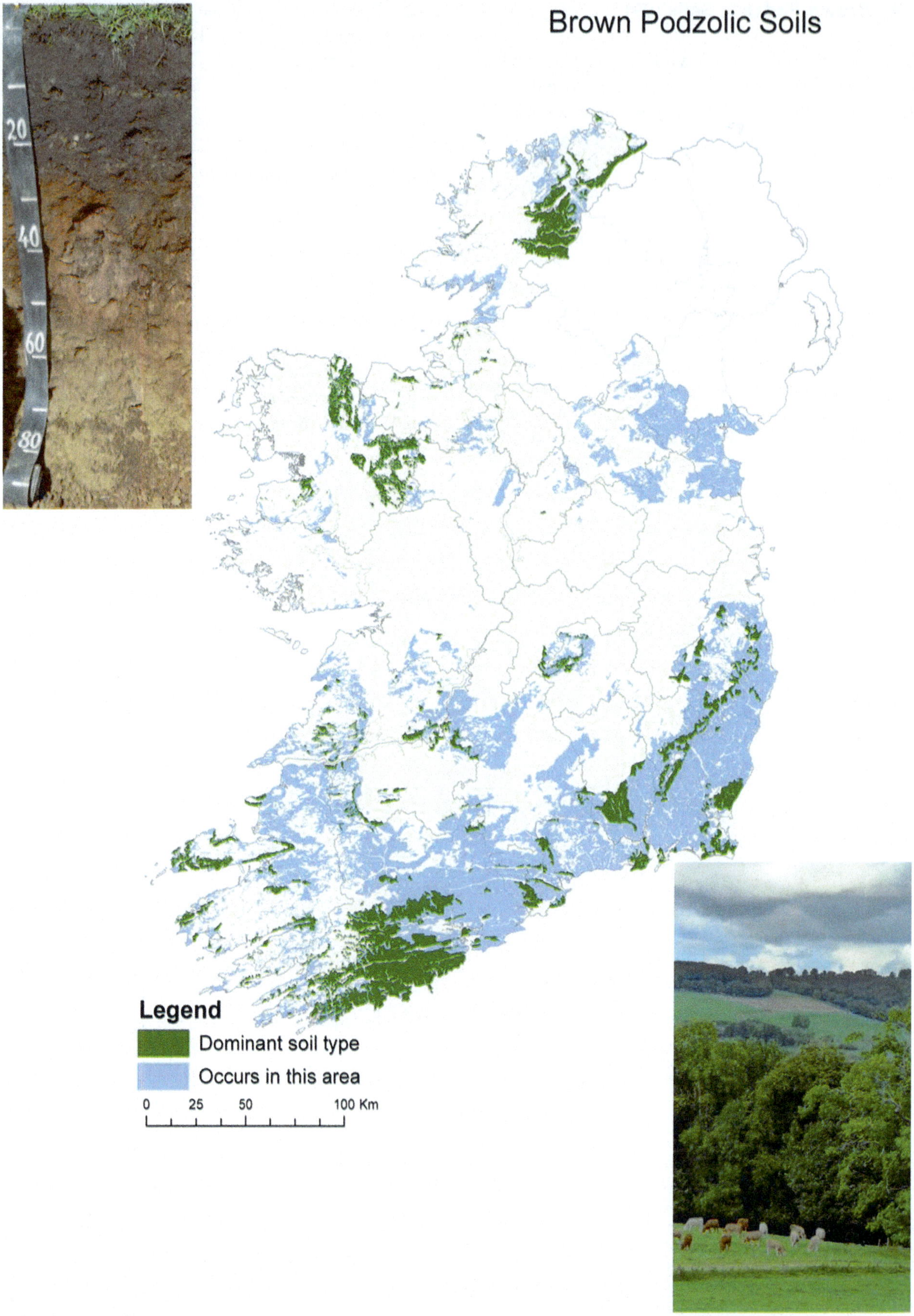

Fig. 4.17 Brown Podzolic: profile pit (left), geographic distribution (centre) and landscape image (right)

REPRESENTATIVE PROFILE DESCRIPTION

SERIES: ROSS CARBERY

Reference profile: RPW39BR01
County: Cork
Weather: Overcast
Elevation: 38

TOPOGRAPHY
Position: Lower slope
Slope degree: 11
Slope Form: Concave
Aspect: SE

PARENT MATERIAL
Substrate Type: Drift
Substrate Subgroup: Siliceous stones

TEXTURAL CRITERIA
Textural Class: Coarse loamy
Texturally contrasting:

LAND USE
Land use: Grassland improved
Human technologies: Fertilizer applications, Slurry applications, Seeded/reseeded
Vegetation: Grassland, Grassland
Species: Bramble/Briars

WATER TABLE None

ROCK OUTCROPS None

SURFACE STONE None

IRISH CLASSIFICATION (2013)
Soil subgroup: 09.0.0 Typical Brown Podzolic

National Soil Series: Ross Carbery
Coarse loamy compact drift with siliceous stones

WRB CLASSIFICATION (2006)
Haplic Entic Podzol (Anthric Novic)

DESCRIPTION

0 - 35 cm **Ap**

MATRIX COLOR: 10YR33. **STONES (%):** Few, 2-6 mm, Angular, Siliceous stones; Few, 2-6 cm, Angular, Siliceous stones. **TEXTURE:** Sandy loam. **STRUCTURE:** Weak, Crumb, Fine. **CONSISTENCY:** Very friable. **PLASTICITY:** Non-plastic. **STICKINESS:** Non-sticky. **ROOTS:** Many, Medium. **PACKING DENSITY:** Very Low **POROSITY:** Very High. **MACROPORES:** Medium. **BOUNDARY:** Clear, Smooth.

35 - 70 cm **AB**

MATRIX COLOR: 10YR43. **STONES (%):** Common, 2-6 mm, Angular, Shale; Common, 2-6 cm, Angular, Shale. **TEXTURE:** Coarse, Sandy loam. **STRUCTURE:** Moderate, Sub-angular blocky, Fine to Medium. **CONSISTENCY:** Friable. **PLASTICITY:** Non-plastic. **STICKINESS:** Non-sticky. **ROOTS:** Common, Fine. **PACKING DENSITY:** Medium. **POROSITY:** High. **MACROPORES:** Fine. **BOUNDARY:** Abrupt, Wavy.

70 - 100 cm **Bs buried**

MATRIX COLOR: 75YR46. **STONES (%):** Common, 2-6 mm, Angular, Shale; Common, 2-6 cm, Angular, Shale. **TEXTURE:** Sandy Loam. **STRUCTURE:** Weak, Sub-angular blocky, Fine. **CONSISTENCY:** Very friable. **PLASTICITY:** Non-plastic. **STICKINESS:** Non-sticky. **ROOTS:** Few, Fine. **COATS:** Clay, Few, Faint, Discontinuous. **PACKING DENSITY:** Medium. **POROSITY:** High. **MACROPORES:** Fine. **BOUNDARY:** Clear, Wavy.

100 - 150 cm **2C buried**

MATRIX COLOR: 25Y62. **VARIEGATION:** 75YR46, Few, Fine, Distinct, Sharp. **STONES (%):** Many, 2-6 mm, Angular, Shale; Many, 2-6 cm, Angular, Shale. **TEXTURE:** Sandy Loam. **STRUCTURE:** Massive. **COMPACITY:** Non-cemented but compacted. **CONSISTENCY:** Firm. **PLASTICITY:** Non-plastic. **STICKINESS:** Non-sticky. **ROOTS:** Very few, Fine. **PACKING DENSITY:** High. **POROSITY:** Low. **MACROPORES:** Fine.

Fig. 4.18 Reference profile pit description of 09.0.0 Typical Brown Podzolic, Ross Carbery Soil Series

LABORATORY ANALYSIS

Horizon	pH	Total (%)		Organic Carbon (%)	Loss-on-ignition (%)
		Nitrogen	Carbon		
1 (Ap)	8.3	0.31	4.26	2.78	
2 (AB)	8.5	0.12	2.62	0.94	
3 (Bs buried)	8.2	0.07	1.19	0.48	
4 (2C buried)	8.5	0.01	0.24	0.16	

Horizon	OXALATE EXTRACTABLE		EXCHANGEABLE COMPLEX					
	Fe ($g\ kg^{-1}$)	Al ($g\ kg^{-1}$)	CEC ($cmol\ kg^{-1}$)	Exchangeable Bases ($cmol\ kg^{-1}$)				Base Saturation (%)
				Na^+	K^+	Mg^{2+}	Ca^{2+}	
1 (Ap)	2.98	0.73	20.60	0.5	8.97	1.18	19.40	Sat.
2 (AB)	4.47	0.83	12.20	0.35	5.00	0.39	11.9	Sat.
3 (Bs buried)	11.35	3.01	13.1	0.51	0.02	0.23	13.20	Sat.
4 (2C buried)	0.93	0.73	1.82	0.26	0.14	0.209	2.27	Sat.

Horizon	PARTICLE SIZE (%)			Textural Class USDA	Bulk Density g/cm^3	Standard Deviation
	Sand 2000-50 μm	Silt 50-2 μm	Clay <2 μm			
1 (Ap)	64	18	18	Sandy Loam	1.19	0.11
2 (AB)	62	23	15	Sandy Loam	0.97	0.06
3 (Bs buried)	54	35	11	Sandy Loam		
4 (2C buried)	57	37	6	Sandy Loam		

Fig. 4.18 (continued)

4.1.10 Luvisols (10)

Luvisols are soils in which the clay material from the topsoil has eluviated and moved down the profile and been deposited further down the subsoil. The clay must disperse in the horizon of eluviation, a process driven largely by pH, before it is transported to the depth of accumulation by percolating waters (ISSS Working Group RB 1998). The clay illuviation results in a Bt horizon with a significant increase of clay (argillic B horizon) compared to the overlying horizons (>20% increase). Base saturation is >35%. An albic E horizon may be present between the A and Bt horizons. These soils are decalcified, thus soils that are calcareous do not have argillic B horizons, unless soils were formed before the carbonate source was derived. Luvisols are good agricultural soils due to the higher clay content which means that they are characterised by good nutrient retention. The increase in clay content down the soil profile means that these soils may be moderately to imperfectly draining depending upon the overall texture of the soil. Luvisols correlate closest with the Luvisol Reference Soil Group of WRB (IUSS Working Group WRB 2006). Six distinct Subgroups occur within this Great Group in Ireland (Fig. 4.19):

10.0.0 ***Typical Luvisol Soils*** have distinct topsoil, without any distinguishing features (Fig. 4.20).

10.2.0 ***Gleyic Luvisol Soils*** display gleyic features due to the presence of a shallow fluctuating groundwater table.

10.2.6 ***Humic Gleyic Luvisol Soils*** have humose topsoil and display gleyic features due to the presence of a shallow fluctuating groundwater table.

10.3.0 ***Stagnic Luvisol Soils*** display stagnic properties as a result of the presence of a slowly permeable subsurface horizon.

10.3.6 ***Humic Stagnic Luvisol Soils*** have humose topsoil and display stagnic properties as a result of the presence of a slowly permeable subsurface horizon.

10.6.0 ***Humic Luvisol Soils*** have a humose topsoil.

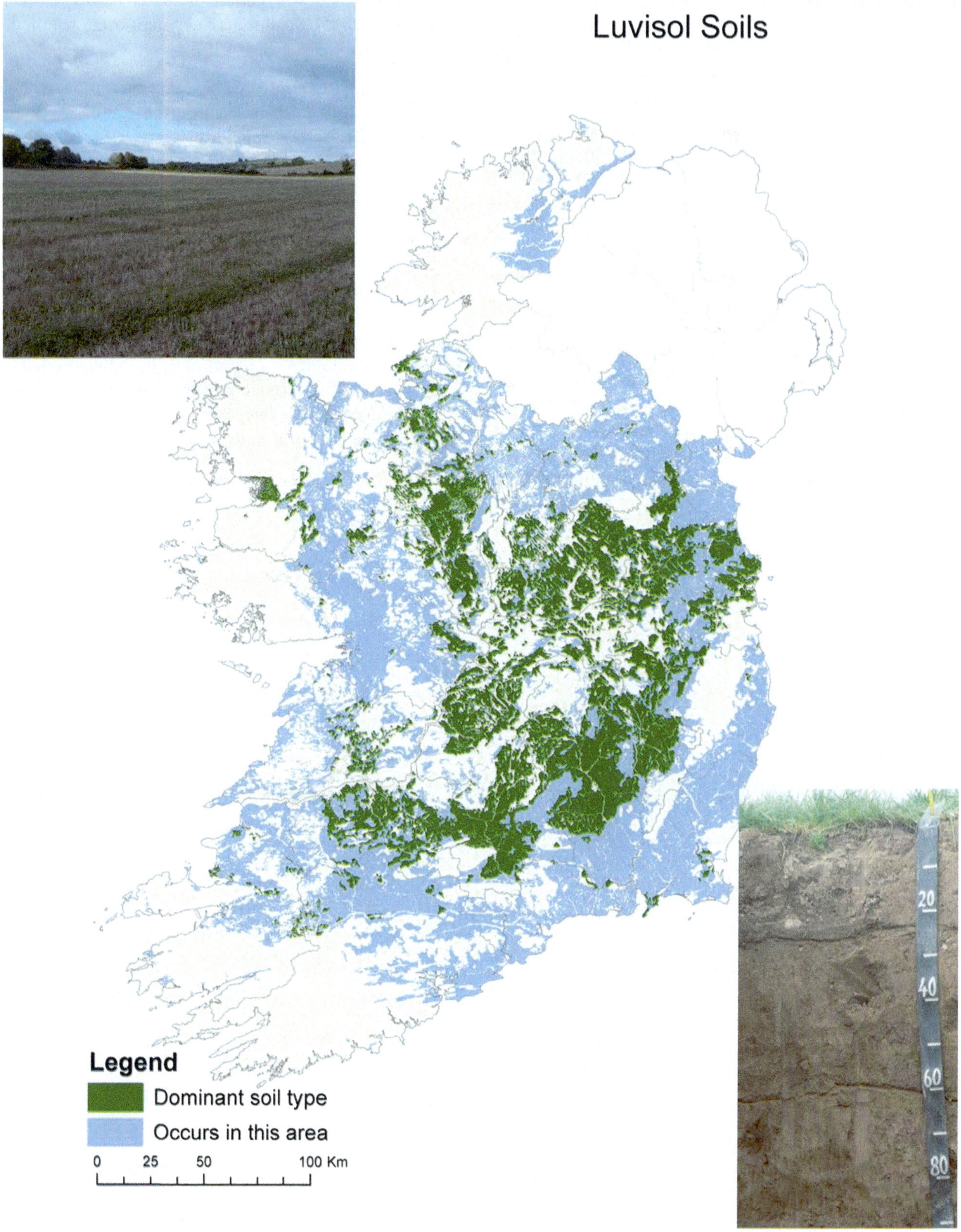

Fig. 4.19 Luvisol: profile pit (left), geographic distribution (centre) and landscape image (right)

REPRESENTATIVE PROFILE DESCRIPTION

SERIES: ELTON

Reference profile: RPS36RC18
County: Kilkenny
Weather: Overcast
Elevation: 201

TOPOGRAPHY
Position: Middle slope
Slope degree: 3
Slope Form: Straight
Aspect: SW

PARENT MATERIAL
Substrate Type: Drift
Substrate Subgroup: Limestones

TEXTURAL CRITERIA
Textural Class: Fine loamy
Texturally contrasting: -

LAND USE
Land use: Grassland Improved
Human technologies: Slurry applications, Fertilizer applications, Ploughing

WATER TABLE None

ROCK OUTCROPS None

SURFACE STONE None

IRISH CLASSIFICATION (2013)
Soil subgroup: 10.0.0 Typical Luvisol

National Soil Series: Elton
Fine loamy drift with limestones

WRB CLASSIFICATION (2006)
Haplic Luvisol (Humic Hypereutric Endoskeletic)

DESCRIPTION

0 - 25 cm **Ap**

MATRIX COLOR: 10YR43. **STONES (%):** Few, 2-6 mm, Sub angular, Limestone; Sandstone. **TEXTURE:** Loam. **STRUCTURE:** Moderate, Sub-angular blocky, Medium. **COMPACTION:** Non-cemented and Non-compacted. **CONSISTENCY:** Firm. **PLASTICITY:** Slightly plastic. **STICKINESS:** Sticky. **ROOTS:** Many, Very fine. **PACKING DENSITY:** Medium. **BOUNDARY:** Abrupt, Smooth.

25 - 60 cm **Bt**

MATRIX COLOR: 75YR44. **STONES (%):** Common, 6mm -2 cm, Sub angular, Sandstone; Limestone. **TEXTURE:** Loam. **STRUCTURE:** Strong, Sub-angular blocky, Medium. **COMPACTION:** Non-cemented and Non-compacted. **CONSISTENCY:** Firm. **PLASTICITY:** Very plastic. **STICKINESS:** Very sticky. **ROOTS:** Few, Very fine. **PACKING DENSITY:** Medium. **BOUNDARY:** Clear, Wavy.

60 - 120 cm **C**

MATRIX COLOR: 10YR54. **STONES (%):** Abundant, 6mm -2 cm, Sub angular, Limestone. **TEXTURE:** Sandy loam. **STRUCTURE:** Moderate, Massive. **CONSISTENCY:** Very firm. **PLASTICITY:** Non-plastic. **STICKINESS:** Slightly sticky. **ROOTS:** Few, Very fine. **HCL REACTION:** Extremely strong (thick foam). **PACKING DENSITY:** High.

Fig. 4.20 Reference profile pit description of 10.0.0 Typical Luvisol, Elton Soil Series

LABORATORY ANALYSIS

Horizon	pH	Total (%)		Organic Carbon (%)	Loss-on-ignition (%)
		Nitrogen	Carbon		
1(Ap)	6.9	0.24	2.21	1.94	
2(Bt)	7.4	0.08	0.78	0.57	
3(C)	8.6	0.01	7.75	0.25	

Horizon	OXALATE EXTRACTABLE		EXCHANGEABLE COMPLEX					
	Fe ($g\ kg^{-1}$)	Al ($g\ kg^{-1}$)	CEC ($cmol\ kg^{-1}$)	Exchangeable Bases ($cmol\ kg^{-1}$)				Base Saturation (%)
				Na^+	K^+	Mg^{2+}	Ca^{2+}	
1(Ap)	4.93	1.68	16.46	0.10	0.17	0.80	14.69	96
2(Bt)	2.90	1.56	9.42	0.08	0.10	0.79	8.87	Sat.
3(C)	0.29	0.18	8.63	0.08	0.05	0.51	32.03	Sat.

Horizon	PARTICLE SIZE (%)			Textural Class USDA	Bulk Density g/cm^3	Standard Deviation
	Sand 2000-50 μm	Silt 50-2 μm	Clay <2 μm			
1(Ap)	40	40	20	Loam	1.24	0.03
2(Bt)	37	37	26	Loam	1.44	0.02
3(C)	57	29	14	Sandy Loam		

Fig. 4.20 (continued)

4.1.11 Brown Earth Soils (11)

Brown Earth soils are well-drained soils possessing rather uniform profiles with little differentiation between horizons (Fig. 4.21). The subsoil is generally influenced by parent rock or glacial drift material on which it was found. They have not been extensively leached or degraded although some leaching may have resulted in the translocation of calcium (Jones et al. 2011). These soils tend to have medium textures, (sandy loam, loam or sandy clay loam), are friable coupled with desirable structure and drainage characteristics, which means that they are the most extensively cultivated soils in Ireland (Gardiner and Radford 1980) (Fig. 4.21). The majority of these soils are found in the tract running from East Cork to County Louth (Fig. 4.21) where the majority of the tillage farming in the country is carried out. These soils possess a Cambic Bw horizon and correlate closest with the Cambisol Reference Soil Group of WRB (IUSS Working Group WRB 2006).

11.0.0 ***Typical Brown Earths*** have distinct topsoil, without any distinguishing features (Fig. 4.22).

11.2.0 ***Gleyic Brown Earths*** display gleyic features due to the presence of a shallow fluctuating groundwater table.

11.2.6 ***Humic Gleyic Brown Earths*** have humose topsoil and display gleyic features due to the presence of a shallow fluctuating groundwater table.

11.3.0 ***Stagnic Brown Earths*** display stagnic properties as a result of the presence of a slowly permeable subsurface horizon.

11.3.6 ***Humic Stagnic Brown Earths*** have humose topsoil and display stagnic properties as a result of the presence of a slowly permeable subsurface horizon.

11.5.0 ***Typical Calcareous Brown Earths*** have a distinct topsoil and a calcareous subsurface horizon starting within 40 cm and are formed in calcareous parent materials.

11.5.3 ***Stagnic Calcareous Brown Earths*** have stagnic properties derived from a slowly permeable subsurface horizon and a calcareous subsurface horizon starting within 40 cm and are formed in calcareous parent materials.

11.5.6 ***Humic Calcareous Brown Earths*** have humose topsoil and a calcareous subsurface horizon starting within 40 cm and are formed in calcareous parent materials.

11.6.0 ***Humic Brown Earths*** have humose topsoil.

11.9.0 ***Anthropic Brown Earths*** soils reflect surface horizons have been altered by significant human activity which has altered the nature of the surface horizon and no other distinguishing features. This human alteration may include; terric applications such as calcareous sea sands and shells, artefacts, importation of soil material, inversion of soils in situ (Fig. 4.22).

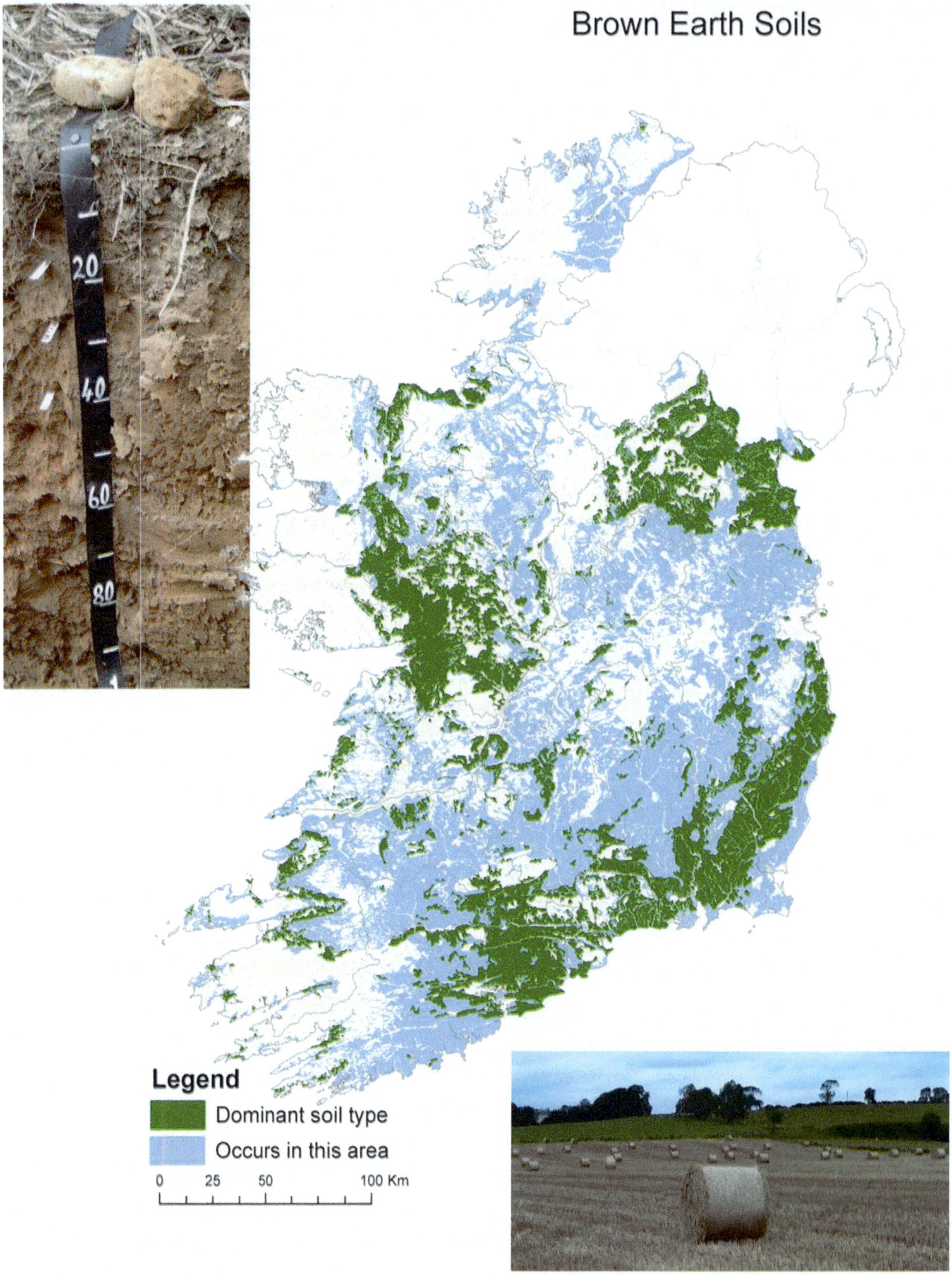

Fig. 4.21 Brown Earth: profile pit (left), geographic distribution (centre) and landscape image (right)

REPRESENTATIVE PROFILE DESCRIPTION

SERIES: CLONROCHE

Reference profile: RPS62RC04
County: Kilkenny
Weather: Overcast
Elevation: 256

TOPOGRAPHY
Position: Lower slope
Slope degree: 1
Slope Form: Straight
Aspect:

PARENT MATERIAL
Substrate Type: Drift
Substrate Subgroup: Siliceous stones

TEXTURAL CRITERIA
Textural Class: Fine loamy
Texturally contrasting:

LAND USE
Land use: Grassland Improved
Human technologies: Fertilizer applications, Ploughing

WATER TABLE: None

ROCK OUTCROPS None

SURFACE STONE None

IRISH CLASSIFICATION (2013)
Soil subgroup: 11.0.0 Typical Brown Earth

National Soil Series: Clonroche
Fine loamy drift with siliceous stones

WRB CLASSIFICATION (2006)
Haplic Cambisol (Humic Eutric)

DESCRIPTION

0 - 21 cm **Ap**

MATRIX COLOR: 10YR43. **STONES (%):** None. **TEXTURE:** Loam. **STRUCTURE:** Moderate, Sub-angular blocky, Medium. **COMPACTION:** Non-cemented and Non-compacted. **CONSISTENCY:** Friable. **PLASTICITY:** Plastic. **STICKINESS:** Slightly sticky. **ROOTS:** Common, Very fine. **PACKING DENSITY:** Low. **BOUNDARY:** Abrupt, Smooth.

21 - 48 cm **Bw1**

MATRIX COLOR: 10YR44. **STONES (% TOTAL):** Common, Angular, Siliceous stones. **TEXTURE:** Loam. **STRUCTURE:** Moderate, Sub-angular blocky, Fine. **COMPACTION:** Non-cemented and Non-compacted. **CONSISTENCY:** Friable. **PLASTICITY:** Plastic. **STICKINESS:** Sticky. **ROOTS:** Common, Very fine. **PACKING DENSITY:** Low. **BOUNDARY:** Clear, Smooth.

48 - 75 cm **Bw2**

MATRIX COLOR: 10YR44. **STONES (%):** Many, 2-6 cm, Angular, Siliceous stones; 6-20 cm, Sub angular, Shale. **TEXTURE:** Loam. **STRUCTURE:** Moderate, Sub-angular blocky, Fine. **COMPACTION:** Non-cemented and Non-compacted. **CONSISTENCY:** Friable. **PLASTICITY:** Plastic. **STICKINESS:** Sticky. **ROOTS:** Very few, Very fine. **PACKING DENSITY:** Low. **BOUNDARY:** Abrupt, Wavy.

75 - 100 cm **BC**

MATRIX COLOR: 25Y54. **MOTTLE:** 25Y66. **STONES (%):** Many, 6mm -2 cm, Flat/platy, Shale; 2-6 cm, Flat/platy, Shale. **TEXTURE:** Sandy loam. **STRUCTURE:** Weak, Angular blocky, Medium. **COMPACTION:** Cemented. **CONSISTENCY:** Firm. **PLASTICITY:** Plastic. **STICKINESS:** Sticky. **COATS:** Manganese. **PACKING DENSITY:** Medium.

Fig. 4.22 Reference profile pit description of 11.0.0 Typical Brown Earth, Clonroche Soil Series

LABORATORY ANALYSIS

Horizon	pH	Total (%)		Organic Carbon (%)	Loss-on-ignition (%)
		Nitrogen	Carbon		
1(Ap)	6.6	0.48	4.27	3.57	
2(Bw1)	6.5	0.29	2.42	1.40	
3(Bw2)	6.5	0.18	1.23	0.80	
4(BC)	6.5	0.06	0.32	0.18	

Horizon	OXALATE EXTRACTABLE		EXCHANGEABLE COMPLEX					
	Fe ($g\ kg^{-1}$)	Al ($g\ kg^{-1}$)	CEC ($cmol\ kg^{-1}$)	Exchangeable Bases ($cmol\ kg^{-1}$)				Base Saturation (%)
				Na^+	K^+	Mg^{2+}	Ca^{2+}	
1(Ap)	8.69	2.94	15.30	0.13	0.59	1.67	12.35	96
2(Bw1)	9.78	3.18	9.63	0.14	0.63	1.26	6.29	86
3(Bw2)	11.82	5.26	8.05	0.10	0.42	1.14	4.19	73
4(BC)	2.02	1.07	4.35	0.08	0.23	0.65	1.82	64

Horizon	PARTICLE SIZE (%)			Textural Class USDA	Bulk Density g/cm^3	Standard Deviation
	Sand 2000-50 μm	Silt 50-2 μm	Clay <2 μm			
1(Ap)	43	34	23	Loam	0.92	0.11
2(Bw1)	40	35	25	Loam	0.96	0.02
3(Bw2)	35	41	24	Loam		
4(BC)	53	33	14	Sandy Loam		

Fig. 4.22 (continued)

4.2 Meeting Future Needs

The development of the 1:250,000 3rd Edition National Soil Map of Ireland and the associated Irish Soil Information Systems (SIS) database are described here followed by a short section on soil quality, sustainability and mapping. This demonstrates the utility of the Irish SIS in meeting the challenges of sustainability through the development of soil property maps based on data held in the soils database. The content of this chapter relies largely on the project reports produced as part of the Irish SIS project, in particular, the Synthesis Report by Creamer et al. (2014) and the Final Technical Report 18 by Creamer et al. (2016).

Prior to the Irish SIS project, systematic pedology in Ireland had largely been related to mapping exercises, such as the previous National Soil Survey (NSS) conducted by An Foras Talúntais (AFT) who mapped 44% of the country at a scale of 1:126,720; developed a General Soil Map of Ireland and a National Peatland map, both at 1:575,000 scale (Creamer et al. 2014). Recognition of the need to increase understanding of soil resources at a global level has been growing due to the interdependence of the supply of food and agricultural production. The issues of land degradation, disparity of production potentials and of population carrying capacities have become international concerns, especially since the 1980s driving the need for enhanced soil information (IUSS 2006). In 1961, the Food and Agriculture Organization (FAO) and the United Nations Educational, Scientific and Cultural Organization (UNESCO) decided to prepare a Soil Map of the World at a scale of 1:5,000,000 (FAO 2016). This project was completed over a period of 20 years and until recently remained the only global overview of soil resources (FAO 2016). The Soil Map of the World subsequently integrated over 15,000 different soil mapping units of regional and national soil information to produce the Harmonized World Soil 30 arc-second raster database Version 1.1 (FAO/IIASA/ISRIC/ISS-CAS/JRC 2009). Since the mid-1990s, a multi-stakeholder collaboration has produced the revised Harmonized World Soil Database, which was published in 2006 at a resolution of 1 km (FAO 2016).

At a European scale, challenges to sustainability were driving the demand for soil protection and highlighted the need for soil data that could be harmonised at that level. Determining the appropriate scale to develop a national soil map is challenging; if too small in scale, it will lack sufficient detail, potentially making it limited in its application (Fig. 4.23) and conversely, a larger scale (more detailed) is likely to incur higher costs. Consequently, scale will typically be constrained by budgetary considerations. The European Soil Bureau Network (ESBN) Technical Working Group responsible for soil monitoring and harmonisation determined that a scale of 1:250,000 was both an economically feasible intermediate scale, sufficiently equipped to identify specific problems at a regional scale (Montanarella and Jones 1999). On that basis, the 1:250,000 scale was determined as the appropriate for the development of the 3rd Edition National Soil Map of Ireland.

4.3 Legacy Data

Prior to the Irish SIS project, substantial soil survey work had already been carried out in Ireland between the 1950s and 1980s by AFT and the National Soil Survey of Ireland (Chap. 1). As part of the development of the 1:250,000 soils map for Ireland and in order to construct a comprehensive database of soil information in Ireland, it was necessary to digitise and develop a supporting database of all available legacy soil survey data. In addition, the Irish SIS also incorporates more recent map products including the Indicative Soil Map (1:150,000) and the Subsoil mapping (1:50,000) described by Fealy et al. (2009).Within the Irish SIS project, the previously surveyed county-level soil survey work is collectively referred to as *Terra Cognita,* a Latin term used in cartography to describe 'known land' (Fig. 4.24).

A key methodological aim was to use existing detailed mapping work to develop a harmonised soil map for previously surveyed areas. Once developed, this could then be used as the basis for the modelling of areas where detailed soil surveys had not been carried out. The AFT survey detailed mapping was used to generate soil-landscape or soilscape models of *Terra Cognita* (Fig. 4.24) which served as a baseline for the statistical modelling methodologies to map *Terra Incognita* (Fig. 4.24), the 'unknown land'. Validation of this methodology was carried out over a 2.5-year field survey, in which 11,000 locations were assessed for soil type using an auger survey approach. Where previously unidentified combinations of soilscape units were found, soil profile pits were excavated, sampled and described at representative locations across the country to define this new soil information. These 225 pits were described and sampled in detail and were used to support the development of a new soil classification system for the country described in Chap. 3. The AFT soil classification included ten Great Soil

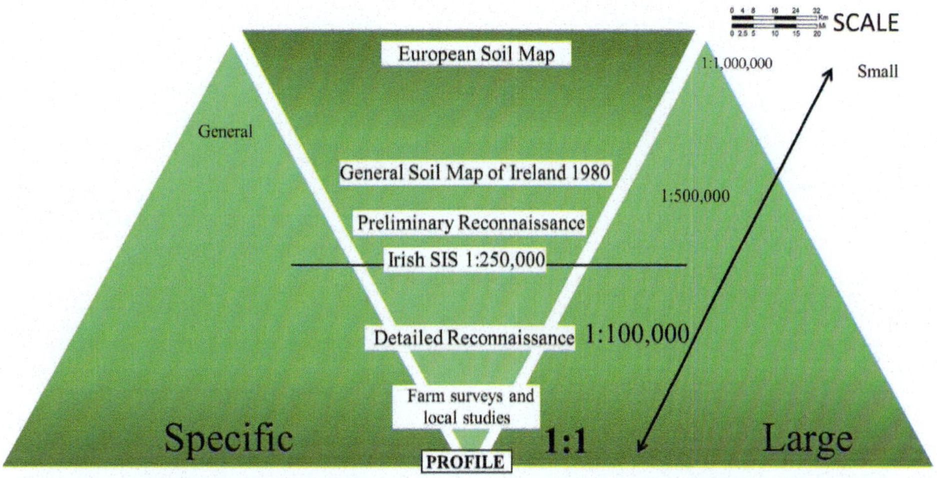

Fig. 4.23 Visualisation of small-scale general European Soil Map to large-scale profile descriptions along with the scale of Irish SIS mapping project shown at 1:250,000

Fig. 4.24 Shows the distribution of Terra Cognita with the coverage of previously published national soil survey work by An Foras Talúntais and Terra Incognita, the areas to be modelled across Ireland from Jones et al. (2014)

Table 4.1 Previous Great Group classification and Irish SIS classification 2014

Great Soil Group (AFT classification 1980)	General properties	Great Soil Group (Irish SIS Classification 2014)
Podzols	Formed by leaching of nutrients (podzolisation process), acidic, poorly drained	Podzols
Brown Podzolics	Less depleted of nutrients than Podzols, good physical characteristics	Brown Podzolics
Grey Brown Podzolics	Usually formed from calcareous parent material which counteracts the effects of leaching, can be light to heavy textured	Luvisols
Brown Earths	Most occur on lime-deficient parent materials, therefore acidic in nature, relatively mature and well drained	Brown Earths
Gleys	Developed under influence of permanent or intermittent waterlogging, impervious with poor physical structure, unsuitable for cultivation or intensive grazing	Surface-water and Groundwater Gleys
Rendzina (Shallow Brown Earth)	Shallow soils, usually no more than 50 cm depth, usually derived from limestone parent material, use limited by shallow depth	Rendzina
Lithosols	Skeletal stony soils usually overlying solid or shattered bedrock, use limited to rough grazing and forestry	Lithosols
Regosols	Alluvial soils found in low lying flat areas along river course and estuaries, texture can vary between sands and clays, can be acid or alkaline depending on the material from which they are derived, wide range of uses	Alluvial Soils
Basin Peat	High organic matter content (>30% down to at least 30 cm depth) formed in depressions	Mineratrophic peat
Blanket Peat	High organic matter content (>30% down to at least 30 cm depth) formed under conditions of high rainfall and humidity	Ombrotrophic peat

Groups (Table 4.1). Within Great Soil Groups, a further subdivision was made into Groups (where the Great Group had relatively contrasting soil types within a Great Soil Group, for example, Calcareous Brown Earths and all other Brown Earth soils). The final detailed classification level was defined as soil series, describing a sequence of horizons with distinct chemical–physical characteristics. The Irish SIS developed a 'Subgroup' category which facilitated a systematic classification approach largely based on field descriptions. The final map and database products have been created using a unique combination of new and traditional methodologies utilising soils data from both the historical AFT survey archive and new data collected for the Irish SIS project.

4.4 Development of the Soil Map

The development of a soil map requires the soil surveyor to identify soil series in the context of the landscape and to name them according to specified criteria (Creamer et al. 2014). The soil mapping unit is the basic geographic component on the soils map and scale determines the level of detail that is cartographically possible. For detailed large scale maps, for example, 1:10,000, the distribution of soil series might be delineated with considerable detail. At smaller scales, such as at 1:250,000, soil mapping units rarely comprise of a single soil type. Instead, soil series are combined into groupings which are referred to as 'soil associations' and which are

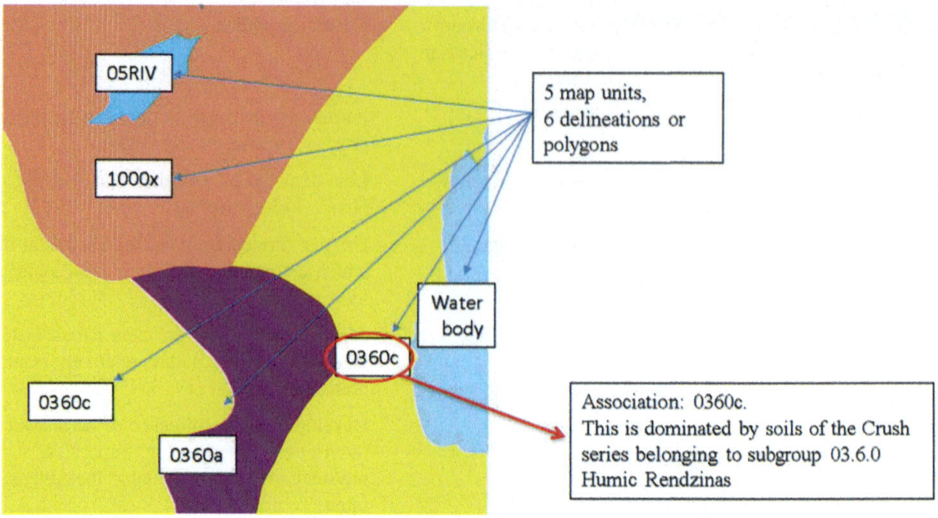

Fig. 4.25 Delineations of soil units with their soil associations

composed of a dominant soil series along with associated soil series. Each soil association refers to a range of soil series that occur together and that are associated with a particular landscape (Creamer et al. 2014). Soil associations may have between two and ten soil series co-occurring in a particular landscape pattern with the soil association named after the most dominant soil series (Fig. 4.25).

The 1:250,000 national soil map of Ireland is a soil map composed of soil associations of the 213 soil series, and altogether a total of 58 soil associations, excluding peat, alluvium, urban, rock and marsh are defined. On a soil map, the soil mapping units or polygons representing a soil association are attributed with the same colour and labelled with the same alphanumeric code. The numbers of each code indicate the predominant Great Group and Subgroup for that association (Fig. 4.25). Figure 4.25 shows the delineation of polygons at a zoomed in scale, with association 0360c highlighted, representing a Humic Rendzina dominated by soils of the Crush series.

A key aspect of the methodology deployed in the Irish SIS project was the digitisation and harmonisation of legacy data of the *Terra Cognita* areas. To do so, information was extracted from the detailed published AFT county bulletins and was entered into the Irish SIS database. Representative soil profile data were also captured for each soil series. To ensure consistency, the existing county maps were harmonised at a scale of 1:250,000. The large scale of the county soil maps meant that a great number of soil series were recorded. In addition, these surveys were carried out over a period spanning several decades with the work conducted by a number of different surveyors. As a result, many of the recorded soil series were similar in description but were named differently on different county soil maps. Furthermore, the updates to classification that were published in more recent AFT maps were not in place for the development of some of the older maps. All of these factors meant that a review of the national soil classification system was required. This included a process of harmonisation and rationalisation to allow generalisation of the county maps from the 1:126,720 to the 1:250,000 scale (Fig. 4.26). To harmonise the county maps in *Terra Cognita,* small polygons ($<0.075\ km^2$) were eliminated and polygons, where soil series had been amalgamated, were combined. Mapping at county borders was realigned so that adjacent soil county maps could be joined together (Fig. 4.26).

Once soil series were harmonised and rationalised, these were then integrated into soil associations to develop the final mapping units for the 3rd Edition National Soil Map of Ireland. During the development phase of the map, a draft national legend was compiled to take the amalgamation of soil series into consideration. This draft legend was used as the basis for the predictive mapping and fieldwork data collection programme. Figure 4.27 shows the Terra Cognita map of Ireland, with the interim legend displayed.

4.5 Predictive Mapping

Globally, digital soil mapping research has increased markedly in recent years. Today, there are large numbers of soil maps that are generated by quantitative statistical methods (Brus et al. 2011). This process is referred to as 'predictive mapping'. In Ireland, once *Terra Cognita* had been stratified into soilscape units, the information was used to predict similar soilscapes in *Terra Incognita.* Soilscapes are groups of soils typically formed from similar substrates that are linked to large-scale landscape features and derived from expert judgement as part of the generalisation of the AFT county soil maps (Creamer et al. 2014). The well-established SCORPAN method of McBratney et al. (2003) after Zakharov (Eq. 4.1) was used in the Irish SIS to derive soil information from other spatially available datasets:

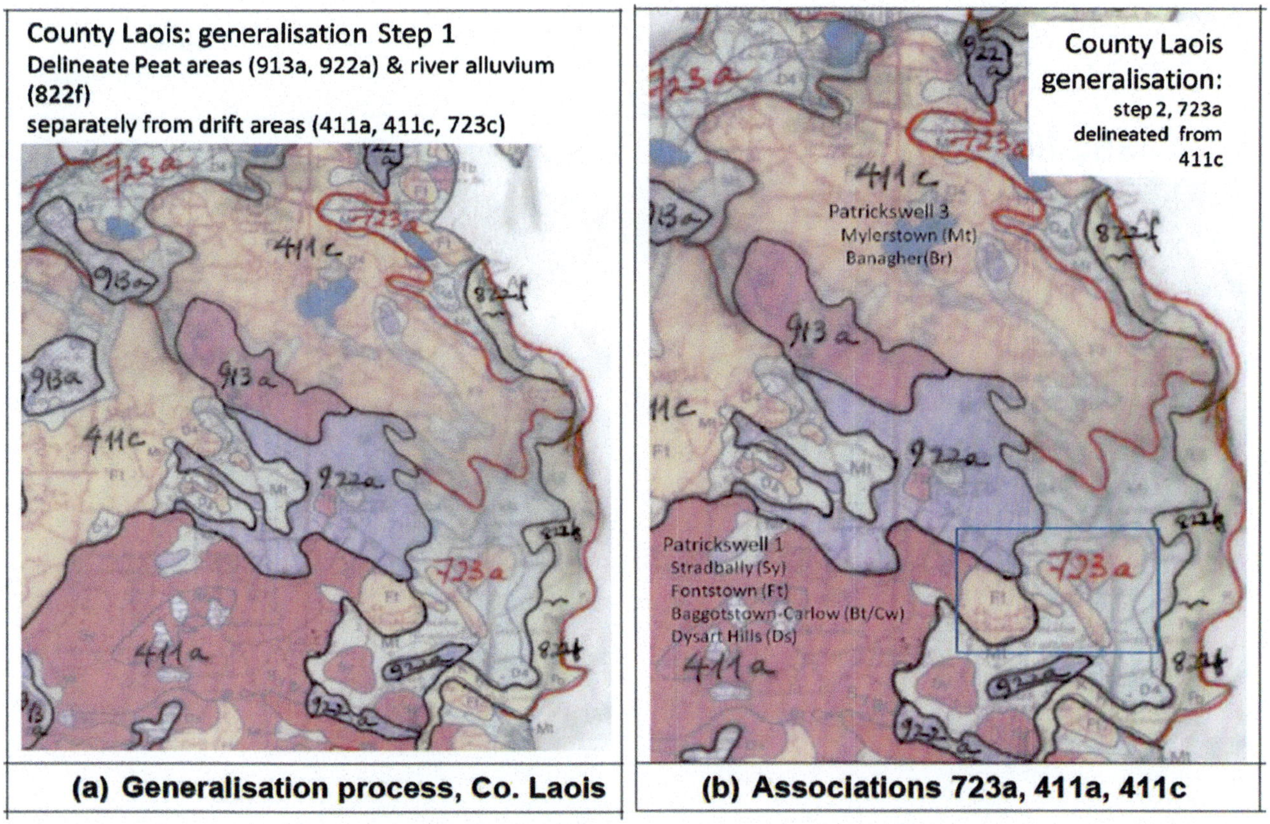

Fig. 4.26 **a** and **b** Example of county map generalisation for Co. Laois from Jones et al. (2014). Examples show generalisation decisions made in County Laois, where (**a**) shows the inclusion of smaller polygons of 723 Mylerstown, a Gley soil, within the 411c Patrickswell association. Also, small polygons of the Fontstown series, shown as a 411 are rationalised to Patrickswell series, occurring within the Mylerstown Association (723a). (**b**) Shows the Fontstown map units, reallocated to the 411a Patrickswell association and elsewhere to the 723a Mylerstown association

$$\mathrm{Sc}, \mathrm{Sa} = f(\mathrm{s}, \mathrm{c}, \mathrm{o}, \mathrm{r}, \mathrm{p}, \mathrm{a}, \mathrm{n}) \tag{4.1}$$

where

- Sc soil class;
- Sa soil attribute;
- s soil;
- c climate, local climatic properties;
- o organisms, vegetation, fauna and human activities;
- r topography;
- p parent material, lithology;
- a age and time, and
- n space and spatial position.

Terra Cognita data was used to derive predictive models for each soilscape, shown in Fig. 4.28, which were then applied to *Terra Incognita* to predict soil types as represented by soil associations. Two modelling approaches were applied to extrapolate soilscapes into the *Terra Incognita*. The first approach was based on similarity to the reference areas (*Terra Cognita*), referred to as feature space (FS) analysis. The second approach, random forest (RF), was a rule-based approach wherein the model inferred a relationship between environmental covariates and soil distribution. Two soilscape maps were developed based on these two approaches. At this stage, a series of models were developed, tested and optimised. Statistical modelling using inference systems based upon statistical concepts including RF, Bayesian belief networks (BBN) and neural networks were applied to determine the soil associations in *Terra Incognita*. Five model realisations or mapped outputs were developed on this basis to represent the *Terra Incognita* areas. To achieve that highest level of accuracy, a method to 'fuse' the modelled outputs so that the best performing areas were selected was developed. This resulted in a single predictive map that represented the highest accuracy score for each of the predicted soil associations. An auger survey was conducted from 2010 to 2013 to ascertain the main soil associations across Ireland. This survey compared the soil observation at an auger borehole to the predicted soil and the composition of the predicted soil association. The survey area was divided into 10 × 10 km map sheets and the major soil series found within each map sheet were recorded.

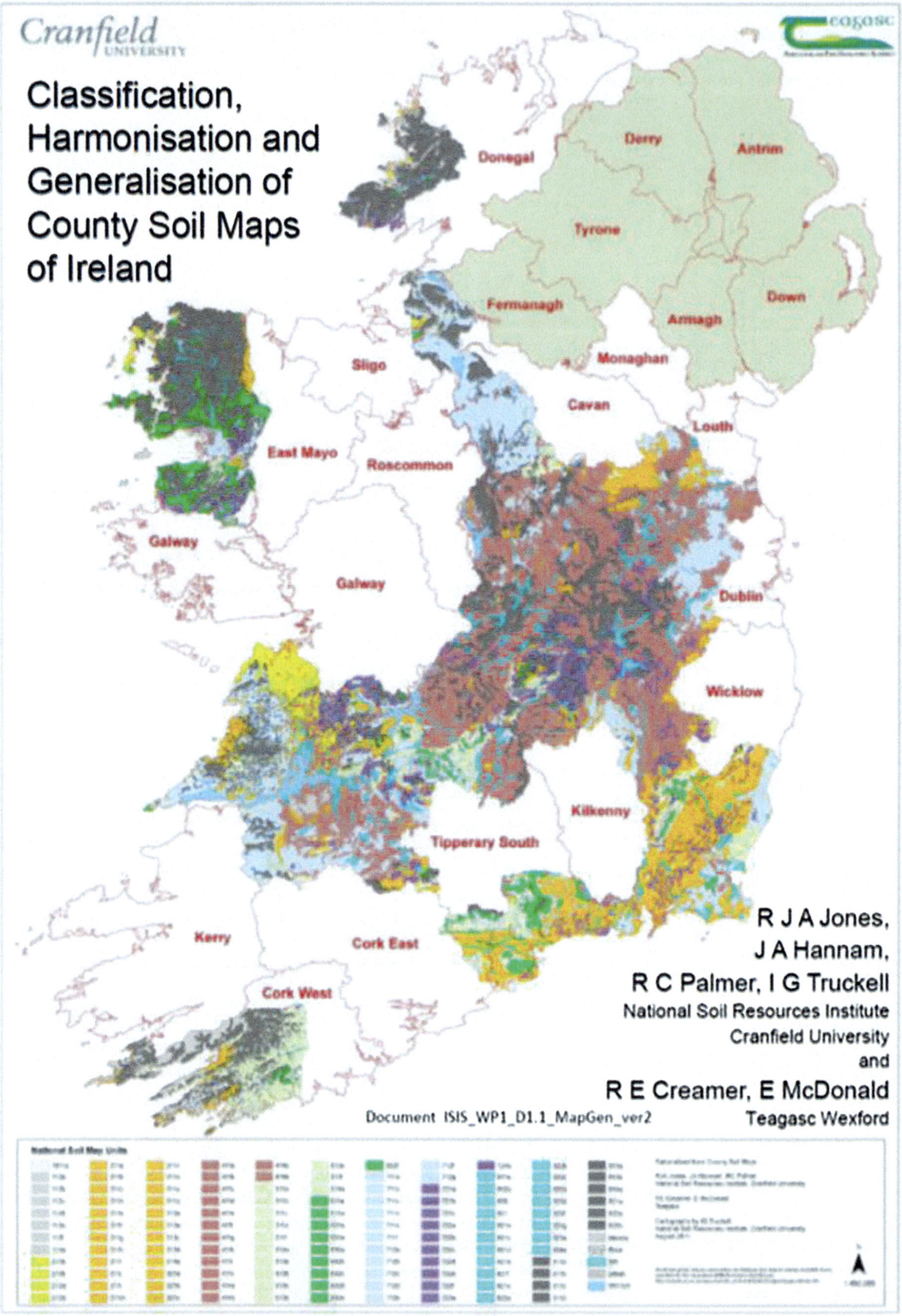

Fig. 4.27 Classification, Harmonisation and Generalisation of County Soil Maps of Ireland by Jones et al. (2014). This represents the Terra Cognita following a review of classification, along with harmonisation of the soil series to allow generalisation of the county maps from 1:126,720 to the 1:250,000 scale

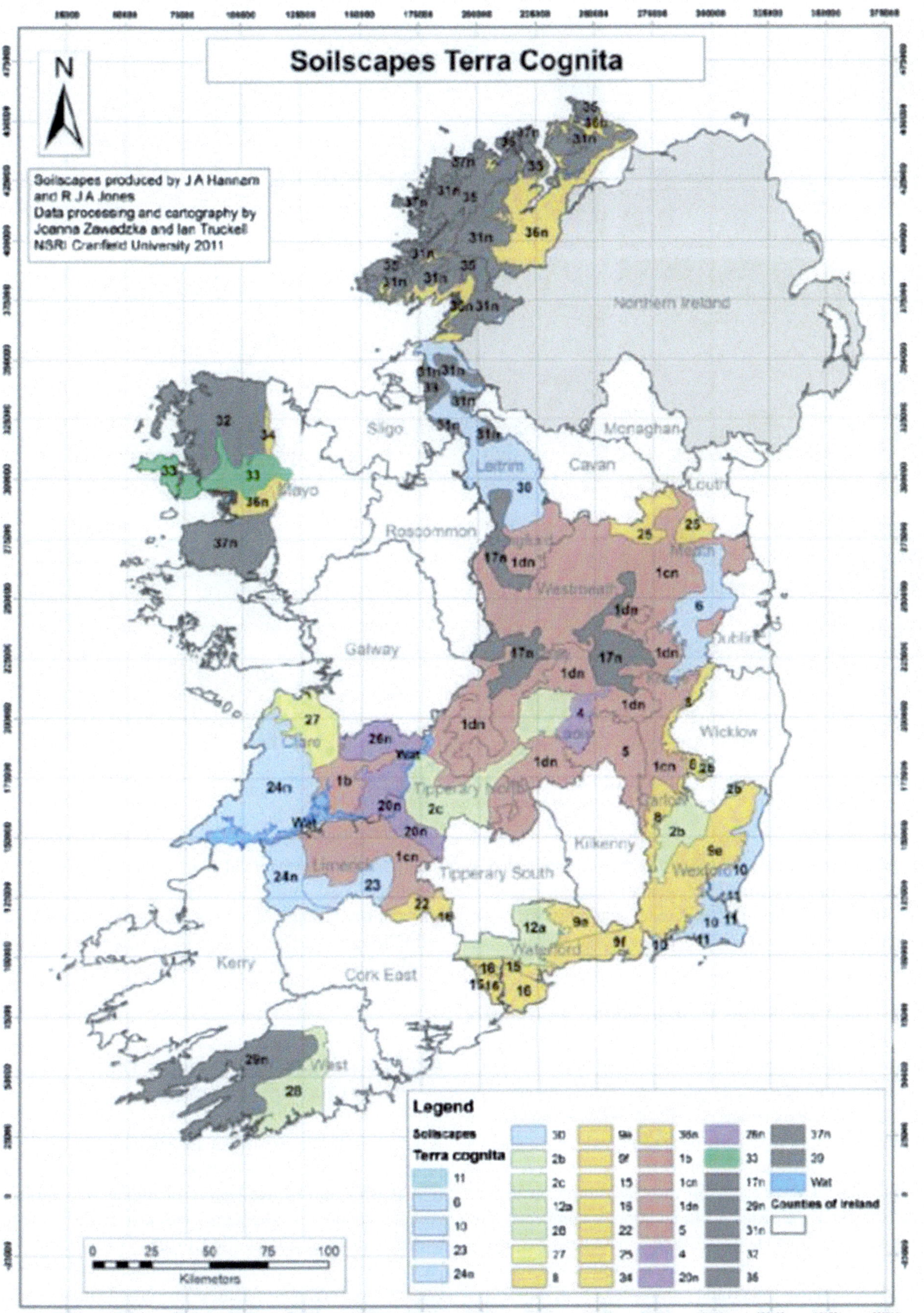

Fig. 4.28 *Terra Cognita* soilscapes developed as a baseline to model *Terra Incognita* from Jones et al. (2014) Soilscapes for Terra Cognita (2007-S-CD-1-S1) Irish Soil Information Systems Final Technical Report 3

Fig. 4.29 Auger borehole samples from validation survey of predictive mapping taken from 2010 to 2013. During this survey approximately 11,000 records were collected across Ireland from *Terra Incognita* areas

Altogether approximately 11,000 records were collected as part of this survey, a sample of which is seen in Fig. 4.29. Once the 'fusion process' was complete, this map was amalgamated with the *Terra Cognita* generalised map.

4.5.1 New Soil Series to the Legend and Matrix

Following the validation of the fusion map with the 11,000 augers and original Terra Cognita data, it became apparent that new soil associations (i.e. soil series occurring in specific landscapes types) had been identified, which had not been captured previously in the Terra Cognita survey. The identification of these new soil associations meant that an updated legend was required. In addition, a number of new series were also found to occur. Where these were frequently found, they were described as modal profiles which meant that a representative soil profile pit was dug to approximately 1 m depth or bedrock (see Chap. 3). These series were incorporated into the final map legend. This final legend formed the basis of the 3rd Edition Soil Map of Ireland and was validated using the same approach described above. The 3rd Edition National Soil Map of Ireland (Fig. 4.30) shows the final mapping units, represented as soil associations on the map.

4.6 Development of SIS the Digital Soil Information System

A lot of soil data existed in Ireland prior to the Irish SIS project. This soil data was typically held in map or paper format with limited availability of digital soil information. Although county monographs were scanned into PDF format and soil maps were digitised in the 1990s, there was no harmonised system that was publicly accessible with capacity for data interrogation. The introduction of a digital system was recognised as crucial towards greater

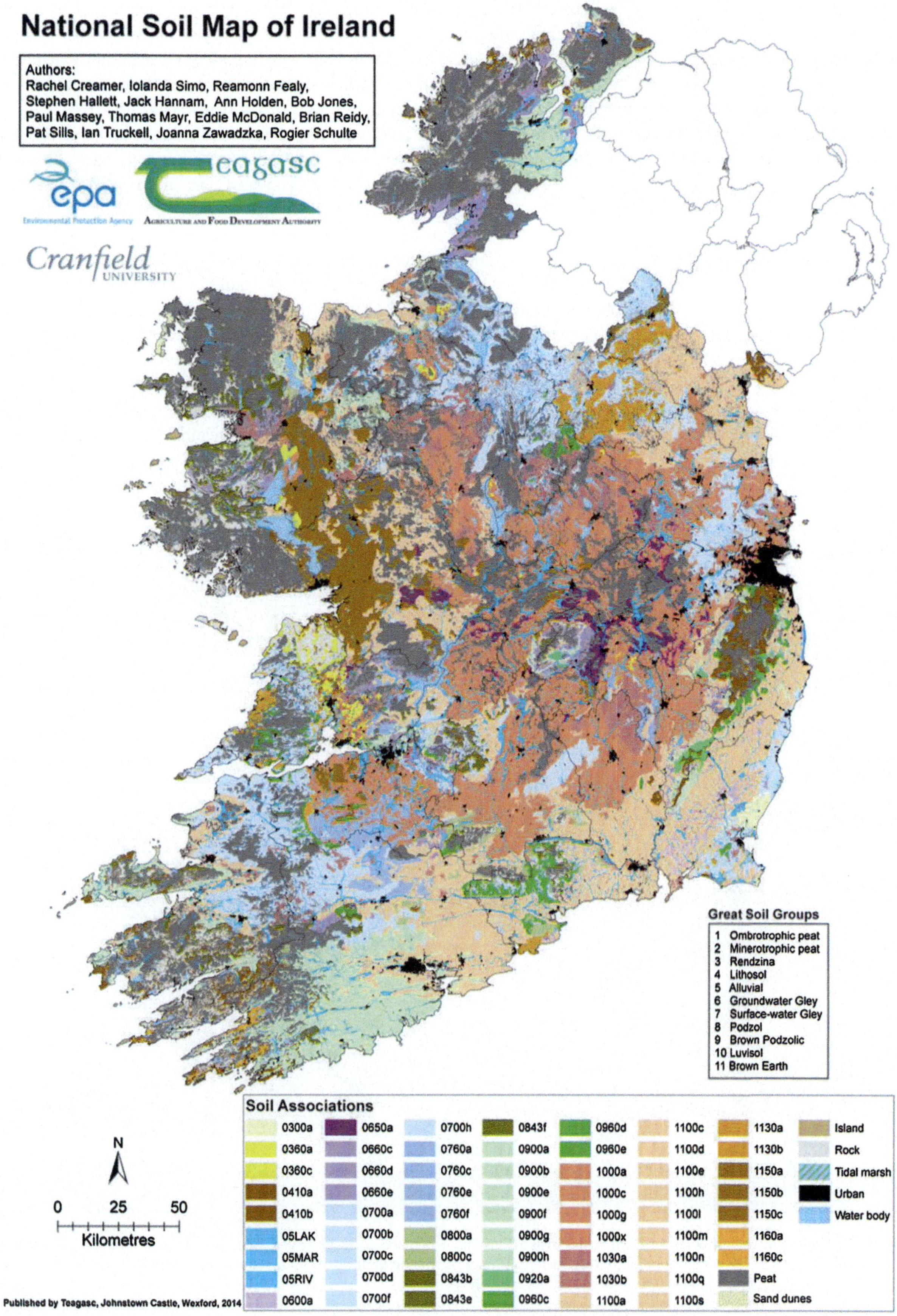

Fig. 4.30 The finalised 3rd Edition National Soil Map of Ireland by Creamer et al. (2014) with soil association legend

management of the soil data resource in Ireland (Daly and Fealy 2007). As well as serving a fundamental role in allowing a better understanding of the natural environment, such a system could support process-based and socio-economic modelling for policy appraisals. Along with a publicly available 1:250,000 soils map of Ireland, the associated soil information system was a key output of the project.

4.6.1 Interactive Soil Map

The spatial components of the SIS database are built on the ArcGIS Server GIS platform from Environmental Systems Research Institute (ESRI). The platform provides a range of application advantages including the scope to develop future extensions. It also incorporates advanced potential for data modelling and interrogation. The system includes a web-enabled GIS server that provides access to the public through a web browser. The new digital 1:250,000 scale soil association map was imported, cleaned and converted into an ArcGIS file geodatabase (FGDB) deployed on the public-facing web server (Creamer et al. 2014).

The map service hosted on the web server allows web browsers access to the map and underlying database and is available at http://gis.teagasc.ie/soils/map.php. The interface is interactive, user-friendly and freely available, and enables users to understand more about the soil associations that occur in their area of interest. Users are able to zoom and pan across the map using standard web mapping tools. Zooming is limited to a certain scale as a reminder to users that the map was constructed for use at a scale of 1:250,000 (SIS 2014).

Figure 4.31 shows a screenshot of the web portal with the selected polygon highlighted by the black outline. The soil association information of the highlighted polygon is brought up to the right, indicating that this polygon contains

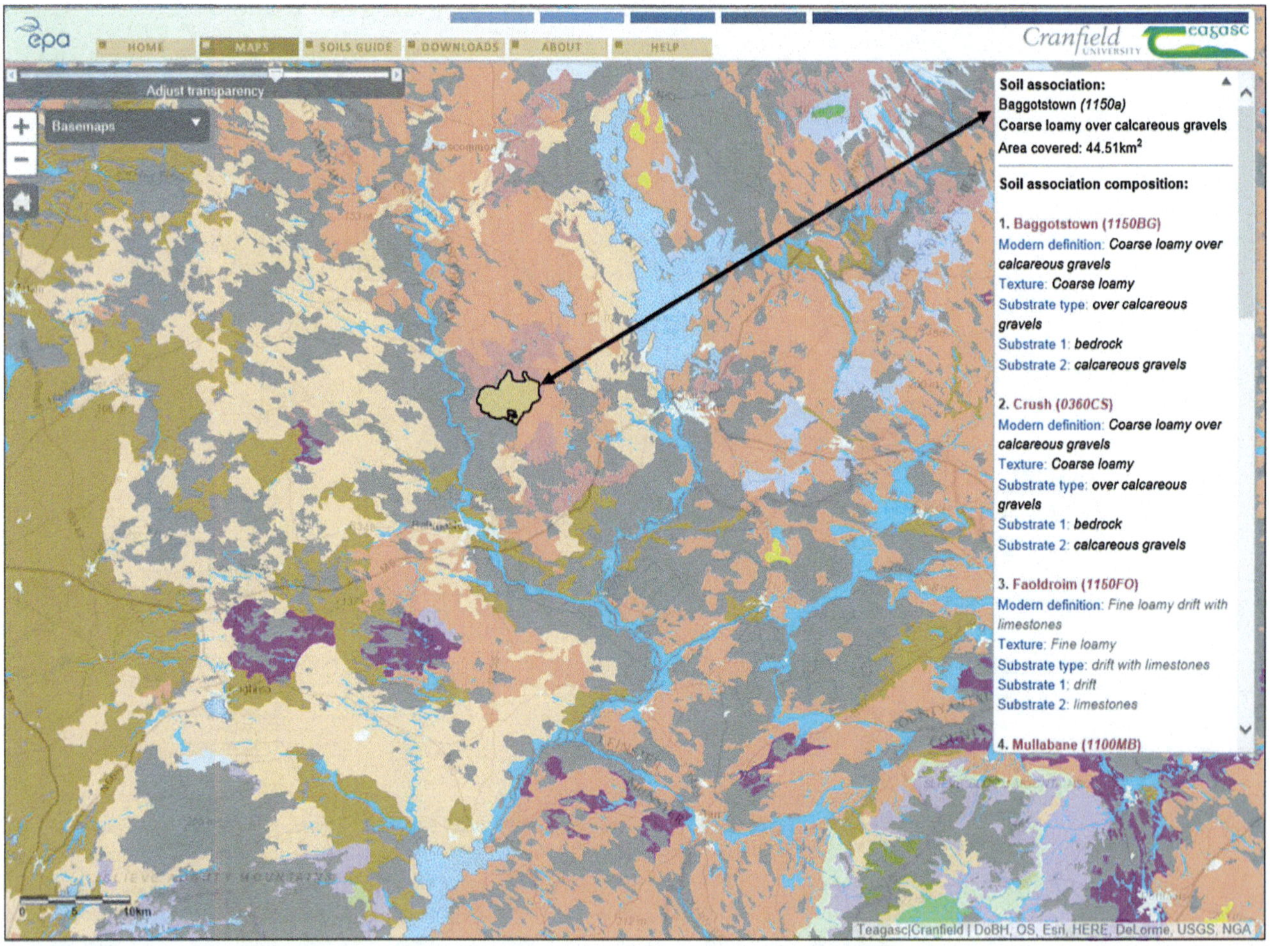

Fig. 4.31 Screenshot of web browser interface of the Irish SIS taken from the freely available web portal at http://gis.teagasc.ie/soils/map.php. The figure shows a zoomed in selection of the 1:250,000 map with one polygon highlighted. The information on soil association related to the highlighted mapping unit is also displayed on the interface

the Baggotstown (1150a) soil association. While the Baggotstown series represents the lead of the association, it also includes ten other series, including Crush (0360CS), Faoldroim (1150FO) and Mullabane (1100MB) (Fig. 4.31). Further information for each of the constituent soil types is shown, along with the functionality to link to the 'soils guide' tool where more detailed soil series attribute information can be accessed.

4.6.2 Soils Guide

Together with the interactive map, the website also includes a soils guide that displays the hierarchical groups of soil with a detailed description of the members of any selected element of a soil group shown on the right. The opening guide shows all of the soil associations which are grouped and displayed using colours that are consistent with the national soil legend. The 'accordion box' design of the constituent soils guide sub-windows allows the user to switch between the Associations, Great Soil Groups, Soil Subgroups and Soil Series levels (Fig. 4.32). All soil groups belong to a Soil Subgroup and a Great Soil Group which forms the taxonomic classification system explained in Chap. 3. The soil association is the cartographic grouping of soils that commonly occur together in the landscape. The database linked to the map allows the user to 'drill down' to learn more information about the soil types that make up a given association and their relative ranking in terms of typical extent (SIS 2014). Figures 4.32 and 4.33 show the related database information associated with Bantry series (0410BA). The information describes the profile and site and provides detailed horizon level information.

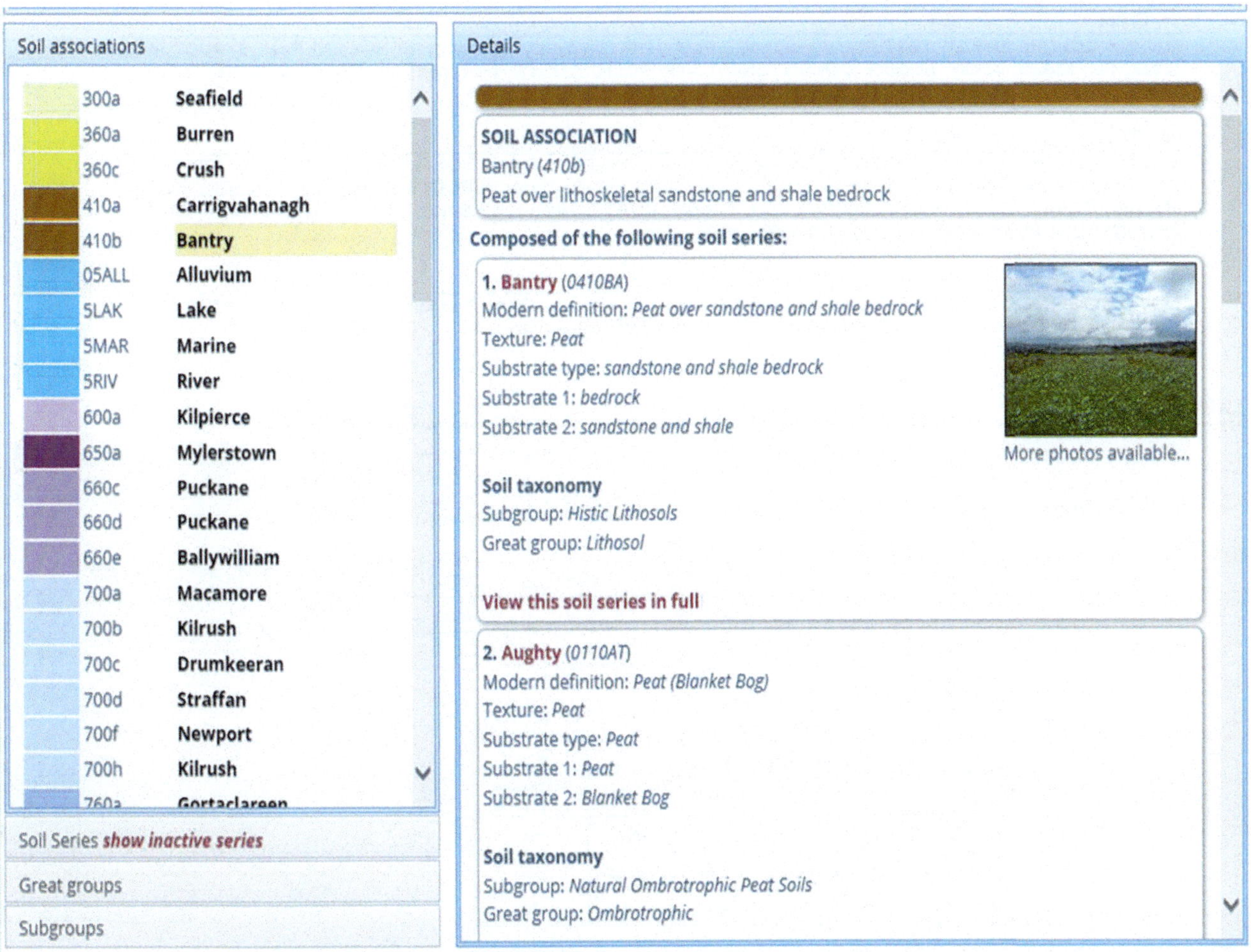

Fig. 4.32 Screenshot of Soils Guide associated with SIS database showing the accordion box display on the left with hierarchy of soil taxonomy options. The right panel describes the selected element from the left with further capacity to 'drill down' to explore the soil series property data in full

REPRESENTATIVE PROFILE DESCRIPTION

SERIES: BANTRY

Reference profile: RPR02BR01
County: Kerry
Weather: Rain
Elevation: 128

TOPOGRAPHY
Position: Middle slope
Slope degree: 5
Slope Form:
Aspect: N

PARENT MATERIAL
Substrate Type: Bedrock
Substrate Subgroup: Sandstone

TEXTURAL CRITERIA
Textural Class:
Texturally contrasting:

LAND USE
Land use: Grassland Unimproved
Human technologies: Artificial drainage, Levelling, Sprayed (herbicides)
Vegetation: Rushes
Species: Ragwort

WATERTABLE None

ROCK OUTCROPS None

SURFACE STONE None

IRISH CLASSIFICATION (2013)
Soil subgroup: 04.1.0 Histic Lithosol

National Soil Series: Bantry
Peat over sandstone and shale bedrock

DESCRIPTION

0 - 10 cm **Oh**

MATRIX COLOR: 10YR21. **TEXTURE:** Amorphous, Peat, H10 Completely decomposed, Humified Peat. **STRUCTURE:** Weak, Sub-angular blocky, Medium. **CONSISTENCY:** Very friable. **ROOTS:** Few, Medium. **PACKING DENSITY:** Low. **POROSITY:** High. **BOUNDARY:** Clear, Smooth.

10 - 30 cm **Omf**

MATRIX COLOR: 10YR22. **STONES (%):** Common, 2-6 cm, Angular, Siliceous stones; 6-20 cm, Flat/platy. **TEXTURE:** Semi-fibrous peat, Peat, H6 Well decomposed, Mesic (semi-fibrous) Peat. **STRUCTURE:** Moderate, Angular blocky, Medium. **CONSISTENCY:** Friable. **ROOTS:** Few, Coarse. **COATS:** Humus, Common, Distinct, Continuous. **PACKING DENSITY:** Low. **POROSITY:** Many. **BOUNDARY:** Abrupt, Wavy.

30 cm + **Cr**

MOTTLE: 75YR58, Few, Medium, Distinct, Diffuse. **STONES (%):** Many, 6-20 cm, Angular, Siliceous stones. **STRUCTURE:** Massive. **ROOTS:** Very few, Fine. **POROSITY:** Very low. **MACROPORES:** Very fine.

LABORATORY ANALYSIS

Horizon	pH	Total (%)		Organic Carbon (%)	Loss-on-ignition (%)
		Nitrogen	Carbon		
1(Oh)	5.7	1.43	19.57	19.57	39.1
2(Omf)	5.9	1.10	23.67	24.33	41.3
3(Cr)	4.8	0.04	0.40	0.27	

EXCHANGEABLE COMPLEX					
CEC (cmol kg^{-1})	Exchangeable Bases (cmol kg^{-1})				Base Saturation (%)
	Na^+	K^+	Mg^{2+}	Ca^{2+}	
39.09	0.27	0.58	7.57	30.74	Sat.
45.78	0.34	0.29	4.54	53.64	Sat.
4.12	0.09	0.18	0.48	2.44	78

Fig. 4.33 Shows the detailed database information available for soil series Bantry 0410BA. The information is downloadable into a pdf report format

4.7 Soil Quality, Sustainability and Mapping

Soil quality refers to the capacity of the soil to function within natural or managed ecosystem boundaries, and to sustain plant and animal productivity, maintain or enhance water and air quality, and support human health and habitation (Karlen et al. 1997; Doran 2002). Soil quality varies due to natural factors, such as climate or slope and due to anthropogenic factors, such as management. Soil has both inherent and dynamic qualities. Inherent qualities refer to the intrinsic capacity of the soil to function and these characteristics do not readily change, for example, soil texture, cation exchange capacity (CEC), or depth to bedrock. Conversely, dynamic properties can change over time in response to land use and soil management. Dynamic properties include organic matter, pH and infiltration rate. Although dynamic properties can be changed through land use and management practices, the extent of change may be determined/constrained by the inherent soil properties. Knowledge of soil properties and qualities is therefore essential in relation to understanding soil quality which can, in turn, support both agricultural and environmental management.

Meeting the challenge of sustainability relies in part on better management of the soil resource. This can only be achieved if there is a comprehensive inventory with adequate soil data to understand the soil resource sufficiently. In this regard, the Irish SIS provides a vital advance in soil information at a national level in Ireland. It represents a significant national resource for research, policy-making and land use practices. Soil is an elemental driver of many agri-environmental processes and the delivery of meeting agri-environmental objectives relies to a great degree on soil properties and hence soil type (Creamer et al. 2014).

In Ireland, an inventory of the most relevant soil attribute data essential to support national objectives related to agriculture and the environment was compiled by a range of stakeholders in 2013 (Creamer et al. 2016). This inventory proposed the following 'indicative' soil attribute maps:

- Soil Drainage (shown in Chap. 14)
- Soil Depth
- Soil Texture
- Soil pH
- Bulk Density
- Soil Organic Carbon (SOC) (shown in Chap. 17)

Soil attribute maps are developed to spatially describe soil properties based upon point/profile data. As part of the Irish SIS project, modal profile pits were selected from over 700 profile descriptions to represent the best available data of the main soil types in Ireland. Soil mapping is complicated due to the inherent high spatial variability of soil properties that may occur at varying scales and ranges across a landscape. Accordingly, the use and interpretation of such maps require careful consideration in relation to suitability which should be considered on a case by case basis (Creamer et al. 2016). Field soil surveys assume high costs that increase in line with scale. The 3rd Edition Soil Map of Ireland is based on a scale of 1:250,000, which means that while this map provides a good overview of the trends and types of soils found at catchment or landscape scale, it is not appropriate for detailed information at farm or field level.

To derive the soil attribute maps three methods were employed which are as follows: (1) kriging, (2) assessment based on the dominant soil series within a soil association and (3) proportional approach to soil series within an association.

1. Kriging refers to a family of estimators within DSMM that use least squares regression algorithms to interpolate spatial data. Ordinary kriging (OK) and universal kriging (UK) are widely used. OK is based solely on the data values from the soil profile pits whereas UK considers some influence from other environmental and management factors such as land use and topography (Creamer et al. 2016). The soil pH, bulk density and soil organic carbon (SOC) indicative maps have been developed using the kriging methodology which is described in detail in Creamer et al. (2016).
2. Dominant soil series characteristics defining the association.
3. Proportional approach to soil association mapping. This was used for the soil depth map.

4.7.1 Indicative Soil Depth Map

Soil depth, or the thickness of the soil cover, is a reflection of both soil parent material and the length of time for which soil forming processes are active. Soil depth is important in determining the potential rooting depth of plants and any restrictions related to depth. The effective soil depth can be reduced by the presence of bedrock or impervious layers such as hardpans (Brady 1974; Creamer et al. 2016). All other factors being equal, deep soils will have greater available moisture-holding capacity than shallow soils (Brady 1974). Root systems anchor plants in the soil and also supply plants with nutrients and water from the soil. The supply of resources to the above-ground parts of the plant depends upon the length and distribution of the soil root which will be modified by soil depth (Killham 1994). The soil solum refers to the upper and most weathered part of the soil profile, the A and B horizons. Root systems are most

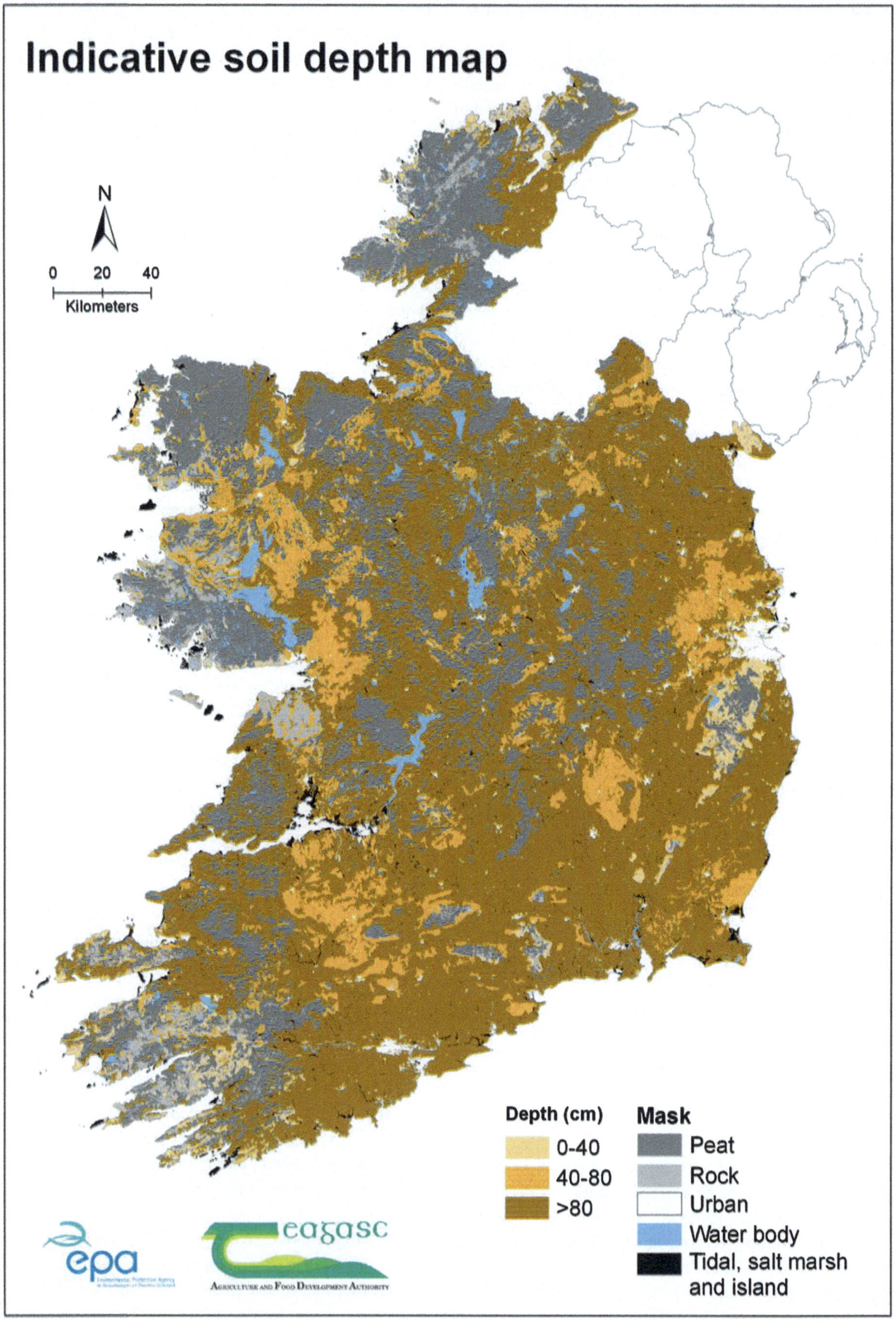

Fig. 4.34 Indicative soil depth map from Creamer et al. (2016) developed using above method in ArcGIS from ESRI and data from the Irish SIS

dense in the upper solum and plants derive the majority of their water requirements from this area but rooting depth will vary due to the plant physiology, the type of soil and the water availability (Creamer et al. 2016).

Soil depth varies considerably across the landscape and defining soil depth accurately is challenging. Figure 4.34 shows the indicative soil depth map which was developed based upon the 3rd Edition National Soil Map of Ireland.

The procedure to determine soil depth included a comparison of auger point data at soil series level to the soil series constituent member of the association. Auger points were matched and summarised to soil series within each association. The maximum depths of the modal profiles for each soil series of the soil association were checked and the most common depth of the modal profile was selected (Table 4.2). Based upon the Irish classification system three depth classes were recognised; 0–40, 40–80 and >80 cm. Therefore, in the example shown in Table 4.2, a depth of >80 cm is the most frequently found, occurring in 32 of the 50 auger bores for the association.

Figure 4.34 shows that shallower soils are found in mountainous areas, along the west coast of West Cork and Mayo and the northern coast of Donegal and the karst landscape of the Burren in Co. Clare. The rest of the country has deeper soils which are shown in two depth classes; soils where bedrock occurs within 80 cm and soils found mostly over glacial till where deep subsoil are present and a soil depth in excess of 1.5 m is often found.

4.7.2 Indicative Bulk Density Map

Bulk density refers to the space or mass of dry soil including porosity per unit of bulk volume. The bulk volume is determined before drying to a constant weight at 105° C (Brady 1974). As bulk density considers pore space as well as soil solids, loose porous soils will have low bulk densities (weight per unit volume). Accordingly, where two soils share the same particle density, if one soil is more compact it will have a higher bulk density. Soil structure refers to the shape, size and arrangement of soil particles into larger soil structural units or 'peds' in the soil. Soil structure reflected by bulk density and porosity, are critical in determining the provision of nutrients, water and air in soil. Bulk density reflects the ability of the soil to function for structural support, water and solute movement, and soil aeration. It allows concentration data of elements to be presented on a mass and/or volume basis. This is important for expressing soil physical, chemical and biological measurements on a volumetric basis to allow the assessment of ecosystem services and for comparisons between management systems.

Bulk density has been found to increase with depth due to lower organic matter content, compaction due to the weight of the overlying layers, and less aggregation and rooting (Foth 1978). While a low bulk density is considered optimum for the movement of air and water through the soil, a high bulk density can restrict root growth and can result in poor movement of water and air through the soil. The bulk density of most cultivated soils is about 1.3 g/cm^3 but values may range in the extreme from 0.5 g/cm^3 in ash soils to 2.0 g/cm^3 for very compacted lower horizons (Fitzpatrick 1986).

During the field campaign of the Irish SIS project, bulk density data were collected from 246 profile pits at horizon level. In total 2950 horizons across 809 soil profiles across Ireland were described. Some data were not available for the previous AFT survey data and a comparison of datasets was conducted to develop pedotransfer functions (PTF) for the estimation of bulk density where these values were absent (Reidy et al. 2016). Universal kriging was applied in R software to produce the final indicative maps (Figs. 4.35 and 4.36). An interpolation approach based on data points to mapping bulk density was therefore preferred over mapping according to soil type.

Table 4.2 Example of calculations and criteria for the selection of the depth for association 1030b from Creamer et al. (2016) where auger bore observations are intersected with soil map to obtain the National Soil Association delineated for the site of the auger bore observation

Association Unit 1030b						
	Series 1	Series 2	Series 3	Series 4	Series 5	Depth (cm)
Series component	Rathowen	Ladestown	Straffan	Elton	Banagher	>80
Modal profiles	RPO17BR01	AFT_Legacy_253	RPO19BR02	RPS36RC18	RPH80BR01	
Maximum depth of modal profile	120	60	75	120	80	
Num. auger bores	12	0	18	11	9	

The auger bore soil series is compared to the soil series member of the association. The maximum depth of the modal profiles for each soil series of association is checked

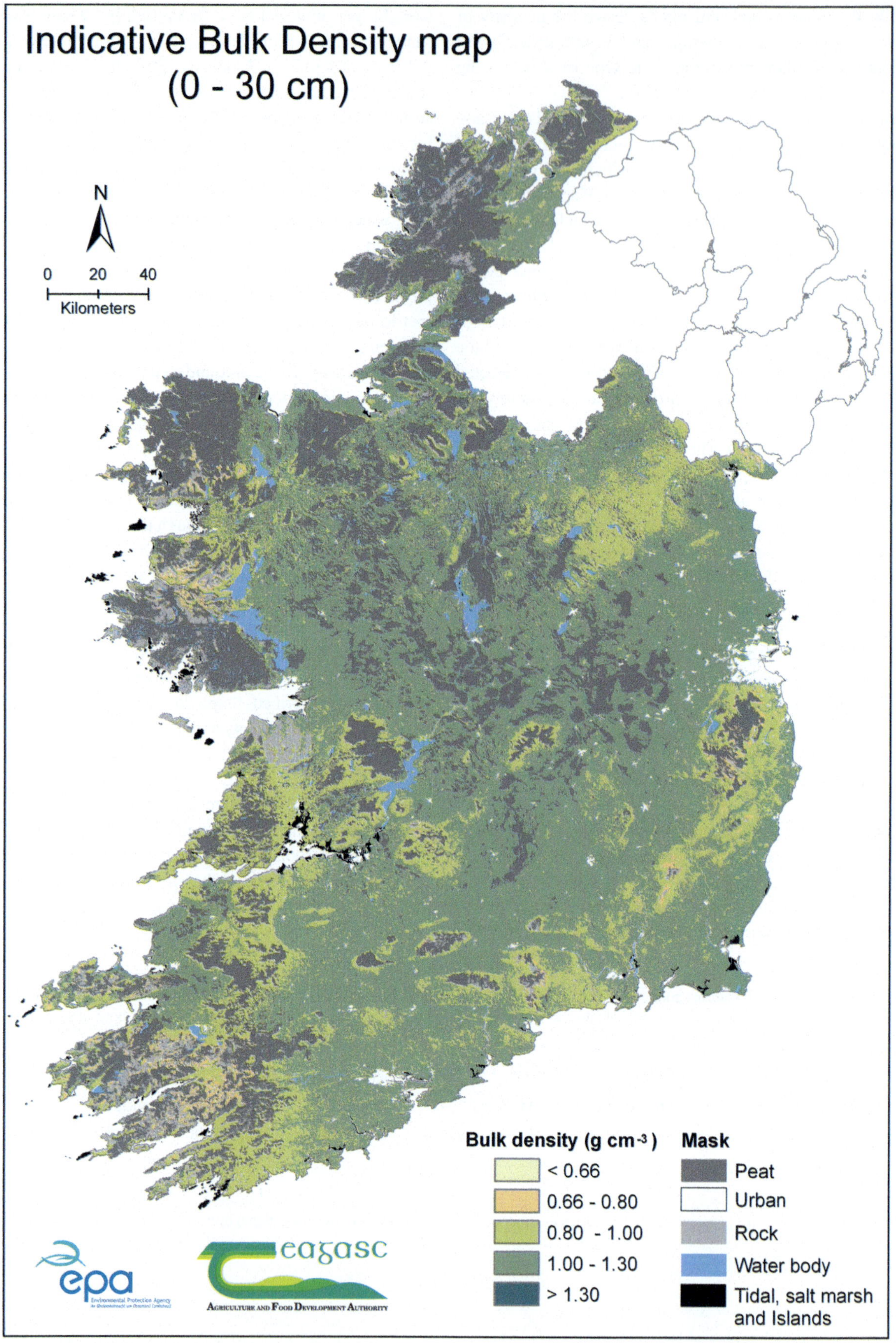

Fig. 4.35 Indicative bulk density map 0–30 cm from Creamer et al. (2016) developed using DSMM including universal kriging. Data based on measured and PTF derived bulk density data measured and calculated as part of the Irish SIS project

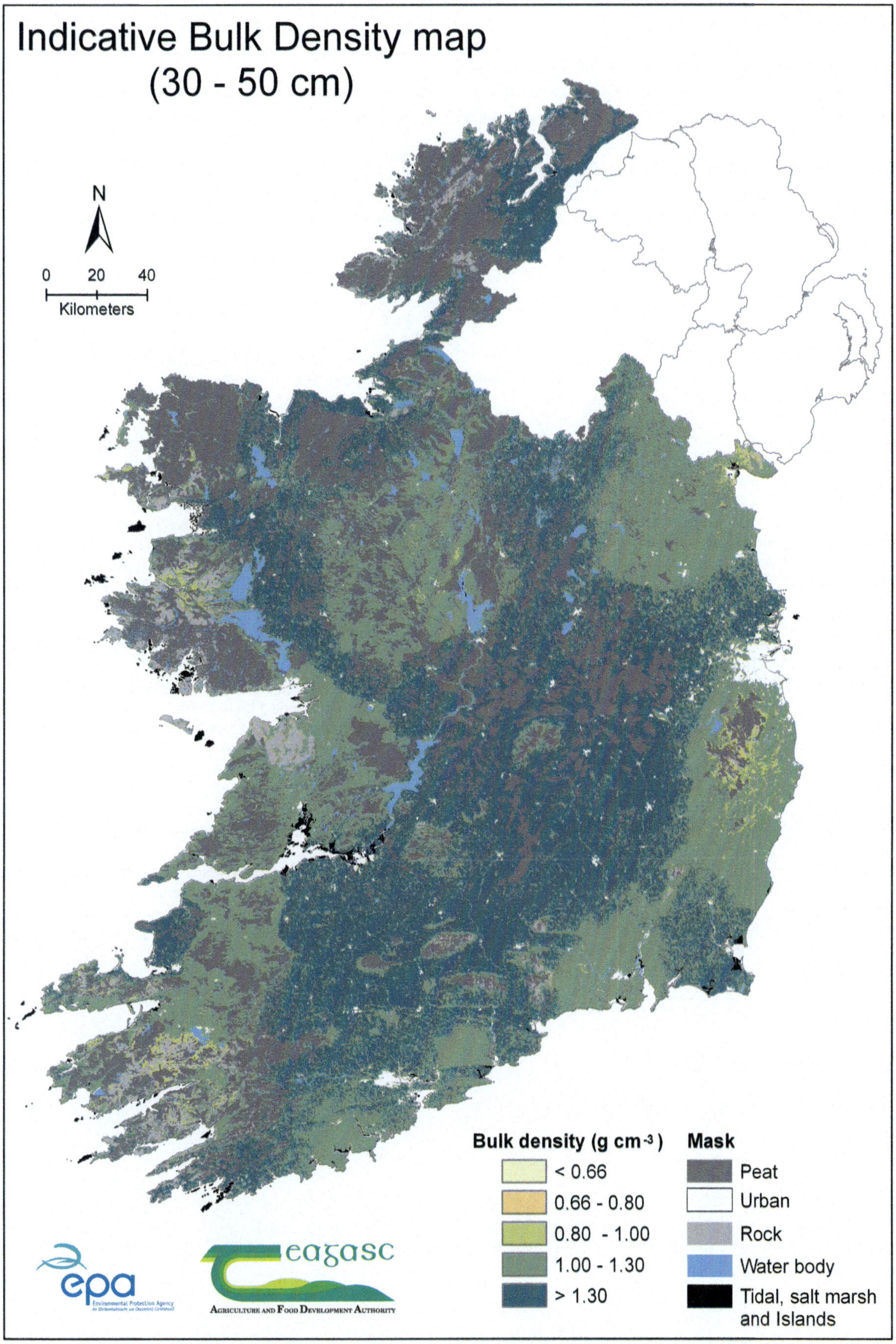

Fig. 4.36 Indicative bulk density map 30–50 cm from Creamer et al. (2016) developed using DSMM including universal kriging. Data based on measured and PTF derived bulk density data measured and calculated as part of the Irish SIS project

4.8 Conclusion

The 3rd Edition National Soil Map of Ireland was developed at a scale of 1:250,000 using a novel approach while also incorporating the valuable detailed legacy soil survey work. Allied with the national soil map, the Irish SIS online database represents the first national soils database with functionality to be interrogated electronically. This accessibility has undoubted potential, particularly in enabling better understanding of the natural environment in Ireland.

References

Brady N (1974) The nature and properties of soils, 8th edn. Macmillan Publishing Co., Inc., New York

Brus DJ, Kempen B, Heuvelink GBM (2011) Sampling for validation of digital soil maps. Eur J Soil Sci 62:394–407

Creamer RE, Simo I, Reidy B, Carvalho J, Fealy R, Hallett S, Jones R, Holden A, Holden N, Hannam J, Massey P, Mayr T, McDonald E, O'Rourke S, Sills P, Truckell I, Zawadzka J, Schulte RPO (2014) Irish Soil Information Systems. Synthesis Report (2007-S-CD-1-S1). EPA STRIVE Programme, Wexford. Available for download from http://erc.epa.ie/safer/reports

Creamer RE, Simo I, O'Sullivan L, Reidy B, Schulte RPO, Fealy RM (2016) Irish Soil Information Systems: Soil properties map Final Technical Report 18. EPA STRIVE Programme, Wexford. Available for download from http://erc.epa.ie/safer/reports

Daly K, Fealy R (2007) Digital soil information system for Ireland (2005-S-DS-22), Final Report. Environmental RTDI programme 2000–2006, Environmental Protection Agency, Johnstown Castle, Wexford

Doran JW (2002) Soil heath and global sustainability: translating science into practice. Agr Ecosyst Environ 88:119–127

ESRI (2016) Polygon. http://support.esri.com/en/knowledgebase/GISDictionary/term/polygon. Last accessed: 19 Apr 2016

FAO/IIASA/ISRIC/ISS-CAS/JRC (2009) Harmonized World Soil Database (version 1.1). FAO, Rome, Italy and IIASA, Laxenburg, Austria

Fealy RM, Green S, Loftus M, Meehan R, Radford T, Cronin C, Bulfin M (2009) Teagasc EPA soil and subsoils mapping project-final report, vol 1. Teagasc, Dublin

Fitzpatrick EA (1986) An introduction to soil science, 2nd edn. Wiley, New York

Food and Agriculture Organization (FAO) FAO/UNESCO (2016) Soil Map of the World. http://www.fao.org/soils-portal/soil-survey/soil-maps-and-databases/faounesco-soil-map-of-the-world/en/. Last accessed: 20 Apr 2016

Foth HD (1978) Fundamentals of soil science. Wiley, New York

Gardiner MJ, Radford T (1980) Soil associations of Ireland and their land use potential explanatory bulletin to soil map of Ireland 1980. National Soil Survey of Ireland, An Foras Talúntais, Dublin

Hammond RF (1981) The Peatlands of Ireland. Soil Survey Bulletin 35. An Forás Talúntais, Dublin, Ireland

Hodgson JM (1976) Soil Survey Field Handbook: Describing and Sampling Soil Profiles. Volume 5 of Soil Survey Technical monograph, Soil Survey of England, Harpenden, UK 98pp

International Union of Soil Sciences (IUSS) (2006) World Reference Base for Soil Resources 2006. World Soil Resources Reports (103). FAO, Rome

ISSS Working Group RB (1998) World Reference Base for Soil Resources: Atlas (Bridges EM, Batjes NH, Nachtergale FO, Eds). ISRIC-FAO-ACCO, Leuven

Jones RJA, Hannam JA, Palmer RC, Truckell IG, Creamer RE, McDonald E (2011) Irish SIS Final Technical Report 1: Interim classification, Harmonisation and Generalisation of County Soil Maps of Ireland. EPA STRIVE Programme, Wexford. Available for download from http://erc.epa.ie/safer/reports

Jones RJA, Hannam JA, Palmer RC, Truckell IG, Creamer RE, McDonald E (2014) Irish SIS Final Technical Report 3: Soilscapes for Terra Cognita Ireland. EPA STRIVE Programme, Wexford. Available for download from http://erc.epa.ie/safer/reports

Karlen DL, Mausbach MJ, Doran JW, Cline RG, Harris RF, Schuman GE (1997) Soil quality: a concept, definition, and framework for evaluation. Soil Sci Soc Am J 61:4–10

Killham K (1994) Soil ecology. University Press Cambridge, United Kingdom

McBratney AB, Santos MLM, Minasny B (2003) On digital soil mapping. Geoderma 117(1–2):3–52

Montanarella L, Jones RJA (1999) The European Soil Bureau. In: Bullock P, Jones RJA, Montanarella L (eds) Soil resources of Europe. EUR 18991 EN 202 pp. Office for Official Publications of the European Communities, Luxembourg

Reidy B, Simo I, Sills P, Creamer RE (2016) Pedotransfer functions for Irish soils—estimation of bulk density (ρb) per horizon type. Soil 2:25–39

SIS—Irish Soil Information Systems (2014) Online resource. http://gis.teagasc.ie/soils/index.php. Last accessed: 22 Apr 2016

Thomasson AJ (1975) Soil properties affecting drainage design. In: Soils and Field Drainage. Soil Survey Technical Monograph No. 7, 80pp

Landscapes: An Introduction

5

Lilian O'Sullivan and Rachel Creamer

Abstract

The approach to describing the landscapes of Ireland is based upon earlier works to describe landscapes of Ireland. The delineation of the landscapes is developed using a digital elevation model in tandem with the Irish Soil Information Systems data, using ArcGIS, geographic information system software. Mountain areas are those that occur above an elevation of 500 m above sea level, while lowlands, are those areas found at an elevation of less than 150 m. Unlike other volumes in this World Soil Series, Ireland is a relatively small island with a strong influence of glaciation, which has resulted in significant changes to the landscape. These glacial influences are clearly defined across a range of physiographic and altitudinal zones. In the lowland areas the physiographic influences are related to the influence of parental material (either limestone or siliceous) or as a result of the influence of groundwater table. Peatlands are an important soil across the Irish landscape, and while these soils appear in other chapters, these soils have a dedicated chapter. Owing to rapid urbanisation in Ireland in the period 2000–2006 the final landscape chapter explores urban soils as urbanisation represents both direct and indirect impacts for the soil resource.

L. O'Sullivan (✉) · R. Creamer
Teagasc Agriculture and Food Development Authority, Johnstown Castle Environment Research Centre, Wexford, Ireland
e-mail: Lilian.OSullivan@teagasc.ie

L. O'Sullivan
Irish Environmental Protection Agency, Johnstown Castle, County Wexford, Ireland

R. Creamer
Wageningen University and Research Atlas—Wageningen Campus, Wageningen, The Netherlands

Keywords

Landscapes • Mountain • Hill • Peat • Lowlands
Urban

5.1 An Introduction to the Landscapes of Ireland

In the following Chaps. (6–12), we apply the same physiographic regions as applied in earlier works to describe the landscapes of Ireland. As with Gardiner and Radford (1980) and Creamer et al. (2014), elevation is used as a key physiographic feature upon which the soils are grouped. Accordingly, mountain areas are described as those that occur at an elevation above 500 m with lowlands as those areas below an elevation of 150 m. Delineation of landscapes has been achieved by use of a digital elevation model in tandem with the Irish Soil Information Systems using ArcGIS, a geographic information system software. The mountain landscapes are mostly found in the western seaboard counties; however some exceptions include the Wicklow Mountains in the east of Ireland. While the spatial extent of mountain landscapes in Ireland is low, much lower than landscapes of similar geological origin in Great Britain or Scandinavia, these soils differ distinctly from soils found at lower elevations requiring a differentiated landscape description. These implicit differences are associated with differences in pedogenesis and as a result, modern day functionality can be associated with differences in land use and management. The colder, wetter conditions found at higher elevations favour peat formation. While blanket peats are the dominant soil at this elevation, peatlands are an extremely important soil across the Irish landscapes generally. The Blanket Peats of the western seaboard are often referred to as climatic peats with high rainfall and humidity driving their accumulation (Gardiner and Radford 1980). More generally, peatlands in Ireland have enormous societal importance for the many social, cultural and ecosystem

R. Creamer and L. O'Sullivan (eds.), *The Soils of Ireland*, World Soils Book Series,
https://doi.org/10.1007/978-3-319-71189-8_5

services that they provide and so, while they appear across many of the landscape chapters, they are described separately in Chap. 8. These conditions also favour leaching, driving podzolisation processes in this landscape. Hill landscapes described here refer to the upland areas between 150 and 500 m above sea level with the majority of the hill locations concentrated in the southern half of the country, particularly in the uplands of Munster.

Ireland is mostly comprised of lowlands (<150 m) and the midlands of Ireland are made up of a central plain of carboniferous limestone giving rise to the limestone soils that dominate the central lowlands of Ireland. Altogether, the limestone lowlands of Ireland account for approximately 1.6 million hectares (M ha), almost one-quarter of the land area (23.2%). Peat is a prominent feature at this lowland elevation and is readily found interspersed with these limestone soils where the flat gradient and soils characterised by a higher clay percentage result in a slower drainage gradient inducing conditions that are favourable for peat formation.

Unlike other volumes in this World Soil Series, Ireland is a relatively small island with a strong influence of glaciation, which has resulted in significant changes to the landscape. These glacial influences are clearly defined across a range of physiographic and altitudinal zones. In the lowland areas the physiographic influences are related to the influence of parental material (either limestone or siliceous) or as a result of the influence of the groundwater table. The drumlin belt of the north midlands of Ireland that stretches across parts of Northern Ireland into the Republic of Ireland is one of the most extensive drumlin belts in Europe (described in Chap. 11). Glaciation has had a profound influence on the soils of Ireland with the majority of Irish soils derived from glacial deposits. As a result, mapping and describing Irish soils represents a complex task, for example, in one soil survey, of the 11 Great Soils Groups in Ireland, nine were found in one field in Co. Cork.

Finally, the landscape chapters conclude with a description of urban soils (Chap. 12). Urbanisation has been rapid in Ireland in the period 2000–2006 and so attention must be paid to the impact that this represents both directly and indirectly for the soil resource.

References

Creamer RE, Simo I, Reidy Carvalho J, Fealy R, Hallett S, Jones R, Holden A, Holden N, Hannam J, Massey P, Mayr T, McDonald E, O'Rourke S, Sills P, Truckell I, Zawadzka J, Schulte RPO (2014) Irish soil information systems. Synthesis report (2007-S-CD-1-S1). EPA STRIVE Programme, Wexford. Available for download from http://erc.epa.ie/safer/reports

Gardiner MJ, Radford T (1980) Soil associations of Ireland and their land use potential explanatory bulletin to soil map of Ireland 1980. National Soil Survey of Ireland, An Foras Talúntais, Dublin

6 Mountain Landscapes

Lilian O'Sullivan, Brian Reidy, and Rachel Creamer

Abstract

The landmass of Ireland is low-lying, particularly the midlands, with mountain areas mostly found around the fringes of the country. The mountain landscape, with its main soil associations and soil landscape units (Gardiner and Radford 1980; Creamer et al. 2014), is defined as that which exceeds 500 m elevation above sea level. It occupies approximately 58,000 hectares (ha) which is much lower than in landscapes of similar geological origin in Great Britain or in Scandinavia. Peats, especially blanket peats are widespread in this landscape and account for half of the mountain soils (51.1%). Blanket peat, podzols and brown podzolics occur as the typical catena sequence at this elevation. Extensive areas of rock outcrops occur (almost 30%) which can be attributed to the increased exposure to erosion and climate associated with this elevation. Rock outcrops are interspersed with histic lithosols and podzols that frequently occur in this landscape. Agricultural land use is typically limited to extensive livestock grazing of sheep and cattle. Although limited from an agricultural perspective, these landscapes represent an important resource providing a habitat for a wide variety of species. Within the mountain landscapes, five of Ireland's six national parks are represented.

Keywords

Podzol • Brown Podzolic • Lithosols • Uplands

6.1 Mountain Landscapes

In general, the landmass of Ireland is low-lying, particularly the midlands, with mountain areas mostly found around the fringes of the country. The mountain landscape, with its main soil associations and soil landscape units (Gardiner and Radford 1980; Creamer et al. 2014), is defined as that which exceeds 500 m elevation above sea level. It occupies approximately 58,000 hectares (ha) (Fig. 6.1) which is much lower than in landscapes of similar geological origin in Great Britain or in Scandinavia. Peats, (Chap. 8) especially blanket peats are widespread in this landscape and account for half of the mountain soils. Blanket peat, Podzols and Brown Podzolic soils occur as the typical catena sequence at this elevation. Extensive areas of rock outcrops occur which can be attributed to the increased exposure to erosion and climate associated with this elevation. Rock outcrops account for almost 30% of the area and are found interspersed with Histic Lithosols and Podzols that frequently occur in this landscape.

County (Co.) Wicklow is home to the most extensive mountain area above 500 m in the country, accounting for almost one-third (29.62%) of this landscape. The geology of these mountains includes granite, gneiss and schist sometimes with glacial drift. Over two-thirds of the mountains in Co. Wicklow are characterised by blanket peat. Histic Lithosols are the next most commonly found soils, and while some rock outcrop occurs, it is not as prevalent as in Co. Kerry. After Co. Wicklow, Co. Kerry has the second most extensive mountain region in the country. In particular, the south of the county is characterised by rugged hills and mountains, including Carrauntoohil which is the highest mountain in Ireland standing at 1038 m above sea level. Carrauntoohil is the central peak of the Macgillycuddy's Reeks range on the Iveragh Peninsula. Other mountain

L. O'Sullivan (✉) · R. Creamer
Teagasc Agriculture and Food Development Authority, Johnstown Castle Environment Research Centre, Wexford, Ireland
e-mail: Lilian.OSullivan@teagasc.ie

L. O'Sullivan
Irish Environmental Protection Agency, Johnstown Castle, County Wexford, Ireland

B. Reidy
Teagasc Moorepark, Animal and Grassland Research and Innovation Centre, Fermoy, County Cork, Ireland

R. Creamer
Wageningen University and Research
Atlas—Wageningen Campus, Wageningen, The Netherlands

R. Creamer and L. O'Sullivan (eds.), *The Soils of Ireland*, World Soils Book Series,
https://doi.org/10.1007/978-3-319-71189-8_6

Fig. 6.1 Images of upland areas taken in Co. Donegal. The presence of ericaceous vegetation is seen in the bottom two images, with the low growing perennial *Calluna vulgaris* (heather) obvious in the foreground of the bottom image. The presence of frequent rock outcrop is also seen which is typical at higher elevations on steeper slopes where erosion and exposure limit potential for soil development

landscapes in the county are found along the Dingle Peninsula. Further south, few of the upland areas reach an altitude in excess of 500 m, and so the Shehy Mountains on the border between counties Cork and Kerry represent the mountain landscape of Co. Cork. The mountains in counties Cork and Kerry date from the Devonian Period and are mainly comprised of Old Red Sandstone (Pracht and Sleeman 2002). Other mountain landscapes along the western seaboard can be found in counties Mayo, Galway, Sligo, Leitrim and Donegal. In Co. Mayo, the mountain areas occur in the west of the county and include Mweelrea Mountain, which is the highest mountain in Connacht. In Co. Sligo, the Dartry Mountains, found on the border with Co. Leitrim, are generally formed of cherty limestones on top of mudstones dating from the Carboniferous Period when Ireland was covered in shallow seas (MacDermott et al. 1996). In Co. Donegal, the Derryveagh Mountains are the main mountain landscape and make up a large portion of the county. There, the area is underlain with granite geology found with a band of quartzite schists (Fay et al. 2007), including the scree covered Errigal which is the highest of the 'Seven Sisters' mountain chain. The highest sea cliffs in Europe are also found in the county. In the south-east, Mount Leinster, the highest peak of the Blackstairs Mountains in Co. Wexford, and the Comeragh Mountains in Co. Waterford can be found. Other mountains are scattered across the country, such as the Slieve Blooms on the Laois–Offaly border and Slievenamon and the Galtymores (The Galtees) in Co. Tipperary (Fig. 6.2).

6.2 Soils of the Mountain Landscape

As described in Chap. 2, soils occur along a landscape and reflect a catena sequence. As such, the majority of Great Soil Groups found in mountain landscapes are blanket peats, Lithosols or Podzols as shown by the areas above the dashed line in Fig. 6.3. Further downslope, Brown Podzolic soils occur but can also be found at high elevations.

In Ireland, over half of the mountain areas are characterised by rock outcrops which are generally found in association with histic lithosols and podzols. With the exception of peatscapes, the Stagnic Iron-pan Podzol Subgroup that belongs to the Podzol Great Soil Group is the most commonly found soil subgroup across the mountain landscapes (53.6%, Table 6.1). These are soils with peaty topsoil above a gleyed albic (leached) horizon (Eg) that rests directly on a cemented, continuous thin iron-pan (Bf) that impedes vertical water movement (Simo et al. 2014). Higher rainfall coupled with the lower temperatures typical at altitude are conducive to the formation of a surface peaty layer. Based on Van't Hoff's temperature rule, every 10 °C rise in temperature increases the velocity of chemical reactions by a factor of two to three (Jenny 1941). Conversely, every 10 °C drop in temperature will decrease the velocity of chemical reactions and at colder altitudes, microbial activity is slowed down, thereby allowing the accumulation of organic matter. The development of the peaty layer relies on Ericaceae vegetation, which is acidifying in its nature. Ericaceae vegetation, including the heath or heather family, is often found in acid, less fertile growing conditions. This peaty layer releases humic acids which contribute to intensive leaching, which is the washing of iron (Fe), aluminium (Al) and organic matter down the soil profile. Combined, this process produces podzolised soils. The term 'podzol' or ashy soil was defined by the Russian scientist Dokuchaev at the end of the 1800s as a zonal soil of the boreal zone of the coniferous forests with Ericaceae vegetation (Duchaufour 1982). Although Podzols are more frequently found in the colder boreal climates, they can be found further south, but distribution will become more reliant on special site conditions and vegetation (Duchaufour 1982). Full podzol profile development will only occur under acidifying vegetation such as coniferous forestry or Ericaceae (Duchaufour 1982).

Within the Stagnic Iron-pan Podzol Subgroup, the most commonly found associations are 0843f and 0843b. *Glenary* is the series lead of association 0843f which is primarily composed of blanket peat and Podzols with rock outcrops and Humic Lithosols on sandstone bedrock. These may be interspersed with Groundwater Gleys and Surface-water Gleys found on drift occurring with siliceous stones. These are found in the mountain landscapes of counties Kerry and Mayo, and to a lesser extent in Co. Wicklow. Association 0843b dominated by the *Knockastanna* series is comprised of Podzols and Brown Podzolics over shale and slate bedrock and peat. Humic Lithosols on bedrock can also be found within this association, as well as Humic Brown Podzolic soils. The association is found to occur mostly in counties Wicklow, Limerick and Tipperary. The local climate found at this elevation is responsible for the intensively leached nature of these soils. The Knockastanna series is typically characterised by an organic horizon overlying a soil that is generally of low base status with a distinct iron accumulation in the B horizon. The rooting depth of these soils is limited due to the presence of an iron-pan but below this layer the profile is free draining (Finch and Ryan 1966). Association 0843e is another Stagnic Iron-pan Podzol, also with the Glenary lead series. Similar to association 0843f, association 0843e also includes Podzols and Brown Podzolics however, this association does not

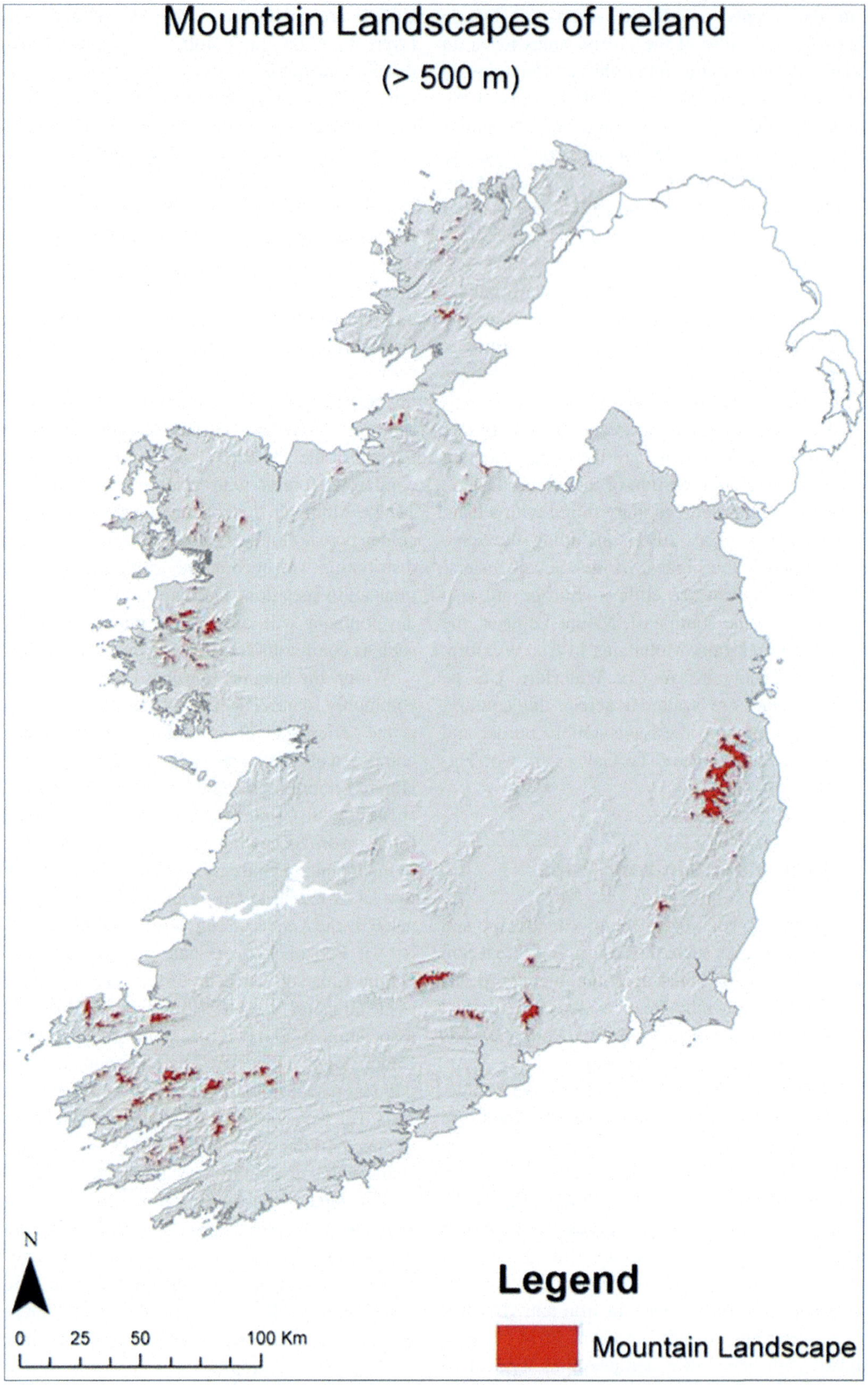

Fig. 6.2 The mountain landscapes of Ireland (in red) in excess of 500 m above sea level. Data based on digital elevation (DEM) by Preston and Mills (2002) and the Irish Soil Information Systems by Creamer et al. (2014). A majority of the higher elevations are found along the western seaboard but some are found in counties Wicklow, Waterford and Tipperary also

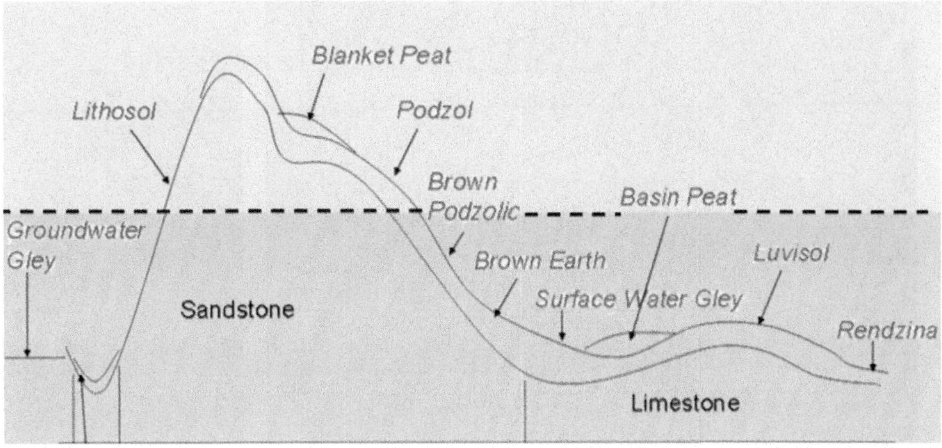

Fig. 6.3 Catena sequence and soil great groups along a landscape adapted from Simo et al. (2014)

Table 6.1 Main associations of the mountain landscapes, association description and the main series included in the association[a]

Main associations (prop. excl. peat, rock outcrops and mask) c11 K ha	Mainly occurs (proportion of association)	Association description	Main series (lead series first)
0843f (40.38%)	Kerry (54.6%), Mayo (37.3%)	Blanket peat and podzols in mountainous areas with outcropping rock and humic lithosols (on sandstones bedrock); interspersed with stagno-podzols and surface water gleys, on drift with siliceous stones	Glenary (0843GL), Puckane (0660PU), Aughty (0110AT), Knockeyon (0460KY), Driminidy (0760DY), Aughty drained (0170ATD)
0410a (31.4%)	Wicklow (87.3%), Dublin (2.95%)	Peat associated with lithosols, brown podzolic, podzols and brown earths over igneous and metamorphic stones, with inclusions of groundwater gleys	Carrigvahanagh (0410CV), Tiknock (0900TN), Knocksquire (1100KQ), Aughty (0110AT)
0410b (12%)	Cork (50.7%), Kerry (45.25%)	Peat associated with lithosols on sandstones and shale bedrock with inclusions of podzols and brown podzolic on sandstones and shale bedrock	Bantry (0410BA), Aughty (0110AT), Knockshigowna (0460KW), Knockeyon (0460KY), Figar (0400FG), Brackloon (0820BK)
0843b (8.42%)	Wicklow (31.9%), Tipperary (22.1%), Limerick (22.1%)	Podzols and brown podzolic over shale and slate bedrock and Peat, with inclusions of humic rendzina and brown earths on bedrock. Loamy over shale bedrock	Knockastanna (0843KS), Cupidstownhill (0900CU), Aughty (0110AT), Borrisoleigh (0960BE), Knockeyon (0460KY)
0843e (4.78%)	Waterford (100%)	Podzols and brown podzolics on drift with siliceous stones and peat. Loamy drift with siliceous stone	Glenary (0843GL), Kiltealy (0900KY), Ross Carbery (0900RO), Knockboy (0960 KB), Aughty (0110AT), Cooga (0900CO), Monavullagh (0843MV)

[a]Full soil descriptions are available at soil series level through the online soils database at http://gis.teagasc.ie/soils/soilguide.php

Fig. 6.4 The top-left shows a Knockastanna series soil profile belonging to the Stagnic Iron-pan Podzol Subgroup that is widely found in the mountain landscapes. The peaty topsoil overlying a leached E horizon is seen sitting on a thin iron-pan layer that impeded water movement. The top right is the Cappacorcoge series, which is a Typical Lithosol with soil profile limited to 30 cm from the surface. The bottom images show the stoney, often shallow nature of soils found at high elevations

include Lithosols, blanket peat or rock outcrops which are more typical at the highest elevations.

Thereafter, Histic Lithosols are the most commonly found Soil Subgroup, and belong to the Lithosol Great Group. These are non-calcareous shallow soils (<30 cm depth) overlying hard rock or skeletal and gravelly material. These are stony soils found frequently with rock outcrops and have a peat surface horizon (>20% organic carbon) that is less than 40 cm thick (Simo et al. 2014). Association 0410a is the most extensively found and has the *Carrigvahanagh* series lead. The majority of this association can be found in the Wicklow Mountains and while it can be found elsewhere, it is to a much lesser extent. This association can be described as peat associated with Lithosols, Brown Podzolics, Podzols and Brown Earths over igneous and metamorphic stones. Wet soils, Groundwater and Surface-water Gleys may occur where the slopes are less steep. Association 0410b is the most commonly found with *Bantry* as the series leader. These are also Histic Lithosols found with Podzols and Brown Podzolics but these differ from the Carrigvahanagh series in relation to parent material. These Lithosols are found on sandstones and shale bedrock, and are almost exclusively found in counties Cork and Kerry (Fig. 6.4).

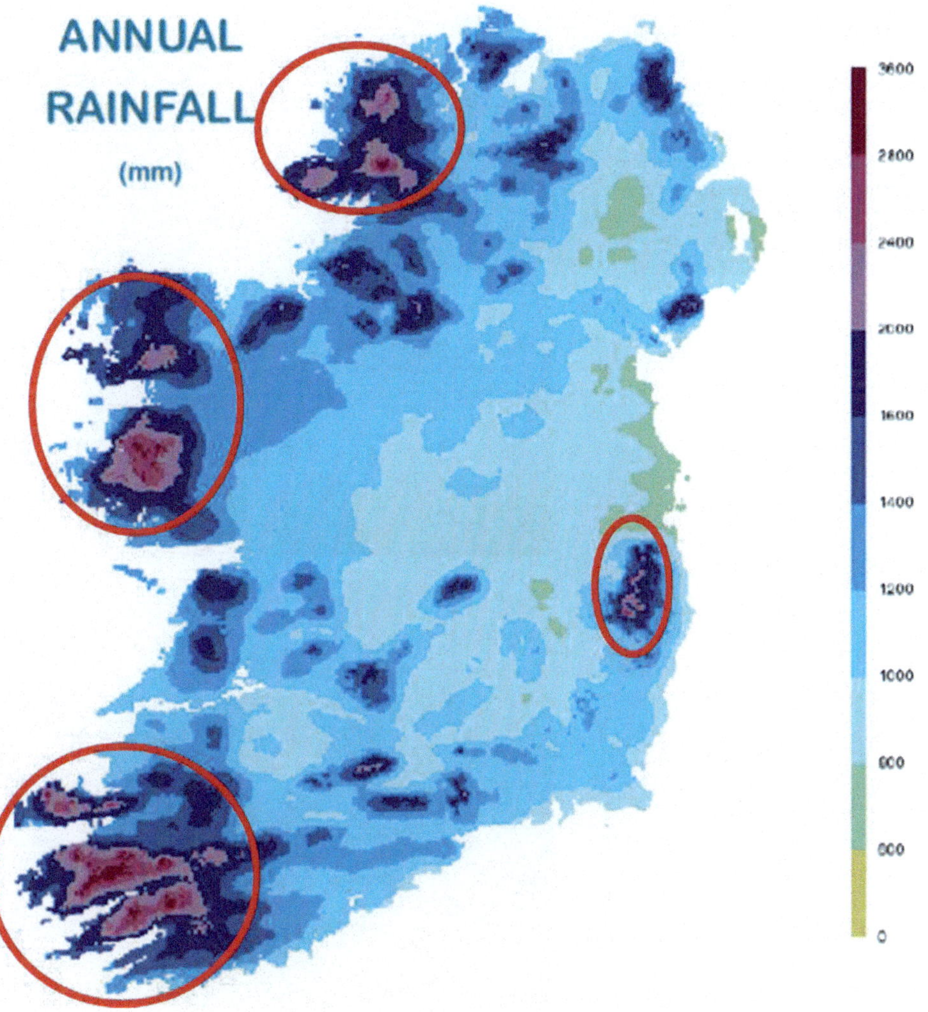

Fig. 6.5 Mean annual rainfall ranges for Ireland from 1981 to 2010 by Walsh (2012) based on data from Met Éireann, the Irish Meteorological Service. The average annual rainfall averaged over all Ireland is approximately 1230 mm. Rainfall increases significantly with height, with amounts over 3000 mm experienced on the highest mountains in the south-west, west and north-west. The areas that experience highest rainfall are circled in red, and occur in the western half of the country and on high ground in the east, on the Wicklow mountains

6.3 Land Use and Management

Mountain soils at high elevations tend to have extremely limited use range from an agricultural perspective. Rough grazing represents the major agricultural land use with some coniferous forestry plantations, where possible. Although mountain areas have been used as rough grazing for centuries, the soils of the mountain landscape produce vegetation that is of low nutrient status for livestock. These soils often lack many minerals including major elements such as nitrogen, phosphorus, potassium and calcium, and micronutrients including manganese, cobalt, copper and molybdenum (Feehan and O'Donovan 1996). In Soil Survey Bulletin No. 39 related to the Slieve Bloom Mountains, Conry and Hammond (1984) indicate annual rainfall increases of about 200 mm per 100 m rise in elevation, resulting in an increase of between 12 and 17% more wet days in the mountain areas than on the surrounding lowlands. Also, while the Slieve Blooms are extremely wet, these amounts are considerably less than those found on similar mountains in the west of Ireland (Conry and Hammond 1984). There is a large range in annual precipitation in Ireland, as low as 800–900 mm per annum (pa) along the south-east, to over 2000 mm pa in the mountainous areas along the western seaboard (Fig. 6.5) (Creamer et al. 2014). As a result of topography and climate, the soils of the mountainous areas are intrinsically shallow, wet and acidic. The prevalence of rock outcrops and blanket peat interspersed with other soils, further limits the land use potential for agriculture in these areas. Another limitation of the soils in mountain landscapes relates to the slopes found at these elevations. The steepness of the slopes renders the use of tractors or other mechanised equipment as dangerous with slopes in excess of 12° frequently found across the mountain landscape. Slopes in this range are considered high, at or above which the use of tractors or other agricultural equipment is less likely (Mockler et al. 2013). While excess soil moisture related to increased precipitation might reduce the number of days at which soil is trafficable, the limited access

Fig. 6.6 Land use at higher elevations is typically limited to rough grazing such as sheep or cattle production as seen in the above images from Co. Kerry. The limited soil development is particularly obvious in the top and central images where frequent rock outcrops are likely interspersed with shallow Lithosols. *Photo* Karen O'Neill

of these sites and the danger related to slope gradient, means that the use of tractors in these areas is atypical, regardless of ground conditions (Fig. 6.6).

Although the mountain landscapes of Ireland might highlight limited agricultural potential, they represent a vital resource in relation to the provision of habitats for wildlife and biodiversity. Of the mountain landscapes, approximately 85% are designated areas of protection under the Natura 2000 network. The European Union (EU) Natura 2000 network includes a Habitats Directive (92/43/EEC) (EU 1992), the Birds Directive (2009/147/EC) (EU 2009) and the EU EIA Directive (2011/92/EU) (EU 2012) that seek to halt the loss of biodiversity through the protection of Europe's most valuable and threatened species and habitats (EC 2016). Natura 2000 also encompasses human activities that focus on conservation through sustainable management (EC Commission (EC) 2016). Within the mountain landscapes, five of Ireland's six national parks are represented.

Study Box. 6.1 Uplands of the National Parks—A Varied and Valuable Landscape

The term national park was reserved by the International Union for the Conservation of Nature (IUCN) in 1969 for areas that share certain characteristics including ecosystems where plant and animal species, geomorphological sites and habitats are of special scientific, educational and recreational interest (NPWS 2016). In Ireland, mountain landscapes (as defined) are represented in five of the six national parks, namely Glenveagh, Connemara, Ballycroy, Killarney and Wicklow National Parks. The highest point of the Burren, the sixth national park, only reaches 207 m at Knockanes and does not include mountain landscapes as classified here.

Blanket bogs are a common feature of upland areas in the national parks. Natural blanket bogs are dominated by cotton species (*Eriophorum*), purple moor grass, lousewort, bell heather (*Calluna vulgaris*) and bryophytes including bristly swan neck moss (*Campylopusatrovirens*) and liverwort (*Pleuroziapurpurea*) (Renou-Wilson et al. 2011). In Ireland, the blanket bog habitat is one of the most important habitats and while raised bogs are more at risk in Ireland, blanket bogs are considered a rarer ecosystem at the international level with Ireland having the largest coverage of blanket bogs in Europe (Renou-Wilson et al. 2011). The Liffey Head Bog within Wicklow National Park is one example of an actively growing blanket bog. Overall, few plants and species survive in the blanket bog habitat due to the low levels of oxygen and the acidic nutrient poor conditions. Research by Professor J. J. Moore in the 1950s studied the fungi in two different peats of the Dublin-Wicklow Mountains, a heather peat and a deep deergrass peat. For both, the most common fungi found were various species of penicillium (Feehan and O'Donovan 1996). Penicillium, with its antibiotic properties, is of major importance to the food and pharmaceutical industries today.

In Co. Donegal, the Glenveagh National Park landscape is dominated by peats, with the highest peaks and hilltops exposed due to peat erosion and extensive weathering. Only certain wildlife can endure these harsh conditions with species such as the arctic mountain hare found at the Glenveagh and the Ballycroy National Parks. Bird species are found in these parks that are otherwise rare outside the north-west of Ireland. The golden eagle was reintroduced to Glenveagh National Park through a project that was initiated by the National Parks and Wildlife Service (NPWS) that began in 1989 (O'Toole et al. 2002). Deer carrion is an important food source when grazing is scarce, supporting a variety of species. Red deer is the only native deer species in Ireland. The impact of large herbivores, such as the red deer, in relation to overgrazing of semi-natural woodlands was assessed at Glenveagh, Killarney and Wicklow National Parks (Newman et al. 2014). A significant change in ground flora composition and abundance was found in response to large herbivore exclusion over approximately a 40-year period, as well as a shift towards homogenisation of the vegetation community highlighting the need for management programmes that promote the conservation of a diverse woodland ecosystem (Newman et al. 2014).

These landscapes, while inhospitable for many wildlife species, provide a unique and valuable habitat for other species. Erosion represents an inherent threat for mountain and hill landscapes (Chap. 7) and is one that can be promoted through increased recreational activities or overgrazing. As these areas are evermore recognised for their tourism and recreational values, it is important that they are managed in a sustainable way.

References

Conry MJ, Hammond RF (1984) Soil resources and potential of the slieve bloom mountains and foothills. Soil Surv Bull No. 39. An Foras Talúntais, Dublin

Creamer R, Simo I, Reidy B, Carvalho J, Fealy R, Hallet S, Jones R, Holden A, Holden N, Hannam J, Massey P, Mayr T, McDonald E, O'Rourke S, Sills P, Truckell I, Zawadzka J, Schulte R (2014) Irish soil information synthesis Report (2007-S-CD-1-S1) prepared for the environmental protection agency by Teagasc. Cranfield and University College Dublin, EPA, Wexford

Duchaufour P (1982) Pedology pedogenesis and classification. Translated by Paton TR, Allen &Unwin (Publishers) Ltd, London UK

EU (1992) Council directive 92/43/EEC of 21 May 1992 on the conservation of natural habitats and of wild fauna and flora. European Union. Available at: http://eur-lex.europa.eu/legal-content/EN/TXT/PDF/?uri=CELEX:31992L0043&from=EN. Accessed 10 May 2016

EU (2009) Directive 2009/147/EC of the European parliament and of the council of 30 November 2009 on the Conservation of wild birds (Codified Version). European Union, Available at: http://eur-lex.europa.eu/LexUriServ/LexUriServ.do?uri=OJ:L:2010:020:0007:0025:en:PDF. Accessed 10 May 2016

EU, 2012.Directive 2011/92/EU of the European parliament and of the council of 13 December 2011 on the assessment of the effects of certain public and private projects on the environment (Codification) (Text with EEA relevance). European Union, Available at: http://eur-lex.europa.eu/LexUriServ/LexUriServ.do?uri=OJ:L:2012:026:0001:0021:EN:PDF. Accessed 10 May 2016

European Commission (EC) (2016) Environment natura 2000. Available at: http://ec.europa.eu/environment/nature/natura2000/index_en.htm. Accessed 10 May 2016

Fay D, Kramers G, Zhang C, McGrath D, Grennan G (2007) Soil geochemical atlas of Ireland. In: Kramer G (ed) Teagasc and the Environmental Protection Agency

Feehan J, O'Donovan G (1996) The bogs of Ireland an introduction to the natural, cultural and industrial heritage of Irish peatlands, UCD, The environment institute, Walsh Printers, Roscrea, Co. Tipperary, Ireland

Finch TF, Ryan P (1966) Soils of Co. limerick, Soil Surv Bulletin No. 16, An Foras Talúntais, Dublin 4

Gardiner MJ, Radford T (1980) Soil Associations of Ireland and their land use potential explanatory bulletin to soil map of Ireland 1980. National Soil Survey of Ireland, An Foras Talúntais, Dublin

Jenny H (1941) Factors of soil formation a system of quantitative pedology. Dover Publications Inc., New York

MacDermott CV, Long CB, Harney SJ (1996) A geological description of Sligo, Leitrim, and adjoining parts of Cavan, Fermanagh, Mayo and Roscommon, to accompany the Bedrock Geology 1:100,000 Scale Map Series, sheet 7, Sligo-Leitrim, with contributions by K. Claringbold, D. Daly, R. Meehan and G. Stanley. Geological Survey of Ireland

Mockler E, Bruen M, Desta M, Misstear B (2013) Pathways project final report volume 4: catchment modelling tool. STRIVE Report No. 56, Irish Environmental Protection Agency, Johnstown Castle, Ireland

National Parks and Wildlife Service (NPWS) (2016) National Parks and Wildlife Service, National Parks and Wildlife Service and Department of Arts, Heritage, Regional, Rural and Gaeltacht Affairs. Online: https://www.npws.ie/national-parks. Accessed 10 Oct 2016

Newman M, Mitchell FJG, Kelly DL (2014) Exclusion of large herbivores: Long-term changes within the plant community. For Ecol Manage 321:136–144

O'Toole L, Fielding AH, Haworth PF (2002) Re-introduction of the golden eagle into the Republic of Ireland. Biol Cons 103:303–312

Pracht M, Sleeman AG (2002) A geological description of west cork and adjacent parts of Kerry to accompany the Bedrock Geology 1:100,000 Scale map series, sheet 24, West Cork, With contributions by G. Wright (Groundwater) and W. Cox (Minerals). Geological Survey of Ireland

Preston R, Mills P (2002) Generation of a hydrologically corrected digital elevation model for the Republic of Ireland. Unpublished report submitted to Environmental Protection Agency of Ireland by Compass Informatics as part of the 2000-LS-2.2.2 Fourth Progress Report

Renou-Wilson F, Bolger T, Bullock C, Convery F, Curry J, Ward S, Wilson D, Muller C (2011) BOGLAND: Sustainable management of peatlands in Ireland. STRIVE Report No. 75, Irish Environmental Protection Agency, Johnstown Castle, Ireland

Simo I, Creamer RE, Reidy B, Hannam JA, Fealy R, Hamilton B, Jahns G, Massey P, McDonald E, Schulte RPO, Sills P and Spaargaren O (2014) Irish SIS final technical report 10: soil profile handbook. Associated datasets and digital information objects connected to this resource are available at: Secure Archive For Environmental Research Data (SAFER) managed by Environmental Protection Agency Ireland. http://erc.epa.ie/safer/resource?id=a1b872a2-3b8c-11e4-b233-005056ae0019. Last accessed 09 May 2016

Walsh S (2012) A summary of climate averages for Ireland 1981–2010. *Climatological Note No. 14*. Met Éireann Meteorological Service of Ireland, Dublin

7 Hill Landscapes

Lilian O'Sullivan, Brian Reidy, and Rachel Creamer

Abstract

Hill landscapes described here refer to the upland areas between 150 and 500 m above sea level with majority of the hill locations concentrated in the southern half of the country, particularly in the uplands of Munster. The area captured within the hill landscapes accounts for approximately 1.4 million (M) hectares (ha), of which about 0.5 M ha can be classified as peat soils. The range in altitude defined here results in a much broader portion of the catena being captured, such that the soil associations being described will range from upland hill soils to those found in the foothills. Accordingly, soil types include the Lithosol Great Soil Group typical of upland areas, to Brown Earths and Luvisols that occur further down the catena that may progress naturally into Surface-water Gleys. Histic Lithosols represent one of the main Soil Subgroups found and belong to the Lithosol Great Soil Group which are non-calcareous shallow soils (<30 cm depth) overlying hard rock or skeletal and gravelly material. These stony soils are found frequently interspersed with rock outcrops and have a peaty surface horizon (>20% organic carbon) that is less than 40 cm thick. Typical Brown Earths are commonly found, as are the wetter Stagnic Luvisols that lead naturally into the even wetter Surface-water Gleys. Hill soils capture a wide range in relation to categories of soil suitability and land use. Histic Lithosols exhibit extremely limited agricultural land use potential in comparison to Typical Brown Earths that exhibit few limitations in that regard. Wetter soils, such as Surface-water Gleys, are considered heavy from an agricultural perspective. As they are frequently found on heavy glacial till, they are often characterised by heavy textured subsoil. Typically, grass production is the optimum agricultural use for these soils but good management for grazing is essential to reduce the risk of poaching.

Keywords

Hill landscapes • High nature value • Biodiversity

7.1 Introduction

The hill landscapes described here refer to the upland areas between 150 and 500 m above sea level. Based on this delineation, altogether, the area captured within the hill landscapes accounts for approximately 1.4 million (M) hectares (ha), of which about 0.5 M ha can be classified as peat soils which are described in more detail in Chap. 8. The range in altitude defined here, results in a much broader portion of the catena being captured, such that the soil associations being described will range from upland hill soils to those found in the foothills. Accordingly, soil types include the Lithosol Great Soil Group, typical of upland areas, to Brown Earths and Luvisols that occur further down the catena that may progress naturally into Surface-water Gleys (Fig. 7.1).

L. O'Sullivan (✉) · R. Creamer
Teagasc Agriculture and Food Development Authority, Johnstown Castle Environment Research Centre, Wexford, Ireland
e-mail: Lilian.OSullivan@teagasc.ie

L. O'Sullivan
Irish Environmental Protection Agency, Johnstown Castle, County Wexford, Ireland

B. Reidy
Teagasc Moorepark, Animal and Grassland Research and Innovation Centre, Fermoy, County Cork, Ireland

R. Creamer
Wageningen University and Research
Atlas—Wageningen Campus, Wageningen, The Netherlands

R. Creamer and L. O'Sullivan (eds.), *The Soils of Ireland*, World Soils Book Series,
https://doi.org/10.1007/978-3-319-71189-8_7

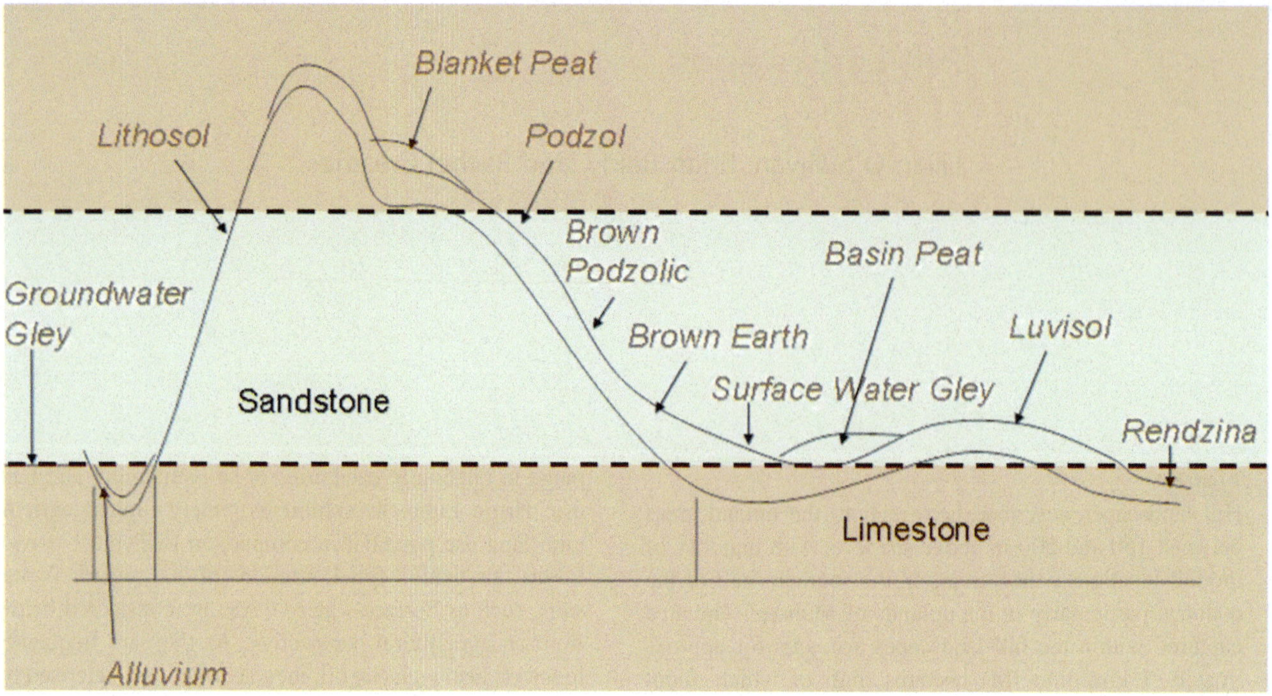

Fig. 7.1 Catena sequence of soil great groups of Ireland across a landscape adapted from Simo et al. (2014) with areas between the dashed lines most typical soil great groups found on the hill landscapes

7.2 Provincial Distribution of the Hill Landscapes of Ireland

The majority of the hill locations are concentrated in the southern half of the country, particularly in the uplands of Munster (Fig. 7.2). County (Co.) Cork has the most extensive hill area, and together with Co. Kerry accounts for almost one-third of the hill landscapes of Ireland. This area includes the Caha Mountains along the Beara Peninsula north of Bantry Bay, the upland areas of the Iveragh and Dingle peninsulas and the Sheehy Mountains that border counties Cork and Kerry. Further north, the Derrynasaggart Mountains provide a natural border between these two counties, while to the east, the Boggeragh Mountains can be found. The geology of these areas is comprised of sandstones, Old Red Sandstone, siltstones and mudstones. The hill relief of Co. Waterford includes the Comeragh and the Knockmealdown Mountain ranges which are also comprised of sandstone bedrock and sedimentary glacial drift. These areas still form part of the Devonian Munster Basin that stretches from Co. Kerry to the Comeragh Mountains in Co. Waterford (Sleeman and McConnell 1995). The hill relief in Co. Waterford reflects elongated east–west valleys with ridges made up of sandstones and mudstones with poorly exposed carboniferous limestones on the valley floors (Sleeman and McConnell 1995). Elsewhere in Munster, the Galtee Mountains can be found spanning three counties: Cork, Limerick and Tipperary, and are characterised by shale bedrock. The Devil's Bit and the Silvermine Hills in Tipperary represent a break in the limestone lowlands of the county and are comprised of upland sandstone, shale and slate bedrock with drift in siliceous stones (Finch and Gardiner 1993). The hill areas of Co. Clare include the Slieve Aughties, Slieve Bernaghs and the Cratloe Hills which can be described as sandstone shale hills associated with poor acid soils and peat (Finch 1971).

In Connacht, counties Mayo and Galway are home to the majority of the hill landscapes in this province. In both counties, upland blanket peat that is commonly cutover can be found interspersed with Stagno-Podzols. While more than 90% of the Sligo–Leitrim area is underlain with sedimentary rocks, the igneous and metamorphic rock groups are also represented (MacDermot et al. 1996). Shared between counties Mayo and Sligo, the upland bog covered Ox Mountains cut by valleys occupied by Lough Talt and Easky Lough are underlain with schists and granites, with rock outcrops less common in the gneiss area to the east (MacDermot et al. 1996). The relief of Co. Leitrim can almost be divided into half by Lough Allen with the table-like mountains alternating with scenic valleys found in the northern half (Walsh 1973). Here, the plateau hills are characterised by shales and sandstones, capped by grits (Dhonau 1973).

In Ulster, the Co. Donegal hill landscape is also predominantly peat, interspersed with some shallow Histic

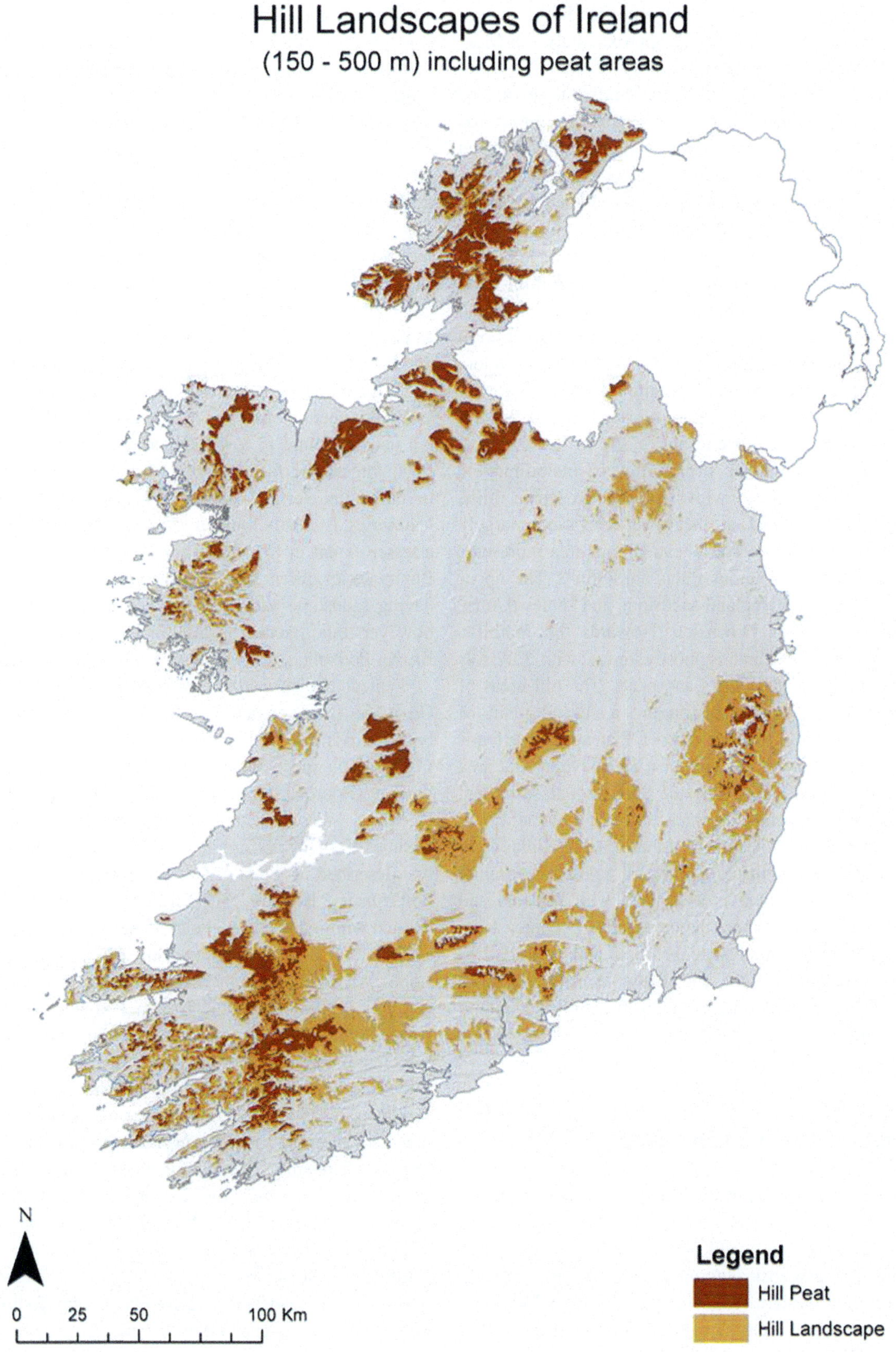

Fig. 7.2 Hill landscapes of Ireland (150–500 m above sea level) including the peat areas at this height range. Data based on 20 m DEM from Preston and Mills (2002) and the Irish Soil Information Systems at 1:250,000 by Creamer et al. (2014). Hill areas are more widespread in the southern half of the country but are also found along the western seaboard fringing the mountain landscapes described in Chap. 6

Lithosols (shallow peat soils) which are often found in rock outcrops. While peat soils occupy approximately 0.5 M ha of the hill landscapes, over one-quarter of this is concentrated in the northwest, in Co. Donegal. Furthermore, outside of Co. Kerry, Co. Donegal has the most extensive proportion of rock outcrops at this altitudinal range. Here, the blanket peat is found mostly over granite and metamorphic rocks. The hill landscapes of counties Cavan and Monaghan belong primarily to the drumlin sequence, where some drumlin features can be found at hill elevation (see Chap. 11).

Similarly in Leinster, in Co. Louth, some drumlin areas can be found at this altitudinal range and are associated with wet soils. In Co. Louth, the hill areas are found in the north of the county at West Feede and in the south around Collon, and are characterised by freely draining shale hills with siliceous drift at the end of the drumlin belt. In Co. Meath, there are few areas that extend above 150 m but the most notable can be found in the northwest including the Loughcrew Hills. Along with well-drained shale hills, limestone can be found at the lower elevations largely associated with Luvisols. The Slieve Bloom area represents the hill landscape of counties Laois and Offaly. The upper areas are dominated by upland sandstone and shales (Conry and Hammond 1984). Downslope, towards the foothills, limestone is prevalent and is mostly found with Luvisols with some Groundwater Gleys and peats. The hill areas of Co. Kilkenny can be found fringing a central plain of limestone lowlands. The Castlecomer Plateau that extends eastwards into Co. Carlow is the most extensive upland area in the county and it, along with the Slieveardagh Hills that extend into Co. Tipperary, is characterised by glacial till of shale, sandstone and limestone. Further east, the uplands of Co. Wicklow are dominated by gneiss, schist and granite bedrock with drift, which is associated with Podzols and Brown Podzolic soils. Where the topography becomes more undulating, the shale and slate bedrock with mixed drift deposits becomes associated with Brown Earths and Luvisols. The Blackstairs Mountains in Co. Wexford that extend into Co. Carlow are similarly characterised by upland gneiss, schist and granite glacial drift.

7.3 Soils of the Hill Landscape

The dominance of peat is reflected along the western seaboard but is particularly obvious in the northwest in counties Donegal, Leitrim, Sligo and Mayo (Figs. 7.2 and 7.3). Excluding peat and rock outcrops, which are also concentrated along the western seaboard, Surface-water Gleys are the most commonly found Great Soil Group. At the higher elevations, Histic Lithosols represent one of the main Soil Subgroups found and belong to the Lithosol Great Soil Group which are non-calcareous shallow soils (<30 cm depth) overlying hard rock or skeletal and gravelly material. These stony soils are found frequently interspersed with rock outcrops and have a peaty surface horizon (>20% organic carbon) that is less than 40 cm thick (Simo et al. 2014). The soil associations with Histic Lithosol lead series include association 0410a that has the 'Carrigvahanagh' series leader and association 0410b that has the 'Bantry' series leader that are described in Chap. 6. The Carrigvahanagh is the more commonly found of these associations and is principally found in Co. Wicklow, with some distributions found in counties Donegal and Dublin also. This association can be described as peat associated with Lithosols, Brown Podzolics, Podzols and Brown Earths over igneous and metamorphic stones. Wet soils, including Groundwater and Surface-water Gleys, can be found where the slopes are less steep. Groundwater Gleys are usually found in low-lying terrain towards the base of hills, whereas Surface-water Gleys tend to be on foot slopes or undulating terrain. Association 0410b, with the Bantry lead series, is mostly found in counties Cork and Kerry and to a lesser extent in Co. Tipperary. The parent material of the Bantry series differs from that of the Carrigvahanagh series, having sandstone and shale bedrock as the parent material; however, this association also includes peats, Podzols and Brown Podzolic soils (Fig. 7.3; Table 7.1).

Typical Brown Earths that belong to the Brown Earths Great Soil Group are amongst the most commonly found soil types across the hill landscapes. Associations 1100a and 1100s are the most extensively occurring. Association 1100a has the 'Clonroche' lead series and can be found in counties Wicklow, Wexford and Kilkenny and to lesser extents in counties Cork, Waterford and Carlow. The Clonroche series, first described in 'Soils of Co. Wexford' which was the first Soil Survey Bulletin, by Gardiner and Ryan (1964), is defined within the Irish soil classification system by Creamer et al. (2014) as a fine loamy soil, over drift with siliceous stones. Overall, these are well-drained soils that occur on undulating land with some Brown Podzolics on the upper slopes. Water-affected soils can also be found within this association including Surface-water Gleys and Groundwater Gleys, the latter being found in landscape depressions. After 1100a, association 1100s is the next most commonly found soil association and is, for the most part, found in counties Cork and Tipperary. Association 1100s with the 'Broomhill' lead series is comprised of Brown Earths that are found to occur with Humic Brown Podzolics mainly on sandstone bedrock or in drift with siliceous stones. Podzols may occur at the higher elevations, while Surface-water Gleys can also occur in this association, typically in the foothills. The Broomhill series is characterised by a friable horizon with a moderate structure, with moderate to deep rooting depth (Diamond and Sills 2011).

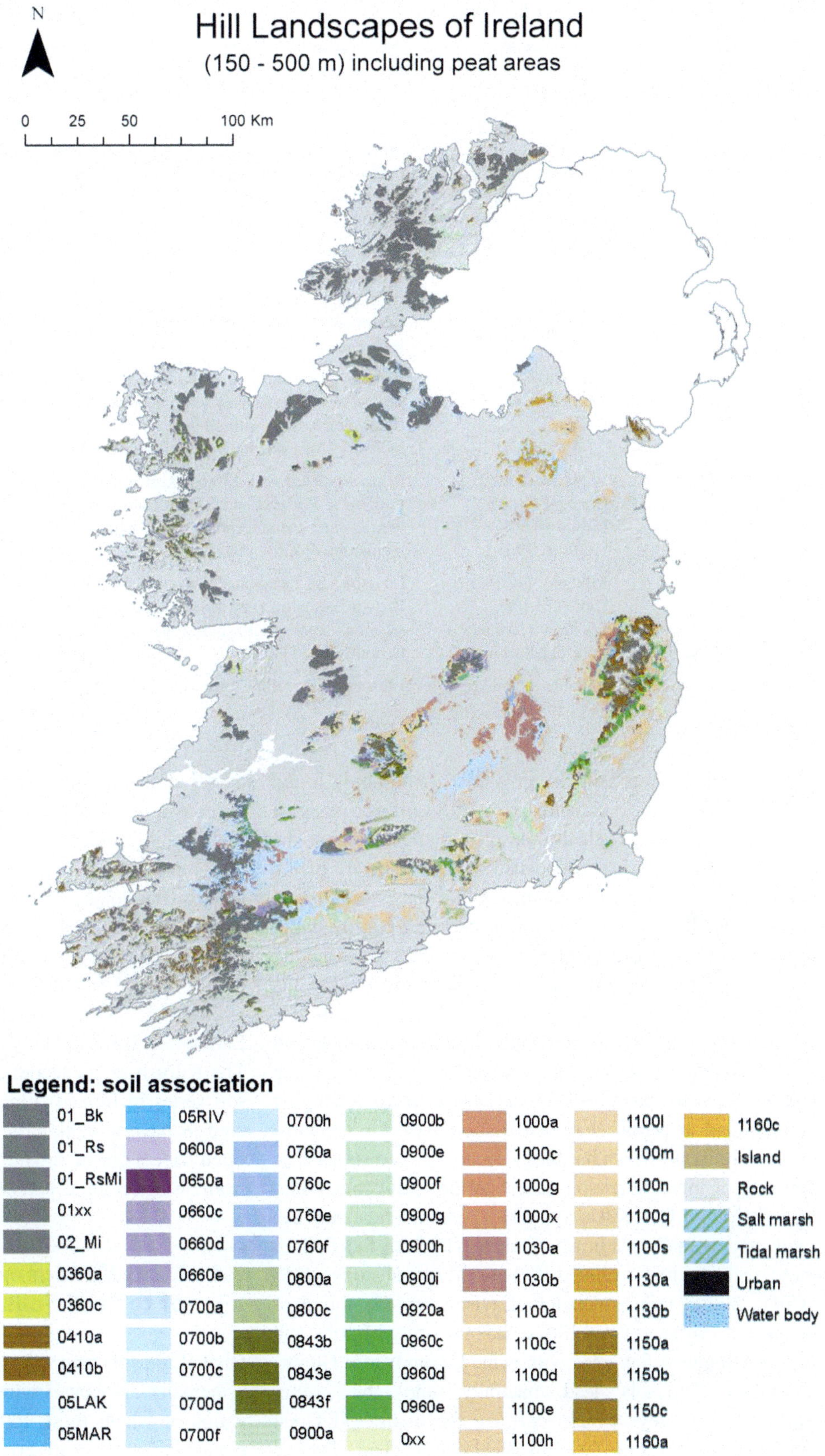

Fig. 7.3 Hill landscapes of Ireland (150–500 m above sea level) including the peat landscapes (01_Bk, 01_Rs, 01_RsMi, 01xx, 02_Mi on legend). Data based on the Irish Soil Information Systems (SIS) 1:250,000 by Creamer et al. (2014) shown with national legend intersected with a 20 m digital elevation model (DEM) from Preston and Mills (2002)

Table 7.1 Main associations of the hill landscapes, association description and the main series included in the association[a]

Main associations (prop. excl. peat, mask, rock outcrop) c800 K ha	Mainly occurs (proportion of association)	Association description	Main series (lead series first)
0700b (15.54%)	Cork (43%), Tipperary (21.2%), Kerry (15.4%), Limerick (8.9%)	Surface-water Gleys, Brown Earths and Brown Podzolics on (fine loamy) drift with siliceous stones and inclusions of Groundwater Gleys	Kilrush (0700KR), Gortaclareen (0760GN), Ballylanders (1100BY), Cupidstownhill (0900CU), Moord (1130MO)
1100a (7.25%)	Wicklow (46.1%), Kilkenny (16.5), Wexford (15.6%), Cork (5.45%)	Well-drained Brown Earths on (fine loamy) drift with siliceous stones in an undulating land with some Brown Podzolics on upper slopes. Surface-water Gleys and Groundwater Gleys found in depressions	Clonroche (1100CL), Ballylanders (1100BY), Kilrush (0700KR), UNnamed_29 (0900UN29), Dunboyne (1000DB)
1100s (7.2%)	Cork (50%), Tipperary (45.39%)	Brown Earths and Brown Podzolics mainly on sandstones bedrock and some occurring on (fine loamy) drift with siliceous stones, with inclusions of Podzols and Surface-water Gleys	Broomhill (1100BL), Clonroche (1100CL), Rathkenny (0960RK), Drumslig (0800DS), UNnamed_29 (0900UN29)
0410a (6.67%)	Wicklow (57.9%), Donegal (10.8%), Dublin (8.69%), Louth (6.9%)	Peat associated with Lithosols, Brown Podzolics, Podzols and Brown Earths over igenous and metamorphic stones, with inclusions of Groundwater Gleys	Carrigvahanah (0410CV), Tiknock (0900TN), Knocksquire (1100KQ), Aughty (0110AT)
1030a (5.9%)	Kilkenny (47.1%), Laois (19.1%), Wicklow (13.3%), Cork (11.8%)	Luvisols, Surface-water Gleys and Stagnic Brown Earths on (fine loamy) drift with siliceous stones, with inclusions of Groundwater Gleys	Crosstown (1030CO), Crossabeg (1020CB), Johnstown (1020JH), Laughil (1030LG), Dungarvan (1000DG)

[a]Full soil descriptions are available at soil series level through the online soils database at http://gis.teagasc.ie/soils/soilguide.php

The Luvisol Great Soil Group is also a commonly found soil type across the hill landscapes. Luvisols are soils in which clay material from the topsoil has moved down the soil profile and has been deposited in deeper layers of the soil. The higher clay content means that these soils retain nutrients well; however, the concentration of clay at depth means that they are moderate to imperfectly draining depending upon the soil texture. Stagnic Luvisols are the most commonly found Soil Subgroup found within this Great Soil Group in hill landscapes. These are Luvisols that display stagnic properties, due to water stagnation occurring below 40 cm for parts of the year, as a result of the higher clay content at depth. Here, association 1030a is the most commonly found which has the 'Crosstown' lead series. The majority of this association occurs in the Castlecomer Plateau region of Co. Kilkenny, which also extends into Co. Laois and a small area in Co. Carlow. The Irish SIS defines these soils as fine loamy drift with siliceous stones (SIS 2014). Within this association, Luvisols can be found to occur with Surface-water Gleys, Stagnic Brown Earths and in lower lying or flat areas with some Groundwater Gleys.

Stagnic Luvisols lead naturally into the wetter Surface-water Gleys which represent the main Soil Great Group found across the hill landscapes. The Typical Surface-water Gley Subgroup occurs in counties Cork, Tipperary, Kerry and Limerick. Surface-water Gleys are poorly draining soils due to a slowly permeable subsurface layer as a result of higher clay content or dense material in the subsoil. This results in waterlogging of the top 40 cm for prolonged periods of the year. These soils tend to have fine texture (fine loamy or clayey) in the lower subsoil horizons and typically show prominent mottling above 40 cm (Jones et al. 2011). The most commonly found soil association is 0700b which has the 'Kilrush' lead series. Within this association, Surface-water Gleys can be found with Brown Earths and Brown Podzolics on drift with siliceous stones. The Kilrush series accounts for one of the most widespread soil series in Ireland (Finch 1971). These soils are considered to have a weak structure, with slow permeability and often have a plastic or sticky consistency described as heavy from an agricultural perspective (Finch 1971).

7.4 Land Management and Suitability of the Hill Soils

Histic Lithosols tend to have a very limited use range. This is principally associated with the slope, topography and elevation at which these soils occur. These factors limit the potential for mechanised agriculture and as a result, the spreading of nutrient amendments and fertilisers is restricted. The shallow nature of these soils means that they have limited rooting depth. As a result, these soils are generally limited to grass production (Finch and Gardiner 1993). These soils mostly occur on steeper slopes in the hill areas of counties Cork, Kerry and Wicklow.

Overall, Typical Brown Earths have a friable consistence, are well drained and have relatively good nutrient retention. The free-draining nature of these soils means that they enjoy an extended grazing season with limited risk of poaching or compaction. They tend to have a high available water capacity and altogether have a wide use range proving suitable for a range of arable crops and grass production systems. The Clonroche soil series has a wide use range and is very suited to tillage with a medium pH. This soil has a friable consistence and strong structure making it one of the best tillage soils in Ireland (Diamond and Sills 2011). Their good available water capacity contributes to their high yield potential. These soils were originally classified by An Foras Talúntais (AFT) in 1964 in Clonroche in Co. Wexford in the first Soil Survey Bulletin (AFT 1964) where they are analogous to the soils of north Co. Wexford where a range of soft fruits, root crops and cereals are grown (Diamond and Sills 2011). The Broomhill soil series is also considered very suitable for arable and grass production with the potential grass yield estimated as being very high (Diamond and Sills 2011). The consistence of the soil allows cultivation over a long period in spring and the soil is suited for both winter and spring sown crops (Diamond and Sills 2011). Finally, in this Soil Subgroup, the Ballylanders soil series is similar to the Clonroche series, observing a wide use range with no particular limitations. The natural lime and nutrient status of these soils is considered low but these soils respond well to nutrient amendments (Finch and Ryan 1966). These soils, while suitable for grass and arable production, are considered particularly favourable for grass production, especially for short-term highly productive leys, and if managed well can sustain a high stocking density with limited hazard of poaching damage (Finch and Ryan 1966). Slope is not found to be particularly steep for the majority of the Typical Brown Earths found across the hill landscape. Tractor safety becomes an issue at slopes greater than 12° (Mockler et al. 2013). Accordingly, these soils are overall suitable for agricultural production based on their structure and physical location. Irish agriculture is dominated by grasslands and the majority of the Brown Earths at this elevation are typically used for grasslands.

Although Stagnic Luvisols tend to have conditions similar to those of Surface-water Gleys, they tend to not be as wet for long periods, and so management of these soils is slightly easier than that of Surface-water Gleys. These soils are usually found mid-slope and with their fine loamy texture tend to have a wider use range than Surface-water Gleys. As these soils are mostly found on the Castlecomer Plateau, it is unsurprising that the relatively flat upland terrain of this area is not limited by slope from a soil management perspective. These soils can be useful for a number of tillage crops and pasture, responding well to management. The majority of these soils are used for grass-based farming systems.

Surface-water Gleys are considered heavy from an agricultural perspective. As they are frequently found on heavy glacial till, they are often characterised by heavy textured subsoil. This results in poor drainage making them difficult to work with, which means that these soils have a limited use range. Typically, grass production is the optimum agricultural use for these soils but good management for grazing is essential to reduce the risk of poaching. The appropriate timing of operations is essential to ensure that the use of farm machinery does not result in damage such as compaction to these soils given their weak structure. The Kilrush series are considered to have high clay content but their poor internal drainage means that these soils are generally not suited for tillage (Finch and Ryan 1966). With good management, these soils can support good pasture production. Overall, poor drainage, weak structure and heavy texture represent the main limitations associated with these soils (Finch 1971). Again, these areas are mostly used for grass-based production systems, but pockets of coniferous forestry (Figs. 7.4 and 7.5) can be found, as well as areas designated as Natura 2000 sites.

Case Study 7.1 Hill Landscapes—Policy and Management Interactions

Soils in upland areas, (mountain and hill landscapes) are more exposed to soil erosion, especially along the western seaboard. As well as climatic conditions and topography, historical land management activities have been associated with a loss of vegetation, soil degradation and subsequent erosion. Although erosion occurs naturally, erosional features such as teracettes (ridges that appear on slopes) are invariably exploited by livestock disturbance, in particular by sheep (Clavin 2000; Guinan 2005). In Ireland, agriculture on hill and mountain landscape areas is generally characterised by extensive livestock production systems. Notably, these areas are not limited to agricultural use and can accommodate multiple land uses including conservation, tourism, fuel harvesting and forestry.

In 1975, under EEC Directive 268/75, headage payments were introduced for mountain and hill farming, along with other certain less-favoured areas (LFA). The aim of the scheme was to compensate farmers in LFAs to ensure a reasonable level of income through payments for livestock including sheep (Heritage Council 1999). In line with the LFA Directive (268/75 EEC), the scheme aimed to

Fig. 7.4 Coniferous plantations are often found on hill slopes such as can be seen in the top image. Hilly areas are widely used for extensive production systems, in particular, livestock production of cattle and sheep as seen in the bottom image

'conserve the countryside by the prevention of further depopulation of rural areas' (Government of Ireland 1995). An unintended consequence of the scheme was an increase in overgrazing which accelerated erosion, thereby resulting in the deterioration of the landscape with some negative environmental impacts. An investigation into the aspects of physical degradation of selected Irish hillsides highlighted widespread physical degradation in these environments that had intensified over the 25-year study period, much of which could be related to the livestock management systems then practiced (Guinan 2005). In 2003, a restructuring of payments resulted in a major decline in sheep numbers from 4.7 to 2.5 million ewes over the period 1992–2015 (CSO 2016). At one hill site in the west of Ireland grazed by sheep, trends in relation to ground cover including vegetation and soil were examined over a 14-year period from 1995 to 2008 (Walsh et al. 2016). The reduced ewe stocking rate (0.8 ewes/ha) with appropriate management both facilitated growth of the semi-natural vegetation and increased soil stability (Walsh et al. 2016).

Importantly, these areas often represent an important value from a biodiversity perspective and provide unique habitats for a diversity of species of flora and fauna. Measures, including the designation of EU Natura 2000 sites, can provide some protection and sustainable management. Increasingly, it is recognised that alone the designation of sites will not be sufficient to meet European biodiversity goals and that targeting

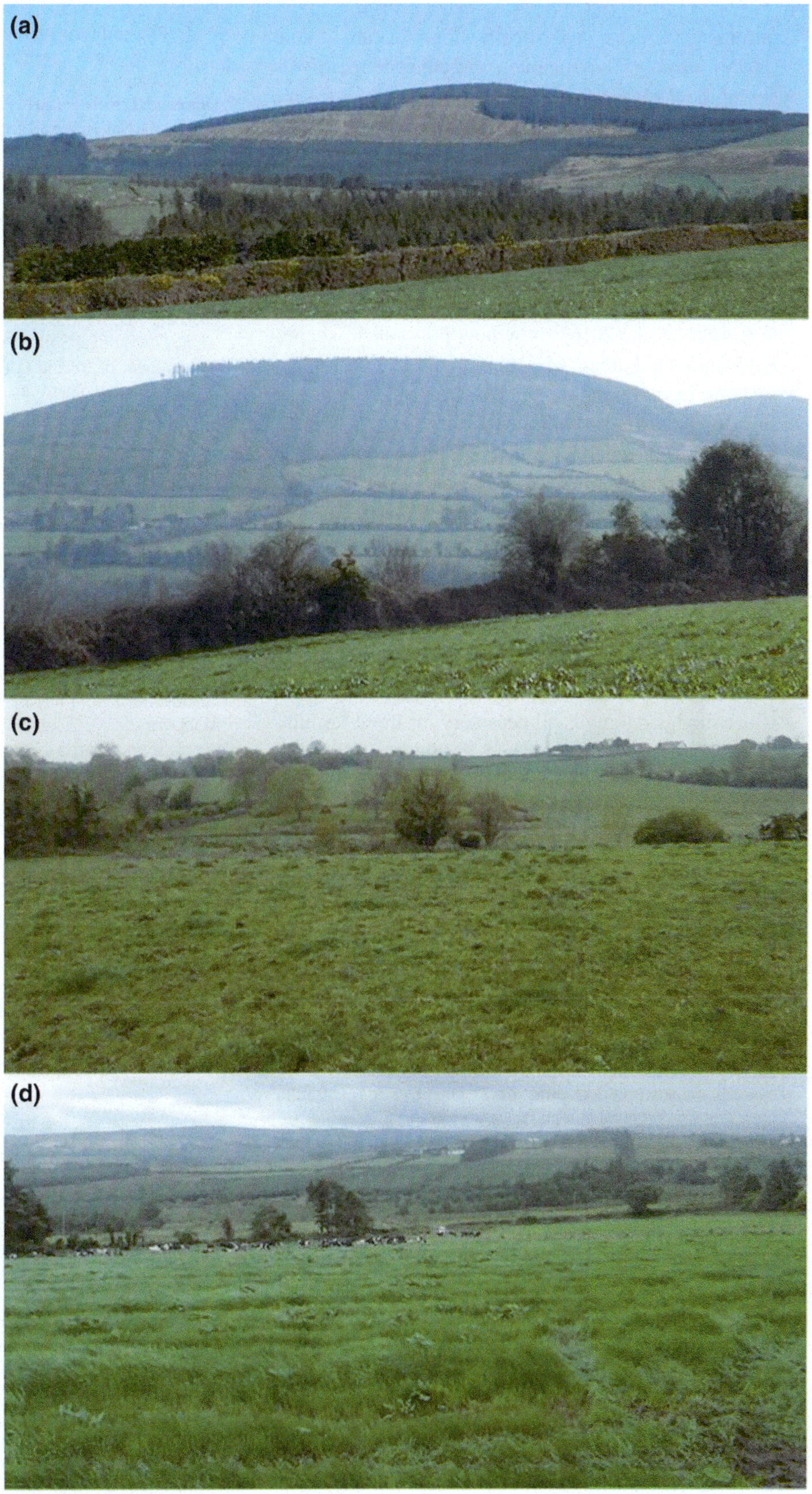

Fig. 7.5 **a** Hill slopes of the Wicklow Mountains visible in the background. These soils may be low in pH due to gneiss and schist parent materials and soils are best suited to forestry which can be seen as the major land use in the background. In many upland areas, commonage is typical and sheep grazing is employed. **b** Also in Co. Wicklow, these granitic hills similarly suffer from low pH and again forestry is prominent in the upper reaches. The grassland in the foreground requires dock (*Rumex obtusifolius*) control measures. **c** Image of limestone drift, bedrock and outcrops from Co. Galway. The hill highlights a complex shape and as a result restricts management due to extremes in small areas. Beef grazing is employed in the area and heavy poaching can be seen in the foreground with some parts effectively scrub. **d** This image from the Boggeragh Mountains in Co. Cork shows improved grassland facilitated by the shale and sandstones parent materials with low inclines. However, high rainfall can promote leaching and nutrient-poor soils could develop on steeper inclines resulting in more afforested areas without intense management effort. Some mechanical compaction is visible in right foreground

farming practices, specifically high nature value (HNV) farming, that maintain and enhance the diversity of the landscapes, habitats and species is necessary (Beaufoy and Cooper 2009; O'Rourke et al. 2016). The concept of HNV farming goes beyond the classic opposition between agricultural production and ecological richness and emphasises the interdependence and synergies of nature and society (Plieninger and Bieling 2013; O'Rourke et al. 2016). Using case studies, including one of the uplands of Co. Kerry, O'Rourke et al. (2016) show how pastoralism shapes HNV in upland areas. Also, how traditional pastoralist systems have responded to EU CAP highlighting a shift towards single-species specialisation and intensification, all of which have marked impacts on vegetation, similarly described above. Despite extensification measures of EU agri-environment policies, a dual process of livestock farming intensification and extensification often on the same farm or abandonment is found to be occurring in many upland areas of Europe (O'Rourke et al. 2016). To meet EU biodiversity targets, a better targeting, evidence base for HNV farmland, as well as better integration with the wider rural economy is all necessary for these farming practices to be maintained into the future (O'Rourke et al. 2016). Active management and monitoring of socio-economic and socio-ecological factors are vital to the development of HNV systems in this fragile environment (Walsh et al. 2016).

References

Beaufoy G, Cooper T (2009) The application of the high nature value impact indicator 2007e2013. (Guidance Document). European evaluation network for rural development. http://ec.europa.eu/agriculture/rurdev/eval/hnv/guidance_en.pdf

Central Statistics Office (CSO) (2016) Livestock and farm numbers. Databases Statbank, online resource: http://www.cso.ie/px/pxeirestat/Statire/SelectVarVal/Define.asp?maintable=AAA01&PLanguage=0. Accessed 08 Oct 2016

Clavin DJ (2000) Terracette studies in the hill and mountain environment of Western Ireland with emphasis on the role of livestock. M. Agr. Sc. Thesis (unpublished). University College Dublin

Conry MJ, Hammond RF (1984) Soil resources and potential of the Slieve Bloom Mountains and Foothills. Soil survey bulletin No. 39, An Foras Talúntais, Dublin

Creamer R, Simo I, Reidy B, Carvalho J, Fealy R, Hallet S, Jones R, Holden A, Holden N, Hannam J, Massey P, Mayr T, McDonald E, O'Rourke S, Sills P, Truckell I, Zawadzka J, Schulte R (2014) Irish Soil Information Synthesis Report (2007-S-CD-1-S1) EPA STRIVE Programme, Wexford. Available for download from http://erc.epa.ie/safer/reports

Dhonau NB (1973) Chapter IV Geology solid geology in county legitim resource survey Part 1—Land use potential (soils, grazing capacity and forestry) soil survey bulletin no. 29, An Foras Talúntais, Dublin 4

Diamond J, Sills P (2011) Soils of Co. Waterford soil survey bulletin no. 44 National Soil Survey of Ireland, Teagasc

Finch TF (1971) Soils of county clare soil survey bulletin no. 23 National Soil Survey of Ireland, Report compiled by Finch, TF, Culleton, E and Diamond S, An Foras Talúntais

Finch TF, Gardiner MJ (1993) Soils of tipperary north riding soil survey bulletin no. 42. National Soil Survey of Ireland, Teagasc

Finch TF and Ryan P (1966) Soils of co. Limerick soil survey bulletin no. 16. National Soil Survey of Ireland, An Foras Talúntais

Gardiner MJ, Ryan P (1964) Soils of Co. Wexford soil survey bulletin no. 1. National Soil Survey of Ireland, An Foras Talúntais, Dublin 4

Government of Ireland (1995) Operational programme for agriculture, rural development and forestry, 1994–1999. Stationery Office, Dublin

Guinan LM (2005) Studies investigating the physical degradation of selected Irish hillsides: plant-herbivore interactions, digital photogrammetry and climate. A dissertation submitted to the National University of Ireland, University College Dublin, in fulfilment for the degree of Doctor of Philosophy, Dublin, 390 pp

Heritage Council (1999) Impact of agriculture schemes and payments on aspects of Ireland's heritage, The Heritage Council

Jones RJA, Hannam JA, Palmer RC, Truckell IG, Creamer RE, McDonald E (2011) Irish SIS final technical report 1: interim classification, harmonisation and generalisation of county soil maps of Ireland. EPA STRIVE Programme, Wexford. Available for download from http://erc.epa.ie/safer/reports

MacDermot CV, Long CB, Harney SJ (1996) A geological description of Sligo, Leitrim, and adjoining parts of cavan, fermanagh, mayo and roscommon, to accompany the bedrock geology 1:100,000 scale map series, sheet 7, Sligo—Leitrim, with contributions by Claringbold K, Daly D, Meehan R and Stanley G. Geological Survey of Ireland

Mockler E, Bruen M, Desta M, Misstear B (2013) Pathways project final report volume 4: catchment modelling tool. STRIVE report no. 56. Irish Environmental Protection Agency, Johnstown Castle, Ireland

O'Rourke E, Charbonneau M, Poinsot Y (2016) High nature value mountain farming systems in Europe: case studies from the Atlantic Pyrenees. Fr Kerry Uplands Irel J Rural Stud 46:47–59

Plieninger T, Bieling C (2013) Resilience-based perspectives to guiding high-nature-value farmland through socioeconomic change. Ecol Soc 18(4):20

Preston R, Mills P (2002) Generation of a hydrologically corrected digital elevation model for the Republic of Ireland. Unpublished report submitted to environmental protection agency of Ireland by Compass Informatics as part of the 2000-LS-2.2.2. Fourth progress report

Simo I, Creamer RE, Reidy B, Hannam JA, Fealy R, Hamilton B, Jahns G, Massey P, McDonald E, Schulte RPO, Sills P, Spaargaren O (2014) Irish SIS final technical report 10: soil profile handbook. Associated datasets and digital information objects connected with this resource are available at: Secure Archive For Environmental Research Data (SAFER) managed by the Environmental Protection Agency Ireland

SIS—Irish soil information systems (2014) Online resource. http://gis.teagasc.ie/soils/index.php. Last accessed 22 Apr 2016

Sleeman AG, McConnell B (1995) Geology of East Cork—waterford a geological description of East Cork, waterford and adjoining parts of tipperary and limerick to accompany the bedrock geology 1:100,000 scale map series, sheet 22, East Cork—waterford with contributions by Claringbold K, O'Connor P, Warren WP and Wright G. Geological Survey of Ireland

Walsh M (1973) Chapter III General description of the area in county leitrim resource survey part 1—land use potential (Soils, grazing capacity and forestry) Soil survey bulletin no. 29, An Foras Talúntais, Dublin 4

Walsh M, Hanrahan JP, O'Malley L, Moles R, (2016) Trends, over 14 years, in the ground cover—vegetation, bare soil and boulder/rock—on an unimproved western hill grazed by sheep and associated animal performance. Irish J Agri Food Res 55(1):47–62

Peatlands

8

Florence Renou-Wilson

Abstract

Peatlands form part of the iconic landscape of Ireland and account for nearly a quarter of the land area. Raised bogs, blanket bogs and fens are Ireland's last great area of wilderness, hovering between land and water, providing unusual habitats for its unique and specialist flora and fauna. At the same time, the peat itself is a significant carbon store (C) and a repository of treasures as well as critical information to understand past climates and land uses. Historically, all natural peatlands have been impacted by humans through our efforts to 'utilise' this land, through draining, burning, afforestation, grazing and other 'improvements'. Nowadays, peatlands are found mainly drained either for agriculture, forestry or peat extraction or utilised for wind farms or recreational purposes. All of these land uses dry the bog and result in a loss of biodiversity at all levels; genetic, species, habitats and landscapes; a loss of water and mineral storage capacity leading to increased flooding and water pollution. Also, a loss of C sequestration capacity and the loss of stored C to the atmosphere and waterways, as well as an increased risk of bog burst and fires and the spread of invasive species can result. There is now a growing national and global interest in peatland restoration and in ending non-sustainable uses of peat by bringing back the 'sustainable' services and benefits that peatlands provide to society as a whole.

Keywords

Peatland • Organic soil • Raised bog • Blanket bog
Fen • Sustainable management • Greenhouse gas emissions

F. Renou-Wilson (✉)
School of Biology and Environmental Science,
University College Dublin, Belfield, Dublin 4, Ireland
e-mail: Florence.renou@ucd.ie

8.1 Introduction

Peatlands have been in the Irish landscape since the last Ice Age and, together with a remnant of primaeval forests, they form our oldest natural heritage. Irish peatlands are the country's last great area of wilderness, hovering between land and water, providing unusual habitats for its unique and specialist flora and fauna. Peatlands cover a large area of the land surface, occurring as raised bogs, blanket bogs or fens and forming cultural landscape icons in many parts of the country (e.g. Connemara, Ox Mountains and Slieve Bloom). Peatlands have accumulated peat over millennia, creating an important economic raw material, on which the livelihoods of certain rural populations have critically depended. This accumulated peat mass makes peatlands a fascinating historical archive of past environmental and cultural change. More importantly in view of recent climate change, peatlands store very large amounts of carbon (C) that would be released to the atmosphere should the peatland degrade, for instance, when a peatland is drained. Peatlands are the most space-effective C stores of all terrestrial ecosystems (Dise 2009). Over centuries, peatlands slowly remove and store more C than they release and, therefore, they exert a net cooling effect on the global climate (Frolking et al. 2006). Once degraded, this process is reversed. Despite the many benefits provided by peatlands, these ecosystem services have generally remained unnoticed. This has resulted in a lack of appreciation of the need for cautious management to maintain this precious resource.

Peatlands and Irish people have been closely connected by a long history of cultural and economic development. In the distant past, peat landscapes were both feared and respected as wilderness areas and often linked to traditional culture, rituals and worship (Feehan et al. 2008). In modern times, peatlands have been afforested, cutover by domestic turf-cutting, cutaway by industrial peat extraction, eroded by overgrazing and agricultural reclamation, damaged by infrastructural developments and invaded by alien

R. Creamer and L. O'Sullivan (eds.), *The Soils of Ireland*, World Soils Book Series,
https://doi.org/10.1007/978-3-319-71189-8_8

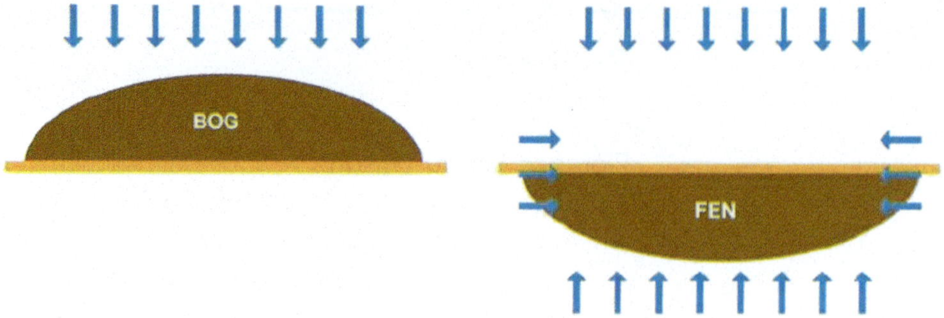

Fig. 8.1 Schematic representation and classical different between 'bog' and 'fen'. Brown = peat; arrow = water flow. Adapted from Joosten (2008)

non-native species. Moreover, climate change poses a further threat to the survival of these natural ecosystems (Belyea and Malmer 2004; Jones et al. 2006).

This chapter will review how Irish peatlands are a significant element in Irish nature and a valuable ecological resource before assessing their past and current utilisation from a land management and environmental perspective.

8.2 Peatlands, Peat Soils and Peat

Peatlands are wetland ecosystems that are characterised by the accumulation of organic matter called peat which derives from dead and decaying plant material under high water saturation conditions. Peat accumulates where the production of plant material exceeds decay. Water is the most important factor limiting decay. Irish peatlands were formed in the last 10,000 years (since the last ice age) and have developed under climatic conditions whereby precipitation exceeded evapotranspiration or in areas of impeded drainage. In natural (those with little human impact) peatlands, peat accumulates at a rate of about 0.5–1 cm per year (or 5–10 m over 10,000 years) with strong local and temporal variations (Clymo 1992, Charman 2002). Peat is defined as material that has passively accumulated consisting of at least 30% (dry mass) of dead organic material. A peat soil contains organic material (peat) over a depth of at least 40 cm (30 cm if undrained); the depth requirement does not apply in the event that the peat layer is directly over bedrock (Renou-Wilson et al. 2011a, b; Simo et al. 2014). A peatland is further defined as a geographical area (with or without vegetation) where peat soil occurs naturally (i.e. all classes of peat-covered terrain). For mapping purposes, a peatland should cover a minimum spatial extent of 1 ha. The properties of Irish peat soils vary widely due the range of peat type encountered throughout the island and their associated anthropogenic disturbances. Surface peats have been found to have a pH as low as 3.3 and bulk density as low as 0.05 g cm^3 (Hammond 1981, Renou-Wilson et al. 2018a). However basal fen peat found at the bottom of raised bog, above a layer of marl [1] can reach pH above 6.

There are different types of peatlands depending on geographic region, terrain and vegetation type. In Ireland, the major distinction is between bogs and fens. Bogs only receive nutrients from precipitation and thus are nutrient-poor whereas fens (which receive nutrients from the surface or groundwater as well as precipitation and tend to be more calcium and nutrient rich, although poor fens also exist). The difference between a bog and a fen is represented in Fig. 8.1. The key characteristics of a peatland include waterlogged conditions, development of specific vegetation and the consequent formation and storage of peat (defined below). The interconnection and interdependencies between water, plants and peat is critical to the survival of the peatland and, therefore, make them vulnerable to a wide range of disturbances.

8.3 Extent and Profiles of Fens, Raised Bogs and Blanket Bogs

While Hammond (1981) estimated that peatlands covered a total area of 1.17 million ha or 17% of the area of the Republic of Ireland, a multi-source data map of peat soils (i.e. which may or may not have the original peat vegetation cover) estimated the national cover at 1.46 million ha or 20.6%. This map excluded small areas (<7 ha) and fens and, therefore, is still likely to be an underestimate (Connolly and Holden 2009) and is being constantly refined as new remote sensing platforms become available. More recent estimates based upon the Irish Soil Information Systems indicate that peatlands occupy as much as 1.7 million ha, roughly 25% of the area (SIS 2014). There have been several different schemes proposed for the classification of peatlands in Ireland. The accepted one is Hammond (1981), in which three types are distinguished: fens, raised bogs and blanket bogs (Fig. 8.2). Blanket bogs can be further subdivided into two categories: Atlantic blanket bogs and mountain blanket bogs.

[1]Marl: a calcareous limnic deposit formed in fresh water lakes by Charophyceae (Stone-worts) permeated by variable quantities of shells of fresh water molluscs; found, beneath the raised bogs in the Irish Midlands.

Fig. 8.2 The common frog (*Rana temporaria*) on a raised bog in the Irish Midlands with the carnivorous sundew (*Drosera rotundifolia*), bog asphodel (*Narthecium ossifragum*), bog cranberry (*Vaccinium oxycoccos*) and ling heather (*Calluna vulgaris*)

The Irish habitat classification (Fossitt 2000) includes an additional type identified as 'cutover bogs' and used in situations where part of the original mass of peat has been removed through turf-cutting or other forms of peat extraction. Cutover bogs can be associated with all peat-forming systems, including fen peats, raised bogs and blanket bogs. However, the terms 'peatland' and 'bog' are not interchangeable.

8.3.1 Fens

Fens are peatlands that are formed from vegetation receiving a constant influx of base-rich groundwaters and, therefore, can be described as minerotrophic (fed by groundwater). Fen peats in Ireland usually have a relatively high pH but some remain acidic with a pH ranging from 4.5 to 8.0 (Doyle and Críodáin 2003). Fen peats are mineral-rich with relatively high ash content (10–20%) and a relatively shallow peat depth (*c.* 2 m). The vegetation is generally species rich and largely dominated by sedges (*Carex* species), rushes (*Juncus articulatus* and *J. effusus*) and tall herbs (*Epilobium palustre* and *Galium palustre*) and with brown mosses a feature of the ground layer. *Schoenus nigricans* forms a characteristic sward in one particular type of rich fen still found across the country. There is a notable absence of *Sphagnum* species. In the Irish climate, fen can be seen as a transitional ecosystem *en route* to a raised bog. However, they are rarely seen to progress in this natural direction due to human-induced disturbances, be it reclamation for agriculture, road works or land-filling. Natural fens are rare, as 97% of the country's fens have been drained for agriculture (Foss et al. 2001). While fens of conservation importance still occur right across the country, their current extent is estimated at 20,180 ha only (Foss 2007).

8.3.2 Raised Bogs

Raised bogs are found mainly in the Irish midlands under a rainfall regime of between 750 and 1000 mm per year. The peat surface is strongly acid, but beneath this layer, glacial alkaline soils of limestone origin can be found. Raised bogs are dominated by *Sphagnum* mosses, with heather (*Calluna vulgaris* [L.] Hull), bog cotton (*Eriophorum angustifolium* Honckeny), and several species of sundew (*Drosera* spp.) and orchids (e.g. *Dactylorhiza incarnata*) can also be found. Raised bog development started at the end of the last glacial period 10,000 years ago, when shallow lakes left by the retreating ice covered much of central Ireland. Nutrient-rich groundwater derived from calcareous glacial drift, fed these lakes (Mitchell and Ryan 1997). Reeds and sedges encroached around the lake edges; their remains fell into the water where they were only partly decomposed, in time forming a thick layer of reed peat. The upward growth of sedges and other plants (brown moss spp. as well *Sphagnum* mosses) and the accumulation of their undecayed remains 'raised' the peat surface above the influence of groundwater, becoming 'rain-fed' and, therefore, ombrotrophic. This gave rise to raised bogs up to 14 m deep (averaging 6–7 m), a feature that made them particularly suitable for peat exploitation. However, Ireland is still home to some of the finest examples of raised bog in Western Europe and its bogs have been recognised as being of national and international conservation importance.

8.3.3 Blanket Bogs

Blanket bogs developed about 4000 years ago; they are found chiefly along the western seaboard and on mountaintops throughout the rest of the country. Most of the blanket bogs were formed through paludification which is a process whereby peat accumulates directly over a formerly dry mineral soil that is now waterlogged. This occurred as a result of changed climatic conditions, and the distribution of blanket bogs is strongly related to climatic factors: they are most widespread in areas where annual rainfall levels are greater than 1200 mm and the number of rain days exceeds 225. These bogs are shallow and form a blanket-like layer averaging 2.5 m in depth over an underlying acidic mineral soil or bedrock. In their natural condition, these areas are dominated by *Eriophorum* species, black bog rush (*Schoenus nigricans* in Atlantic blanket bog), and purple moorgrass (*Molinia caerulea*), lousewort (*Pedicularis sylvatica*), bell heather (*Erica cinerea*) and the bryophytes *Campylopus atrovirens* and *Pleurozia purpurea.*

> ***Study Box 8.1 Blanket Bog Sub-types***
> *Atlantic blanket bogs* are found in low-lying coastal plains and valleys in mountainous areas of western counties, below 200 m ordnance datum (O.D.). They are particularly well developed in counties Donegal, Mayo, Galway, Kerry, Clare and Sligo. Their vegetation is clearly distinct from raised bogs and mountain blanket bogs (White and Doyle 1982). Vegetation is characterised by a grassy appearance due to the occurrence of purple moor grass (*Molinia caerulea*) and black bog rush (*Schoenus nigricans*). Patterned surfaces of pools, hummocks, flats and flushes are important features of these blanket bogs. *Mountain blanket bogs* occur on relatively flat terrain (across mountain plateaux and gentle slopes) in the higher Irish mountains above 200 m O.D. and are distributed more widely than Atlantic blanket bog. The vegetation of these blanket bogs is characterised by the presence of ericoid shrubs and in particular ling heather (*Calluna vulgaris*), crowberry (*Empetrum nigrum*) and bilberry (*Vaccinium myrtillus*). Natural water channels, lakes and flushes are features of mountain blanket bogs.

8.4 Irish Peatlands: A Valuable Ecological Resource

Irish peatlands are mainly open, windswept, wilderness areas virtually devoid of tree cover due to the wet conditions that have prevailed in particular in recent centuries. High precipitation did not favour the spread of trees and a stagnant water table favoured wetland communities over woody shrubs and trees that are typically found in peatland ecosystems elsewhere in the world (Charman 2002). Peatlands are exceptional natural entities. Composed of a unique combination of habitats, they can form a diversity of ecosystems with a unique biodiversity, at species and genetic level. They represent a considerable national biodiversity resource with some species being endemic and rare at a global scale. No less than three bog habitats, two fen habitats and six other habitats associated with peatlands are listed in Annex I of the Habitats Directive (EU Directive on the Conservation of Habitats, Flora and Fauna 92/43/EEC) because they are particularly threatened in Europe. At the species level, peatlands are home to flora and fauna of highly significant conservation value with species new to Ireland still being discovered (Renou-Wilson et al. 2011a, b), for example an extinct moss refound in a restored raised bog at Clara Bog in Co. Offaly (Smith 2014).

From an ecosystem point of view, natural peatlands act as sinks where peat deposits (and its C) accumulate and the peat, together with water and minerals are stored. Natural peatlands are unique ecosystems because they are generally net sinks for carbon dioxide (CO_2 uptake) and sources of methane (CH_4 emission). Therefore, their climate footprint depends on the magnitude of the land-atmosphere exchange of these two major greenhouse gases (GHG) while nitrous oxide (N_2O) becomes significant only in nutrient-rich fens and when wetlands are drained to agriculture or forested. Peatlands are large C stores globally (Yu 2012), and are estimated to contain between 53 and 75% of total soil organic C stocks in Ireland (Tomlinson 2005; Renou-Wilson et al. 2011a, b). The accumulation of these vast quantities of C occurs over many thousands of years and results from the slow accumulation of partly decomposed plant remains (C-rich organic material) under the water-saturated, oxygen-depleted conditions that prevail in natural peatlands. While the net annual GHG budget of natural peatlands is spatially and temporally variable (Koehler et al. 2011; Wilson et al. 2016), it is sensitive to natural and anthropogenic perturbations. Peat extraction, drainage and cultivation for agricultural and forestry purposes have changed nearly 90% of the original peat soils in Ireland from C sinks to sources (Wilson et al. 2013).

8.5 Historical Review of Peatland Utilisation and Stakeholders

Peat has been a source of fuel in Ireland since prehistoric times. Seventh- and eighth-century law-texts contain references to turf-cutting although these are few and ambiguous. The clearest is in the text on mill races: here the ditch of a turf-bog (clad fótbaig móna) is included among the seven ditches exempt from liability in the case of accidental drowning (Binchy 1955). Reading between the lines, one can assume that the ditch was made during the cutting of turf for use as fuel. The first explicit mention of turf as fuel is in the early twelfth-century tale Aislinge Meic Con Glinne (Jackson 1990) where it is mentioned that the miserable hospitality of the monastery of Cork included two sparks of fire in the middle of a wisp or oat-chaff and two sods of wet turf (dá fót do úr-mónaid). Before the Anglo-Normans arrived, wetlands were not only drained for turf-cutting but they may also have been used as a source of reeds for thatching, basket-making etc. (Kelly 2000). From these mediaeval sources (e.g. Calendar of Ormond Deeds), it is understood that some raised bogs around Dublin seem to have been cut-away as early the thirteenth century. Despite traditional hand-cutting, Irish peatlands remained largely untouched up to modern times. In the nineteenth century, peat was the main fuel available to a population that had expanded to more than 8.2 million (Feehan et al. 2008). According to Mitchell (1976), the disappearance of peatlands in the eastern part of Ireland is related to this period of intense peat-cutting due to high population density. Furthermore, peatlands were extensively used for the summer grazing of animals and the cultivation of crops. Of the different peatland types, fens suffered most from agricultural reclamation. Cole (1984) reviewed the history of agricultural development on peatland in Ireland, which intensified in the 1950s at the same time as peat extraction became an industry with the establishment of Bord na Móna, the main producer of peat energy and supplier of peat products in Ireland (Fig. 8.3). Mechanisation of the process and utilisation of adapted vehicles has allowed for more peat to be extracted over a wider area of bog, in less accessible terrain and also on a semi-commercial basis (for domestic use). This was not solely confined to raised bogs in the Midlands, as Conoghan (2000) reported the dramatic rise in the use of the excavator method and hopper method of peat extraction on blanket bogs since the mid-1980s. Since the 1990s with the increasing demand for horticultural peat, peat extraction for horticultural products has particularly impacted smaller raised bogs, which had been so far spared from industrial peat extraction for energy use.

At the same time as industrial cutaway peatlands appeared for the first time in the Irish landscape, considerable areas of bogs were being afforested with coniferous species, such as Sitka spruce (*Picea sitchensis*), lodgepole pine (*Pinus contorta*) and Norway spruce (*Picea abies*) (Fig. 8.3). Most planting was carried out on lowland and

Fig. 8.3 Examples of natural mountain blanket bog (Glenveagh National Park, Co. Donegal), Atlanticblanket bog (Owenirragh, Co. Mayo) and fen (Scragh, Co. Westmeath). *Photos* F. Renou-Wilson

montane blanket bogs, with some financial and technical difficulties (waterlogging, nutrient-poor soil, exposure to name but a few) and more significantly unsustainable outlooks (Renou and Farrell 2005; Renou-Wilson and Byrne 2015). Although the afforestation of blanket bogs is officially declining, private afforestation may continue to occur on non-designated blanket bogs.

Agriculture is the oldest land use of peat soils, and reclamation and drainage of organic soils were intensified in the last two centuries as a result population pressures. In addition, several Acts and schemes, including the 1945 Arterial Drainage Act, the Farm Improvement Programme and the Programme for Western Development have supported the reclamation and drainage of organic soils for agricultural purposes. In the 1980s and early 1990s, the European Union (EU) Headage grant scheme intensified grazing patterns with an upsurge in sheep numbers. In recent years, the threat from overgrazing has somewhat receded with the introduction of agri-environmental schemes (Green Low Carbon Agri-environment Scheme (GLAS), formerly the Rural Environmental Protection Scheme (REPS)) and the decoupling of Common Agricultural Policy subsidies from production (translated into the National Farm Plan Scheme and Commonage Plans).

More recently, due to the fact that peatlands are generally uninhabited, they have become attractive for a wide variety of new land use options including urban, industrial and infrastructural development as well as dumping of waste. Fens have been mostly subject to dumping or landfill because of their low-lying nature. Fens have often been filled in so as to create drier ground conditions and to make a site more suitable for subsequent developments such as housing (Foss 2007). Other industrial developments: quarries, industrial infrastructure or road constructions have impacted peatlands in a very limited way (thousands of hectares), but at an increasing number of sites. The Irish climate is ideally suited for both peatlands and wind farms and the vast majority of wind farms are located on upland and cutaway peatland areas (Renou-Wilson and Farrell 2009). Peatlands have also provided space for recreation, tourism and even military exercises. In the last decades, the number of tourists interested in outdoor activities has increased and with it, the impact of trampling on blanket bogs, which was until recently confined to National Parks (Wicklow, Glenveagh and Connemara), and has started to spread onto other peatland areas of conservation importance. Upland blanket bogs, some designated, have been unacceptably degraded, with the impact likened to overgrazing (MacGowan and Doyle 1997).

8.6 Current Peatland Land Use Categories

While the land area covered by peatlands deemed worthy of conservation has been estimated at 269,270 ha (Malone and O'Connell 2009), it is likely to be an overestimation of the area of natural peatlands in Ireland given that it probably includes areas that have been degraded with some having the potential to be restored. Given the large remaining area of 'managed' peatland (Table 8.1), decisions regarding their management are of critical importance for their future sustainability.

Table 8.1 Distribution of the main land use categories of peatlands

Natural peatlands	269,270[a]
Cutover peatlands (affected by domestic turf-cutting)	612,380[a]
Afforested peatland	301,700[b]
Farmed peatland (grassland and cropland)	295,000[c]
Industrial cutaway peatlands	70,000[d]
Rehabilitated cutaway	18,000

[a]Malone, S, and C O'Connell. 2009. Ireland's Peatland Conservation Action Plan 2020—Halting the Loss of Biodiversity. Irish Peatland Conservation Council, Lullymore

[b]Black, K., O'Brien P., Redmond J., Barett F. & Twomey M. 2008. The extent of recent peatland afforestation in Ireland. Irish Forestry 65 (1&2): 71–81

[c]CRF Table 5.C, National Inventory Report 2007–2009 (Environmental Protection Agency)

[d]57,000 ha (including hard surfaces and fringes) belongs to Bord na Móna and the remainder is an estimate of the area that is cutaway by private companies extracting peat mainly for horticulture

8.6.1 Agriculture

Current agricultural activity on peat soils is largely confined to grassland production and the grazing of cattle or sheep. An estimated 300,000 ha of grassland is on organic based soils (Wilson et al. 2013) with a negligible area used for cropland (Donlan and Byrne 2015). In Ireland, grasslands over organic soils are used for direct livestock (domestic herbivores) grazing with a small area used for forage production (either hay or silage). Some typical grassland over organic soils may be considered unimproved and includes large expanse of blanket bog/heath mosaic with various degree of human-made drainage. Most grasslands, however, have been managed to some degree by grazing, mowing, fertilisation application or intensive drainage. Management of grasslands over organic soils is described in detail by Renou-Wilson et al. (2015). The more nutrient-rich peat soils used for agriculture are hotspots of CO_2 emissions, and the rewetting of agricultural peat soils has been shown to be an important GHG mitigation tool (Renou-Wilson et al. 2016). Land use management related to sheep grazing on the hill and mountain peatlands (and heaths), remains a complex issue that needs further investigation.

8.6.2 Forestry

Peatland forests were not part of the natural assemblage of wild habitats in Ireland. In 2008, 43% or *c.* 301,700 ha of the total forest estate was located on peat soils with the majority located on blanket bog (218,850 ha) and the remainder on raised peat (74,000 ha) and cutaway peatlands (8850 ha) (Black et al. 2008). The twenty-first century has seen the near cessation of the afforestation of natural peatlands but is ongoing on already drained organic soils (cutover and cutaway bogs and some used for agriculture). The main impact of afforestation of peatlands is the drying out of the site which leads to (1) loss of characteristic flora and fauna and (2) direct loss of CO_2 to the atmosphere and indirectly as dissolved organic C losses via hydrological pathways. Cultivation, fertilisation and other management operations also leads to nutrient leaching and degradation of pristine water quality. Nowadays, several controversial challenges, such as more efficient timber production, increased multiple use of forests, mitigation of the effects of climate change, and enhancement of the quality of surface waters, put pressure on the practical forest management of peatlands (Renou-Wilson and Byrne 2015). Specific forest management methods have been developed to account for the specific soil conditions and the interactions between terrestrial and water ecosystems. Coillte, the national state-owned forest company and largest peatland landowner, published a strategy for the 'Future Management of Low Production Forests' detailing a protocol agreed for 43,000 ha of western peatland forests deemed uneconomic and unsustainable (Tiernan 2008). The strategy requires that these areas be replanted with minimal inputs while others will be managed with the aim of restoring a bog ecosystem. Coillte has also been actively involved in the restoration of 2500 ha of afforested blanket bogs and raised bogs through the LIFE programme (Delaney 2008).

8.6.3 Peat Extraction

Peat is currently extracted for (1) electricity generation in condensing power plants, (2) fuel for domestic heating (briquettes and turf), (3) horticultural products and (4) as raw material for chemical products, bedding material, filters and absorbent material. It has been estimated that a total of 100,000 ha of peatland is currently utilised for industrial peat extraction in Ireland; half of which is in active production (Fitzgerald 2006). While the total area of peatlands currently affected by domestic peat extraction (mechanical and hand-cutting) remains unknown, it is estimated that domestic turf-cutting has affected more than half a million hectares (Malone and O'Connell 2009; Wilson et al. 2013).

Thousands of hectares of drained peatlands and organic soils have been 'abandoned' because they have such low productivity for grazing or because turf-cutting is not practiced anymore. The old drainage systems continue to work in many cases while in some cases rewetting may occur naturally. Many peatlands designated as part of the Natura 2000 network (European Union coordinated network of protected areas) include such degraded land where turf-cutting took place on the outskirts of a bog and has not been restored. In the absence of proper management, these sites typically retain an oxic layer that is continually mineralised. These areas remain drier than natural peatlands and are often prone to wildfires.

Industrial cutaway peatlands present a challenging environment in regard to restoration to bring back some of the bog functions and associated benefits (C sequestration, biodiversity etc.). Therefore, rehabilitation or rewetting (drain blocking and damming) has been implemented in most cases with wide range consequences for climate and/or biodiversity benefits depending on the initial degradation status (Renou-Wilson et al. 2018b). Rewetting of bogs used for horticultural peat is even more problematic due to the significant depth of the drains and the type of peat remaining at the surface, which often coincides with a woody fen peat layer that displays high porosity. In addition, Bord na Móna has rewetted 1175 ha of raised bog (drained but not cut) since 2009 using drain blocking, informed by detailed topographic mapping (Bord na Móna 2016).

8.7 Condition of Irish Peatlands

The human-related impact on peatlands is nothing new and Irish peatlands can be described as deeply humanised landscapes that have evolved, indeed sometimes originated, in close association with land use systems. For example, tree clearance by humans has been implicated in the initiation of peatland growth in the western part of the country (Moore 1993; Molloy and O'Connell 1995; Caulfield 2004). Nevertheless, anthropogenic disturbances of peatlands have been mainly destructive with significant environmental degradation along the way. These disturbances resulted in a loss of biodiversity at all levels; genetic, species, habitats and landscapes. Also, the loss of water and mineral storage capacity leading to increased flooding and water pollution, the loss of C sequestration capacity and the loss of stored C to the atmosphere and waterways, increased risk of bog bursts and fires and the spread of invasive species. Research has shown that past and current management of the majority of Irish peatlands has been unsustainable and Irish peat soils will deteriorate further without intervention (Renou-Wilson et al. 2011a, b). The increasing scale and rate of disturbance that occurred in the twentieth century have clearly contributed to the current situation whereby only a small amount of peatlands remains in a natural or undamaged condition. Of the Holocene extent of fens and bogs in Ireland, only 15% of the original peatland cover is in a near-intact condition reflecting a low level of degradation with an even smaller proportion acting as fully functioning mires, which are active peat-forming ecosystems (Wilson et al. 2013). The area of undisturbed fens is very small as they have long been drained and reclaimed for agricultural use. Similarly, most of the bogs have been affected by peat-cutting, grazing or fire to one extent or another. It was estimated that only 10% of the original raised bog and 28% of the original blanket peatland resource are deemed suitable for conservation (Malone and O'Connell 2009). As the vast majority of peatlands are degraded to various extents, they are critically at risk of future disturbances such as climate change. Predicted climate changes are likely to affect low Atlantic blanket bogs in the west of Ireland the least, while the areas showing greatest changes in precipitation and temperature are the areas containing basin peat in the Midlands.

8.8 Ecosystem Services of Peatlands

Natural peatlands are considered one of the most important ecosystems of the world, because of their key value for biodiversity, water filtration and supply, their role in climate regulation via the sequestration and storage of C from the atmosphere, as an important support for human welfare: for the cultural services associated with the amenity and recreational value of this 'wild land', for providing a source of well-being and knowledge (Fig. 8.4). Therefore, peatlands provide a wide range of ecosystem services to the global society, but these services are often 'hidden' as they are not traded in markets. All of these benefits may be reduced in supply as a result of certain land management interventions, such as drainage for agriculture, afforestation or peat extraction. These uses of bogs produce 'goods' (turf used for energy, domestic heating or horticultural products, livestock, timber production) that have known existing markets. However, these uses are 'unsustainable' as they involve disturbances that are irreversible on any time scale relevant to society (Fig. 8.5).

8.9 Towards a Sustainable Management of Irish Peatlands

Peatlands are a key part of Ireland's natural heritage and provide many such ecosystem services, including reducing C emissions, assisting with flood regulation, providing a home for our most threatened habitats and species and a place to improve the mind and body. In the first decades of the twenty-first century, the dilemmas facing the peatland resource have been heightened with only a few 'near-intact' or natural peatlands remaining in the Irish landscape and

Fig. 8.4 Examples of peatlands used for turf cutting (Co. Offaly) and industrial peat extraction (Co. Offaly) andafforestation (Co. Sligo). *Photos* F. Renou-Wilson

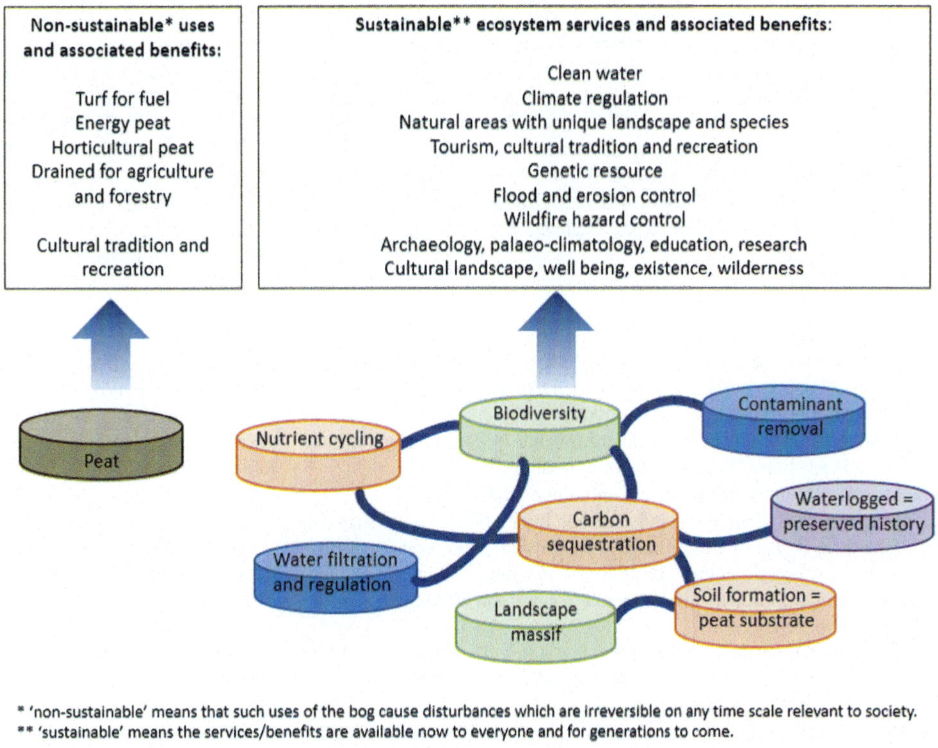

Fig. 8.5 Peatland ecosystem functions, uses, services and associated benefits

which are likely to be further damaged, be it directly by human actions or by global changes.

There is growing national and global interest in peatland restoration and in ending non-sustainable uses of peat by bringing back the 'sustainable' services and benefits that peatlands provide to society as a whole. In Ireland, the first Peatland Strategy was published in 2015 and represents the political will to manage this huge natural resource in a more sustainable way, with the commitment to give direction to Ireland's approach to peatland management and how best to optimise the benefits derived from our vast peatland resource now and for future generations. The main objective of the Peatland Strategy is to guide Government policy in relation to all peatlands across various sectors (agriculture, forestry, energy, environment etc.) but in particular to those nominated as Natura 2000 designated sites. Together with remnants of primaeval forests, these represent the oldest surviving wild landscape on this island, a source of beauty at many levels (aesthetic, cultural, educational), that should be nurtured for future generations.

References

Belyea LR, Malmer N (2004) Carbon sequestration in peatland: patterns and mechanisms of response to climate change. Glob Change Biol 10:1043–1052

Binchy DA (1955) Coibnes Uisci Thairidne. Ériu 17:52–85

Black K, O' Brien P, Redmond J, Barrett F, Twomey M (2008) The extent of recent peatland afforestation in Ireland. Ir Forest 65(1&2):71–81

Caulfield S (2004) The different effects of climate change on human impact on Ireland's native woodland development. Ireland's Native Woodlands, Galway, Woodlands of Ireland

Charman D (2002) Peatlands and Environmental Change. John Wiley & Sons, Chichester, p 301

Clymo RS (1992) Models of peat growth. Suo 43:127–136

Cole AJ (1984) The history of agricultural development on peatland in Ireland. In: Proceedings of the 7th international peat congress, International Peat Society, Dublin, Ireland

Connolly J, Holden NM (2009) Mapping peat soils in Ireland: updating the derived Irish peat map. Irish Geogr 42(3):343–352

Conoghan J (2000) Evaluation of blanket bogs for conservation in the Republic of Ireland. A synthesis of the reports on survey to identify blanket bog sites of scientific interest commissioned by the wildlife service 1987, 1989–1991. Unpublished report to NPWS. Department of the Environment, Heritage and Local Government, Dublin

Delaney M (2008) Bringing the bogs back to LIFE. In: 13th international peat congress: after wise-use: the future of peatlands, IPS, Tullamore, Co. Offaly, Ireland

Dise NB (2009) Peatland response to global change. Science 326 (5954):810–811

Donlan J, Byrne KA (2015) Assessment of agricultural activity on drained organic soils. EPA, Environmental Protection Agency, Wexford, Ireland, p 22

Doyle GJ, Críodáin CÓ (2003) Peatlands—fens and bogs. In: Otte ML (ed) Wetlands of Ireland, University College Dublin Press, Dublin, pp 79–108

Feehan J, O'Donovan G, Renou-Wilson F, Wilson D (2008) The bogs of Ireland—an introduction to the natural, cultural and industrial Heritage of Irish peatlands, 2nd edn. Digital Format. University College Dubli, Dublin

Fitzgerald P (2006) Energy peat in Ireland. In: Farrell CA (ed) Peatland utilisation and research in Ireland, Walsh Printers, Roscrea, Co. Tipperary, pp 20–23

Foss P (2007). Study of the extent and conservation status of springs, fens and flushes in Ireland 2007. Internal Report for the National Parks and Wildlife Service, Department of the Environment, Heritage and Local Government, Dublin, Ireland, p 142

Foss PJ, O'Connell CA, Crushell PH (2001) Bogs and fens of Ireland conservation plan 2005. Irish Peatland Conservation Council, Dublin, Ireland

Fossitt JA (2000) A guide to habitats in Ireland. Dublin, The Heritage Council

Frolking S, Roulet NT, Fuglestvedt J (2006) How northern peatlands influence the Earth's radiative budget: sustained methane emission versus sustained carbon sequestration. J Geophys Res 111:G01008. https://doi.org/10.1029/2005JG000091

Hammond RF (1981) The peatlands of Ireland. In: Soil survey bulletin no. 35, An Foras Taluntais, Dublin, Ireland

Jackson K (1990) Aislinge Meic Con Glinne, Dublin

Joosten H (2008) Peatlands and Carbon. In: F. Parish AA, Sirin D, Charman H, Joosten T, Minayeva M, Silvius (eds) Assessment on Peatlands, Biodiversity and Climate Change. Kalalumpur and Wageningen, Global Environmental Centre and Wetlands International, pp 99–117

Jones MB, Donnelly A, Albanito F (2006) Responses of Irish vegetation to future climate change. Biol Environ Proc R Irish Acad 106B(3):323–334

Kelly F (2000) Early Irish farming. School of Celtic Studies, Dublin Institute for Advanced Studies, Dublin

Koehler A-K, Sottocornola M, Kiely G (2011) How strong is the current carbon sequestration of an Atlantic blanket bog? Glob Change Biol 17:309–319

MacGowan F, Doyle GJ (1997) Vegetation and soil characteristics of damaged Atlantic blanket bogs in the west of Ireland. In: Tallis JH, Meade R, Hulme PD (eds) Blanket mire degradation: causes, consequences and challenges, British Ecological Society and the MAcaulay Land USe Research Group, Aberdeen, pp 54–63

Malone S, O'Connell C (2009) Ireland's peatland conservation action plan 2020—halting the loss of peatland biodiversity. Irish Peatland Conservation Council, Rathangan, Co. Kildare, p 152

Mitchell F (1976) The Irish landscape. Collins, London

Mitchell F, Ryan M (1997) Reading the Irish landscape. Town House, Dublin

Molloy K, O'Connell M (1995) Palaeo-ecological investigations towards the reconstruction of environment and land use changes during prehistory at Céide fields, western Ireland. Probleme der Küstenforschung im südlichen Nordseegebiet 23:187–225

Moore PD (1993) The origin of blanket mire, revisited. In: Chambers BJ, Chapman H, Hall AV (eds) Climate change and human impact on the landscape, London, pp 217–224

Renou F, Farrell EP (2005) Reclaiming peatlands for forestry: the Irish experience. In: Stanturf JA, Madsen PA (eds) Restoration of boreal and temperate forests. CRC Press, Boca Raton, pp 541–557

Renou-Wilson F, Byrne KA (2015) Irish peatland forests: lessons from the past and pathways to a sustainable future. In: Stanturf JA (ed) Restoration of boreal and temperate forests. Taylor & Francis Group, Boca Raton, pp 321–335

Renou-Wilson F, Farrell CA (2009) Peatland vulnerability to energy-related developments from climate-change policy in Ireland: the case of wind farms. Mires and Peat 4(Article 08):1–11, Online at: http://www.mires-and-peat.net

Renou-Wilson F, Bolger T, Bullock C, Convery F, Curry JP, Ward S, Wilson D, Müller C, (2011a). BOGLAND—Sustainable management of peatlands in Ireland. STRIVE report no 75 prepared for the Environmental Protection Agency (EPA), Johnstown Castle, Co. Wexford, p 157

Renou-Wilson F, Bolger T, Bullock C, Convery F, Curry JP, Ward S, Wilson D, Müller C (2011b) BOGLAND: a protocol for the sustainable management of Irish peatlands. STRIVE report no 76 prepared for the Environmental Protection Agency (EPA), Johnstown Castle, Co. Wexford, p 24

Renou-Wilson F, Wilson D, Barry C, Foy B, Müller C (2015) Carbon loss from drained organic soils under grassland—CALISTO. EPA research report no 141, Environmental Protection Agency, Wexford, Ireland

Renou-Wilson F, Müller C, Moser G, Wilson D (2016) To graze or not to graze? Four years GHG balances and vegetation composition from a drained and a rewetted organic soil under grassland. Agric Ecosyst Environ 222:156–170

Renou-Wilson F, Wilson D, Rigney C, Byrne K, Farrell C, Müller C (2018a) Network Monitoring Rewetted and Restored Peatlands/Organic Soils for Climate and Biodiversity Benefits (NEROS project) EPA Research Report. Synthesis Report. Environmental Protection Agency. Wexford, Ireland p 49

Renou-Wilson F, Moser G, Fallon D, Farrell, CA, Mueller C, Wilson D (2018b) Rewetting degraded peatlands for climate and biodiversity benefits: results from two raised bogs. Ecological Engineering, in press

Simo I, Creamer RE, Reidy B, Hannam JA, Fealy R, Hamilton B, Jahns G, Massey P, McDonald E, Schulte RPO, Sills P, Spaargaren O (2014) Irish SIS final technical report 10: soil profile handbook. Associated datasets and digital information objects connected to this resource are available at: secure archive for Environmental Research Data (SAFER) managed by Environmental Protection Agency Ireland

Smith GF (2014) Extinct moss re-found in raised bog. Retrieved 6/12, 2015, from http://www.blackthornecology.ie/general/extinct-moss-re-found-raised-bog/

Soil Information System (2014) Online resource at: http://gis.teagasc.ie/soils/index.php

Tiernan D (2008) Redesigning afforested western peatlands in Ireland. In: 13th International peat congress: after wise-use: the future of peatlands, IPS, Tullamore, Co. Offaly, Ireland

Tomlinson RW (2005) Soil carbon stocks and changes in the Republic of Ireland. J Environ Manage 76:77–93

White J, Doyle GJ (1982) The vegetation of Ireland: a catalogue raisonné. J Life Sci Roy Dublin Soc 3:289–368

Wilson D, Müller C, Renou-Wilson F (2013) Carbon emissions and removals from Irish peatlands: current trends and future mitigation measures. Irish Geogr 46(1–2):1–23

Wilson D, Farrell C, Fallon D, Moser G, Muller C, Renou-Wilson F (2016) Multi-year greenhouse gas balances at a rewetted temperate peatland. Glob Change Biol. https://doi.org/10.1111/gcb.13325

Yu ZC (2012) Northern peatland carbon stocks and dynamics: a review. Biogeosciences 9:4071–4085

Limestone Lowlands

9

Lilian O'Sullivan, Brian Reidy, and Rachel Creamer

Abstract

Ireland is mostly comprised of lowlands (<150 m) and the midlands of Ireland are made up of a central plain of carboniferous limestone giving rise to the limestone soils that dominate the central lowlands of Ireland. Altogether, the limestone lowlands of Ireland account for approximately 1.6 million hectares (M ha), almost one-quarter of the land area (23.2%). Peat is a prominent feature and is readily found interspersed with these limestone soils where the flat gradient and soils characterised by a higher clay percentage result in a slower drainage gradient, inducing conditions that are favourable for peat formation. Luvisols are the more prevalent soils found and account for 60.6% of the limestone lowland soils. Brown Earths are the next most commonly found soil type, with the larger portion of these being Typical Calcareous Brown Earths. The Typical Calcareous Brown Earths are spatially concentrated in the west of Ireland where the limestone bedrock is found at shallower depths. Wetter soils, such as Surface-water Gleys are found in counties Meath, Dublin and Kildare with the even wetter Humic Surface-water Gleys also common across this landscape. Calcareous Groundwater Gleys (2.37%) are found in counties Kildare, Laois and Offaly which are soils that include both gleying within the top 40 cm and a calcareous sub-surface horizon starting within 40 cm. While Humic Rendzinas account for a relatively small proportion of the lowland limestone soils (<2%), the majority of these (76.9%) are found in the Burren region of Co. Clare. The Burren is one of the largest karst landscapes in Europe and is home to an extensive assemblage of flora and fauna.

Keywords

Limestone • Carboniferous • Karst • Burren

9.1 Introduction

Ireland is mostly comprised of lowlands (<150 m) and the midlands of Ireland are made up of a central plain of carboniferous limestone. It is this carboniferous limestone parent material that gives rise to the limestone soils that dominate the central lowlands of Ireland. Altogether, the limestone lowlands of Ireland account for approximately 1.6 million hectares (M ha), almost one-quarter of the land area (23.2%).

Peat is a prominent feature across the lowlands of Ireland and this is particularly true of the midland counties where peat is readily found interspersed with the limestone lowland soils. The flat landscape and heavier (>clay percentage) soils lead to slower drainage for much of the midland topography which has induced the wet conditions that are favourable for peat formation (Hammond 1981).

The solid geology of the midlands is largely characterised by limestone. One example is the solid geology in the south of Co. Kildare that includes a full limestone succession consisting of lower bedded limestones, middle oolites and dolomites to upper pure bedded limestone (Conry et al. 1970). Although the central plain is where the limestone lowland soils are most readily found (Fig. 9.1), these soils are found in almost every county in Ireland. The largest extents are found in counties Galway (16.1%), Tipperary (10.8%), Meath (9.1%), Limerick (7.7%), Roscommon

L. O'Sullivan (✉) · R. Creamer
Teagasc Agriculture and Food Development Authority, Johnstown Castle Environment Research Centre, Wexford, Ireland
e-mail: Lilian.OSullivan@teagasc.ie

L. O'Sullivan
Irish Environmental Protection Agency, Johnstown Castle, County Wexford, Ireland

B. Reidy
Teagasc Moorepark, Animal and Grassland Research and Innovation Centre, Fermoy, County Cork, Ireland

R. Creamer
Wageningen University and Research
Atlas—Wageningen Campus, Wageningen, The Netherlands

R. Creamer and L. O'Sullivan (eds.), *The Soils of Ireland*, World Soils Book Series,
https://doi.org/10.1007/978-3-319-71189-8_9

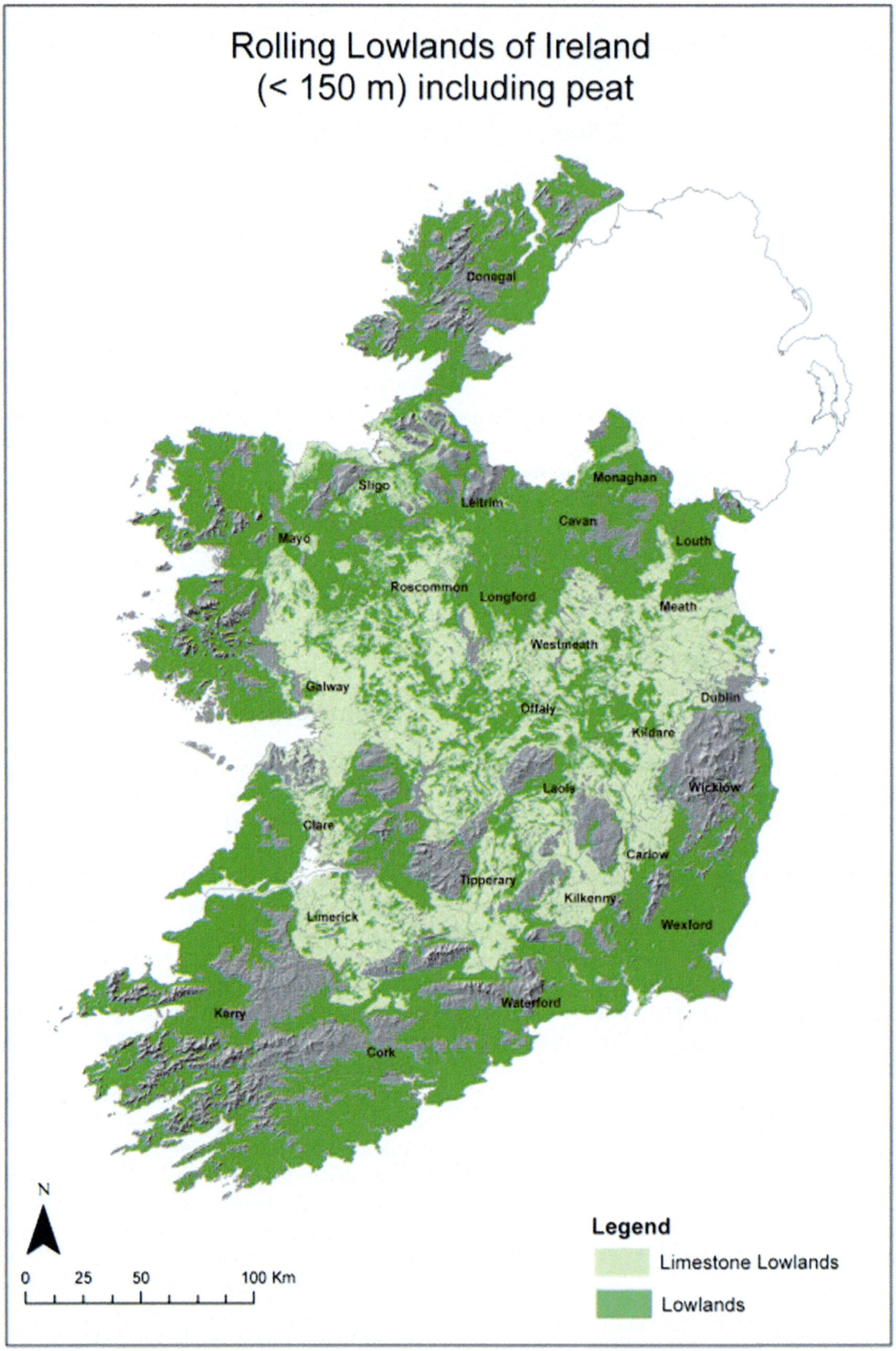

Fig. 9.1 Lowland landscapes of Ireland (<150 m above sea level). Data based on digital elevation (DEM) by Preston and Mills (2002) and the Irish Soil Information Systems by Creamer et al. (2014). The limestone lowlands of the midlands are shown to occupy the central plain of Ireland

(7.3%), Westmeath (7.1%), Kildare (6.5%), Mayo (5.1%), Offaly (5.0%), Laois (4.5%) and Kilkenny (4.5%). In many parts of the country, soils are found on glacial deposits that contain a lot of limestone material and so they tend to be calcareous in nature. In Co. Kildare, this glacial limestone till mantle is associated with Carboniferous limestone that was scraped off the north side of the Slieve Bloom Mountains as the glacier pushed eastwards (Conry et al. 1970).

The Luvisol Great Soil Group is the most prevalent Great Soil Group occurring across this landscape. These soils account for 60.6% of which 56.3% are Typical Luvisols with the remaining 4.3% belonging to the Stagnic Luvisol Soil Subgroup. The latter soils reflect stagnic properties due to the presence of a slowly permeable sub-surface horizon. Brown Earths are the next most commonly occurring soils representing almost a quarter of the lowland limestone soils. The larger portion of these is comprised of Typical Calcareous Brown Earths (15.9%), while the remainder is classified as Typical Brown Earths. The Typical Calcareous Brown Earths are spatially concentrated in the west of Ireland in counties Galway and Mayo where the limestone bedrock is found at shallower depths. These soils have a calcareous sub-surface horizon starting within 40 cm and are formed in calcareous parent material. Typical Surface-water Gleys (4.8%) are found in counties Meath, Dublin and Kildare. The even wetter Humic Surface-water Gleys (5.97%) are more widely distributed across the limestone lowlands while Calcareous Groundwater Gleys (2.37%) are found in counties Kildare, Laois and Offaly which are soils that include both gleying within the top 40 cm and a calcareous sub-surface horizon starting within 40 cm. While Humic Rendzinas account for a relatively small proportion of the lowland limestone soils (<2%), the majority of these (76.9%) are found in the Burren region of Co. Clare. The Burren is one of the largest karst landscapes in Europe and is home to an extensive assemblage of flora and fauna (see Case Study 9.1). Humic Rendzinas are shallow (<30 cm depth), calcareous (>2% $CaCO_3$) soils with hard rock or skeletal material comprising >80% coarse fragments with a humose top layer (Simo et al. 2014).

9.2 Soil Associations of the Limestone Lowlands

The distribution of soil associations across the limestone lowlands of Ireland is shown in Fig. 9.2. Associations 1000a, 1000x and 1000c are the most extensively found, all of which have the Elton (1000ET) lead series (Table 9.1). Combined these associations account for 56% of the soils across the limestone lowlands. Association 1000a is comprised of Typical Luvisols found with Surface-water Gleys, Stagnic Brown Earths and Calcareous Brown Earths on drift with limestones. Altogether, the association is classified as fine loamy drift with limestone which can be found in many counties in Ireland but the greatest extents are found in counties Westmeath, Roscommon, Meath, Offaly and Tipperary. Luvisols are soils where clay has moved down the soil profile. The higher clay content at depth means that these soils can be moderately to poorly draining depending upon the soil texture. The Elton series is defined as fine loamy drift with limestones which means that these soils are naturally moderately draining. Altogether these are considered to have good agricultural potential being friable deep soils with plentiful, well-developed roots, a high base saturation with good nutrient retention (SIS 2014).

Association 1000x also has the Elton lead series, but this association is characterised by Luvisols found with Surface-water Gleys on drift with mixed limestone and siliceous stones. The key difference between association 1000a and 1000x is that the latter association is characterised by a combination of wetter soils. In Co. Meath, the Elton series is widespread in the river catchment stretching from Oldcastle in the west, through Navan and Trim. Here, an eastward trend of increasing wetness is observable, with the 1000a association found to the west of the county, transitioning to association 1000x which are bordered to the east by a dominance of Surface-water Gleys. Spatially, while this association is found widely across the lowlands, it is more predominantly found in the south central region of the country particularly in counties Tipperary, Limerick and Kilkenny.

The third most extensively found soil association is 1000c which also has the Elton lead series. This association is composed of Luvisols associated with histic and humic Groundwater Gleys and Calcareous Brown Earths on drift with limestones and basin peat. The occurrence of histic and humic soils within this association is an indication that the soils of this association are wetter with the exception of the Calcareous Brown Earths which will be naturally more free draining. Equally, the presence of basin peats and Groundwater Gleys indicates that these soils are found in low-lying topography as both are likely to occur in landscape depressions. This association is predominantly found in the midlands, in particular in counties Tipperary, Laois, Carlow, Offaly, Kildare, Westmeath and Kilkenny.

Next, soil association 1100q with the Mullabane lead series 1100 MB (Fig. 9.3b) can be found. This is comprised of Typical Brown Earths, found with Calcareous Brown Earths on drift with limestones associated with Luvisols, Rendzinas and peat. This soil association is predominantly found in the west and north-west of Ireland, primarily in Co. Galway but also to a lesser extent in counties Sligo and Roscommon. In Co. Galway, these soils are found towards the low-lying eastern part of the county in the areas around Loughrea. The association is classified as a coarse loamy

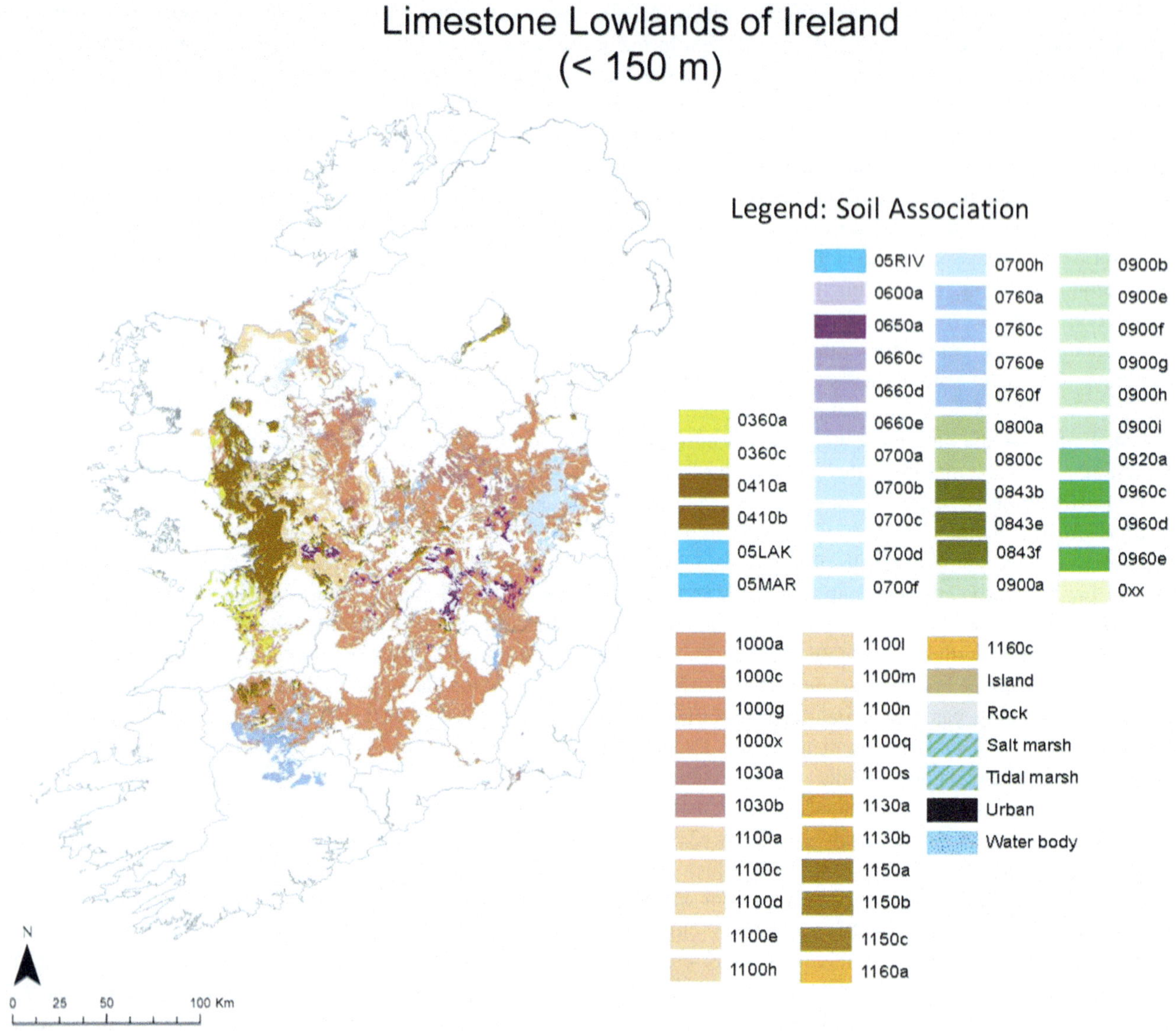

Fig. 9.2 Soil associations of the limestone lowlands of Ireland highlighting the dominance of the Luvisols across the midlands and the geographical concentration of the Typical Calcareous Brown Earths in the West of Ireland

drift with limestones. Many of the soils in this association are shallow in nature including the Burren series (0360BU), the Crush series (0360CS) and the Ballydeloughly series (0300BD) which all belong to the Rendzina Great Soil Group. The classification of these soils is based on their shallow calcareous nature. The shallow nature of many of the soils in this association represents a limitation in relation to agricultural use range owing to their intrinsic rooting depth limitation. This is further compounded by the likelihood of rock outcrop preventing trafficability for fertiliser and cropping. The peat series found in this association is the Banagher (0270BR) which is a drained mineratrophic fen peat. Fen peats occur under the influence of base-rich groundwater and as a result are characterised by a pH > 4.0. The majority of fen peats have been drained for agricultural purposes in Ireland or they have already evolved into bogs as a natural progression.

Associations 1150b is the next most extensively found with the Ballinacurra lead series, followed by the 1150c association with the Faoldroim lead series. These soils are concentrated in the west of Ireland, in particular, in the counties Galway and Mayo. The Ballinacurra series is classified as a fine loamy drift over limestone bedrock. This soil type correlates with Eutric Leptic Cambisol (Humic) in the WRB classification (IUSS 2006). As well as Typical Calcareous Brown Earths, this association includes Typical Brown Earths and Luvisols on drift with limestones, with Rendzinas and decalcified Lithosols on limestone bedrock and Peat. The calcareous nature of these soils means that they have a high pH and a high base saturation. They have a

Table 9.1 Main soil associations of the rolling lowlands, association description and the main series in the association[a]

Main associations 1.67 M ha	Mainly occurs (proportion of association)	Association description	Main series (lead series first)
1000a (22.7%)	Westmeath (18.4%), Roscommon (17.2), Meath (16.6%), Offaly (8.56%), Tipperary (7.6%) …	Luvisols associated with Surface-water Gleys, Stagnic Brown Earths and Calcareous Brown Earths, on fine loamy drift with limestones	Elton (1000ET), Faoldroim (1150FO), Musicfield (1130MF), Mullabane (1100 MB), Straffan (0700ST) …
1000x (22.2%)	Tipperary (29.9%), Limerick (20.3%), Kilkenny (14.4%), Meath (9.6%), Kildare (9.03%), Dublin (6.04%) …	Luvisols and Surface-water Gleys on fine loamy drift with mixture of limestones and siliceous stones	Elton (1000ET), Dunboyne (1000DB), Howardstown (0760HN), Straffan (0700ST), Kilrush (0700KR) …
1000c (11.1%)	Tipperary (20.7%), Laois (18.6%), Carlow (16.8%), Offaly (16.3%), Galway (11.9%) …	Luvisols associated with histic and humic Groundwater Gleys and Calcareous Brown Earths, on fine loamy drift with limestones on Basin Peat	Elton (1000ET), Kellistown (1000KL), Baggotstown (1150BG) …
1100q (8.4%)	Galway (67%), Sligo (18.9%), Roscommon (10.6%), Louth (2.9%)	Mostly Brown Earths and Calcareous Brown Earths on coarse loamy drift with limestones, associated with Luvisols and some inclusions of Rendzinas and Peat	Mullabane (1100 MB), Burren (0360BU), Crush (0360CS), NBE17 (1160NBE17) …
1150b (5.74%)	Galway (47.1%), Mayo (25.5%), Limerick (14.85%), Monaghan (7.4%)…	Calcareous Brown Earths and Luvisols associated with Rendzinas and decalcified Lithosols, fine loamy over limestone bedrock	Ballincurra (1150BC), Elton (1000ET), Kilfenora (1150KF), Faoldroim (1150FO) …
1150c (5.4%)	Galway (81.1%), Clare (10.1%), Mayo (6.76%) …	Calcareous Brown Earths, Brown Earths and Luvisols on fine loamy drift with limestones, associated with Rendzinas and decalcified Lithosols on limestones bedrock and Peat	Faoldroim (1150FO), Kilfenora (1150KF), Elton (1000ET), Mullabane (1100 MB), Burren (0360BU) …

[a]Full soil descriptions are available at series level through the online soils database at http://gis.teagasc.ie/soilsguide.php

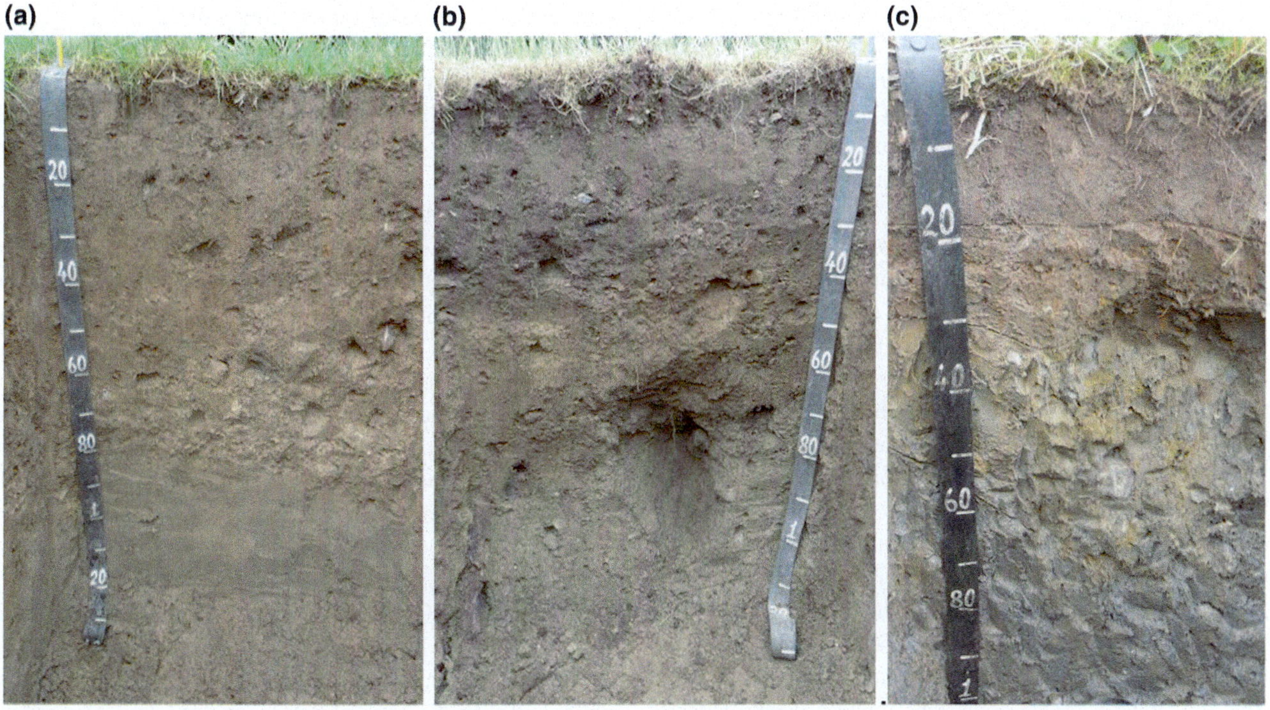

Fig. 9.3 **a** Elton Typical Luvisol. **b** Mullabane Typical Brown Earth. **c** Howardstown Humic Surface-water Gley

fine to crumbly texture, a friable consistency with plenty of evidence of rooting found. Association 1150c with the Faoldroim lead series is also a fine loamy, over drift with limestones. Also found, albeit to a lesser extent is association 1150a with the Baggotstown lead series which is texturally different, and is classified as a coarse loamy over calcareous gravels (SIS 2014). Although not as shallow as Rendzinas, Typical Calcareous Brown Earths tend to have a comparatively shallow profile compared to Typical Brown Earths meaning that their use for the production of arable crops is typically limited due to shallow rooting depth.

Association 0760c with the Howardstown lead series (0760HN) is classified as clayey drift with limestones and the presence of a humic surface horizon means that these soils are even wetter than Typical Surface-water Gleys. Along with Humic Surface-water Gleys, the association includes Luvisols in clayey lowlands with inclusions of Typical Calcareous Brown Earths associated with drier areas. The majority of these heavy wet soils are found in north Cork and in Co. Limerick. These soils are characterised by humose topsoil with evidence of gleying within the first 40 cm (Fig. 9.3c). These soils may benefit from land drainage for the purpose of agricultural production.

Association 0650a with the Mylerstown lead series (0650MT) is a fine loamy drift soil with limestones. This series is largely found in counties Kildare, Laois, Galway, Offaly and Tipperary. The series is characterised by a high pH that increases with depth. The soils have a friable consistency with a moderate to average structure in the upper horizons The profile is characterised by a greyish brown colouring, with a Munsell 10YR42 dark greyish brown typical in the upper horizon. Down the profile, the presence of a Bg horizon will be found which is characterised by more clay and less calcium carbonate ($CaCO_3$) than the underlying horizon. This association consists of poorly drained Calcareous or Humic Calcareous Groundwater Gleys with Luvisols on drift with limestones.

Association 0360a with the Burren series lead (0360BU) is predominantly found in the Burren region of Co. Clare. The Burren series are classified as fine loamy soils over bedrock. These soils correlate with Eutric Leptic Regosol (Humic) soils in World Reference Base classification (IUSS 2006). By definition, Rendzinas are shallow calcareous soils but these soils are further classified as Humic characterised by humose topsoil indicating high organic matter content. This means that these soils have a low pH in the upper horizon coupled with a high carbon content, which is reversed down the profile as the soil comes closer to the limestone bedrock. Although the top horizon is characterised by many roots, the shallow nature of these soils means that rooting depth is a general inherent limitation of these soils. The soils have a friable consistency with a weak-moderate structure becoming massive down the profile.

9.3 Soil Suitability of the Lowland Limestone Landscapes

Ireland has a global reputation as being a lead producer of high performing thoroughbred horses. It is the calcium rich soil and temperate climate of the Irish limestone lowlands that are considered to provide the ideal conditions for the development of young horses. Horse breeding is an industry of national importance in Ireland in terms of exports and tourism, with Irish bred horses in demand worldwide. The limestone soils are considered to provide ideal grazing conditions for healthy foal bone development. In many midland counties, including Westmeath and Kildare, mares are put into foal in spring to take advantage of the calcium-rich grass plains found there.

Overall, Luvisols dominate the limestone lowlands. Luvisols are considered good agricultural soils due to their higher clay content which means that they can retain nutrients well. They tend to be deep soils with good moisture holding capacity and therefore exhibit a wide agricultural use range. Depending upon the texture of the soil, Luvisols will be moderately to imperfectly draining but the Luvisols of this landscape are mostly moderately draining. The inherent characteristics of these soils mean that they make excellent grassland soils. Careful management in relation to stocking rates and prudent management under wet conditions means that these soils can be grazed over a long season to achieve optimum grass utilisation (Conry et al. 1970; Finch and Ryan 1966). Although the majority of these soils are used for grass production, they also make good tillage soils and can be used for a variety of tillage crops and vegetable production. In Co. Dublin, Luvisols are the most productive soils in the county and horticulture production in the Fingal area benefits from proximity to such a large domestic market. Although a high yield of tillage crops can be obtained with Luvisols, lodging with cereal crops can occur. Lodging refers to the process by which small-grained cereals can snap due to a weak stem (Berry et al. 2004). Along with potentially lowering yields (Berry et al. 2004), this can also result in a deterioration of soil structure associated with tillage production, in which case a short rotation is important (Finch et al. 1971). Luvisols tend to have a moderate soil structure which represents the main challenge in relation to the productive capacity of these soils as compaction and poaching can occur if not sufficiently well managed particularly after periods of prolonged wetness. Although not widely used for forestry, most forest species would be successful on these soils.

Surface-water Gleys are characterised by poor drainage associated with a slowly permeable sub-surface layer. The Howardstown series, a Humic Surface-water Gley is an even wetter soil than the Typical Surface-water Gley. As a result,

Fig. 9.4 Limestone pavement, a priority habitat (8240) under the EU Habitats Directive, consists of blocks of limestone bedrock, known as clints, separated by fissures called grikes forming a complex mosaic with calcareous grasslands and heath (Burren Programme, 2015). *Photo* Habitats of the Burren (Burren Programme 2015)

these soils have a limited use range. Despite their limitations, these do, however, make good grassland soils and are suitable for forestry (AFT 1973; Finch et al. 1971). Controlled grazing and proper use of lime and fertilisers will give a high level of grass production but the poor soil structure of these soils requires optimum management to reduce the risk of compaction or poaching (AFT 1973; Finch et al. 1971).

Typical Calcareous Brown Earths are concentrated in the west of Ireland, in counties Galway and Mayo. The free-draining nature of these soils allows for grazing over a long season with limited risk of poaching but there is a potential risk of moisture deficit in dry periods (Finch et al. 1971). These soils tend to be stony and shallow thereby limiting the potential for tillage production. The soil is not generally suited for forestry.

Groundwater Gleys are water-affected soils and have a limited use range with a high susceptibility to poaching. The poor drainage of the Mylerstown series coupled with their landscape position means that these soils are limited from an agricultural perspective. For grass production, artificial drainage with good management practices will optimise the agricultural productivity of these soils. Otherwise, the soils offer some potential for forestry.

Case Study 9.1—The Burren

Within the European Union, Ireland has the largest limestone pavement area, with over 32,187 ha compared to less than 3000 ha in the UK (Wilson and Fernández 2013). The Burren, from the Gaelic word *Boireann* meaning 'stoney place', located in the north-west of Co. Clare and south Co. Galway, represents one of the finest karst landscapes in Ireland and indeed the world. As rain falls through the atmosphere, it picks up molecules of carbon dioxide (CO_2) to form a weak carbonic acid (H_2CO_3) which can dissolve limestone overtime. H_2CO_3 dissolves the calcite in limestone to produce calcium bicarbonate resulting in many limestone features specific to karst landscapes. These features include limestone pavements (Fig. 9.4), clints and grikes (Fig. 9.4), dolines, gorges, springs, swallowholes and turloughs. Karst landscapes can deliver many ecosystem benefits and represent an important source of water. Some of Ireland's largest springs emerge from karst which is particularly important due to the absence of surface-water courses where these springs are found (GSI 2016). Also, karst landscapes, such as the Burren, are home to a very diverse and species-rich landscape that includes an extensive assemblage of flora and fauna (Figs. 9.4 and 9.5). Many plant and animal species survive in the Burren that are otherwise rare in Ireland and Europe.

Association 0360a with the dominant Burren series (Humic Rendzina) represents the main soil type found across this landscape. These are shallow calcareous

Fig. 9.5 Dry calcareous grassland is a priority habitat (6210) for conservation under the EU Habitats Directive. Widespread in the Burren, these include unimproved or semi-improved dry grasslands on thin soils that are rich in calcium. They are normally species-rich and include classic 'Burren' species such as spring gentian, mountain avens and bloody cranesbill (Burren Programme 2015). *Photo* Habitats of the Burren Programme (2015)

soils over limestone bedrock with an accumulation of organic matter in the surface horizon. These soils have a limited use range due to their shallow nature which results in inadequate rooting depth. In many places, the lack of soil depth precludes tillage which is compounded further by the frequent rock outcrops found. However, the soils are naturally free draining and are suited for grazing. Their free-draining nature means that these soils produce good winter grazing pastures (Finch et al. 1971). Despite the challenges of farming in this unforgiving terrain, several 100 farm families continue to live and farm in the Burren, maintaining practices that are essential to support the diverse Burren landscape (Burren Programme 2015). The Burren enjoys a rich tradition of farming and is considered a pastoral landscape. The ancient tradition of transhumance is practiced whereby in-calf mothers are put back onto the Burren hills in October to graze the rough limestone grasslands and heaths over winter (Burren Programme 2015). To support the sustainability of the Burren, the Burren Life research project (EU LIFE Nature fund with three local partners: the National Parks and Wildlife Service, Teagasc and the Burren branch of the Irish Farmers Association) that commenced in 2005 established a blueprint for sustainable farming in this area. This was subsequently rolled out between 2010 and 2015 under the 'Burren Farming for Conservation Programme' and was further expanded in 2016 through 'The Burren Programme'. The practical (paying for actions and results) hybrid approach to farming and conservation has delivered a range of proven environmental benefits on about 14,500 ha of prime Burren habitat and has fostered a positive attitude towards conservation among farmers which will support the sustainability of one of Ireland's most extraordinary landscapes into the future (Burren Programme 2015) (Fig. 9.6).

Fig. 9.6 Farming in the Burren is challenging in the difficult terrain. Here, scrub has been removed and walls have been repaired in an area of limestone pasture, enabling the restoration of winter grazing which has been shown to be the key driver of biodiversity in such pastures. *Photo* Burren Programme (2015)

Acknowledgements This chapter gratefully acknowledges Dr. Brendan Dunford and the Burren Programme for the use of the images contained in the Burren Case Study in this chapter.

References

An Foras Talúntais (1973) County Leitrim resource survey part 1—land use potential (Soils, Grazing Capacity and Forestry) Soil Survey Bulletin No. 29, An Foras Talúntais, Dublin

Berry PM, Sterling M, Spink JH, Baker CJ, Sylvester-Bradley R, Mooney SJ, Tams AR, Ennos AR (2004) Understanding and reducing lodging in cereals. Adv Agron 84:217–271

Burren Programme (2015) The Burren, online: http://burrenprogramme.com/. Accessed 12/09/2016

Conry MJ, Hammond RF, O'Shea T (1970) Soils of County Kildare. National Soil Survey of Ireland, AnForasTalúntais (The Agricultural Institute), Dublin 4

Creamer R, Simo I, Reidy B, Carvalho J, Fealy R, Hallet S, Jones R, Holden A, Holden N, Hannam J, Massey P, Mayr T, McDonald E, O'Rourke S, Sills P, Truckell I, Zawadzka J, Schulte R (2014) Irish soil information synthesis report (2007-S-CD-1-S1) EPA STRIVE Programme, Wexford. Available from download from http://erc.epa.ie/safer/reports

Finch TF, Ryan P (1966) Soils of Co. Limerick Soil Survey Bulletin No. 16. National Soil Survey of Ireland AnForasTalúntais (The Agricultural Institute), Dublin 4

Finch TF, Culleton E, Diamond S (1971) Soils of Co. Clare Soil Survey Bulletin 23, National Soil Survey of Ireland. AnForasTalúntais, Dublin

Geological Survey of Ireland (GSI) (2016) Importance of limestone and karst. Karst Booklet, Geological Survey of Ireland

Hammond RF (1981) The Peatlands of Ireland to accompany Peatland Map of Ireland. 1978 Soil Survey Bulletin No. 3, 2nd edn.An ForasTalúntais, Dublin, Ireland

IUSS Working Group WRB (2006). World Reference Base for Soil Resources—a framework for international classification, correlation and communication. World Soil Resources Reports 103. ISSN 0532-0188, FAO Rome, 128 pp

Preston R, Mills P (2002) Generation of a hydrologically corrected digital elevation model for the Republic of Ireland. Unpublished report submitted to Environmental Protection Agency of Ireland by Compass Informatics as part of the 2000-LS-2.2.2 Fourth Progress Report

Simo I, Creamer RE, Reidy B, Hannam JA, Fealy R, Hamilton B, Jahns G, Massey P, McDonald E, Schulte RPO, Sills P, Spaargaren O (2014) Irish SIS Final Technical Report 10: Soil Profile Handbook. Associated datasets and digital information objects connected to this resource are available at: Secure Archive For Environmental Research Data (SAFER) managed by Environmental Protection Agency Ireland. http://erc.epa.ie/safer/resource?id=a1b872a2-3b8c-11e4-b233-005056ae0019. Last accessed: 08/09/2016

SIS—Irish Soil Information System (2014) Online resource. http://gis.teagasc.ie/soils/index.php

Wilson S and Fernández F (2013) National survey of limestone pavement and associated habitats in Ireland, Irish Wildlife Manuals No. 73 prepared by Ecologic Environmental & Ecological Consultants Ltd. for Irish Wildlife Manuals Series (EdsMarnell F and Jeffrey R) for National Parks and Wildlife Service (Department of Arts, Heritage and the Gaeltacht)

Rolling Lowlands 10

Lilian O'Sullivan, Brian Reidy, and Rachel Creamer

Abstract

Ireland is mostly comprised of lowlands. The midlands of Ireland are made up of a central plain of carboniferous limestone with elevation generally less than 150 m above sea level, giving rise to limestone soils which are described in Chap. 9. In this chapter, we focus on the remaining lowland areas that occur around this central plain. Lowlands are typically associated with Brown Earths and Luvisols. Brown Earths are found associated with sandstone bedrock whilst Rendzina and Luvisols (although not always) are typically associated with limestone bedrock. Alluvial soils are widely found along waterways, estuarine or marine environments. In general, Brown Earths are considered good agricultural soils with a wide use range potential. In particular, their natural free drainage means that they do not require additional drainage and their ability to retain nutrients means that they have good agricultural potential. Humic Brown Earths, such as the Ashgrove series with high organic matter content in the topsoil may be moderately acidic and may therefore benefit from periodic liming. Groundwater Gleys are water-affected soils, typically found in depressions and so have a limited use range with a high susceptibility to poaching. Alluvial soils reflect similar characteristics to Groundwater Gleys and are limited due to their very high risk of flooding. The main limitation with Brown Podzolics is associated with the nutrient status of these soils, which require frequent liming to neutralise the acidity and other amendments to increase fertility.

Keywords

Lowlands • Gneiss • Schist • Alluvial

10.1 Introduction

Ireland is mostly comprised of lowlands. Here, lowlands are classified as the areas that are <150 m above sea level. The midlands of Ireland are made up of a central plain of carboniferous limestone giving rise to limestone soils which are described in more detail in Chap. 9. In this chapter, we focus on the remaining lowland areas which are found to occur around this central plain (Fig. 10.1). Altogether, the lowlands of Ireland account for approximately 79.7% of the area of the country. Excluding non-soil areas, and the limestone lowlands [1.670 million (M) hectares (ha)], the remaining lowlands (including peats 1.2 M ha), account for almost 3.5 M ha. This equates to approximately half of the land area of Ireland.

Lowlands are typically associated with Brown Earths and Luvisols (Fig. 10.2). On upper slopes, some Brown Podzolic soils may be found, whilst the lower slopes tend to be characterised by Surface-water Gleys and Groundwater Gleys. Alluvial soils can be found along waterways, estuarine or marine environments. A simplified distinction between limestone and non-limestone lowland soils can also be seen in Fig. 10.2. Brown Earths are found associated with sandstone bedrock whilst Rendzina and Luvisols (although not always) are typically associated with limestone bedrock. Basin peat is a common feature of the lowlands of Ireland, often forming in landscape hollows in the midlands, or as Atlantic blanket bogs along the western seaboard. Whilst mountain blanket bogs occur at higher elevations, Atlantic

L. O'Sullivan (✉) · R. Creamer
Teagasc Agriculture and Food Development Authority, Johnstown Castle Environment Research Centre, Wexford, Ireland
e-mail: Lilian.OSullivan@teagasc.ie

L. O'Sullivan
Irish Environmental Protection Agency, Johnstown Castle, County Wexford, Ireland

B. Reidy
Teagasc Moorepark, Animal and Grassland Research and Innovation Centre, Fermoy, County Cork, Ireland

R. Creamer
Wageningen University and Research
Atlas—Wageningen Campus, Wageningen, The Netherlands

R. Creamer and L. O'Sullivan (eds.), *The Soils of Ireland*, World Soils Book Series,
https://doi.org/10.1007/978-3-319-71189-8_10

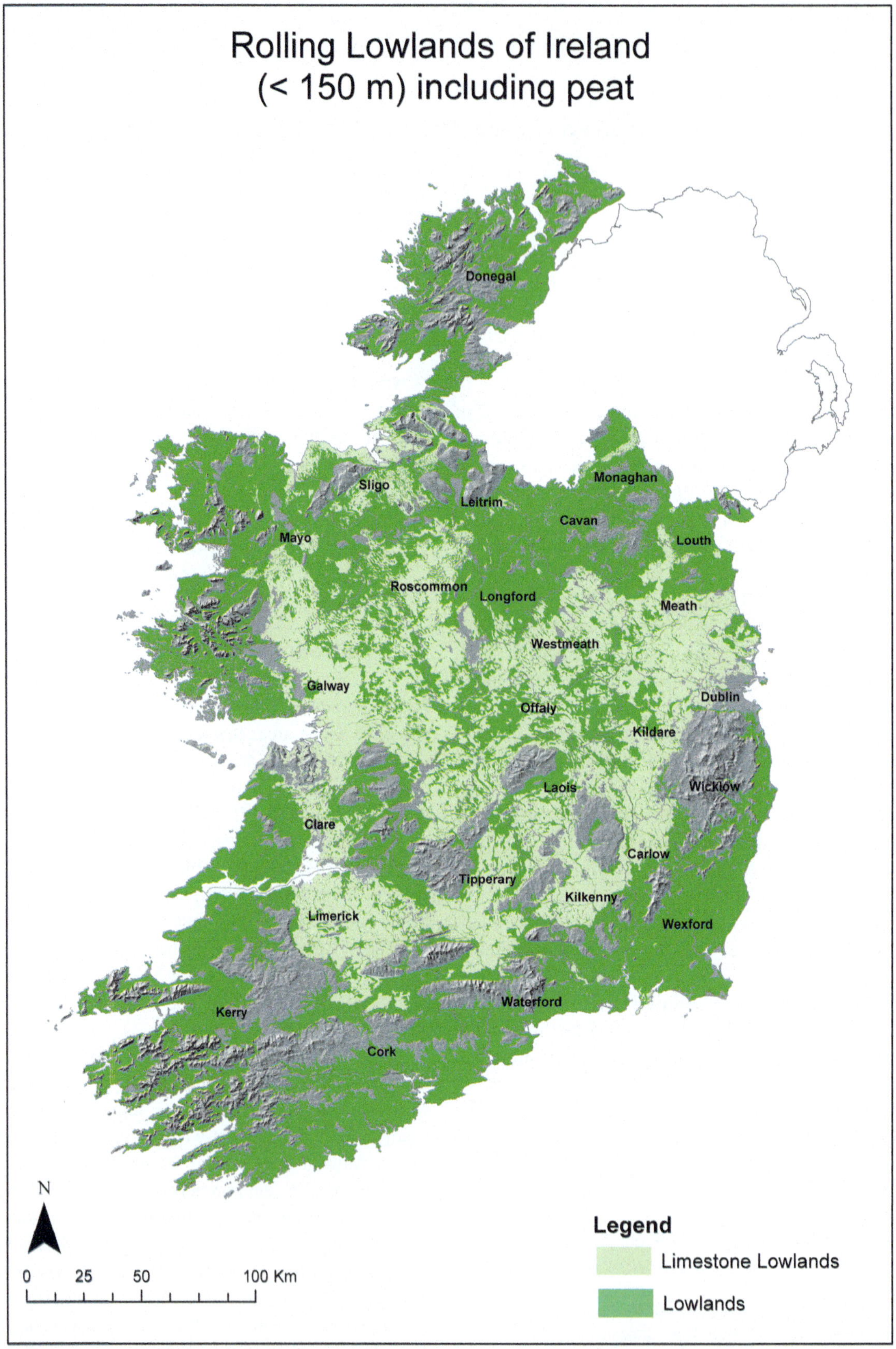

Fig. 10.1 Lowland landscapes of Ireland (<150 m above sea level). Data based on digital elevation (DEM) by Preston and Mills (2002) and the Irish Soil Information Systems by Creamer et al. (2014). The rolling lowlands frame the limestone lowland soils of the midlands that occupy the central plain of Ireland

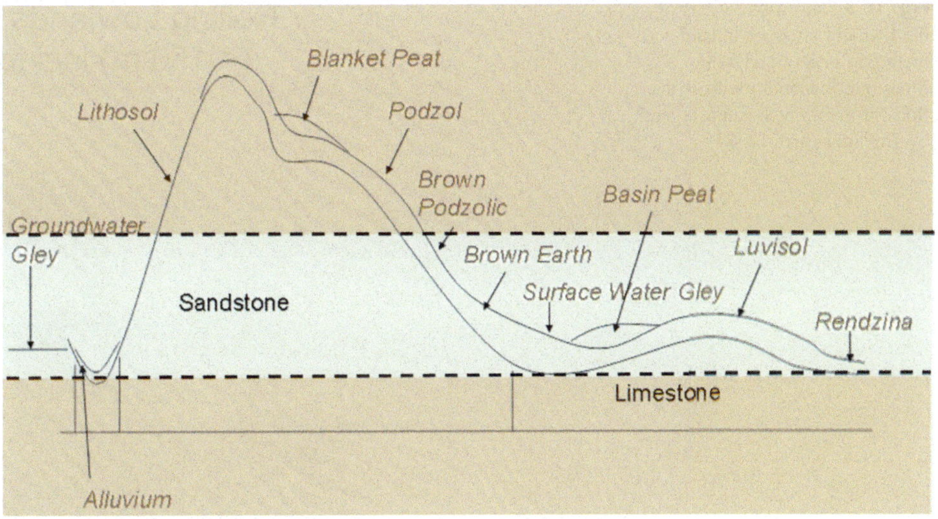

Fig. 10.2 Catena sequence of Soil Great Groups of Ireland across a landscape adapted from Simo et al. (2014a, b) with areas between the dashed lines the most typical Soil Great Groups found on lowlands landscapes

blanket bogs associated with high rainfall and acid peat substrate can be found to extend to sea level (IPCC 2016) (Case Study 10.1).

Of the rolling lowlands, the largest extent is unsurprisingly found in Co. Cork, which is the largest county in Ireland. This area is part of the Munster Basin of Old Red Sandstone geology and is associated mainly with peats, Typical Brown Podzolics and Typical Brown Earths, occurring in this order from west to east across the county. Siliceous lowland soils are also found in counties Kerry, Waterford and Tipperary which are also part of the Munster Basin dating from the Devonian Period. Further east, Co. Wexford has the largest extent of lowland soils outside of Co. Cork, accounting for almost 10%. Here, the soil parent material is glacial drift predominantly derived from Ordovician shale with some granite influence (Gardiner and Ryan 1964). Elsewhere, lowland soils are found across every county in Ireland but other larger extents can be found in Co. Mayo in the west and in Co. Donegal in the north-west. North-west of Co. Mayo is dominated by peat. The blanket bog here is low-level Atlantic type some of which has been man modified (Hammond 1981). This area of the landscape is generally characterised by igenous bedrock associated with Brown Podzolic soils. Further eastwards, some parts of the county are characterised by limestone glacial till lowlands. Co. Donegal is a landscape dominated by peat but in the east of the county, the landscape is comprised of valleys and depressions with glacial drift deposits from acid igenous rocks formed through the cooling and solidification of magma or lava resulting in the slightly more acidic Typical Brown Podzolic soils. Lowland soils are also commonly found in counties Leitrim, Cavan, Monaghan and Louth but these areas form part of the drumlin belt of Ireland which are described in more detail in Chap. 11.

10.2 Peatlands of the Rolling Lowlands

The soil associations found across the lowlands, including the peat soils are shown in Fig. 10.3. Peat soils account for approximately 1.2 M ha of the lowlands of Ireland. Much of this peat is found in the counties of the western seaboard, in particular, counties Mayo, Galway, Donegal, Kerry, Clare and Sligo. Large parts of the midlands are also characterised by peat, with counties Offaly, Roscommon, Westmeath, Longford, Tipperary and Kildare home to peatlands, much of which is found interspersed with the limestone lowlands. In general, the bogs of the midlands are raised bogs and have developed under a rainfall regime of 750–1000 mm/year (Renou-Wilson et al. 2011). The low-lying undulating topography of the midlands has resulted in a low drainage gradient meaning that conditions favour peat formation (Hammond 1981). Although the peat surface is acid, the basis is derived from nutrient-rich groundwater derived from calcareous glacial drift, that fed the lakes remaining at the end of the last glacial period (Mitchell and Ryan 1986). Their upward growth overtime resulted in their becoming Ombrotrophic rain-fed peats as the domed surface became raised above the influence of groundwater. The raised appearance of these bogs means that they have historically been suitable for peat exploitation (Chap. 8).

Case Study 10.1: Atlantic Blanket Bogs

Blanket bogs are climatic peats that develop under cool humid conditions of high rainfall exceeding 1250 mm/year. The underlying rock structure is important with nutrient-poor acidic rocks favouring peat formation, particularly obvious in the Ox Mountains in Co. Sligo and the Derryveagh Mountains in Co. Donegal (Feehan and O'Donovan 1996). Along

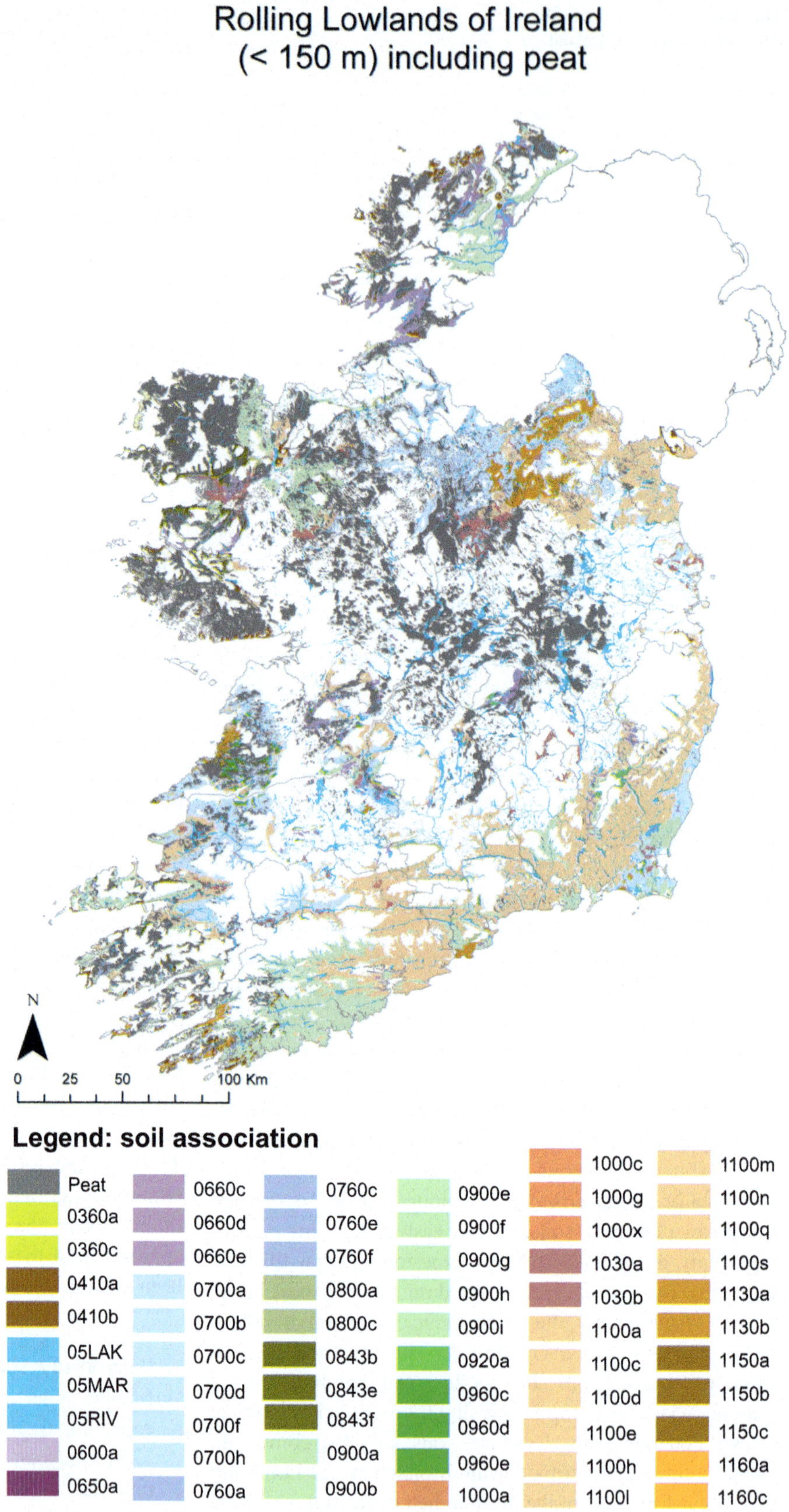

Fig. 10.3 Soil associations of the lowland areas of Ireland including peats (excluding limestone lowlands) with Irish Soil Information Systems legend by Creamer et al. (2014)

with climate, geology and podzolisation, human influence is considered to have played an important part in the growth of blanket bogs. Bog initiation in many areas coincides not only with the deterioration in climate but also with the advent of metallurgy and the general opening up of woodland areas from about 4000 B.P. onwards, associated with the early Bronze Age time (O'Connell 1990).

Unlike fen peats, blanket bogs rarely form in depressions and are found to undulate in accordance with the underlying mineral rock. Blanket bog constitutes a major landscape feature in Ireland, in particular along the western seaboard counties of which over one-third can be classified as lowland or Atlantic blanket bog (O'Connell 1990). This blanket bog type occurs in counties Galway and Mayo and to a lesser extent in Cork, Donegal and Kerry on flat to gently undulating topography below 150 m ordnance datum (O.D.) (Hammond 1981). The 150 m contour is an important divide, as *Schoenus nigricans* represents a key species in the botanical composition of the lowland Atlantic bogs subtype, above which the floristic composition changes (Hammond 1981). A floristic gradient between mountain and lowland bogs, as well as from east to west in Ireland is observed whereby the proximity to the ocean increases the nutrient status of the lowland bogs along the west coast (Feehan and O'Donovan 1996). This produces a mixture of botanical species that includes both hard-core ombrotrophic plants and some that are more mineratrophic in nature. These include both species of cottongrass, deergrass and bog asphodel, black bog-rush and purple moor grass (Feehan and O'Donovan 1996).

10.3 Associations of the Rolling Lowlands

Across the rolling lowlands, 1100a is the most extensively found soil association (Table 10.1; Fig. 10.3) and accounts for almost 14% of the rolling lowland soils. Association 1100a with the Clonroche lead series (1100CL) is not only found primarily in Co. Wexford (30.2%) but also in counties Cork, Wicklow, Kilkenny, Tipperary, Waterford, Louth and elsewhere to a lesser extent. Association 1100a is comprised of Brown Earths which are well-drained soils found in association with Brown Podzolics on upper slopes. The association also includes wetter soils such as Surface-water Gleys but in landscape depressions, some Groundwater Gleys can be found. Typical Brown Earth soils possess a rather uniform profile with little differentiation between horizons and therefore do not exhibit extensive signs of leaching or degradation. The Clonroche soil series was initially described and classified in the first Soil Survey Bulletin 1, 'Soils of Co. Wexford' by Gardiner and Ryan (1964). The series is a fine loamy drift soil with siliceous stones found on undulating land. The soil is well drained and is characterised by a friable consistence, a medium pH ($\sim$6.5) with very fine roots commonly found, particularly in the upper horizons. Whilst the Clonroche soil series have moderately good water holding capacity, drought conditions may occur in shallow profiles in dry seasons (Gardiner and Ryan 1964).

Wetter soils within this soil association include the Kilpierce series (0600KP), a Typical Groundwater Gley. This soil is classified in World Reference Base (WRB) classification as a Haplic Gleysol (Humic Eutric) (IUSS 2006; Reidy et al. 2014). In Co. Wexford, these soils are described as forming an intricate pattern with the Clonroche series (Gardiner and Ryan 1964). The Kilpierce series is predominantly formed of Ordovician shale parent material, and like the majority of series within this association is classified as a fine loamy drift soil with siliceous stones. These soils are prone to a high groundwater table associated with their low-lying topography. Accordingly, these are poorly drained soils limited to unfavourable slopes or landscape depressions. The soil is recognisable by a uniform grey colour throughout the profile, with extensive mottling in a Bg horizon attributable to reduction of iron caused by periodic saturation with water in the presence of organic matter (Simo et al. 2014a). The Knockroe series (0600KO) is another Groundwater Gley found in this association and although still a Typical Groundwater Gley, the texture classification for this soil is coarse loamy drift with siliceous stones.

Other associations with a Brown Earth lead series include 1100n and 1100c, which are texturally different from association 1100a, with both classified as coarse loamy drift associations with siliceous stones with the Clashmore (1100CM) lead series (Fig. 10.4a). The Clashmore lead series is a Typical Brown Earth characterised by brown topsoil with a reddish hue, with a Munsell colour matrix value of 7.5YR. The soil has a friable consistence with a low packing density graduating to a medium packing density down the profile. Also, 1100e fine loamy over shale and slate bedrock with the Ballylanders (1100BY) lead series (Fig. 10.4b). Ballylanders is also a Typical Brown Earth found over shale bedrock which enhances the free draining nature of the soil. These soils are slightly acid, with moderate base saturation, a moderate structure and with a friable consistency (SIS 2014). Although this series tends to have a low pH, the soil is considered to respond well to fertiliser amendments (Finch and Ryan 1966).

After association 1100a, association 05RIV is the next most prevalent soil association across the rolling lowlands

Table 10.1 Main soil associations of the rolling lowlands, association description and the main series in the association[a]

Main associations (prop. excl. peat and mask) 2.2 M ha	Mainly occurs (prop. of association)	Association description	Main series (lead series first)
1100a (13.6%)	Wexford (30.3%), Cork (17.2%) Wicklow (13.65%), Kilkenny (12.15%), Tipperary (9.03%), Waterford (6.69%), Louth (6.5%) …	Well-drained Brown Earths on fine loamy drift with siliceous stones in an undulating land with some Brown Podzolics on upper slopes. Surface-water Gleys and Groundwater Gleys found in depressions	Clonroche (1100CL), Ballylanders (1100BY), Kilrush (0700KR), UNnamed_29 (0900UN29), Dunboyne (1000DB), Kilpierce (0600KP) …
05RIV (11.7%)	Tipperary (9.18%), Cork (7.89%), Limerick (7.68%), Meath (7.37%), Kerry (6.76%), Offaly (5.19%) …	Alluvial and drained alluvial soils on river floodplain with base-rich and medium to coarse texture	Boyne (0500BO), Finisk (0570FI), Aherlow (0570AH), Clohamon (0570CN), Suir (0570SU) …
0900e (9.2%)	Cork (80.1%), Kerry (17.4%), Mayo (2.45%)	Brown Podzolics and Brown Earths predominate with altitudinal sequence of upland Peat, Lithosols and Podzols on coarse loamy drift with igneous and metamorphic stones	Ross Carbery (0900RO), UNnamed_29 (0900UN29), Ballylanders (1100BY), Driminidy (0760DY) …
0700b (9.1%)	Kerry (27.9%), Clare (23.6%), Roscommon (11.7%), Cork (11.5%), Wexford (9.01%), Limerick (7.26%) …	Surface-water Gleys, Brown Earths and Brown Podzolics on fine loamy drift with siliceous stones and inclusions of Groundwater Gleys	Kilrush (0700KR), Gortaclareen (0760GN), Ballylanders (1100BY), Cupidstownhill (0900CU) …
1100n (7.2%)	Cork (45.5%), Kerry (29.2%), Waterford (7.97%), Cavan (5.6%) …	Brown Earths, Luvisols and Surface-water Gleys on coarse drift with siliceous stones	Clashmore (1100CM), Kilrush (0700KR), Dunboyne (1000DB), Clonroche (1100CL), Crosstown (1030CO) …
0700 h (6.5%)	Cavan (30.5%), Leitrim (25.6%), Monaghan (17.3%), Meath (10.6%), Longford (10.2%) …	Surface-water Gleys associated with Luvisols, on fine loamy drift with siliceous stones and Basin Peat	Kilrush (0700KR), Gortaclareen (0760GN), Banagher (0270BR), Aughty (0110AT), Heathtown (0630HT), Corlea (0636CL) …
1100e (4.5%)	Louth (37.1%), Monaghan (25.4%), Wexford (9.5%), Meath (8.42%), Clare (8.1%) …	Brown Earths on lower slopes, Brown Podzolics on steeper slopes, Luvisols and Surface-water Gleys in depressions related to fine textures on shale bedrock and on drift with siliceous stones	Ballylanders (1100BY), Ballyphilip (0700BY), Kilrush (0700KR), Kells (1100KE) …
1100c (3.6%)	Mayo (44.5%), Waterford (25.3%), Kilkenny (16.65%), Wexford (12%) …	Brown Earths and Luvisols on upper slopes; Surface-water Gleys and Groundwater Gleys in depressions. Coarse loamy drift with siliceous stones	Clashmore (1100CM), Clonroche (1100CL), Broomhill (1100BL), Ballylanders (1100BY).…

[a]Full soil descriptions are available at series level through the online soils database at http://gis.teagasc.ie/soilsguide.php

accounting for almost 12% of the area. In Ireland, two main types of alluvial soils are found. First, alluvial soils that formed in alluvial deposits and remain poorly drained due to the influence of a shallow groundwater and second, drained alluvial soils formed on alluvial deposits that have been drained but are well to imperfectly drained (Simo et al. 2014a). These soils occur along flat topography or river floodplains and so are mostly found in lowland areas (Fig. 10.5). These soils display similar characteristics to Groundwater Gleys. However, alluvial soils typically display distinct layers of soil material. The layers are present from a new layer of soil or sediment which is added to the surface after each major flooding event of the river. Typically, alluvial soils reflect gleyic properties, however, if there is a significant increase in clay in the B horizon overlying a coarse substratum, it can give rise to stagnic properties (Simo et al. 2014a). Overall, alluvial soils are poorly draining soils that have a weak soil structure (Fig. 10.6).

Alluvial soils are found in every county in Ireland. Tipperary accounts for the largest extent of the alluvial soils but with an extensive river network that includes the River Suir Basin in the south and the River Shannon to the north, this is expected. The Camoge Series, classified as a Humic Alluvial Gley, can be found in the lowlands to the north of Nenagh and areas around Templemore and Thurles where a shale

Fig. 10.4 Modal profile pits Typical Brown Earths **a** Clashmore series coarse loamy soil with siliceous stones. **b** Ballylanders series fine loamy soil with siliceous stones. Images from the Irish Soil Information Systems field campaign

and sandstone influence can be found in the alluvium (Finch and Gardiner 1993). This series is also prevalent in Limerick, which accounts for another large portion of alluvial soils. The Suir series is widely found in Co. Waterford in the floodplains of the major rivers, Suir, Blackwater and Bride (Fig. 10.7a, b). The parent material is till composed predominantly of sandstone and some limestone (Diamond and Sills 2011). Co. Cork has three main river networks, Lee, Bandon and Blackwater and whilst these provide the natural drainage for the county, a considerable portion of alluvial soils can be found along their river floodplains.

Association 0900e with the Ross Carbery lead series (0900RO) classified as a coarse loamy drift association with siliceous stones is the next most commonly found association accounting for almost 10% of the rolling lowlands. This association is composed mostly of Brown Podzolic soils with Brown Earths, with some inclusions of water-affected Surface-water and Groundwater Gleys. The most distinguishing feature of Brown Podzolic soils is the reddish subsoil associated with the enrichment of leached iron or humus from upper horizons (Fig. 10.8). Some leaching may have occurred but as yet, these soils have not developed the poor acidic mineral topsoil or the bleached sandy layer characteristic of Podzols. Even so, these soils still have lost much of their nutrients, particularly iron and aluminium overtime due to acidic conditions and high rainfall. These soils can be very good agricultural soils benefitting from periodic liming and other nutrient amendments. 0900e with Ross Carbery lead series is almost entirely found in Co. Cork, but with some found in counties Kerry and Mayo also. The series is classified as a coarse loamy compact drift with siliceous stones. The soil has a friable consistency with high porosity with rooting common in the upper horizons (SIS 2014).

Association 0900h is another association that has a Brown Podzolic lead series, NBP4 (0900NBP4) albeit less extensively found (approximately 3%). This association is predominantly Brown Podzolics, Brown Earths and Lithosols on gneiss and schist bedrock and on drift with igenous and metamorphic stones. Some inclusions of Surface-water Gley and Groundwater Gleys can be found. This association is concentrated in the north-west, primarily in Co. Donegal

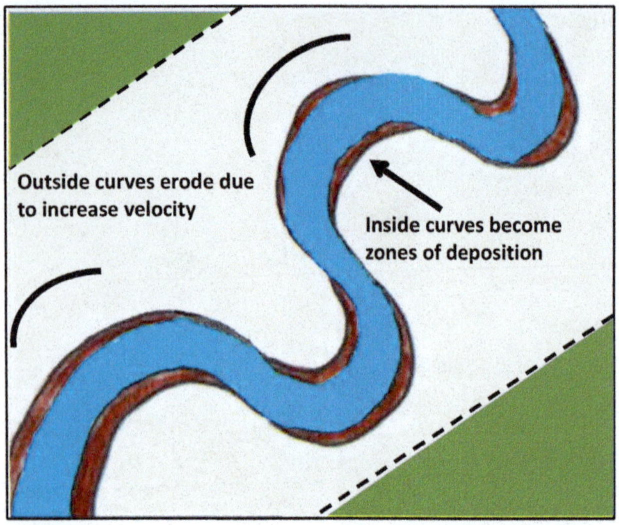

Fig. 10.5 Land that is almost horizontal and found bordering a river can result in an alluvial flat or floodplain. During times of flooding, this area will receive the alluvium deposited by the river (shown between the dashed lines). The faster flow of the water in the outer concave curves (meander) of the river water results in erosion. Conversely, the lower friction and slower flowing water on the inner convex bend (slip-off slope) results in deposition

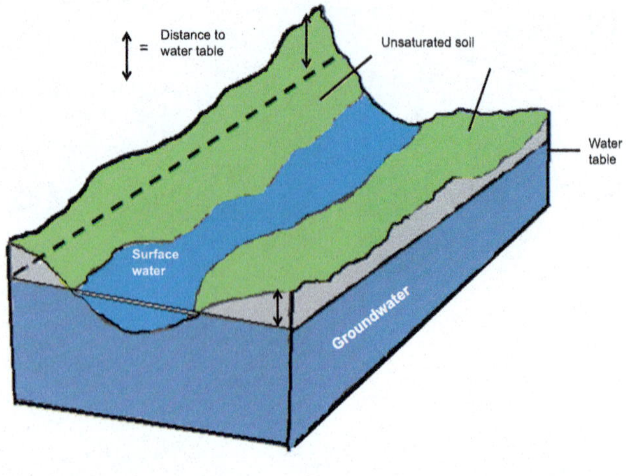

Fig. 10.6 Cross section of land showing water table which occurs at the top of the groundwater. Groundwater comes mostly from precipitation (referred to as recharge), that has infiltrated the soil zone and has percolated downwards until a zone of saturation where all soil spaces/pores are filled with water. As shown, across a landscape, the depth of the groundwater table varies, being closer to the surface in some areas over others depending upon the landscape topography and relief

(90%) and in Co. Mayo (7.7%) with the remainder in Co. Sligo.

After association 0900e, association 0700b led by the 'Kilrush' soil series is the next most extensively found soil association across the rolling lowlands. The poor drainage of Surface-water Gleys is typically due to high clay content or the presence of dense material in the subsoil. These soils exhibit a gleyed argillic B horizon (Btg). They have a medium to high packing density and reflect increasing plasticity in the soil profile. The soil tends to have a fine texture in the lower subsoil horizon and prominent mottling is observable above 40 cm (Simo et al. 2014a, b). In WRB classification, these soils correlate with Stagnosols (Reidy et al. 2014). Association 0700b is found in 16 counties, but is most extensively found in counties Kerry (27.9%), Clare (23.6%), Roscommon (11.7%) and Cork (11.5%). The association is found in counties Wexford, Limerick, Tipperary, Monaghan and elsewhere, but to a lesser extent. The Kilrush soil series texture class is defined as fine loamy over drift with siliceous stones, with consistency that inclines towards being plastic and sticky at depth (SIS 2014). Overall, this association is characterised mostly by wet soils. As well as the Kilrush soil series, 11 other soil series are classified in this association, including three Typical and two Humic Surface-water Gleys and one Typical and two Humic Groundwater Gleys. The presence of the humose A horizon is an indication of very wet conditions. The humose topsoil usually qualifies as a mollic horizon if it has high base saturation (FAO 2006) or umbric horizon in WRB if it is base depleted (IUSS 2006). Humose soil materials will have a moist rubbed colour value/chroma of 3/2 and in the field, this soil will feel soapy and slippery due to the organic matter. This makes it more difficult to detect in silty soils as organic and silty materials have a similar feel. Some Brown Earths are found in the association, including the Ballylanders series which will have better natural drainage compared to the other soils in this association.

Association 0700a with the Macamore lead series is another wet association found in the lowlands. Although this association accounts for only a small areal extent, this soil series has a unique spatial context associated with soil formation. The parent material of this series consists of dense, calcareous, marine muds and oozes which were removed from the Irish Seabed and deposited during the Midlandian ice age. The Irish Sea ice deposited till that contained sand and gravel, mixed with marine shell fragments and local erratics including Leinster granite (Culleton 1978). As a result, the occurrence of this association is limited to the east coast, mostly Co. Wexford with some distribution found in Co. Wicklow. The Macamores are defined as fine loamy soils over calcareous Irish Sea till, correlating with Luvic Stagnosol (Abruptic Humic Eutric Clayic) in WRB classification (Reidy et al. 2014). These tend to have a sandy clay loam texture in the upper horizons with a medium base status. They have a weak to coarse prismatic structure further down the soil profile with strong mottling throughout.

Fig. 10.7 **a** Alluvial soil profile for Suir (0570SU) series, a silty river alluvium highlighting evidence of depositional layering and **b** the associated above ground landscape

10.4 Soil Suitability of the Soil Associations

In general, Brown Earths are considered good agricultural soils with a wide use range potential. In particular, their natural free drainage means that they do not require additional drainage and their ability to retain nutrients means that they have good agricultural potential. The Clonroche series is considered to have a very wide use range and the friable consistency and durable structure of the soil means that they are among the best tillage soils in Ireland (Diamond and Sills 2011). In Co. Wexford, these soils are used for the production of a range of crops such as cereals, root crops and soft fruits. If not used for tillage, these soils are very good for grass production. The free draining soil conditions mean that potential grass utilisation is high with limited risk of damage due to poaching. Although easy to cultivate, the natural nutrient status of these may be low but they respond well to management and nutrient amendments. The Ballylanders series also has a wide use range. They are often likened to the Clonroche series and exhibit no particular limitations. They can be very productive grasslands particularly for short-term highly productive leys (Finch and Ryan 1966). Again, the natural nutrient status can be low but these soils respond well to management. These soils have good water holding capacity but shallow profiles may be prone to drought in dry seasons. Humic Brown Earths, such as the Ashgrove series with high organic matter content in the topsoil may be moderately acidic and may therefore benefit from periodic liming.

Groundwater Gleys are water-affected soils and have a limited use range with a high susceptibility to poaching. The Kilpierce series are often found in depressions which may render them unsuitable for land drainage operations. These soils are typified by a low lime and nutrient status, however, these soils may still support pasture production but careful management to limit the risk of poaching must be applied (Gardiner and Ryan 1964). Alluvial soils reflect similar characteristics to Groundwater Gleys. Alluvial soils are limited due to their very high risk of flooding. They have poor soil drainage, weak structure and are very vulnerable to poaching when grazed (Finch and Gardiner 1993).

The main limitation associated with Surface-water Gleys also relates to wetness due to their poor internal drainage. Their wetness results in a limited use range with grassland or forestry representing the most suitable agricultural land use options. Poor drainage of the Macamore series is attributed

Fig. 10.8 Ross Carbery series modal profile with evidence of a reddish subsurface horizon indicating historical leaching of iron/humus from upper horizons. Brown Podzolics make good agricultural soils once managed appropriately

to the weak structure coupled with the heavy texture of the subsoil. As with all wet soils, controlled grazing, appropriate nutrient management and controlled timing of operations are all essential to enhance the productive capacity of these soils, whilst also limiting the risks of soil structural damage, such as compaction or poaching associated with machine operations or grazing. Many of these soils will benefit from drainage operations to improve their productive capacity.

Brown Podzolic soils are good agricultural soils due to their favourable physical properties. Poaching is not a particular problem as these have good natural drainage and so an extended grazing season is possible. The main limitation is associated with the nutrient status of these soils, which require frequent liming to neutralise the acidity and other amendments to increase fertility.

Case Study 10.2—A Varied Landscape: Drainage Measures in Lowlands

To overcome the limitation of wetter areas in lowlands, management practices have evolved over the years. Open drains used to be common measures, along with subsoil stone drains. These can have limited success if not installed in the best locations. They also require long-term maintenance to prove effective. Many of these old systems become blocked with debris and clay and silt particles over time. Subsurface drains correctly spaced, are the most effective method used at the field scale.

In Fig. 10.9a1, we can see a good sward growing in a drained field. Despite a heavy clay loam A and B horizon, this Humic-Stagnic Surface-water Gley can be trafficable and grazed for a longer season as the soil structure is improved in the upper horizon (Fig. 10.9 a2). Figure 10.9b1 indicates how improved land conditions can result in new problems with weed species such as ragwort becoming established, that would normally not have become established on this Humic Surface-water Gley. This productive sward is growing on an A horizon containing 35% clay (Fig. 10.9b2).

A consequence of draining peat in lowland areas can be shrinkage. As the water table lowers, the peat dries out and increases in density. The drained field in Fig. 10.9c1 has lowered 30 cm in two years of drainage. The peat in these upper horizons has compacted and become earthy (Fig. 10.9c2). Figure 10.9d1 despite a steep incline of 10° still has drainage problems due to its north-facing aspect and subsoil stagnation due to an impediment in the subsoil material preventing lateral flow. The humic material has built up in the A horizon due to the drainage restriction (Fig. 10.9d2)

The history of land drainage and the role of drainage in primary productivity in Ireland are described in more detail in Chap. 14.

Fig. 10.9 (**a1**) Drained and reseeded field Co. Clare displaying humic features in the A horizon (**a2**); Ragwort growing in a drained field Co. Limerick (**b1**) with strong gleyic features in the subsoil (**b2**); Drained peat in Co. Kerry (**c1**) with shrinkage of the A horizon (**c2**); North-facing hill undrained (**d1**) subsurface lateral flow causing humose build up in the A horizon (**d2**)

References

Creamer R, Simo I, Reidy B, Carvalho J, Fealy R, Hallet S, Jones R, Holden A, Holden N, Hannam J, Massey P, Mayr T, McDonald E, O'Rourke S, Sills P, Truckell I, Zawadzka J, Schulte R (2014) Irish Soil Information Synthesis Report (2007-S-CD-1-S1). Associated datasets and digital information objects connected to this resource are available at: Secure Archive For Environmental Research Data (SAFER) managed by Environmental Protection Agency Ireland http://erc.epa.ie/safer/resource?id=1bfe6109-3b92-11e4-b233-005056ae0019. Last Accessed: 2016-10-03

Culleton EB (1978) Characterisation of glacial deposits in south Wexford. Proc R Irish Acad 78B:293–308

Diamond J, Sills P (2011) Soils of Co. Waterford Soil Survey Bulletin No. 44. National Soil Survey of Ireland, Teagasc, Co. Carlow

Feehan J, O'Donovan G (1996) The bogs of Ireland an introduction to the natural, cultural and industrial heritage of Irish peatlands. UCD, The Environment Institute, Walsh Printers, Roscrea, Co. Tipperary, Ireland

Finch TF, Gardiner MJ (1993) Soils of Tipperary North Riding Soil Survey Bulletin No. 42. National Soil Survey of Ireland, Teagasc, Dublin 4

Finch TF, Ryan P (1966) Soils of Co. Limerick Soil Survey Bulletin No. 16, National Soil Survey of Ireland. An Foras Talúntais, Dublin

Food and Agriculture Organization (FAO) (2006) Guidelines for Soil Description. Fourth Edition, FAO Rome, 97pp

Gardiner MJ, Ryan P (1964) Soils of Co. Wexford Soil Survey Bulletin No. 1. National Soil Survey of Ireland, An Foras Talúntais, Dublin 4

Hammond RF (1981) The Peatlands of Ireland to accompany Peatland Map of Ireland, 1978. An Foras Talúntais, Dublin 4

Irish Peatland Conservation Council (IPCC) (2016) Blanket Bogs, Online at: http://www.ipcc.ie/a-to-z-peatlands/blanket-bogs/. Accessed 18/08/2016

IUSS Working Group WRB (2006). World Reference Base for Soil Resources—a framework for international classification, correlation and communication. World Soil Resources Reports 103. ISSN 0532-0188, FAO Rome, 128 pp

Mitchell F, Ryan M (1986) Reading the Irish landscape. Town House, Dublin

O' Connell M (1990) Origins of lowland Irish blanket bog. In: Doyle GJ (ed) Ecology and conservation of Irish peatlands. Royal Irish Academy, Dublin, pp 49–71

Reidy B, Simo I, Spaargaren O, Creamer RE (2014) Irish SIS Final Technical Report 8: Correlation of the Irish Soil Classification System to World Reference Base—2006. Associated datasets and digital information objects connected to this resource are available at: Secure Archive For Environmental Research Data (SAFER) managed by Environmental Protection Agency Ireland. http://erc.epa.ie/safer/resource?id=a1b872a2-3b8c-11e4-b233-005056ae0019 . Last accessed: 25/10/2016

Renou-Wilson F, Bolger T, Bullock C, Convery F, Curry J, Ward S, Wilson D, Müeller C (2011) BOGLAND: Sustainable management of Peatlands in Ireland. STRIVE report series No. 75 prepared for the Environmental Protection Agency by University College Dublin

Simo I, Creamer RE, Reidy B, Hannam JA, Fealy R, Hamilton B, Jahns G, Massey P, McDonald E, Schulte RPO, Sills P, Spaargaren O (2014a) Irish SIS Final Technical Report 10: Soil Profile Handbook. Associated datasets and digital information objects connected to this resource are available at: Secure Archive for Environmental Research Data (SAFER) managed by Environmental Protection Agency Ireland. http://erc.epa.ie/safer/resource?id=a1b872a2-3b8c-11e4-b233-005056ae0019.

Simo I, Creamer RE, Hannam JA, Jones RJA (2014b) Irish SIS Final Technical Report 13: Irish Soil Information Systems Legend. Associated datasets and digital information objects connected to this resource are available at: Secure Archive For Environmental Research Data (SAFER) managed by Environmental Protection Agency Ireland http://erc.epa.ie/safer/resource?id=38a6e9e9-4364-11e4-b233-005056ae0019. Last accessed: 2016-10-03.

SIS—Irish Soil Information Systems (2014) Online resource. http://gis.teagasc.ie/soils/index.php. Accessed 28/09/2016

Drumlin Landscapes

11

Lilian O'Sullivan, Brian Reidy, and Rachel Creamer

Abstract

In Ireland, evidence of two main glacial periods exists, the Munsterian (c 300,000–132,000 years ago) and the more recent Midlandian (c 79,000–13,000 years ago). Today, Ireland has a very rich legacy of glacial deposits and landforms which can be widely found across the Irish landscape. Drumlins, one of the most commonly found glacial landform features, are smooth, oval shaped hills consisting mostly of deposited boulder clay or till. In Ireland, as the glacier retreated, the melting ice left large areas across the country, particularly the central plain, covered in boulder clay deposits. Two different types of drumlins occur, rock and till. Till or drift drumlins are comprised of compact, unstratified glacial till or drift. In contrast, rock drumlins are mostly made up of rock core with a concentric covering of drift. Rock drumlins tend to have a slightly better drainage status than till drumlins. In Ireland, around 25,000 drumlins have been recorded by the Geological Survey of Ireland. Drumlins generally occur in swarms and Ireland is home to one of the most extensive drumlin belts in Europe stretching across parts of Northern Ireland from Co. Down into the Republic of Ireland. Drumlins also occur in the Castlebar—Newport—Clew Bay region of West Mayo, in Co. Clare, and in the Bantry Bay area of West Cork. Of the seven major soil associations across the drumlin landscape, four have a gley lead series meaning that these are predominantly water-affected soils. Peats are also prevalent across the drumlin belt, particularly in counties Donegal and Leitrim. Poor natural drainage represents a major limitation to utilisation and while certain soils might benefit from artificial drainage, drumlin slopes present a major limitation in the use of agricultural equipment to carry out such operations.

Keyword

Boulder clay • Drumlin belt • Till • Drift Glaciation

11.1 Introduction

In Ireland, evidence of two main glacial periods exists, the Munsterian (c 300,000–132,000 years ago) and the more recent Midlandian (c 79,000–13,000 years ago). While more glacial periods may have occurred, such as the Elster glaciation, earlier evidence has largely been removed due to the destructive power of more recent ice sheets. Today, Ireland has a very rich legacy of glacial deposits and landforms which can be widely found across the Irish landscape. Drumlins, one of the most commonly found glacial landform features, are considered to have been first recognised in Ireland (Mitchell and Ryan 1986). The name drumlin comes from the Gaelic word '*dromnín*' which means small hill and was first used by Maxwell Close in 1866 as a general term for hills of this kind (Foster Flint 1957). Drumlins are smooth, oval shaped hills consisting mostly of deposited boulder clay or till.

During interglacial periods, exposed rock surfaces are subject to denudation resulting in a surface layer of disintegrated rock material. During glaciation, the friction between the advancing glacier and the rock bed results in ice melting around the base of the glacier. When the meltwater refreezes, materials on the rock bed freeze to become part of

L. O'Sullivan (✉) · R. Creamer
Teagasc Agriculture and Food Development Authority, Johnstown Castle Environment Research Centre, Wexford, Ireland
e-mail: Lilian.OSullivan@teagasc.ie

L. O'Sullivan
Irish Environmental Protection Agency, Johnstown Castle, County Wexford, Ireland

B. Reidy
Teagasc Moorepark, Animal and Grassland Research and Innovation Centre, Fermoy, County Cork, Ireland

R. Creamer
Wageningen University and Research Atlas—Wageningen Campus, Wageningen, The Netherlands

R. Creamer and L. O'Sullivan (eds.), *The Soils of Ireland*, World Soils Book Series,
https://doi.org/10.1007/978-3-319-71189-8_11

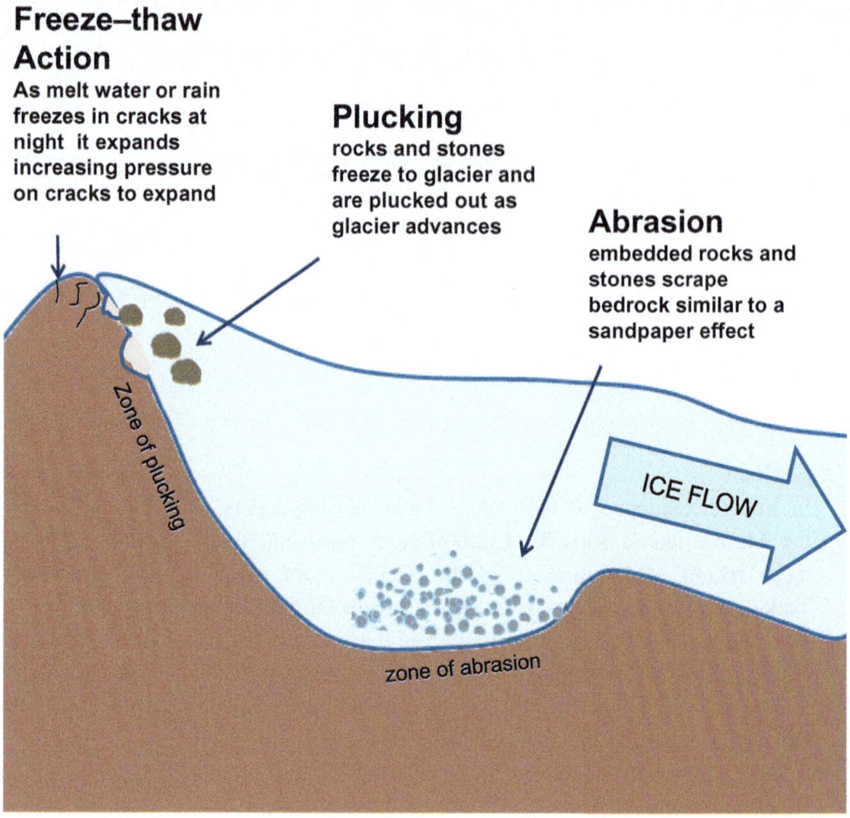

Fig. 11.1 Plucking and abrasion as ice advances resulting in erosion of the bedrock

the glacier so that the newly trapped rock is plucked out (Fig. 11.1). This then becomes part of the abrasion process. Abrasion is similar to a sandpaper effect where debris and material that are embedded in the glacier, erode the rock bed (Fig. 11.1). Abrasion results in glacial striations which serve as an indication of the direction of ice movement. The presence of striations on polished, grooved subangular stones is a leading characteristic of boulder clay. The presence of erratics (non-local stones) is also a differentiating characteristic of boulder clay. In Ireland, as the glacier retreated, the melting ice left large areas across the country, particularly the central plain, covered in boulder clay deposits. Kilroe et al. (1910) explain that the stony portion of the boulder clay deposit is derived from the underlying rocks while the clayey matrix represents an even more broken down condition of the same rock. The friction resulting from stones rubbing against each other and the rock floor smoothed and rounded them, while others were further broken down into fine plastic clay (Kilroe et al. 1910).

Although the exact formation of drumlins is a much debated topic it is generally accepted that drumlins formed as the ice became overloaded with sediment material which was deposited when the strength of the glacier was reduced. Also, material on the ground may have acted as an obstacle resulting in till building up around it. The power of the glacier may have reduced due to the melting of ice or a change in the speed of the glacier. Glaciers slide due to ice melting which occurs under pressure giving rise to a film of water between the bedrock and the ice. This results in a separation between the bedrock and the ice that can promote fast ice flow or 'basal sliding'. Geological research into the drumlin fields of Ireland indicates that the drumlins are a result of high basal water volumes leading to a separation of the glacier from the underlying sediment causing a down draw into the surrounding ocean and a surge in the glacier (Coxon 2001). The streamlining of the drumlin fields in the direction of the major bays to the west coast of Ireland is evidence of such a surge (Coxon 2001). The surging associated with high water pressure is likely to have streamlined the underlying deposits and the large volumes of water, coupled with reduced shear stresses would have allowed for the preservation of the resulting drumlins (Coxon 2001).

While composed mostly of glacial deposits, drumlins may contain a core of bedrock (Catt 1986). Kilroe et al. (1910) provide an in-depth explanation of the drumlins found at the Ballyhaise Agricultural Station in Co. Cavan. In

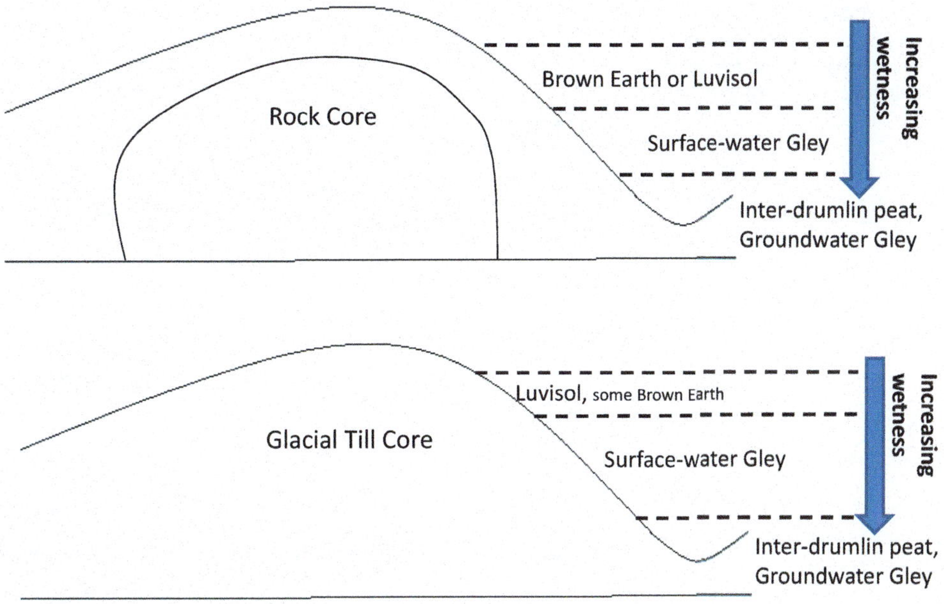

Fig. 11.2 Typical distribution of Soil Subgroups across a drumlin showing core structure, with a greater proportion of wetter soils found on drumlins with a glacial till core

it, the two different types of drumlins that occur, rock and till are described. Till or drift drumlins are comprised of compact, unstratified glacial till or drift. In contrast, rock drumlins are mostly made up of rock core with a concentric covering of drift. Rock drumlins are more likely to be found closer to the centre of the drumlin field (Charlesworth 1957). Although rock drumlins are very similar to true drift drumlins, their shape is generally less symmetrical and they may appear larger and narrower (Charlesworth 1957). Rock drumlins display upper and lower convex slopes with concave slopes intervening while till drumlins have almost uniform convex slopes on all sides (Lee and Ryan 1965). Rock drumlins tend to have a slightly better drainage status than till drumlins. While drainage status is largely determined by soil texture, soils downslope will tend to be more water affected than those further upslope, even where the drumlin has a rock core. Figure 11.2 shows the typical position of Soil Great Groups across a drumlin, highlighting that the wetter soils are more likely to be found further downslope and that drumlins with a rock core tend to have a wider extent of free-draining soils.

The most distinctive characteristic of drumlins is their shape which is half-ellipsoidal meaning that drumlins are often likened to half an egg or the inverted bowl of a spoon. As drumlins are formed beneath moving ice they can indicate the direction of the glacier movement. The long-axis runs parallel to the direction of the ice flow with the higher wider 'stoss' side upstream, tapering down to a pointed 'lee' end (West 1968). The ice flowing from right to left with the long axis of the drumlin running from A to B (Fig. 11.3). The maximum width is indicated by the line running C–D with the highest point at the intersection of A, B, C and D. In Fig. 11.4 a drumlin is visible feeding into Sligo Bay. In Figs. 11.3 and 11.4, the stoss and lee ends are clearly defined but such a distinct difference is sometimes not so readily evident, such as occurs with the drumlins in Co. Clare where many of the drumlins lack clear stoss end development (Finch and Walsh 1973).

Drumlins frequently occur in clusters across the landscape giving rise to the so-called 'egg-basket' topography related to their external morphology. Although several different dimensions are described in the literature, in general drumlins may be up to 1 kilometre long, 300–400 m wide (Meehan 2013) and range in height from a few metres to >100 m (Catt 1986). Figure 11.5 is based upon the Geological Survey of Ireland (GSI) data that builds on the work of Greenwood and Clark (2009) that digitised the subglacial bed forms of the last ice sheet in Ireland (Meehan 2013). The availability of a 10 m Digital Elevation Model (DEM) by the Ordnance Survey of Ireland (OSI) allowed an even greater level of detail to be achieved (Fig. 11.5). There are almost 25,000 drumlins recorded across Ireland with an average

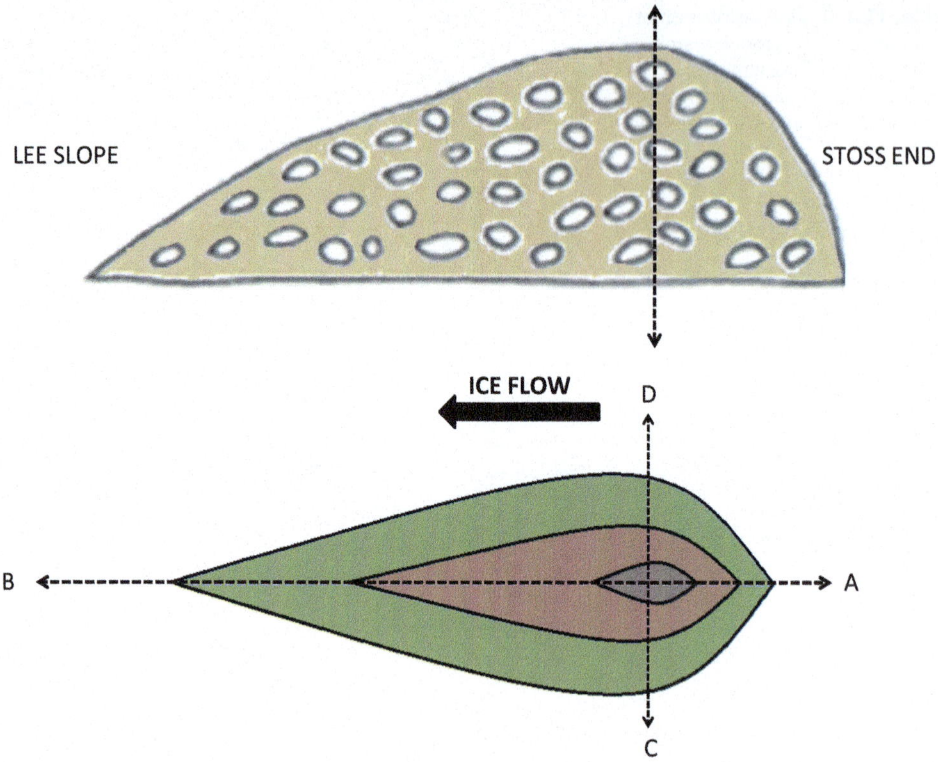

Fig. 11.3 Elevation and plan of a typical drumlin showing direction of ice flow and the steeper stoss end and the gentler sloping lee end (**a**–**b**)

Fig. 11.4 Drumlin feeding into Sligo Bay on the stoss end

drumlin length of 660 m and a height of between 15 and 25 m (Meehan 2013).

Drumlins represent a major physiographic feature across the Irish landscape but this is particularly true of north-central Ireland (Fig. 11.6). Drumlins generally occur in swarms and Ireland is home to one of the most extensive drumlin belts in Europe stretching across parts of Northern Ireland from Co. Down into the Republic of Ireland. Drumlins also occur in the Castlebar—Newport—Clew Bay region of West Mayo (10.15%), in Co. Clare (13.85%) and in the Bantry Bay area of West Cork (Lee and Ryan 1965). Although Co. Cork only represents a small proportion of the drumlin landscape overall, those found in the Bantry Bay area are well developed and represent good examples. Drumlins also occur outside of these major pockets, albeit to a lesser extent, right across the lowlands of Ireland.

The coastal area of Co. Mayo is made up of sharply undulating drumlins with a maze of islands formed from the partially submerged drumlins in Clew Bay. Further east, the terrain of the east-central drumlin spine of Mayo becomes less uniform. The drumlin topography in north Mayo includes both low-lying lakeland drumlins towards the

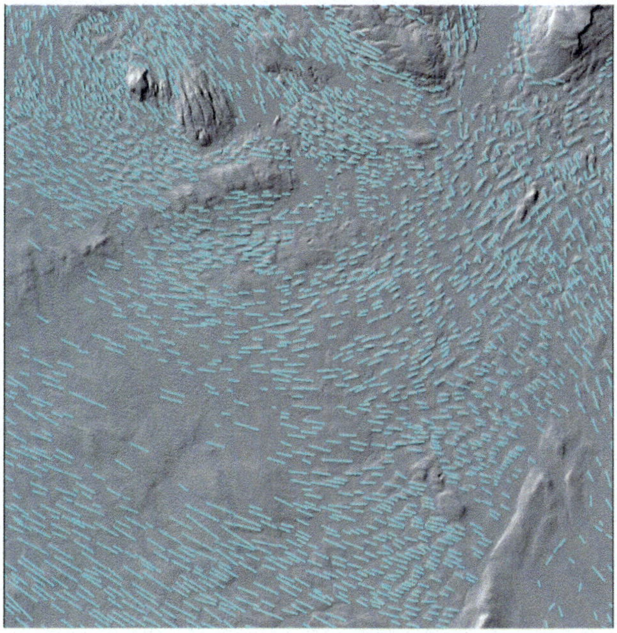

Fig. 11.5 Mapping detail shown by zooming into the map of the drumlin landscape with each individual fleck representing the crest of an individual drumlin. Data published by Geological Survey of Ireland from GSI Quaternary Subglacial Lineations, freely downloadable at http://spatial.dcenr.gov.i.e./GSI_DOWNLOAD/QuaternaryGeology.zip described by Meehan (2013)

southern end and steeper drumlins found around the foothills of mountains in the north-west and the Ox Mountains to the east (Fleming 2003).

In Co. Clare, there are four major concentrations of drumlins found at (1) Crusheen, (2) Scarriff–Sixmilebridge, (3) Ogonnellow–Limerick and (4) West Clare (Finch and Walsh 1973). In general, the drumlins of Co. Clare occur in the lowlands with few occurring on the limestone bedrock of the Burren. Detailed survey work conducted in the 1970s found that the majority of the drumlins are composed of glacial till of uniform geological composition. The majority of drumlins in the county lack stoss end development, with the exception of the Scarriff-Sixmilebridge concentration that is more classically defined (Finch and Walsh 1973). The distribution of drumlins reflects a concentration in the Upper Carboniferous Shale area. These shales weather to produce the heaviest-textured glacial till resulting in soils with poor internal drainage.

In Co. Cork the soils of the drumlins in the Bantry Bay area were described in the 1963 West Cork soil survey as dense, blue-grey drift material deposited characteristically as drumlin shaped hills (AFT 1963). Furthermore, that a rolling topography is found with a slope range of between 6° and 20° with the majority of the slopes between 6° and 15° (AFT 1963). The majority of the drumlins of Bantry Bay are association 1160c which is a Humic Brown Earth. This association is described as Brown Earths found typically in association with Brown Podzolics, Podzols and Surface-water Gleys associated to humic and developed on drift with siliceous stones.

11.2 Soil Associations of the Drumlin Belt of Ireland

Ireland is home to one of the most extensive drumlin fields in Europe stretching across parts of Northern Ireland from Strangford Lough in Co. Down into the Republic of Ireland across north Louth, north Meath, Monaghan, Cavan, Leitrim and south Donegal (Gardiner and Radford 1980). This is described as a drumlin belt and accounts for almost 40% of the drumlins in Ireland.

Across the drumlin belt, the drumlin soils are generally poorly draining, however, some soil types found, such as the 'Ballylanders' which is a Typical Brown Earth, tend to be associated with more free draining soils on the steeper drumlin slopes. Peat soils are regularly found across the drumlin landscape with peat frequently found in the inter-drumlin areas. Often these inter-drumlin areas may be waterlogged or may be occupied by lakes. This is commonly the case in counties Cavan and Monaghan as are fen peats and interlinking streams (Fig. 11.7).

Table 11.1 synthesises the main soil associations found across the drumlin belt of Ireland. Of the seven major soil associations, four have a gley lead series (0700h, 0660e, 0700c and 0760c) which are water-affected soils. Peats are also prevalent across the drumlin belt, particularly in counties Donegal and Leitrim.

Association 0700h is the dominant association found across the drumlin belt, led by the 'Kilrush' soil series. Co. Leitrim typifies the wetter areas of the drumlin belt with these heavy, slowly permeable soils on drumlin topography representing the principle soil of the county. The poor drainage of Surface-water Gleys is typically due to high clay content or the presence of dense material in the subsoil. In Leitrim, both factors are likely to contribute to these waterlogged soils. Even on favourable slopes, the heavy boulder clay of Upper Carboniferous shale composition means that these soils have poor natural drainage (Gardiner and Radford 1980). Moreover, gleying is intensified in areas where soils that are predisposed to permeability issues experience higher rainfall, as in Co. Leitrim. The association is also widely found in Co. Cavan and Co. Monaghan. Although the more freely draining drumlins are found towards the east, association 0700h can still be found to occur in Co. Meath. Elsewhere in Ireland, this soil association is found in the drumlin areas of Longford and Sligo.

The Kilrush lead series of association 0700h is comprised of fine loamy drift with siliceous stones associated with slow permeability resulting in poorly draining soils and waterlogging of the upper 40 cm of the soil profile for prolonged

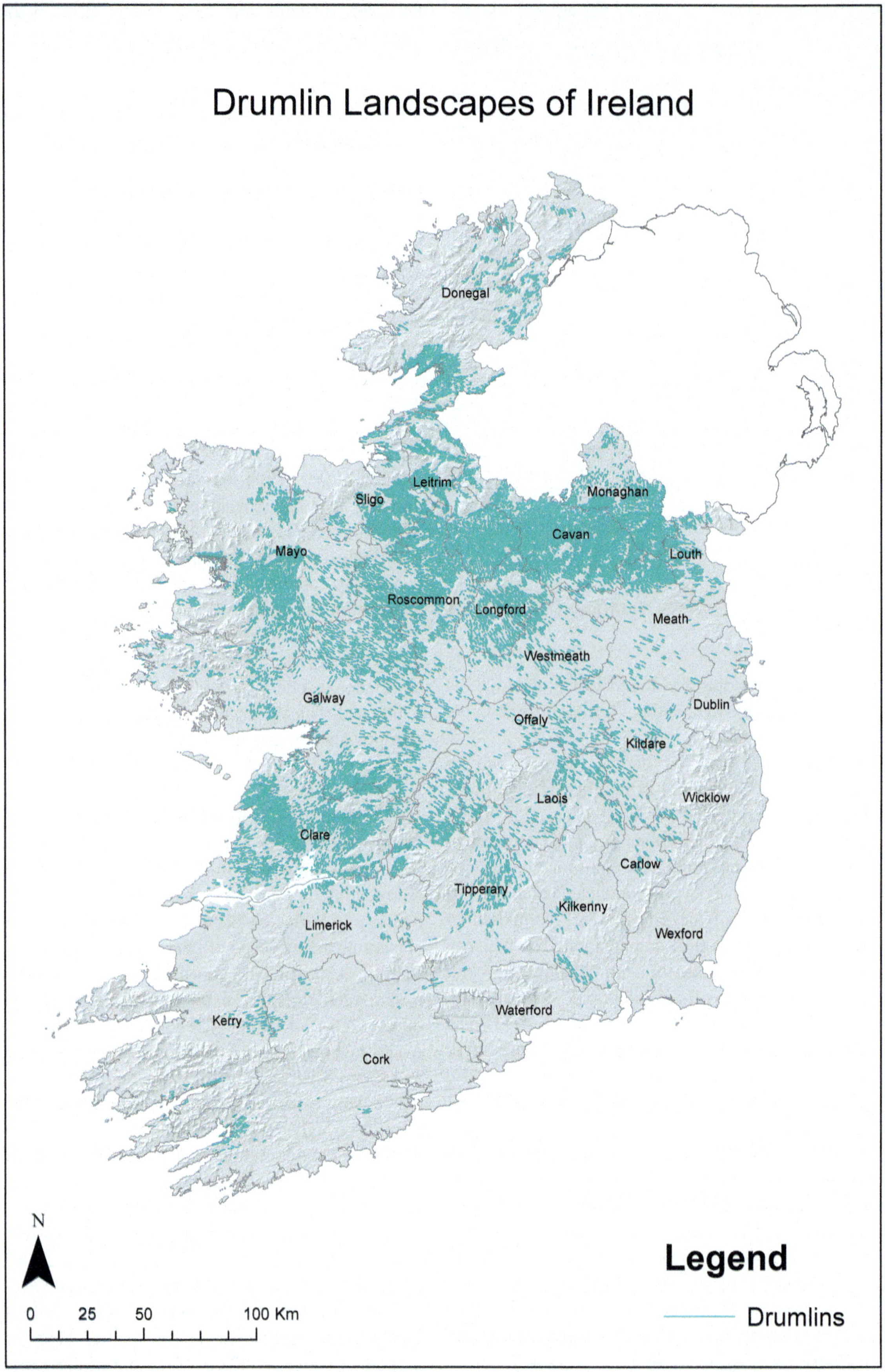
Drumlin Landscapes of Ireland
Donegal
Leitrim
Sligo
Monaghan
Cavan
Mayo
Louth
Roscommon
Longford
Meath
Westmeath
Galway
Dublin
Offaly
Kildare
Laois
Wicklow
Clare
Carlow
Tipperary
Kilkenny
Limerick
Wexford
Kerry
Waterford
Cork
N
0 25 50 100 Km
Legend
Drumlins

◀ **Fig. 11.6** Indication of the distribution of drumlins in Ireland based upon GSI data described by Meehan (2013) showing lines that have been digitised along the features' crests showing almost 25,000 drumlins recorded across Ireland. When shown at national scale the finer detail captured in Fig. 11.5 is lost and so the area shown serves as an indication only and not as spatial extent. Data based on published data from Geological Survey of Ireland, 'GSI Quaternary Subglacial Lineations', freely downloadable at http://spatial.dcenr.gov.i.e./GSI_-DOWNLOAD/QuaternaryGeology.zip described by Meehan (2013)

periods of the year (Simo et al. 2014). The series texture class is defined as fine loamy with consistency that inclines towards being plastic and sticky at depth (SIS 2014).

Within association 0700h, eight other soil series are found to co-occur. These include two Humic Surface-water Gleys; 'Gortaclareen' (0760GN) and 'Driminidy' (0760DY). The Gortaclareen series is characterised by poor drainage, mottling, a weak structure and a high base status (Finch and Ryan 1966; Finch et al. 1971). One Luvisol occurs within the association, the 'Dunboyne' series (1000BD) which is a deep soil with medium to high base status, and unlike the rest of the soils in this association is a moderately well to well-drained soil (Finch et al. 1983). Two peat soils occur within the association. The 'Aughty' series (0110AT) an Ombrotrophic Peat which is a blanket peat that has accumulated under conditions of high rainfall and humidity can be found in central and north Leitrim and in south Donegal. These climatic peats have a thick organic layer and are characterised by a low pH, <4.0. The 'Banagher' series (0270BR), the other peat, is a drained mineratrophic fen peat. Fen peats occur under the influence of base-rich groundwater and as a result are characterised by a pH > 4.0. Fen peats are commonly found in the inter-drumlin hollows and often include peaty soils and intermittent layers of alluvium (AFT 1963). The Banagher series is a reclaimed fen peat that has been drained for agricultural purposes.

Association 1100e is the next most widely found across the drumlin landscape of east Cavan, Monaghan and Louth and is led by the 'Ballylanders' soil series. The Ballylanders soil is classified as having a fine loamy texture over shale bedrock which results in soils that are naturally free draining. These soils reflect a low to medium packing density (SIS 2014). Association 1100e is found in association with Luvisols and Surface-Water Gleys (Table 11.1). This soil tends to have a low pH and nutrient status but is considered to respond well to fertiliser amendments (Finch and Ryan 1966). The association includes four Brown Earths, one Typical Brown Earth and three Stagnic Brown Earths. The 'Kells' series (1100KE) is a Typical Brown Earth which is classified as coarse loamy over shale bedrock. These are well-drained friable soils but reflect a low base status (Finch et al. 1983). So, while easy to cultivate they require adequate nutrient amendments to increase fertility. One Typical Luvisol occurs, the 'Dromkeen' series (1000DK) which is a fine loamy soil over shale bedrock.

Fig. 11.7 Image of a drumlin landscape taken along the drumlin belt of south Ulster taken during the Irish Soil Information Systems field campaign

Table 11.1 Major associations of the drumlin belt, association description and the main series in the association[a]

Assoc.	Occurs (prevalent areas in bold)	Association description	Main series (lead series shown in bold)
0700h	Cavan, Leitrim, Meath, Monaghan	Surface-water Gleys associated with Luvisols, on fine loamy drift with siliceous stones and basin peat	Kilrush 0700KR, Gortaclareen 0760GN, Banagher 0270BR, Aughty 0110AT
1100e	Monaghan, Cavan, Louth, Meath	Brown Earths on lower slopes, Brown Podzolics on steeper slopes, Luvisols and Surface-water Gleys in depressions, related to fine textures on shale bedrock on drift with siliceous stones	Ballylanders 1100BY, Ballyphilip 0700BY, Kilrush 0700KR Kells 1100KE …
1130a	Cavan, Monaghan	Brown Earths and Surface-water Gleys associated with humic and gleyic diagnostic features on fine loamy drift with siliceous stones with Luvisols	Moord 1130MO, Gortaclareen 0760GN, Kilrush 0700KR Gneaves 1136GV, Dunboyne 1000DB
0660e	Donegal	Poorly drained soils in upland areas composed of Humic Groundwater Gleys, Brown Earths and Lithosols on coarse loamy drift with igenous and metamorphic stones. Also, Brown Podzolics and Surface-water Gleys occurring	Ballywilliam 0660BW, Tramore 0700TR, Ballynabreen 0760BB, NBP2 0900NPB2 …
1000a	Meath, Leitrim, Monaghan	Luvisols associated with Surface-water Gleys, Stagnic Brown Earths and Calcareous Brown Earths, on fine loamy drift with limestones	Elton 1000ET, Faoldroim 1150FO, Musicfield 1130MF, Mullabane 1100MB, Straffan 0700ST
0700c	Leitrim, Cavan	Heavy soils with clayey and fine textures. Mostly Surface-water Gleys but some Luvisols and Brown Earths found to a lesser extent	Drumkeeran 0700DK, Cluggin 0760CG, Gortaclareen 0760GN, Kilrush 0700KR, Ballybrannagh 0700BB
0760c	Leitrim	Wetter than Typical Surface-water Gleys with poorly decomposed surface organic matter giving a humic classification. Found with Luvisols in Clayey lowlands on drift with limestones. Some Calcareous Brown Earths	Howardstown 0760HN, Ballyshear 0656BS, Elton 1000ET, Banagher 0270BR, Baggotstown 1150BG, Straffan 0700ST

[a]Full soil descriptions are available at series level through the online soils database at http://gis.teagasc.ie/soilsguide.php

The drumlins of Co. Meath are predominantly associated with calcareous soils. These drumlins are largely found in the north-west of the county, in the area between Moynalty to Castletown, north to the county boundary (Finch et al. 1983). Association 1000a is the most commonly found association in this area, which has the 'Elton' lead series. The Elton series is a Typical Luvisol which can be described as soils in which the clay material from the topsoil has eluviated and moved down the profile where it has been deposited further down resulting in an argillic B horizon (Creamer et al. 2014). The base saturation of these soils is >35%. These soils retain nutrients well and are therefore considered to have good agricultural potential. The higher clay content at depth can result in the slower movement of water through the soil profile depending upon soil texture, however, the Elton series are moderately drained friable deep soils with plentiful, well developed roots. In this association, Luvisol soils are found in association with Surface-water Gleys, Stagnic Brown Earths and Calcareous Brown Earths, on glacial drift with limestones (Table 11.1). Outside of Co. Meath, this soil association is found in drumlin areas in many other counties including; Clare, Kildare, Kilkenny, Offaly, Roscommon, Sligo, Tipperary and Westmeath.

In Co. Leitrim, Humic Surface-water Gleys can be found in areas near Manorhamilton, Kinlough and Ballinamore. Association 0760c with the 'Howardstown' (0760HN) lead series is not typically found elsewhere across the drumlin belt but can be found in the drumlin areas of Limerick and to a lesser extent in Longford and Mayo. These are extremely wet gleyed soils, where poorly decomposed organic matter has accumulated at the surface giving these soils a humic classification. The Howardstown series are classified as soils formed in clayey glacial drift with limestones. The texture of these soils is usually clayey, and clay content generally increases from the A horizon down the profile. A high percentage base saturation is typical for this soil type. Rooting tends to be confined to the A horizon (Finch et al. 1971). This association is predominantly found with Surface-water Gleys and Luvisols in clayey lowlands on drift with limestone, with inclusions of Calcareous Brown Earths. The 'Ballyshear' (0650BS) Humic Calcareous Groundwater Gley is also found to occur and is affected by a high groundwater table associated with its low-lying position in the landscape and consequently has poor drainage.

11.3 Soil Suitability and Land Use of the Drumlin Soils

The emergent soil quality of drumlins is generally determined by parent material but soils on drumlins in Ireland tend to be poorly drained. Where the parent material is entirely composed of drift core it will give rise to wet drumlins, whereas those with a rock core may be considered dry drumlins. Drumlins formed on low ground deposited under heavy ice pressure result in till that is very dense which is likely to produce land that is wet and difficult to drain. Poor natural drainage represents a major limitation to utilisation and while certain soils might benefit from artificial drainage, drumlin slopes present a major limitation in the use of agricultural equipment to carry out such operations. Drumlins frequently present high slopes in excess of 12°, above which the use of agricultural equipment is less likely due to the associated danger (Mockler et al. 2013). Ultimately, drainage operations or other mechanised operations should not be carried out in areas with steep slopes which are inherent in drumlin topography.

The most commonly found soil series across the drumlin landscapes is the Kilrush series. This soil tends to be poorly drained with a clay content of between 40 and 60%. They reflect a heavy texture and weak structure with slow permeability. As a result, they have a limited use range and land use is confined to grassland or to forestry (Finch and Walsh 1973). Vegetation consists of poor rush and sedge dominated swards. The weak structure of these soils limits the potential of mechanised agriculture due to an increased risk of compaction in high soil moisture conditions. The Howardstown soil series found in Co. Leitrim also has a limited use range, highlighting similar limitations associated with permeability issues. Adverse soil physical conditions, high annual rainfall and occasional steep slopes make these soils unsuitable for arable production. They are, however, good grassland soils (AFT 1973; Finch et al. 1971). Controlled grazing and proper use of lime and fertilisers will give a high level of grass production but efficient utilisation of the productive capacity of the soils is hindered due to the long wintering period and poor soil structure (AFT 1973; Finch et al. 1971). The higher rainfall rates in the west of the country compound the challenge of soil and land management. Co. Leitrim experiences adverse soil conditions and the combination of peat soils with wet Surface-water Gleys renders the productive capacity of these soils as extremely limited, even with optimum management practices. In general, for the wetter drumlin soils, forestry or grass production offers the best potential utilisation. Grass production is the main land use, but the susceptibility of these soils to compaction and poaching damage insists that prudent management practices are applied.

Several soils found across the drumlin landscapes, including the Brown Earths, allow for a relatively wide use range. It is important that while some soils offer a wide use range and good natural potential, the extent to which management options are applicable will be determined by the drumlin slope as the better draining soils occur on the steeper slopes. The Ballylanders soil is amongst those found that typically has a wide use range. They are often likened to the 'Clonroche' soils of Co. Wexford and similarly exhibit no serious physical limitation (Finch and Ryan 1966). Grassland can be very productive on these soils especially in short-term highly productive leys (Finch and Ryan 1966). The natural lime and nutrient status is low but these soils respond well to management. Another Brown Earth, the Kells series is typified by a low base status, and while friable and easy to cultivate, this soil type also requires adequate nutrient applications of lime and fertiliser (Finch et al. 1983). The Baggotstown series, a Calcareous Brown Earth, has a moderately wide use range and is only limited by slope, shallowness and the associated drought risk.

Luvisol soils, such as the Dunboyne series, are generally considered good agricultural soils. These soils are more typically found in Co. Meath where rainfall is less, but the excellent water holding capacity of these soils means that they do not suffer from drought. These soils have a moderate to wide range of uses and have a well-drained, medium to high base status (Finch et al. 1983). These soils are excellent for grass production but poaching may occur following periods of prolonged wetness.

Outside of agriculture and forestry, the drumlin landscapes provide many other important ecosystem services to society. In particular, the Clew Bay complex of drumlins in Co. Mayo represents an area of key value for biodiversity. The drumlins here are designated Special Areas of Conservation (SAC) or are proposed National Heritage Areas (pNHA). The maze of interlocking sunken drumlins in Clew Bay represents one of the few landscapes of its kind in the world. These drumlins were submerged at the end of the last ice age due to a rise in sea level associated with melting ice. The drumlin landscape found here is home to a wide variety of terrestrial and marine habitats including seven habitats listed on Annex I and two mammals listed on Annex II of the Habitats Directive (CAAS 2010). Hence, the drumlin landscape in this area makes the region important both nationally and internationally due to the wide range of habitats found here.

References

An Foras Talúntais (1963) West Cork resource survey. An Foras Talúntais, Dublin

An Foras Talúntais (1973) County Leitrim resource survey part 1—land use potential (Soils, Grazing Capacity and Forestry) Soil Survey Bulletin No. 29. An Foras Talúntais, Dublin

CAAS (2010) Environmental report of the Westport Town and environs development plan 2010-2016 Strategic Environmental Assessment. Report by CASS for Westport Town Council and Mayo County Council

Catt JA (1986) Soils and quaternary geology a handbook for field scientists. Clarendon Press, Oxford

Charlesworth JK (1957) The Quaternary Era with special reference to its glaciation in two volumes, vol one. William Clowes and Sons, Limited, London and Beccles

Coxon P (2001) Cenozoic: tertiary and quaternary (until 10,000 years before present). In: Holland CH (ed) Geology of Ireland. Dunedin Academic Press, Edinburgh

Creamer R, Simo I, Reidy B, Carvalho J, Fealy R, Hallet S, Jones R, Holden A, Holden N, Hannam J, Massey P, Mayr T, McDonald E, O'Rourke S, Sills P, Truckell I, Zawadzka J, Schulte R (2014) Irish soil information synthesis report (2007-S-CD-1-S1). Prepared for the environmental protection agency by Teagasc, Cranfield and University College Dublin. EPA, Wexford

Finch TF, Culleton E, Diamond S (1971) Soils of Co. Clare Soil Survey Bulletin 23, National Soil Survey of Ireland. An Foras Talúntais, Dublin

Finch TF, Gardiner MJ, Comey A, Radford T (1983) Soils of Co. Meath National Soil Survey of Ireland. An Foras Talúntais, Dublin

Finch TF, Ryan P (1966) Soils of Co. Limerick Soil Survey Bulletin No. 16, National Soil Survey of Ireland. An Foras Talúntais, Dublin

Finch TF, Walsh M (1973) Drumlins of County Clare. Proc R Irish Acad B Biol Geol Chem Sci 73:405–413

Fleming A (2003) Landscape Appraisal of County Mayo. Report prepared by CAAS Ltd., for Mayo County Council

Foster Flint R (1957) Glacial and pleistocene geology. Wiley, London

Gardiner MJ, Radford T (1980) Soil associations of Ireland and their land use potential explanatory bulletin to soil map of Ireland 1980. National Soil Survey of Ireland, An Foras Talúntais, Dublin

Greenwood SL, Clark CD (2009) Reconstructing the last Irish Ice Sheet 1: changing flow geometries and ice flow dynamics deciphered from the glacial landform record. Quatern Sci Rev 28(27–28):3085–3100

Kilroe JR, Seymour HJ, Hallissy T (1910) The geological features and soils of the agricultural station at Ballyhaise in the County of Cavan. Department of Agriculture and Technical Instruction for Ireland, Dublin

Lee J, Ryan M (1965) A clinosequence of soils associated with Drumlin Landscape. Irish Journal of Agricultural Research 4:1–12

Meehan R (2013) End of Project Report on Project Outputs. Report 1 of 1 on The Merging of Quaternary Map Datasets. Geological Survey of Ireland

Mitchell F, Ryan M (1986) Reading the Irish landscape. Town House Dublin

Mockler E, Bruen M, Desta M, Misstear B (2013) Pathways project final report volume 4: catchment modelling tool. STRIVE Report No. 56, Irish Environmental Protection Agency, Johnstown Castle, Ireland

Simo I, Creamer RE, Reidy B, Hannam JA, Fealy R, Hamilton B, Jahns G, Massey P, McDonald E, Schulte RPO, Sills P, Spaargaren O (2014) Irish SIS Final Technical Report 10: Soil Profile Handbook. Associated datasets and digital information objects connected to this resource are available at: Secure Archive For Environmental Research Data (SAFER) managed by Environmental Protection Agency Ireland. http://erc.epa.ie/safer/resource?id=a1b872a2-3b8c-11e4-b233-005056ae0019. Last accessed: 29/03/2016

SIS—Irish Soil Information Systems (2014) Online resource. http://gis.teagasc.ie/soils/index.php. Accessed 28/09/2016

West RG (1968) Pleistocene Geology and Biology with especial reference to the British Isles. Longmans, Green and Co Ltd., London and Harrow

Urban Soilscapes 12

Thomas Cummins, Franc Myles, Ray Scanlon, Mairéad Glennon, Vincent Gallagher, and Maryann Harris

Abstract

Urban areas have distinct soils, formed through earth-shaping and construction activities, and incorporating manufactured materials and wastes. Urbanisation and urban sprawl have been rapid in Ireland with a high relative increase in artificial surfaces in the period 1990–2006. While land conversion creates a valuable urban environment it also alters the delivery of soil functions. Historically, urban soils have been systematically under-represented in primary resource surveys, leading to critical information gaps. Soil landscapes of urban Dublin are discussed, with evidence of soil sealing, soil material affected by urban activity, contamination of soils, cultural deposits discovered during archaeological excavation, and consideration of the origin of open spaces in the city. Finally, a preliminary soil map for urban Dublin is presented.

T. Cummins (✉)
UCD School of Agriculture and Food Science, University College Dublin, Belfield, Dublin 4, Ireland
e-mail: thomas.cummins@ucd.ie

F. Myles
Archaeology and Built Heritage, Dublin, Ireland

R. Scanlon · M. Glennon · V. Gallagher
Department of Communications, Climate Action and Environment, Geological Survey of Ireland, Beggars Bush, Haddington Road,, Dublin 4, Ireland

M. Harris
UCD School of Biology and Environmental Science, University College Dublin, Dublin 4, Ireland

M. Harris
Parks and Landscape Services Division, Dublin City Council, Dublin, Ireland

Keywords

Urban soils • Anthropogenic soils • Urban landscapes Soil landscapes • Soil contamination • Soil map Technosols • Dublin • Belfield

12.1 Soil Sealing in Ireland

Converting land to urban cover creates a valuable built environment. The functions of structures and pavements, providing shelter, infrastructure, personal and social spaces are all important for society. Urban areas are our most economically valuable lands, and tiny residual open spaces capable of supporting vegetation are the most precious of all—so important that they are withheld from the free market by public authorities and retained for public use. But construction is also often seen as a loss, specifically in relation to the loss of soil functions, and soil sealing is identified as the principal threat to soils in the European Union (EU).

The rate of urbanisation has been very high in Ireland. Land-take in the period 2000–2006 of 0.77% of agricultural land converted to 'artificial surfaces' was described as 'quite alarming' by Tóth (2012), and equates to a permanent loss of annual productive potential of 3 kg of wheat per person. While the loss of soil functions because of urbanisation is seen as a current and long-term concern in Europe, there are no legally binding policy targets for soil protection (EEA 2016). The EU's proposed Soil Framework Directive was withdrawn in May 2014. Despite this, the EU remains committed to soil protection through policies such as the Seventh Environment Action Programme and the Resource Efficiency Roadmap. A similar commitment is reflected by the United Nations in the Sustainable Development Goals.

'Soil Sealing' is defined as 'covering of the soil surface with impervious materials as a result of urban development and infrastructure construction' (EEA 2015). Soil Sealing in Ireland represents a growing problem. Although national policies directly focussed on avoiding soil sealing are

R. Creamer and L. O'Sullivan (eds.), *The Soils of Ireland*, World Soils Book Series,
https://doi.org/10.1007/978-3-319-71189-8_12

lacking (Prokop et al. 2011), the effects of urban sprawl are acknowledged within planning policies. National data on soil sealing in Ireland are unavailable, but analysis of CORINE land cover data for EU member states for 1990–2006 shows Ireland's rate of relative increase in artificial surfaces during the time period as 49.9%, compared to the EU average of 8.8% (Prokop et al. 2011). The area of total land cover of Ireland as artificial surfaces increased rapidly during 1990–2000, from 1.5 to 1.9%, caused by urban sprawl and developments in infrastructure and sports facilities (Prokop et al. 2011). During 2000–2006, a further 0.3% increase coincided with population growth of 11%, characterised by low-density development at the edge of existing urban areas at the expense of agricultural land and green urban areas (Prokop et al. 2011). By 2012, 3.9% of total land cover in Ireland was artificial surfaces (Virto et al. 2015). Importantly, soil sealing not only affects the soil functions in the area that is sealed, but also indirectly impacts the delivery of soil functions elsewhere, for example, through the redirection of water courses or through the disruption of habitats.

However, urban land is neither completely nor uniformly sealed. Pavements retain some permeability and can support some vegetation: low vegetation grows in the cracks, often only temporarily in the face of weed-control practices, while street trees extend their roots into the sub-pavement soil mass. Soil Sealing affects infiltration and runoff during wet weather, increasing peak-flow runoff and puts stress on urban drainage systems, with increasing concern for flooding such as in the River Tolka catchment studied by Verbeiren et al. (2013). Although reversal of sealing is seen as difficult and expensive, vacant lots can become rapidly vegetated in Ireland's moist climate, with paved surfaces and building remains colonised and fragmented after a few years of abandonment, allowing water infiltration and offering new habitats.

The European Environment Agency estimates a mean soil sealing of 41.9% in the Dublin urban morphological zone (Fig. 12.1), equating to around 128 m^2 of sealed land per person (EEA 2010, 2011). The degree of soil sealing is highest in the area of the city centre between the canals, the port and docklands, older harbours such as Dun Laoghaire and Malahide, Dublin Airport and environs, and in industrial zones to the west of the city. The lowest degree of sealing is in the Phoenix Park, the flight paths of Dublin Airport and where there is steep topography.

In addition to Dublin, large urban centres around Ireland share built-up cores, low-density suburbs, concentrated highly paved industrial estates, radial networks and satellite towns. Cork harbour has substantial industrial facilities, including legacies such as a deposit of steel-production wastes at Haulbowline Island. Limerick city and other estuarine settlements have polder reclamations. Airports, isolated industrial facilities, mines, refineries and sites in development have urban qualities against a highly rural hinterland, while a network of transport corridors extends between the major centres. Industrial, transport, military and mining areas occur in both urban and rural areas, while settlements and dwellings have urban-like soils regardless of their size. The history of an individual urban site is likely to include development and redevelopment, dereliction and dumping, site excavation, shaping and capping, and some past long-term use extending beyond living memory. The human influence, even when it dominates soil formation, is a continuum from the built environment outwards, and is not confined to urban areas. Even in remote rural areas, isolated houses, farms and single industrial facilities have sealed or reconstructed soils.

12.2 Anthropogenic Contamination of Urban Soils

Anthropogenic contamination of soils in Ireland can be attributed to four main factors: urbanisation, industry, mining and agriculture. Recent investigations of the geochemical impacts of urbanisation (Glennon et al. 2014) and mining (Stanley et al. 2009; Inter-Agency Group 2000; CDM 2008) have improved understanding of their role in the anthropogenic contamination of soils. National and regional geochemical surveys (Fay et al. 2007; Gallagher et al. 2016) provide a broader picture of the extent and nature of such contamination in the country as a whole.

New materials are constantly entering the soil, as well as the oceans and the atmosphere. These include combustion products (spheroidal carbonaceous particles, glassy ash spheres, black carbon and oxides of nitrogen and sulphur), manufactured materials (concretes, ceramics including brick, plastics, elemental aluminium, glass and composites) and organic molecules (polyaromatic hydrocarbons, polychlorinated biphenyls and pesticide residues), along with tiny traces of radionuclides from weapons testing and nuclear-plant accidents such as occurred in Chernobyl, Ukraine or in Windscale (now Sellafield), Cumbria. Also, increased emissions of carbon dioxide, methane and nitrous oxide among others have entered the atmosphere.

In urban areas, human activities can alter the chemical, physical and biological properties of the soil through construction, industrial activity, fuel combustion, transport emissions and waste dumping. Topsoil, in particular, is a receptor for urban contamination both directly, via emissions to the ground, and indirectly, via diffuse emissions to the air that are eventually deposited on the ground. Acknowledging these pressures on urban soil quality, the Geological Survey of Ireland, in partnership with the Geological Survey of Norway under the EuroGeoSurveys Urban Geochemistry

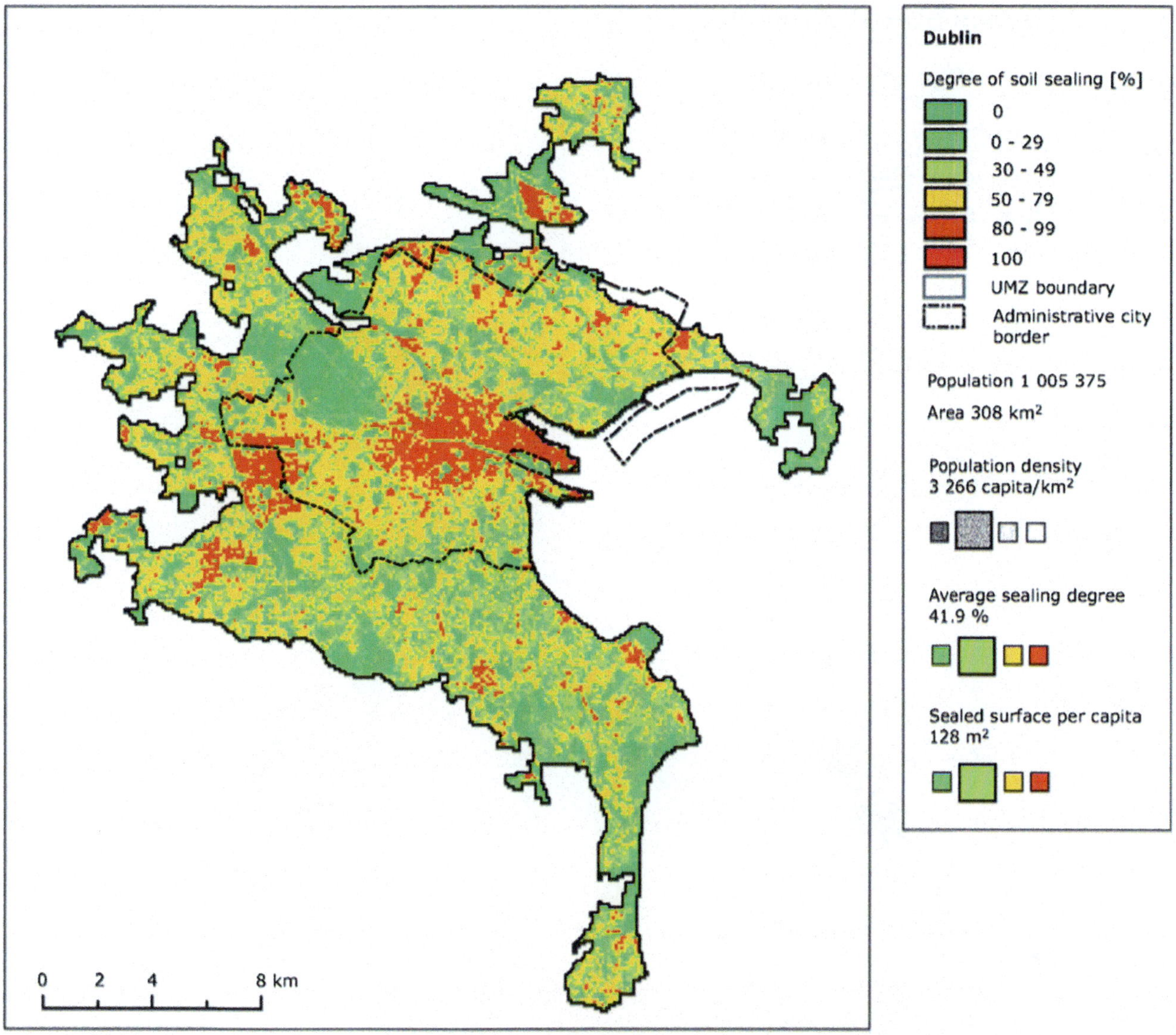

Fig. 12.1 Soil sealing in Dublin urban morphological zone. *Source* European Environment Agency (2011)

initiative, undertook, for the first time, systematic geochemical mapping of topsoils in the greater Dublin area in 2009 (SURGE project, Glennon et al. 2014). Concentrations of 31 inorganic elements (primarily heavy metals) in 1058 samples were measured. A subset of 194 samples was analysed for persistent organic pollutants, including 16 polycyclic aromatic hydrocarbons and seven polychlorinated biphenyls.

Dublin's industrial and settlement history are important considerations in the interpretation of its modern urban soil geochemistry. Geochemical analyses from the SURGE study indicate that the topsoil concentrations of lead (Pb), copper (Cu), zinc (Zn) and mercury (Hg) are strongly influenced by human activities in Dublin (Glennon et al. 2014). The concentrations of these metals were elevated in the made ground, in the docklands, the city centre and areas of heavy industry (Fig. 12.2). Sources of heavy metals in these areas include former industries such as metal and chemical works, domestic and industrial coal burning, the historical use of waste materials and fly ash as fill, and traffic emissions. Similar conclusions were drawn by Nice (2010) for soils in the Belfast urban area and by Zhang and Dao (2011) for Galway city.

Regional geochemical mapping in the northern part of Ireland conducted as part of the Tellus project (Young and Donald 2013; Gallagher et al. 2016) has demonstrated a widespread impact of urbanisation on topsoil composition, even outside the main metropolitan areas. Numerous tin (Sn) anomalies were detected in topsoil around small towns and villages across the region, with domestic coal- and wood-burning a likely source for at least some of them. High Sn concentrations have also been detected in urban soils in Belfast (Nice 2010).

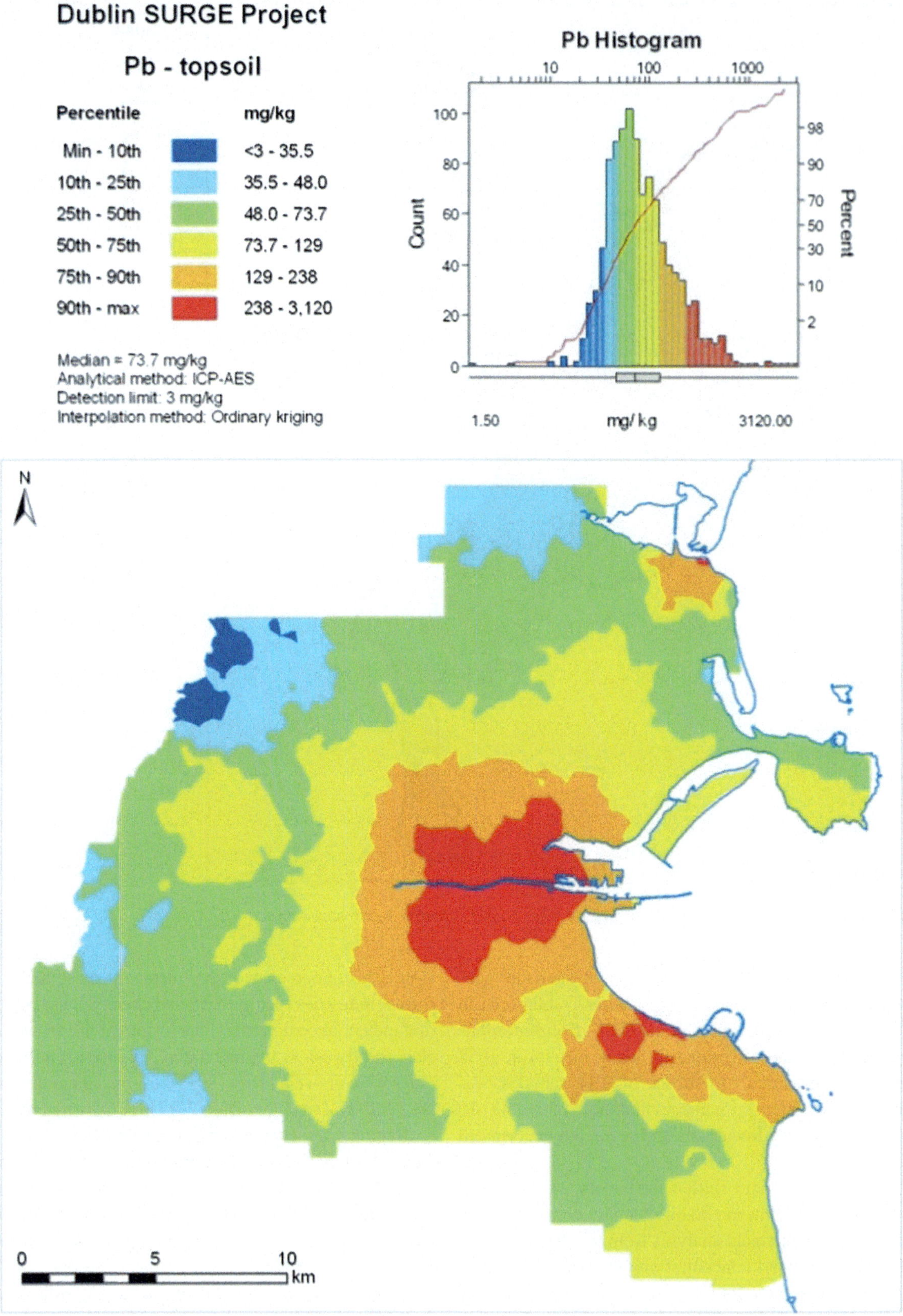

Fig. 12.2 Lead (Pb) concentrations in urban topsoil in Dublin, showing elevated concentrations of Pb in the city centre, with local outliers around Booterstown and Malahide (Glennon et al. 2014)

12.2.1 Health Implications of Contaminated Soil

As a consequence of intensive human interaction with soil in urban and industrial areas, concentrations of contaminants in soils may reach levels that give rise to concern for human health. Worldwide, an increasing proportion of people live in cities; in Ireland, 62% of the population now live in urban areas (CSO 2012). Therefore, it has become increasingly important to understand, monitor, remediate and prevent contamination of urban soils. Urban inhabitants readily come into contact with topsoil in gardens, playgrounds and sports fields, and it is used as the growing medium for crop cultivation in urban domestic gardens. Potential human exposure pathways to contaminants in soil include ingestion of home-grown crops, dermal contact and inhalation of soil dust (Jeffries and Martin 2009). An understanding of the present environmental conditions, i.e. via a baseline survey such as SURGE, backed by regional baseline geochemical surveys such as Tellus, is an essential component of any response to future soil contamination events and for assisting in the establishment of appropriate health criteria for soil (Johnson and Ander 2008). Ireland does not yet have dedicated guideline action values for contaminated land. Even at a European level, there are no uniform standards for soil. Various guidelines have been developed by individual countries over the years but there is no consensus on the establishment of threshold concentrations for various contaminants in soil for the protection of human health (Abel et al. 2010). In the absence of such guideline values, it is important to have a detailed knowledge of local conditions to enable assessment of risks to human health from soil contaminants. Most cities worldwide have well-established monitoring and assessment systems for air and water. In comparison to air and water quality, soils have received little direct attention but the establishment of a geochemical baseline for Ireland is the first step towards science-based management, assessment and monitoring of soil chemical status.

Case Study 12.1: Former Land Uses of Open Spaces in Dublin

The parks of Dublin City have arisen not as a designed system to a master plan but rather as individual sites that became public open spaces mainly if the site was unsuitable for construction due to topography. Other open spaces had previous uses such as graveyards, older waste grounds, natural wetlands reclaimed through refuse disposal, nineteenth and twentieth century quarries, and landfills unsuitable for development (Noone 2016). Other parks have been developed from lands prone to flooding and a small number of Georgian and Victorian squares formerly in private ownership. In the early twentieth century, Dublin Corporation purchased a few private estates which were partly retained as large land banks for recreation (St. Anne's Park, Bushy Park). Although such developments have affected soil formation and can influence park development and management, there has been no comprehensive analysis of the impacts of the previous land uses on soils of public green spaces of Dublin at a city-wide scale. While parks managers and staff have managed open spaces with some awareness of site histories this knowledge is largely undocumented and is being eroded by retirements. According to information from Dublin City Council (DCC),[1] Dublin City is home to 1238 parks and green spaces in state and institutional ownership, including privately owned sports grounds. One hundred and thirty five of these, which are recognised as the main public parks,[2] were selected and studied in detail to identify past land uses, focussing on landfills, historic mapping (Ordnance Survey 1829–41 and 1897–1913), personal communication with DCC Parks staff and historical accounts of individual parks.

Almost one-third (32%) of the 135 public green spaces examined were converted from agricultural lands, found mainly to the north and west of the city. Twenty-nine precent have had some form of landfill, or imported or made ground: 18% are former landfills, 7% are former quarries, 4% are reclaimed wetlands. These values are approximate, due to limitations of available records. Of the public green spaces that have been in continuous use as green spaces since the late seventeenth century, most have come into public use from agriculture, 10% are ecclesiastical and 16% have been purposefully planned as formal green spaces. It should be assumed that any public park which has been redeveloped may have had imported topsoil or other material, while some have had topsoil removed and sold if they were previously housing estate lands, such as at Ballymun. The last remaining farmlands at Pelletstown and Clongriffin were developed by the end of the twentieth century to meet housing demand during a period of economic boom in Ireland. The last farm in the city, Hillside Farm at Glasnevin, is scheduled, at time of writing, as part of the route of the tunnel for the proposed Metro North line.

[1]ArcGIS database of parks and green spaces compiled by Parks and Landscape Services, Dublin City Council 2015.

[2]Sources were Dublin City Council park listings; the Ballymun Regeneration Master Plan (2005) list of parks; the Office of Public Works website http://www.opw.ie/; and Dublin City Council Parks Superintendents.

The ten percent of public green spaces in Council ownership that are ecclesiastical sites (disused graveyards or sites with churches in ruins) seem to be a significant resource in terms of soil history, as they have had limited changes due to their historical and religious importance and unsuitability for construction. They can serve as local biodiversity hotspots (DCC 2008).

Public parks with former industrial land use are mainly in the riparian zones of the Liffey, Dodder, Camac and Tolka rivers. Hydropower was used for ironworks, textile mills, paper mills, sawmills, printing and dyeing factories. Quarries were active throughout the city in the nineteenth and twentieth centuries where sand, gravel, calp limestone and clay were extracted. Active quarries are no longer found in Dublin City (DCC 2011).

Enjoyment of green space was primarily for the upper social classes initially, and the opening of green space began as a means to civilise the lower classes in the 1860s, restricted mainly to permitted days and times at the National Botanic Gardens and the Phoenix Park (Brück and Tierney 2009). Georgian (late eighteenth century) squares, each a fenced ornamental ground faced by four streets of terraced residences, had been constructed as a selling point by developers to serve the residents, who were keyholders. There were market squares where games were played, and a few locations for promenading existed in the seventeenth century, but these were subsequently built over and lost (Costello 2007). Following the passage of the Open Spaces Act (1886), Dublin Corporation was empowered to create public green spaces, supporting the extension of recreation to the lower classes. The first inner city park to be independently created by Dublin Corporation was the Blessington Street Basin which was developed in 1889 (Brück and Tierney 2009) and is still in use.

In the early twentieth Century, Dublin Corporation began to resolve the long-standing social and health problems caused by tenement living conditions. This resulted in a wave of housing clearances, generating building waste to be disposed of and the development of suburban agricultural estates into terraced housing with small communal greens. The building waste was disposed of by the Corporation in various ways: to reclaim wetlands around Dublin Bay, to fill in old quarries, and to build up levels of roads and new green spaces. By the 1950s, Dublin Corporation established a parks service. Following the appointment of parks superintendents in the 1970s, and coinciding with the closure of many landfills due to new European legislation and extensive housing development in Dublin, more municipal lands and landfills were redeveloped into public parks.

Recent housing developments have been mainly to the north and west of the City, including a major regeneration project at Ballymun (DCC 2005). The parks on the north side of Dublin have high water tables and heavy gley soils. The soil here is suitable for manipulation to create constructed wetlands, and this has been done at Tolka Valley (1999 and 2011), Johnstown (2011) and Poppintree Parks (2008–2010) and for private developments at Mayne River Linear Park at Clongriffin (2011–12).

12.3 Urban Soil Materials

The following tentative checklist of the forms and types of materials that occur in the urban soils of Ireland is offered as a starting point for integrating cultural features into soil knowledge.

Materials forming the soil are likely to include members of this list:

1. Loose, aeromorphic (wind-shaped) **surface litter** of deciduous plant leaves, other seasonally shed and fragmentary woody, foliar and floral plant parts, packaging materials, printed papers, cigarette ends, plastic fragments, as well as finer materials from road surfaces, tyres and combustion, and a fine crumb of faunal frass (powdery refuse produced by activity of litter-dwelling animals);
2. **Pavements** of concrete, asphalt composites, masonry, bricks, with sub-layers of sand, angular stones, demolition waste—and existing **buildings**;
3. Soft, dark, crumbly biomantle **topsoil** structured by earthworm-casting and plant-rooting, with visible roots, fungi and animals, formed in situ since underlying material was placed, and containing isolated **artefacts** of all ages, especially fragmentary ceramics, concretes, angular stones, glass, metals and wood;
4. Compacted **subsoil** of yellow–brown, grey or mottled colours; weathered, moved, redeposited and compacted soil material of B- or C-horizon origin, containing subangular stones of mixed lithology;

5. **Remains of** demolished **structures** dominated by lime mortar, with angular stone fragments, concretes and occasional pieces of broken brick and tile;
6. **Dark-coloured soil materials** representing buried former topsoils, or subsurface archaeological features of similar composition, with occasional durable artefacts of ceramic and angular stone, and weathered bone and shells;
7. **Angular stones** from stoneworking during quarrying or construction and unweathered excavated rock in mine spoil;
8. **Organic waste** layers; remains of dung heaps, cesspits, food- and food-processing waste and municipal refuse;
9. Cinders, ash, fused materials, reddened soil and stones—industrial and domestic **mineral waste** from high-temperature processes in small and large open, stove and furnace hearths;
10. At some level, groundwater is likely to occur as a distinct, dynamic surface;
11. The lower limit of artefacts, below which cultural evidence and human action are no longer found;
12. Weathered earth materials, ultimately derived from glacial deposits, slope deposits, alluvium or weathered bedrock; or gravels and sand;
13. Bedrock weathered in upper parts.

Earth movements and earthworks, including land levelling, backfill, inert soil fills and soil displaced from building footprints create new landforms. Installing subsurface drains and communication conduits leaves permanent subsurface features of cut-and-fill. Soils are engineered to improve their function in supporting structures, usually by compaction.

Land that has been reclaimed from the sea through polder construction is used for urban development in many areas. Polder construction dominates flat land in the Shannon and Fergus estuary, and along tidal stretches of most of our larger rivers. Many urban areas on estuaries are built on polders on substrates of former intertidal areas, with the surface raised using dredge spoil, and municipal and demolition wastes.

Heavy industries produce wastes, including ash, cinder and clinker that have been placed on the surface, initially on site, but that are often moved or levelled. Brownfield sites are areas where former land uses leave a perceived obstacle to redevelopment, putting pressure on greenfields for urban expansion.

Archaeological soils are layered deposits of material that retain a stratigraphic sequence from human activities. These can occur as enclosed fills of cavities cut into subsoils, or in urban settlements as thick deposits covering sites opened for construction.

Occupation soils is the archaeological term given to an accumulation of strata relating to construction and habitation sequences. Dublin provides one of the best examples of occupation sequences of any urban Viking settlement. The Wood Quay excavations, synthesised by Wallace (2016) demonstrate the accumulation of house plots and structures characterised by grey 'marl' floors, with post and wattle walls and timber uprights supporting the roof structure. Where these structures were destroyed by fire, the next phase of building was constructed directly over the charred remains of the last. The survival of the distinctive flooring allows the different phases to be identified. Where a high concentration of water in the soil is found (which has been a feature of most of the investigations carried out in the Hiberno-Norse town) there is excellent preservation associated with the organic components of the flooring.

House plots of all ages are associated with deposits that can relate to construction and demolition debris as well as the organic detritus of human occupation. Faecal matter is generally confined to pits that can be timber-lined or simply dug through the underlying deposits and layers, and maintains a distinctive organic odour on well-preserved sites. Ecofacts (non-cultural organic material) extracted from environmental samples are important for interpreting the deposits.

Garden soils tend to be associated with later mediaeval sites and often seal earlier occupation. These soils are characterised by a high organic content derived from the liberal spreading of dung and other organic matter, where the ceramic assemblage tends to be very fragmented as the soil was frequently turned over for cultivation. Outside the city walls (where there is less contamination from later occupation), garden soils tend to form thicker, more homogenous deposits than occupation soils and are generally distinguished from the latter by an absence of structural evidence.

Burial soils are by definition typically confined to burial sites. Where articulated burials have been disturbed, the dark organic soils are characterised by the presence of fragments of human bone. Undisturbed burial sites are recognisable by distinctive grave cuts; however where there are national and local policies in place supporting 'preservation in situ', graveyards are less frequently excavated during commercial developments.

Brickfields are characterised by burnt red clays. Brick is first recorded in Dublin in 1560 where George Burrows leased a 4½ acre plot from the Corporation in 1599 along the eastern bank of the Stein (at the western end of Pearse Street). Other documented brickfields were located on

Merrion Strand and at Cardiffsbridge beyond Cabra. The short functional life of urban brickfields has rendered them invisible on the city's archaeological record. Archaeological investigations have uncovered evidence for dumping in the more-modern disused quarries and brickfields located south of the Grand Canal in Crumlin, surviving as stratified early twentieth-Century deposits up to 10 m thick (Myles 2002, 78; excavations.).

Waterlogged soils are less easy to define in urban contexts and many of the soils described above fall into this category. Extinguished watercourses are often marked by Victorian drainage works above and tend to present as linear extensions of loose silt (in the case of the watercourse depicted by Speed *c.* 1610 which entered the Liffey east of Smithfield) where defining clay banks, artificial or natural, can also be encountered.

Case Study 12.2: Site-Specific Cultural Soils

Writers on the area occupied by mediaeval Dublin make many references to black, peaty material, with buildings either supported on piles, wattles or hurdles, or sinking into the soft earth. This black material is not bog or fen peat, but the accumulated organic remains derived from occupation related to habitation sequences. Stanihurst, writing in 1570, reported marshy ground and that it was necessary 'to fasten the quakemire with hurdels and upon them to build the Citie' while a century later Camden described the ground as 'soft and quaggy'. These are quoted by Charles Haliday, author of *The Scandinavian Kingdom of Dublin* (1884). Haliday's attention was drawn to 'a stratum of boggy soil' which was in turn sealed by a layer of nut-shells, branches and tree trunks. This stratigraphy had been uncovered by Arthur Park Neville, the City Engineer, in sewage-pipe trenches through High Street, Castle Street, Winetavern Street and Fishamble Street in 1880, at depths varying from six to ten feet. Above this, and immediately below the road surface, was a thick mass of soft clay that contained shells (Haliday 1884, 208). Along with mussel shells on Fishamble Street, Haliday more significantly recorded a quantity of squared oak beams resting on a layer of twigs, leaves and shells. These timbers, supported on piles four to five feet in length, sealed the peaty clay. Little (1957, 36) records workmen on Cork Hill, perhaps in the 1920s, coming across a 'thick layer of hazel-nut husks' and further records an observation by Sir Thomas Drew, the architect responsible for the restoration of Christ Church, who encountered a similar deposit of shells when working on the cathedral's foundations.

It is along the shoreline however that the earliest recorded human activity in Dublin city has been identified. This presented as timber fish traps, stake rows and pieces of worked wood, preserved in the waterlogged natural silts of the Liffey delta and dating to the Late Mesolithic period (McQuade 2007). Located under the National Convention Centre at Spencer Dock (Fig. 12.3), this material was recorded well away from the urban mediaeval core that has seen the majority of archaeological excavations. It is in the mediaeval city that the vast majority of archaeological soils have been recorded, which can be found at levels as deep as 10 m below the modern street surface.

It is likely that an Early Christian presence was located on the till ridge south of the Liffey, the significance of which has been eradicated by the development of Christ Church and the later mediaeval parish churches. The embankments were superseded by masonry walls by *c.* 1050, while the city was subsequently extended towards the present waterfront on Wood Quay in the thirteenth century. The archaeological soils encountered in each place are a function of the specific location under investigation and its individual development.

Reclamation soils are a feature of the area between Trinity College and Ringsend and east from the Custom House on the north side of the Liffey. The enclosure of the South Lotts by means of a masonry quay from the early 1700s echoed the enlargement of the mediaeval walled town where refuse was dumped across the natural silts and gravels, building up the ground above the water table and allowing construction to proceed. The most recent investigation undertaken in the south docklands showed 400–700 mm of mixed soils containing numerous ceramic inclusions dating from the period 1650–1750, which were dumped across the area for this specific purpose (Myles 2015). This material came from a number of *laystalls* or dumping grounds, depicted on John Rocque's 1756 *Exact Survey of the City and Suburbs of Dublin*… and shown as hillocks occupying undeveloped ground (Fig. 12.4). In the seventeenth century scavengers were employed to regularly empty the latrine pits located in the back plots of houses; it had previously been ordained that rubbish would be dumped outside the walls in Mullinahack (*Mullach na Caca*, hill of excrement) or at the western end of Hangman's (now Hammond) Lane, both sites that have been investigated (Myles 2003, 114–7).

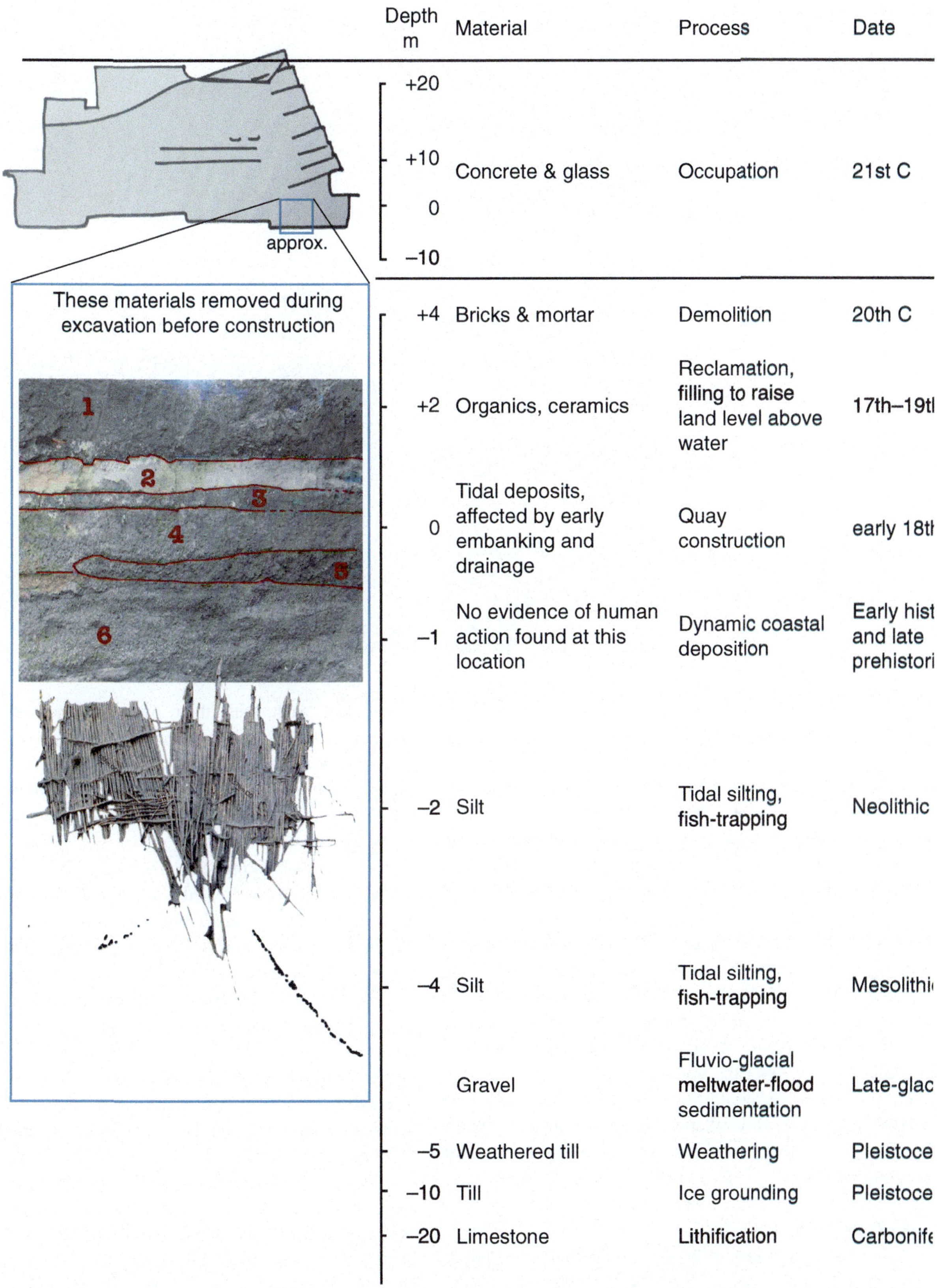

Fig. 12.3 The Convention Centre, Dublin (top) and the soils it has displaced, generalised from archaeological excavations before construction. Estuarine sediments enclosed Mesolithic and Neolithic fish traps, while large-scale reclamation in the eighteenth century created a sequence of drains, sheltered intertidal sediment, windblown sand, thick refuse deposits and buildings which became building remains before all was removed to make way for this latest round of cultural activity. Future deposits at this site will perhaps have remains of all of these, including remains of the Convention Centre, incorporated into and buried under some new structure. Fishtrap images courtesy of Melanie McQuade

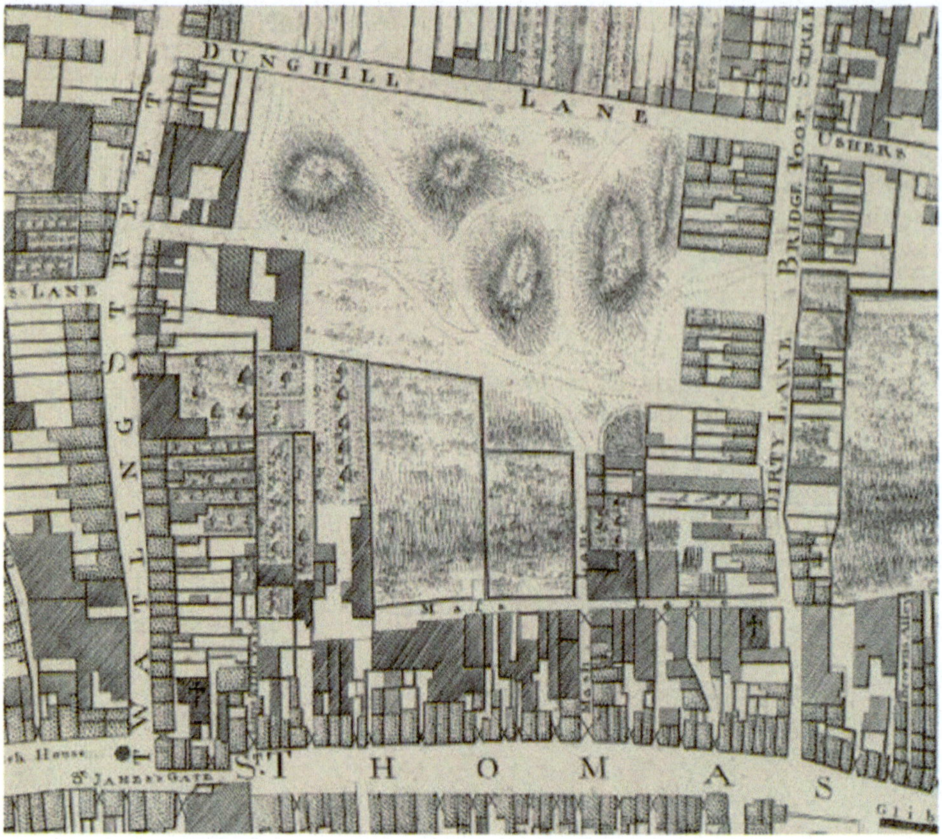

Fig. 12.4 Dunghill Lane (now Island Street), a typical eighteenth century laystall where domestic waste from latrine pits was brought. The accumulated material was later used during reclamation along the Liffey. From John Rocque's *Exact Survey of the City and Suburbs of Dublin*..., 1756

12.4 Soil Landscapes of Urban Dublin

Dublin is underlain mainly by limestone but also by granite that forms the Dublin and Wicklow mountains to the south and by hard metamorphic rocks on the Howth peninsula to the northeast. This bedrock is generally covered by 1–10 m of glacial till interrupted by coarse fluvioglacial gravels formed in the valley floors from meltwater floods. Also, fine estuarine sediments are up to 40 m thick around the converging river mouths in Dublin Bay, accumulated during postglacial sea level recovery (GSI). Eroding cliffs of till have given rise to gravel beaches between hard headlands to the south, while a chain of sand-dune systems alternates with muddy estuaries along the northern coastline. Since glaciation, soil parent materials have weathered for more than 10,000 years to give decalcified, organic-rich, thick loamy lowland soils in the county. Luvisols and Stagnosols are the predominant soils to the north, formed from surface-decalcified limestone till. Centuries of liming have delayed natural acidification.

Long-term nucleated settlement in Dublin city and its constellation of merging villages has provided some consistent large soil-landscape elements: extensive urban land-cover that extends radially from a mediaeval core of archaeological soils, extended by a series of large-scale expansions, each with their own character of land-take depending upon what was removed, whether cellars were built, the extent and material of paving, and whether excavation machinery was used. Gardens and immediate surroundings of long-established houses will have Technic and Hortic Anthrosols.

Stepwise embanking and infill of the Liffey, Dodder and Tolka estuaries have created a large continuous reclamation landform extending into Dublin Bay, formerly Fluvisols and Tidalic Gleysols, latterly Fluvisols (Drainic) and Gleysols (Drainic), occupied by distinctive heavy industry, and now mostly Ekranic Technosols but including Spolic Technosols. The three radially convergent rivers preserve open parkland because of unsuitability for housing, between what were formerly strings of water-powered mills, now derelict.

Buildings and pavements are constructed using a combination of soil displacement and compaction, establishing a suitably stable base for the structures. The displacement of soil in earlier constructions, where manual methods were used, may have been minimised to the exact footprint of foundation beams. Availability of excavation machinery has increased the amount of soil removed. In early mechanised decades, this tended to be exported off-site, for use in coastal reclamation or to be placed in municipal landfills. Landfill charges, waste management planning and building regulations mean that practice has shifted markedly, with displaced soil incorporated into site surfaces for landscaping. This simple change has greatly altered urban landscapes, introducing small, sometimes steep, grassy hills around larger buildings. Sometimes, soil horizons are inverted, either inadvertently, or as a means to suppress weed growth and create lower-productivity, higher-biodiversity swards (NRA, undated: 70).

Viewed in plan, the process of displacing soil from building footprints creates Ekranic Technosols in the constructions and an immediate surrounding zone of Regosols. Regosols are soils with no distinct horizons, in this case because of earth movement. Less-disturbed soils of the pre-development background may survive in peripheral locations. The suburban cover incorporates patches of large institutional campuses. Open amenity grasslands are common, many of them derived from earlier formal and informal municipal landfills, Urbic Garbic Technosols. Truncated soils are parts of many landscapes, including quarries and gravel pits, brickfields, construction sites and cut-and-fill couples in transport corridors.

12.5 Mapping Urban Soils

Historically, urban soils have been systematically under-represented in primary resource surveys, leading to gaps in critical information. In Ireland, the National Soil Survey of An Foras Talúntais (Gardiner and Radford 1980), the National Soil Database survey (Fay et al. 2007), the Irish Soil Information System survey (Creamer et al. 2014) and the Tellus survey underway at time of writing (www.tellus.ie) all generally avoid urban areas, and all for good reasons. The agricultural and environmental motivations of these surveys, the unsuitability of survey methods to urban areas and the practicalities of accessing small heterogeneous private sites all limit their ability to include urban areas.

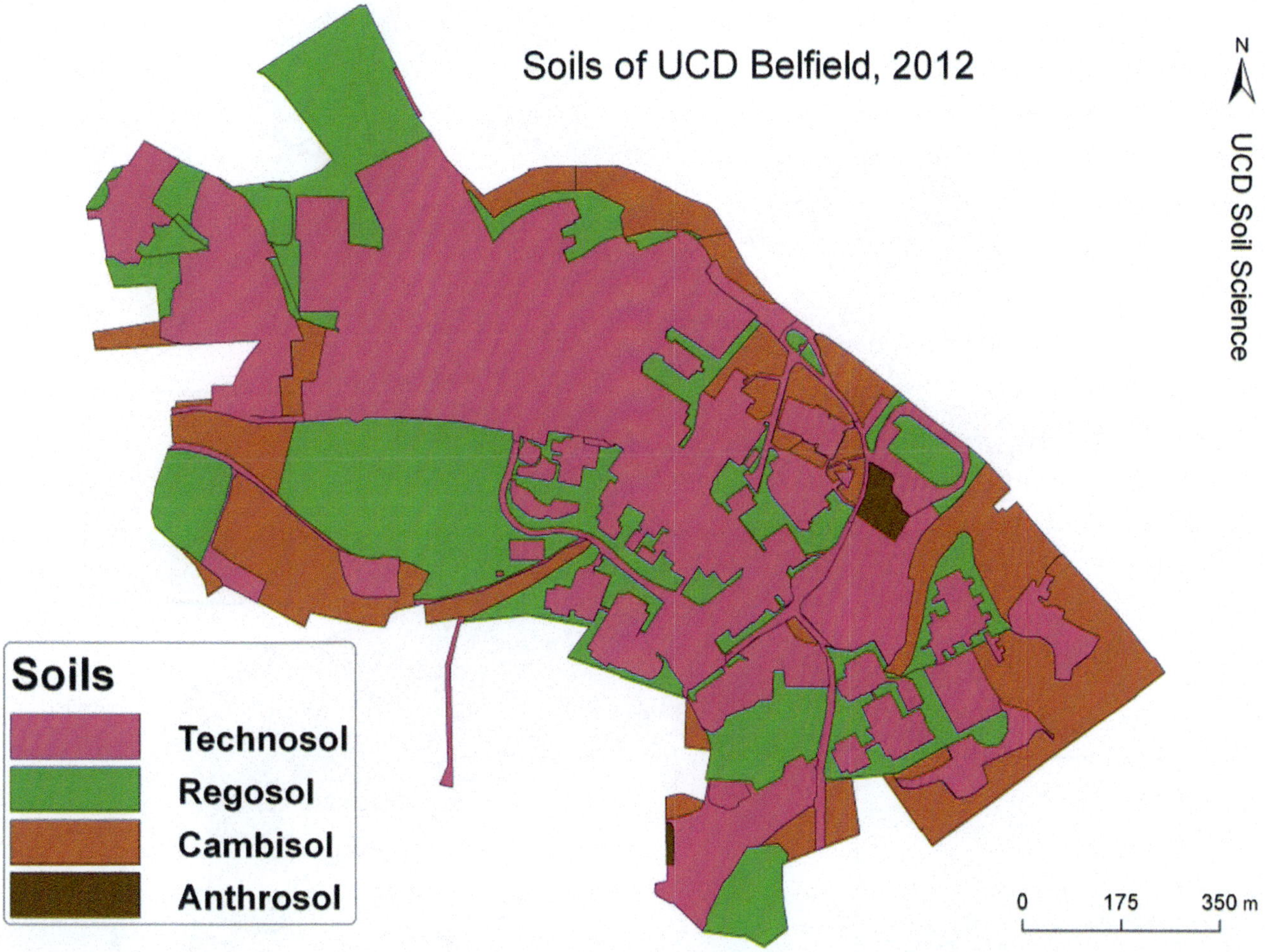

Fig. 12.5 Soils of University College Dublin Belfield campus. Buildings and pavement, shown as Technosols, are in the centre, surrounded by Regosols formed from soil displaced during construction. Less-disturbed peripheral soils (Cambisols) have traces of agricultural modifications, in this case, straightened streams, stonewall fragments, relict hedge banks and ceramic-pipe drains. Anthrosols are found at locations of centuries-long occupation

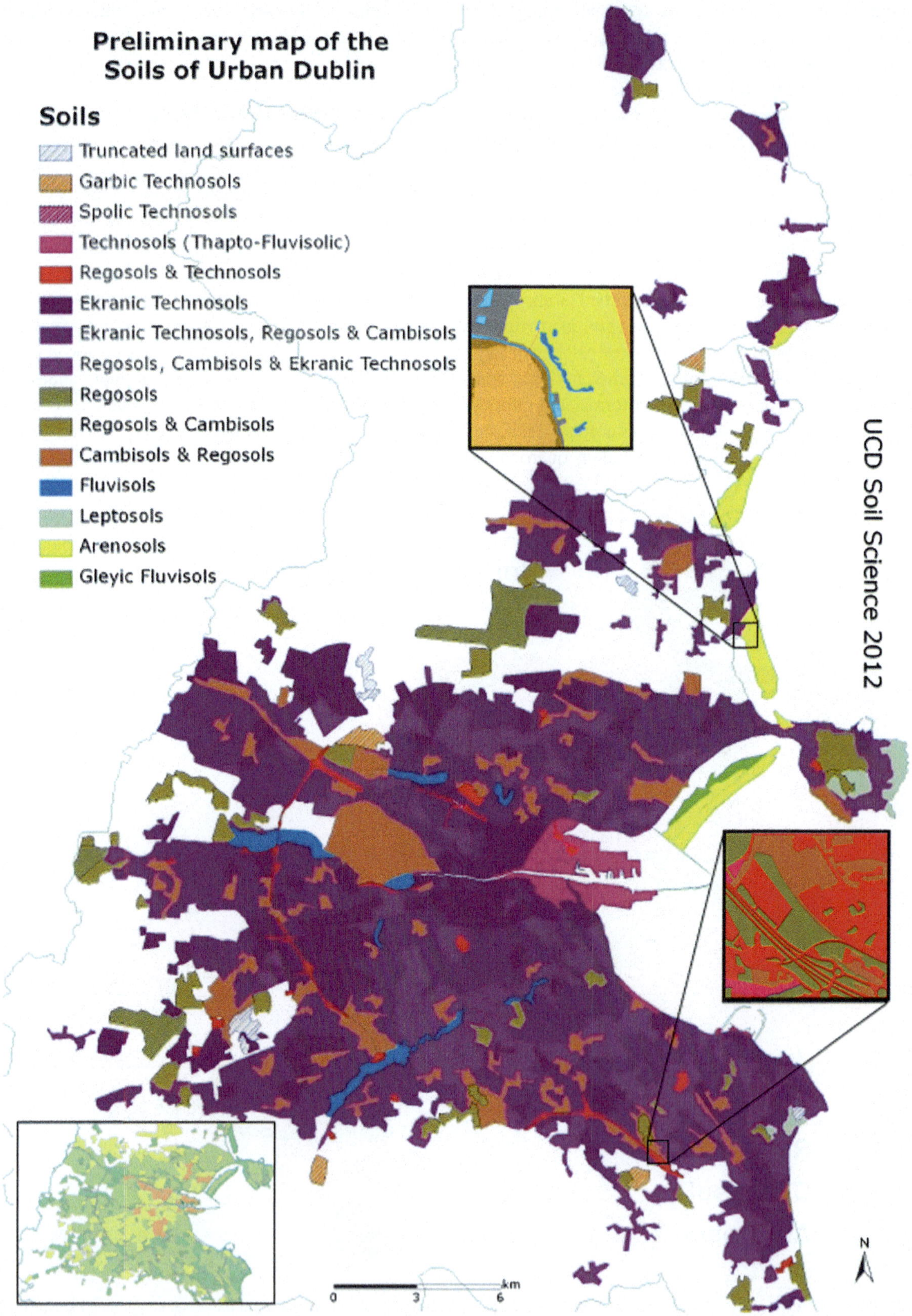

Fig. 12.6 Preliminary map of the soils of urban Dublin (after Cummins 2012). Soil classification is from the *World Reference Base for soil resources* (IUSS Working Group WRB 2015). The administrative boundary is shown to the west, and coastline to the east. Insets are from unpublished student projects, showing ecosystem value of soils (green = high, brown = low), based on stability, sealing, vegetation and toxicity (Ziggiotto 2016), and close-ups of field-surveyed areas (Xie 2014, using a different legend, and including subaqueous soils)

Creating a name for sealed urban soils make it possible for urban areas to be included in soil maps. Soil maps usually showed urban areas with a boundary, and a miscellaneous category, but with no soil information. A ubiquitous hole in the soil map was observed by Denizen (2012) in his graphical archive *The Eighth Approximation.* Early advances in urban soil mapping were made by United States Department of Agriculture surveyors (see Scheyer and Hipple 2005), culminating in the survey of New York City (New York City Soil Survey Staff 2005). There, they took the approach of describing whatever earth-surface material they found, and were able to describe and classify every part of the five boroughs of the city, typical of the most-built environments on earth. This philosophical approach was an advance on their earlier method for mapping the city area of Washington D.C. In the earlier survey, the same surveyors had described city soils as modifications of soils previously known from surveys in the rural hinterland. This method was not sufficient to allow them map areas of sanitary landfill, or land reclaimed using dredge spoil (Smith 1976). In the New York City survey, both of these land types were captured by the map and given full soil-profile descriptions.

The second edition of the *World Reference Base for Soil Resources* (WRB) introduced the Technosol concept in 2006, allowing the built environment to be fully understood as a soil system. Elements of WRB (IUSS Working Group WRB 2015) are discussed in Chap. 3 Classification.

The large University College Dublin Belfield campus shows a pattern of central buildings and pavement—Ekranic Technosols, with peripheral mounds of displaced and reshaped soil, Regosols (Fig. 12.5). Closer inspection, based mostly on surface topography, reveals a complex of landforms created during the work of levelling sports surfaces, from older demolished buildings and by the accumulation of surplus topsoil. Moreover, the core region of longer established buildings has seen multiple generations of soil reshaping and removal since the early 1960s.

Soils are always difficult to observe, and proxy methods are mainly used, rather than physical excavation and examination. Archaeological excavation reports provide detailed accounts of deposits on sites, though in every case these are preserved only by record, since large buildings typically now occupy the space that was once soil. Even if samples can be collected and analysed, as with any soil survey, extrapolating to represent large soil bodies has limitations: in heterogenous urban soils, the assumptions underlying extrapolation often do not hold. In every case, the chief indicator of landscape character, and first hints of its origin come from maps and aerial photographs. Dublin's and Ireland's excellent mapping heritage—with full mapping for the seventeenth century and early nineteenth century at levels of detail achieved nowhere else in the world, provide great scope for desktop investigation. Rapid advances in map availability are happening, with current developments from the Heritage Council, the Environmental Protection Agency, Geological Survey Ireland and Ordnance Survey Ireland, the front-runners.

Engineering properties of soils must be studied on an individual-site basis prior to construction and boreholes are often used for the preliminary investigation. A large number of borehole logs are accessible from Geological Survey Ireland National Geotechnical Borehole Database. In many cases, the upper, human-influenced portions of these logs are recorded by generic terms such as 'made ground' or 'fill'. As data increases, 3D mapping of the substrates—including the 'fill' layer—is increasingly available.

A map of soils of urban Dublin has been prepared, based on proportions of sealing, on evidence for earth movements, and on the background of quaternary sediments and soils in surrounding rural areas. Shown in Fig. 12.6, it identifies areas of coastal reclamation (pink, called 'Thapto-Fluvisolic'), and urban open areas. Much of the evidence for this map is geomorphic or microtopographic rather than based on soil profiles, and represents soil removals, additions and sealing with buildings and pavements. Small settlements outside the continuous urban area are shown; with suitable scales, such urban soils can be seen to occur throughout Ireland, with small farmyards and single isolated houses representing the same soil types.

12.6 Conclusion

The systematic under-representation of the human influence in soils results in an omission of critical information. Primary resource descriptions for urban soils (such as their carbon content) are not available but are urgently required. The classification of urban soils based upon the World Reference Base for Soil Resources is presented in Chap. 3. It is possible to describe urban soils in ways that are compatible with other surveys, group them using soil classification and use multiple sources including field observation, to map them.

Acknowledgements The research leading to these results (MH) has received funding from the Science, Technology, Research and Innovation for the Environment (STRIVE) Programme 2007–2013 under Grant Ref. No. 2365. The programme is financed by the Irish Government under the National Development Plan 2007–2013. It is administered on behalf of the Department of the Environment, Community and Local Government by the Environmental Protection Agency which has the statutory function of coordinating and promoting environmental research.The authors acknowledge the European Environment Agency as the data source for the 'Soil Sealing and population density in the capitals of EEA countries and the Western Balkans' in accordance with copyright (http://www.eea.europa.eu/legal/copyright). The soil map of urban Dublin (TC) was developed with the aid of UCD students Louise Doherty and Agnieszka Piwowarczyk, with contributions to urban-soil studies by Conor Bracken, Brian Burke, Dermot Colreavy, Owen Joe Crilly, Mamadou Diallo, Michael Dillon, Mel

Dermody, Damien Fitzpatrick, Daniel Hayden, Mary Ann Healy, William Keane, Aoife King, Philip Murphy, Lisa Noone, Atsu Norgbedzi, Joseph Onunkwor, Joseph Stones, Togzhan Sultanbayeva, Wen Xie and Eva Ziggiotto.

References

Abel MT, Suedel B, Presley SM, Rainwater TR, Austin GP, Cox SB et al (2010) Spatial distribution of lead concentrations in urban surface soils of New Orleans, Louisiana, USA. Environ Geochem Health 32:379–389

Brück J, Tierney A (2009) Landscapes of desire: parks, colonialism and identity in Victorian and Edwardian Ireland. Report to the Heritage Council for Project No. 1678, pp 48–50

CDM (2008) Feasibility study for the management and remediation of the Avoca Mining Site

Costello V (2007) Public spaces for recreation in Dublin, 1660–1760. Gard Hist 35(2)(Winter):160–179

Creamer R, Simo I, Reidy B, Carvalho J, Fealy R, Hallet S, Jones R, Holden A, Holden N, Hannam J, Massey P, Mayr T, McDonald E, O'Rourke S, Sills P, Truckell I, Zawadzka J, Schulte R (2014) Irish soil information synthesis report (2007-S-CD-1-S1). Prepared for the Environmental Protection Agency by Teagasc, Cranfield and University College Dublin. EPA, Wexford

CSO (Central Statistics Office) (2012) This is Ireland. Highlights from census 2011, part 1. The Stationery Office, Dublin

Cummins T (2012) Preliminary map of the soils of urban Dublin. Poster presented at Environ 2012, the 22nd Irish Environmental Researchers' Colloquium, held at UCD Belfield, 7–9 Mar 2012

Denizen S (2012) There is a hole in the map. Where is the city on the soil survey? The eighth approximation: urban soil taxonomy in the Anthropocene. Visual archive, MLA thesis, University of California Berkeley. https://anthropocenesoil.wordpress.com/

Dublin City Council (DCC) (2005) Ballymun masterplan—frequently asked questions 1998–2005. Ballymun Regeneration Ltd, Dublin, p 28

Dublin City Council (DCC) (2008) Dublin City Biodiversity Action Plan 2008–2012

Dublin City Council (DCC) (2011) Dublin City Council City Development Plan 2011–2017

EEA (2011) Soil sealing and population density in the capitals of EEA countries and the Western Balkans. http://www.eea.europa.eu/data-and-maps/figures/soil-sealing-in-the-capitals/#tab-hard-copy. Reviewed 21 Sept 16

EEA (2015) GIO land (GMES/Copernicus initial operations land) high resolution layers (HRLs)—summary of product specifications, version 11. European Environment Agency, Brussels, p 7

EEA (2016) SOER 2015—the European environment—state and outlook 2015. A comprehensive assessment of the European environment's state, trends and prospects, in a global context. European Environment Agency. http://www.eea.europa.eu/soer

European Environment Agency (EEA) (2010) The European environment state and outlook—land. European Environment Agency. http://www.eea.europa.eu/soer/europe/land-use

Fay D, Kramers G, Zhang C, McGrath D, Grennan E (2007) Soil geochemical Atlas of Ireland. Published by Teagasc and the Environmental Protection Agency

Gallagher V, Knights KV, Carey S, Glennon M, Scanlon R (2016) Atlas of topsoil geochemistry of the Northern Counties of Ireland: data from the Tellus and Tellus border projects. Geological Survey of Ireland

Gardiner MJ, Radford T (1980) Soil associations of Ireland and their land use potential explanatory bulletin to soil map of Ireland 1980. National Soil Survey of Ireland, An Foras Talúntais, Dublin

Glennon M, Harris P, Ottesen RT, Scanlon RP, O'Connor P (2014) The Dublin SURGE project: geochemical baseline for heavy metals in topsoils and spatial correlation with historic industry in Dublin, Ireland. Environ Geochem Health 36(2):235–254. Special issue on the 9th international symposium on environmental geochemistry

Haliday C (1884) The Scandinavian kingdom of Dublin, 2nd edn. Irish University Press, Dublin

Inter-Agency Group (2000) Report of the investigation into the presence and influence of lead in the Silvermines area of County Tipperary. Department of Agriculture, Food and Rural Development, Dublin

IUSS Working Group WRB (2015) World reference base for soil resources 2014, update 2015. International soil classification system for naming soils and creating legends for soil maps. World soil resources reports No. 106. FAO, Rome

Jeffries J, Martin I (2009) Updated technical background to the CLEA model. Scientific report SC050021/SR3. Environment Agency, Bristol

Johnson CJ, Ander EL (2008) Urban geochemical mapping: how and why we do them. Environ Geochem Health 30:511–530

Little GA (1957) Dublin before the vikings: an adventure in discovery. MH Gill, Dublin

McQuade M (2007) 2007:494—Spencer Dock, North Wall Quay, Dublin. Available from http://www.excavations.ie/report/2007/Dublin/0017470/. Accessed on 25 Nov 2016

Myles F (2002) Pearse College, Clogher Road, Crumlin. In: Bennett I (ed) Excavations 2000, Bray, p 78

Myles F (2003) Tram Street/Phoenix Street, Dublin 7. In: Bennett I (ed) Excavations 2001, Bray, pp 114–117

Myles F (2015) Archaeological excavations at Windmill Lane, Dublin 1. Unpublished archaeological report submitted to the Department of Arts, Heritage and the Gaeltacht

New York City Soil Survey Staff (2005) New York City reconnaissance soil survey. United States Department of Agriculture, Natural Resources Conservation Service, Staten Island

Nice SE (2010) Inorganic soil geochemical baseline data for the urban area of the Belfast Metropolitan Area, Northern Ireland. British geological survey open report 08/021. http://nora.nerc.ac.uk/9599/

Noone L (2016) Landfills of Dublin. Unpublished MSc thesis, University College Dublin

Prokop G, Jobstmann H, Schönbauer A (2011) Report on best practices for limiting soil sealing and mitigating its effects. Technical report—2011–050. European Commission, Brussels, pp. 96–99

Scheyer JM, KW Hipple (2005) Urban soil primer for homeowners and renters, local planning boards, property managers, students, and educators. United States Department of Agriculture, Natural Resources Conservation Service, National Soil Survey Center, Lincoln, Nebraska (http://soils.usda.gov/use)

Smith H (1976) Soil survey of District of Columbia. United States Department of Agriculture, Natural Resources Conservation Service, Lincoln, Nebraska

Stanley G, Gallagher V, Ní Mhairtín F, Brogan J, Lally P, Doyle E, Farrell L (2009) Historic mine sites—inventory and classification. EPA and GSI, Ireland

Tóth G (2012) Impact of land-take on the land resource base for crop production in the European Union. Sci Total Environ 435–436:202–214. http://www.sciencedirect.com/science/article/pii/S0048969712009369

Verbeiren B, Van De Voorde T, Canters F, Binard M, Cornet Y, Batelaan O (2013) Assessing urbanisation effects on rainfall-runoff using a remote sensing supported modelling strategy. Int J Appl Earth Obs Geoinf 21:92–102

Virto I, Imaz MJ, Fernández-Ugalde O, Gartzia-Bengoetxea N, Enrique A, Bescansa P (2015) Soil degradation and soil quality in Western Europe: current situation and future perspectives. Sustainability 7:313–365

Wallace PF (2016) Viking Dublin: the wood quay excavations. Merrion Press, Dublin
Xie W (2014) Predictive mapping and ground-truthing of soils in urban Dublin. Unpublished MSc minor thesis, University College Dublin
National Roads Authority (NRA) (Undated) A guide to landscape treatments for national road schemes in Ireland. National Roads Authority, Dublin
Young M, Donald A (eds) (2013) A guide to the Tellus data. Geological Survey of Northern Ireland, Belfast
Zhang C, Dao L (2011) Characterising variation of heavy metal pollution in urban soils. EPA SAFER repository
Ziggiotto E (2016) Rating urban soils based on their ecosystem services: soils of urban Dublin and related ecosystem services. Unpublished undergraduate thesis, University College Dublin

Soil Functions—An Introduction

13

Rogier Schulte, Lilian O'Sullivan, and Rachel Creamer

Abstract

It is now widely recognised that soils not only provide food but in addition, they deliver a wide range of ecosystem services to society. The EU Thematic Strategy on Soils (2006) identified seven 'environmental, economic, social and cultural functions'. In research studies from Ireland, these ecosystem services have been rearranged into the following five main soil functions for agricultural land (1) primary production, (2) water purification and regulation, (3) carbon storage and sequestration, (4) habitat for intrinsic and functional biodiversity and (5) the cycling and provision of nutrients. In addition, soils also provide two ancillary functions, namely: (6) a platform for infrastructure (e.g. roads, buildings) and (7) an outdoor archive of archaeological heritage. In principle, all soils perform each of these functions simultaneously. However, the extent to which each function is delivered depends in the first instance, on land use. For example, the functionality of arable soils is characterised by primary production and nutrient cycling. Second, the functionality of soils depends on soil properties. The dominant soil properties in Atlantic climates relate to soil moisture dynamics, specifically the occurrence of excess soil water. Quantifying soil functions is difficult as each soil function encompasses a set of processes which may be altered through management which could alter the delivery of another function, resulting in potentially a synergy or a trade-off between functions. Therefore, proxy-indicators are used to estimate the extent to which a soil provides the five functions. The functionality of soils worldwide is threatened by unsustainable management practices and in Europe, there are eight main threats to soil functionality.

Keywords

Primary production • Water purification and regulation Carbon storage and sequestration • Habitat for biodiversity • Nutrient cycling

R. Schulte (✉) · R. Creamer
Teagasc, Environment Research Centre, Johnstown Castle, County Wexford, Ireland
e-mail: Rogier.schulte@wur.nl

R. Schulte · L. O'Sullivan · R. Creamer
Wageningen University and Research, Droevendaalsesteeg 4, Building 104, 6708 PB Wageningen, The Netherlands

L. O'Sullivan
Irish Environmental Protection Agency, Johnstown Castle, County Wexford, Ireland

13.1 Soil Functions

In Ireland, the surface 80 cm of the lithosphere is considered to be the managed soil profile. Soil formation is a very slow process evolving over hundreds or thousands of years. This means that soil is both a valuable and a finite resource, as rates of regeneration cannot match those of soil depletion. The majority of materials used by humankind are either mined from the earth or grown in soil. This means that human civilisation is literally built on the outermost skin of our planet. Therefore, sustainable management of land and soil is critical so that they can continue to provide these essential goods and services for humankind. In this context, it is not surprising that five of the 17 Sustainable Development Goals (UNDP 2016) explicitly refer to the role of soil.

It is now widely recognised that soils not only provide food but in addition, they deliver a wide range of ecosystem services to society. These have been described and categorised in a number of studies: Haygarth and Ritz (2009) proposed 18 ecosystem services that are critical for soil and land use in the United Kingdom. Table 13.1 is based upon the work of Haygarth and Ritz (2009) that identifies 18 ecosystem services and functions that are specifically focused on soils and land use.

The EU Thematic Strategy on Soils (2006) identified seven 'environmental, economic, social and cultural functions'. In research studies for Ireland, these ecosystem

R. Creamer and L. O'Sullivan (eds.), *The Soils of Ireland*, World Soils Book Series,
https://doi.org/10.1007/978-3-319-71189-8_13

Table 13.1 Soil based ecosystem services and functions based upon Haygarth and Ritz (2009)[a]

Supporting		Provisioning		Regulating		Cultural	
Service	Function	Service	Function	Service	Function	Service	Function
Primary productivity	Support vegetation	Refugia	Habitat for macrofauna	Water quality	Purified/buffered water	Recreation	Platform for tourism/sport
Soil formation	Soil formation—weathering	Water storage	Retention of water in pore network	Water supply	Regulation of hydrological flows e.g. flood control	Cognitive	Aesthetic
Nutrient cycling	Store/process nutrients; N-fixation, N, P mineralisation	Platform	Supporting structures	Gas regulation	CO_2/O_2 balance, O_3 and SO_x levels.	Heritage	Archaeological archive
		Food supply	Plant growth, crops, livestock	Climate regulation	Greenhouse gas regulation	Heritage	
		Biomaterials	Plant growth such as fuel, fibre, timber.	Erosion control	Soil and colloid retention e.g. Hillslopes/wetlands		
		Raw materials	Topsoil, minerals, aggregates				
		Biodiversity and genetic resource	Unique biological materials including medical products.				

[a]Original table by Philip M. Haygarth and Karl Ritz (2009). The future of soils and land use in the UK: Soil systems for the provision of land-based ecosystem services. Land use policy http://dx.doi.org/10.1016/j.landusepol.2009.09.016. Adapted with permission

services have been rearranged into the following five main soil functions for agricultural land (Schulte et al. 2014):

(1) **Primary production**: this function refers to the speed and capacity of the soil to produce plant biomass. This function provides the basis of human life and refers to agriculture and forestry, specifically the provision of food, feed, fibre (such as wood, paper, cotton or linen) and fuel (biofuel, bioenergy) within natural and managed ecosystem boundaries. Primary productivity is a fundamental ecological variable that measures the energy input to the biosphere and carbon dioxide (CO_2) assimilation. This function also provides an indication of the condition of the land surface area and the status of a wide range of ecological processes. The land area of Ireland is 6.9 million ha, of which 4.5 million ha or about 65% of which is used for agriculture. Some 81% of agricultural land is devoted to grass (silage, hay and pasture), 11% to rough grazing (0.48 million ha) and 8% to crops, fruit and horticulture production (0.38 million ha). About 11% of the total land area or 737,904 ha are used for forestry. Beef and milk production currently account for 67% of agricultural output at producer prices (excluding) forage.

(2) **Water purification and regulation**: soils play an important role in providing clean water for human consumption and for aquatic ecosystems. Soils purify percolating water through either adsorption of potentially harmful compounds (such as metals, excess orthophosphate concentrations) or through transformation (e.g. denitrification of nitrates). Water stored in soil is used for evapotranspiration and plant growth that supply food and fibre (FAO and ITPS 2015). In addition, soils play an important role in regulating water dynamics. Their water storage capacity, controls the rainfall drainage system into rivers, and may thus prevent both flooding and droughts. At the same time, percolating water can replenish groundwater and related streamflow and surface water ecosystems (FAO and ITPS 2015).

(3) **Carbon storage and sequestration**: globally, soils constitute the second largest store of carbon (C). The total quantity of C stored in soil, estimated at 2344 Gt of soil organic carbon (SOC) (Jobbágy and Jackson 2000), is considered to be three times larger than the amount stored in above-ground biomass and twice the amount present in the atmosphere (Eswaran et al. 1993). Soil carbon sequestration can be defined as the net removal of atmospheric CO_2, which is the transfer of CO_2 into pools that are long-lived such as soil organic matter so that CO_2 in the atmosphere is reduced or slowed (McBratney et al. 2014). As well as productivity and litter inputs, soil temperature and water content greatly

influence soil C storage due to their effect on microbial activity. Soil C stocks are positively correlated with mean annual temperatures (Jobbágy and Jackson 2000). Higher soil temperatures increases microbial decomposition of organic matter resulting in less SOC accumulation. Hence, temperature is a major control of SOM storage in soil C models but the decomposition for different SOM fractions is, as yet, an area of uncertainty (Conant et al. 2011). Soil moisture is another key driver that affects the accumulation and storage of C in soils. Soils with larger C stocks are frequently found in latitudinal gradients with moist cold ecosystems and frequently saturated soils (Moyano et al. 2013). These conditions are typical in Ireland and approximately 20–25% of the land area of Ireland is mapped as peats (Connolly and Holden 2009; Irish SIS (Creamer et al. 2014)). Irish peatlands are a huge C store and contain more than 75% of all of the national SOC (Renou-Wilson et al. 2011). Many grassland and peat soils still actively sequester C. In Ireland, near intact peatlands may actively sequester on average 57,402 t C/year over the whole country (Renou-Wilson et al. 2011). Therefore, soils provide an important function in climate regulation, both as a C store and a sink of atmospheric C. Besides climate regulation, the soil C store plays an important role in supporting the other four soil functions. Increasing soil C content improves chemical and physical properties, such as nutrient storage, water holding capacity, aggregation and sorption of organic and inorganic pollutants (Kibblewhite et al. 2008).

(4) **Habitat for intrinsic and functional biodiversity**: soil provides a habitat to all phyla of life. The exact diversity and abundance of species in soils has not yet been quantified, but it is estimated that only 1% of soil microorganism species have been identified (Orgiazzi et al. 2016). Soil biota and soil contribute in the delivery of many other soil functions and support the provision of food, fuel and fibre, the infiltration, storage and delivery of clean water, the suppression of plant pests and the control of nutrient cycles (Orgiazzi et al. 2016). We consider two values of biodiversity, namely:
   a. Intrinsic value: species have their own value, as living beings, regardless of their 'usefulness' to humankind.
   b. Functional value: species are worth protecting because they have a useful role in the delivery of goods and services to humankind;
   c. These two values are not mutually exclusive.

(5) **Cycling and provision of nutrients**: soil provides a 'home' not only to the biotic world, but also to the abiotic world in the form of minerals or nutrients. This function manifests itself in many forms. Soils can be mined to provide fertiliser to other soils, e.g. in the form of rock phosphate or lime. Soils can provide a sustainable receptor of nutrients in waste products such as sewage sludge or animal manures. Or soils may simply act as a temporary store and recycling facility of nutrients applied in the form of chemical fertiliser. In agricultural systems, nutrient cycling is the capacity of the soil to received nutrients in the form of farm, industrial or urban 'wastes', to provide nutrients from intrinsic resources and to effectively carry over these nutrients into harvested crops (Schröder et al. 2016).

In addition to these five functions, soils also provide: (6) a **platform for infrastructure** such as roads or buildings and (7) an outdoor **archive of archaeological heritage**.

13.2 Soils—A Multifunctional Resource

In principle, all soils perform each of these functions simultaneously. However, the extent to which each function is delivered depends in the first instance on land use. For example, the functionality of arable soils is characterised by primary production and nutrient cycling. The capacity of arable soils to accept nutrients is due to the large inputs of nutrients required to replace those rapidly taken up by arable crops (Sullivan et al. 1999). Arable soils also perform the functions water purification, habitat for biodiversity and storage of carbon, but to a much lesser extent. For example, land use type has a large influence on soil biodiversity, with earthworm biodiversity lower in arable production systems than grassland systems due to the injury/death associated with cultivation, soil erosion, compaction, pesticide use, increased predation and lower availability of organic matter (food) (Bardgett 2005; Brussard et al. 2007; Van Eekeren et al. 2008; Turbé et al. 2010). Equally, SOC accumulation in conventional tillage systems is less due to the regular disruption of soil structure through inversion tillage practices, such as ploughing, which aerates the soil and results in increased microbial decomposition of organic matter (Briggs and Courtney 1989; Lal 2004).

Blanket Peats, dominant in the Irish mountain landscape, is at the opposite end of the spectrum: these provide important stores of carbon, a home for biodiversity and supply clean water; in addition, they can provide food (grass), but at a very limited capacity. However, they have reduced nutrient accommodation capacity due to the fact that these soils are already saturated or nearly saturated and in winter will result in overland flow (Gardiner 1986) to surface water (Coyle et al. 2016). Figure 13.1 shows the relative delivery of suites of these five functions across different landscapes, based upon the interaction of land use with soil type.

Fig. 13.1 All soils perform all five functions, but different parts of the landscape provide different 'suites' of soil functions. The top left image represents a typical farmland in Co. Donegal, while the bottom photograph shows a landscape in Co. Wexford. The size of each of the coloured boxes represents the 'supply' of each of the five soil functions (legend top right). Based on Schulte et al. 2014

Second, the functionality of soils not only depends on land use, but also on soil properties. While there are countless properties that impact on each of the five soil functions, the dominant soil properties in the Atlantic climates of Ireland relate to soil moisture dynamics, specifically the occurrence of excess soil water (for two comprehensive reviews, see Schulte et al. 2012; Coyle et al. 2016). In Ireland, soil moisture contents may exceed Field Capacity for a prolonged period throughout the year. Table 13.2 shows how soil moisture dynamics (known as 'Field Capacity Days') has a profound impact on each of the five soil functions.

13.3 Quantifying Soil Functions

It is difficult to quantify soil functionsSoil functions because each of the soil functions encompasses a set of processes. Moreover, management in support of one soil function may alter the delivery of the other functions and can result in a co-benefit (synergy) or a conflict (trade-off) (Robinson et al. 2013; Dominati et al. 2010; O'Sullivan et al. 2015). Therefore, we use proxy-indicators to estimate the extent to which a soil provides the five functions. These proxy-indicators describe the most important aspect(s) of each function that can be measured. Which aspect of each function is considered most important, is highly context-specific and varies from place to place. For example, in their scoping study on the functionality of soils in Ireland, Schulte et al. (2014) used 'grazing capacity' (i.e. the density of animals that a soil can support) as a proxy-indicator for the function 'primary production'. While this indicator appropriately captures the most important aspect of production for grassland soils, other proxy-indicators would obviously be required in regions or countries characterised by other land use such as forestry or tillage (Table 13.3).

Table 13.2 Impact of soil moisture dynamics on each of the five soil functions

Soil function	Impact of soil moisture dynamics
Primary production	Excess soil moisture may induce anaerobicity, possible toxicity, resulting in reduced plant growth Grassland: reduced herbage utilisation as a result of reduced trafficability Tillage: shorter windows of opportunity for tillage operations as a result of reduced trafficability Droughts induce stomatal closure, reducing plant growth
Water purification and regulation	Excess soil moisture contributes to nutrient mobilisation, occurrence of overland flow, erosion Droughts may result in release of organic nitrogen and hence elevated nitrate conditions, following rewetting
Carbon storage sequestration potential	Anaerobic conditions reduce carbon mineralisation and may thus increase carbon sequestration and storage. Conversely, increased oxidation under aerobic conditions can speed up C mineralisation resulting in losses of CO_2 to the atmosphere
Habitat for biodiversity	Soil biodiversity may be impacted upon by both excess soil moisture conditions (anaerobic conditions) and droughts
Nutrient cycling and provision	Excess soil moisture conditions reduce the nutrient accommodation (opportunities for nutrient application), the fertiliser value, harvest index and utilisation value

Sources Schulte et al. (2012) and Coyle et al. (2016)

Table 13.3 Proxy-indicators for each of the five soil functions, used in the Irish case study by Schulte et al. (2014), and alternative candidate proxy-indicators being considered by the European Landmark (Land Management: Assessment, Research Knowledge base) project

Soil function	Proxy-indicators used for Ireland by Schulte et al. (2014)	Alternative optional proxy-indicators
Primary production	Grazing capacity (grassland)	Crop yield (tillage), annual growth rate, biomass (forestry), energetic yield, protein yield, …
Water purification and regulation	Denitrification capacity Phosphorus sorption capacity	Water storage and buffering capacity, resistance to erosion, adsorption capacity for metals, pesticides, metals, organic compounds, mortality rate of pathogens, …
Carbon storage sequestration potential	Soil carbon sequestration rate following conversion to forestry	Carbon stock, soil organic matter content, actual carbon sequestration rate, soil carbon residence times, …
Habitat for biodiversity	Above-ground biodiversity (in absence of data on below ground biodiversity)	Belowground biodiversity: species richness, abundance, biomass, PLFA, genetic diversity, …
Nutrient cycling and provision	Phosphorus sorption capacity	Potential supply of phosphorus, calcium, nutrient accommodation, fertiliser value, harvest index, utilisation value, …

13.4 Areas of Natural Constraint

EU policies have long recognised differences between soils or landscapes in their capacity to produce food. In an effort to reduce land abandonment, farmers located in so-called 'Less Favourable Areas' (LFAs) receive additional EU support to maintain their livelihood. In 2013, the European Commission published a review of soil properties that limit primary production (Van Orshoven et al. 2013). These were collated into seven 'constraints' that will be used to delineate the 'Areas of Natural Constraint' that will supersede the LFAs:

1. Low temperatures throughout the growing season
2. Frequent and prolonged droughts
3. Limited soil drainage and/or prolonged occurrence of excess soil moisture conditions
4. Unfavourable texture and stoniness
5. Shallow rooting depth
6. Poor chemical properties, including soil salinity, sodicity or acidity
7. Steep slopes.

Some of these constraints occur in specific geographical regions. For example, low temperatures throughout the growing season are mainly of relevance to Northern regions in Europe, while frequent and prolonged droughts are most widespread in the Mediterranean South. The third criterion, limited soil drainage and prolonged occurrence of excess soil moisture conditions, is of specific relevance to the Atlantic climate, including Ireland (Study box 12.1). The remaining four criteria are largely independent of climate and may, therefore, be found in any of the climatic zones of Europe.

Case Study 13.1 Soil Moisture Dynamics

In the Irish climate, soil moisture is the dominant soil property controlling the supply of soil functions. Soil moisture can be measured in situ as volumetric water content, but is frequently quantified and expressed using the concept of soil moisture deficits (SMD). SMD is the amount of rainfall that would be required to bring the water content to field capacity (FC), i.e. the amount of water that a soil can hold against gravity. Positive values of SMD refer to a deficit of water. Large deficits may result in droughts and reduce plant growth; a common occurrence in continental and Mediterranean summers. Negative values of SMD refer to conditions where the soil water content exceeds FC: by definition, this results in percolation, i.e. a downward movement of water. If percolation is impeded, or if the precipitation rate exceeds the percolation capacity of the soil, the soil may reach saturation. At this point, all soil pores are fully filled with water, and further rainfall will either result in ponding (standing water) or overland flow. Excess soil moisture conditions are common in Atlantic climates; they commonly occur for prolonged periods of time during winter, when rates of evapotranspiration are low and exceeded by precipitation. However, excess soil moisture conditions may also occur at other times during spells of wet weather.

Teagasc Agriculture and Food Development Authority and Met Éireann, the Irish meteorological service jointly developed the Hybrid Soil Moisture Deficit Model (HSMD), which estimates regional SMD values on a weekly basis for a range of soil drainage categories (Schulte et al. 2005) see: www.met.ie/agmet (Fig. 13.2).

13.5 Threats to Soil Functioning

The functionality of soils worldwide is threatened by unsustainable management practices that reduce the future ability of soils to perform the five functions. Creamer et al. (2010) summarised the main threats to soil quality under Irish agricultural management systems for arable and grassland systems:

Erosion is the threat that has received most attention in Europe, as it results in the permanent removal of topsoil by water or wind. The European Commission (2006) has estimated that over 150 million ha of farmland in Europe are subject to erosion. Under natural conditions, erosion is a slow process that redistributes minerals and can create fertile deposits (e.g. floodplains).

Loss of organic matter can be a precursor to erosion: organic matter typically acts as 'glue' that keeps soil particles together. Frequent ploughing of soils stimulates mineralisation of organic matter. If this is not counterbalanced by inputs of new organic matter, (e.g. stubble, animal manure, slurry, sewage sludge), it may result in a net loss of organic matter. An additional concern associated with loss of organic matter is that it contributes to greenhouse gas emissions and reduces the store of carbon that is 'locked up' in soils worldwide. The European Commission (2006) has estimated that 45% of soils in Europe are now low in organic matter.

Contamination of soils is typically associated with mining operations or historical industrial activities, which can result in elevated or toxic concentrations of, e.g. heavy metals or organic compounds, which may severely restrict the use of soils. In Ireland, the spatial extent of both activities has been limited. However, a 'softer' form of contamination refers to over-application of plant nutrients such as nitrogen and phosphorus, which may exceed the ability of soils to purify water and thus result in eutrophication of groundwater or surface waters (see Chap. 16).

Compaction is a 'hidden threat' as it may occur gradually and can be difficult to diagnose. Compaction is typically caused by trafficking of soils under wet conditions, either by farm machinery or animals. It results in a compression of the pore space, and thus a reduction in the air and water holding capacity; this, in turn, affects all five functions. Considering Ireland's wet climate, compaction is likely to be the most widespread threat to the functionality of the country's soils (Creamer et al. 2010). Chapter 14 on soils and productivity in Ireland describes compaction in greater detail.

Loss of soil biodiversity is a threat that is interlinked with all other threats and functions, as many of the soil processes are performed or mediated by soil organisms. Loss of biodiversity may be triggered by one or more of the other threats and can, in turn, aggravate other threats and reduce functionality. Science is only beginning to assess the current

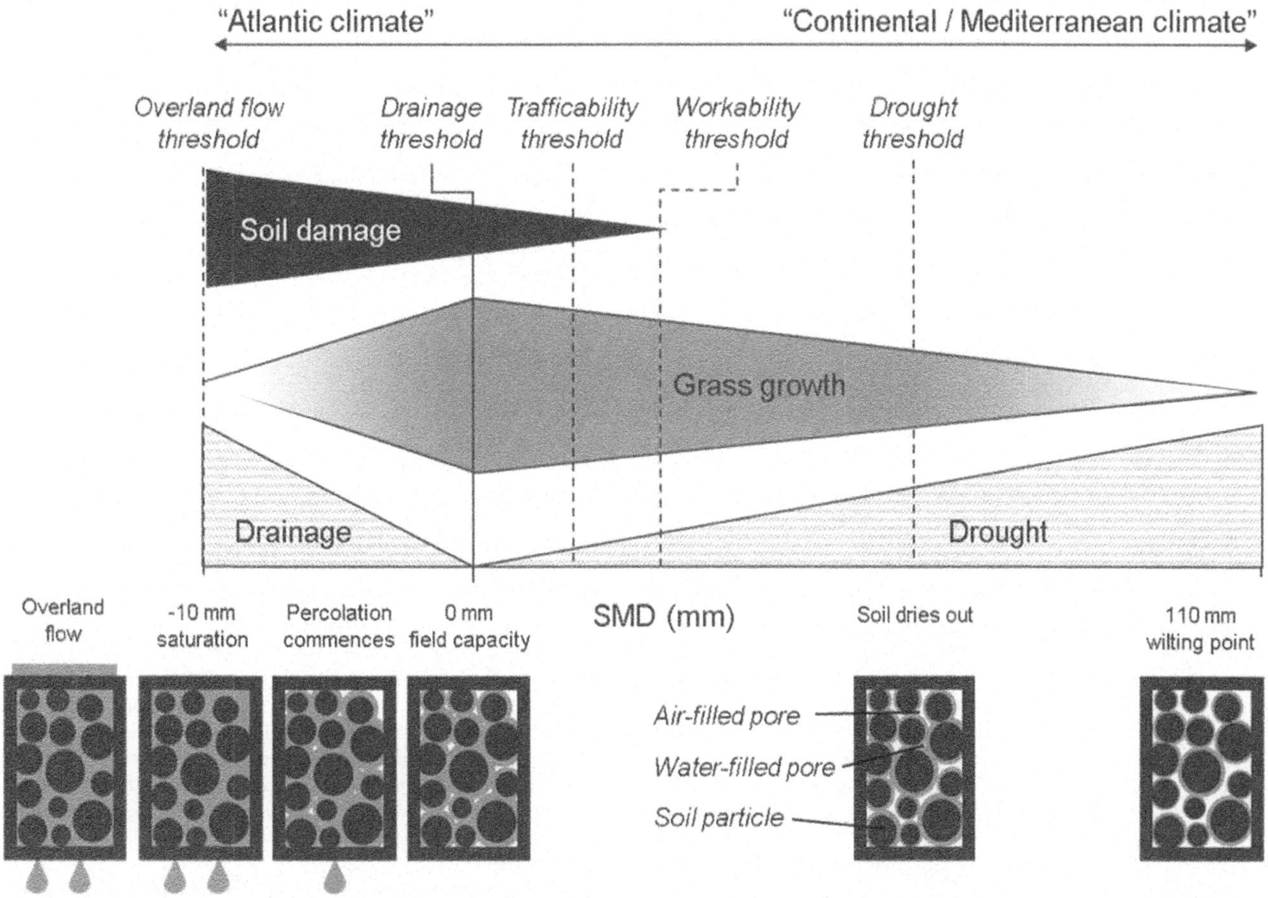

Fig. 13.2 Illustration of the SMD as an indicator of soil moisture dynamics. Positive values of SMD indicate drought conditions, in which plant growth is reduced. The wilting point is defined as the point at which plants have extracted the maximum amount of water from the soil. Excess soil moisture results in percolation and may result in overland flow when the soil reaches saturation. Trafficking or working of wet soil is likely to result in soil damage or compaction *Source* Schulte et al. (2015), Coyle et al. (2016)

state of biodiversity (Orgiazzi et al. 2016) and so the decline of soil biodiversity has yet to be quantified for most parts of Europe, including Ireland.

In the following Chapters, we will examine each of the five soil functions in detail. In Chap. 21, we will explore how soils can be managed in the future to meet the expectations or 'demands' for soil functions by society.

References

Bardgett R (2005) The biology of soil: a community and ecosystem approach. Oxford University Press, Oxford

Briggs D, Courtney F (1989) Agriculture and environment, the physical geography of temperate agriculture systems. Longman Scientific and Technical

Brussard L, de Ruiter PC, Brown GG (2007) Soil biodiversity for agricultural sustainability. Agric Ecosyst Environ 121:233–244

Conant RT, Ryan MG, Agren GI, Birge JHE, Davidson EA, Eliasson PE, Evans SE, Frey SD, Giardina CE, Hopkins FM, Hyvonen R, Kirschbaum M, Lavallee JM, Leifeld J, Parton WJ, Steinweg JM, Wallenstein MD, Wetterstedt JAM, Bradford MA (2011) Temperature and soil organic matter decomposition rates—synthesis of current knowledge and a way forward. Glob Change Biol 17:3392–3404

Connolly J, Holden NM (2009) Mapping peat soils in Ireland; updating the Derived Irish Peat Map. Ir Geogr 3:343–352

Coyle C, Creamer RE, Schulte RPO, O'Sullivan L, Jordan P (2016) A functional land management conceptual framework under soil drainage and land use scenarios. Environ Sci Policy 56:39–48

Creamer RE, Brennan F, Fenton O, Healy MG, Lalor STJ, Lanigan GJ, Regan JT, Griffiths BS (2010) Implications of the proposed soil framework directive on agricultural systems in Atlantic Europe—a review. Soil Use Manag 26:198–211

Creamer R, Simo I, Reidy B, Carvalho J, Fealy R, Hallett S, Jones R, Holden A, Holden N, Hannam J, Massey P, Mayr T, McDonald E, O'Rourke S, Sills P, Truckell I, Zawadzka J, Schulte R (2014) Irish Soil Information System (2007-S-CD-1-S1) EPA STRIVE Programme 2007–2013, Synthesis Report. Available for download from http://erc.epa.ie/safer/reports

Dominati E, Patterson M, Mackay A (2010) A framework for classifying and quantifying the natural capital and ecosystem services of soils. Ecol Econ 69:1858–1868

Eswaran H, van de Berg E, Reich P (1993) Organic carbon in soils of the world. Soil Sci Soc Am J 57:192–194

European Commission (EC) (2006) (COM(2006) 232) Proposal for a Directive of the European Parliament and of the Council establishing a framework for the protection of soils and amending directive 2004/35/EC. European Commission, Brussels

Food and Agriculture Organization (FAO) Intergovernmental Technical Panel on Soils (ITPS) (2015) Status of the World's Soil Resources (SWSR)—main report. Food and Agriculture Organization of the United Nations and Intergovernmental Technical Panel on Soils, Rome, Italy

Gardiner MJ (1986) Soil, site and climatic factors in slurry disposal. Ir J Agric Res 25:273–284

Haygarth PM, Ritz K (2009) The future of soils and land use policy in the UK: soil systems for the provision of land-based ecosystem services. Land Use Policy 26:187–198

Jobággy EG, Jackson RB (2000) The vertical distribution of soil organic carbon and its relation to climate and vegetation. Ecol Appl 10:423–436

Kibblewhite MG, Ritz K, Swift MJ (2008) Soil health in agricultural systems. Phil Trans Royal Soc 363:685–701

Lal R (2004) Soil carbon sequestration to mitigate climate change. Geoderma 123:1–22

McBratney AB, Stockmann U, Angers DA, Minasny B, Field DJ (2014) Challenges for soil organic carbon research. In: Hartemink and McSweeney (eds) Soil carbon, Springer, Switzerland

Moyano FE, Manzoni S, Chenu C (2013) Responses of soil heterotrophic respiration to moisture availability: an exploration of processes and models. Soil Biol Biochem 59:72–85

Orgiazzi A, Bardgett RD, Barrios E, Behan-Pelletier V, Briones MJI, Chotte JL, De Deyn GB, Eggleton P, Fierer N, Fraser T, Hedlund K, Jeffrey S, Johnson NC, Jones A, Kandeler E, Kaneko N, Lavelle P, Lemanceau P, Miko L, Montanarella L, Moreira FMS, Ramirez KS, Scheu S, Singh BD, Six J, van der Putten WH, Wall DH (eds) (2016) Global soil biodiversity Atlas. European Commission, Publications Office of the European Union, Luxembourg, p 176

O'Sullivan L, Creamer RE, Fealy R, Lanigan G, Simo I, Fenton O, Carfrae J, Schulte RPO (2015) Functional land management for managing soil functions: a case-study of the trade-off between primary productivity and carbon storage in response to the intervention of drainage systems in Ireland. Land Use Policy 47:42–54

Renou-Wilson F, Bolger T, Bullock C, Convery F, Curry J, Ward S, Wilson D, Müller C (2011) Bogland: sustainable management of Peatlands in Ireland, STRIVE Report prepared for the Environmental Protection Agency by University College Dublin

Robinson DA, Hockley N, Cooper DM, Emmett BA, Keith AM, Lebron I, Reynolds B, Tipping E, Tye AM, Watts CW, Whalley WR, Black HIJ, Warren GP, Robinson JS (2013) Natural capital and ecosystem services, developing an appropriate soils framework as a basis for valuation. Soil Biol Biochem 57:1023–1033

Schröeder JJ, Schulte RPO, Creamer RE, Delgado A, van Leeuwen J, Lehtinen T, Rutgers M, Spiegel H, Staes J, Tóth G, Wall DP (2016) The elusive role of soil quality in nutrient cycling: a review. Soil Use Manag 32:476–486. https://doi.org/10.1111/sum.12288

Schulte RPO et al. (2005) Predicting the soil moisture conditions of Irish grasslands. Irish J Agricul Food Res: 95–110

Schulte RPO et al. (2015) A note on the hybrid soil moisture deficit model v2. 0. Irish J Agricul Food Res 54(2):126–131

Schulte RPO, Fealy R, Creamer RE, Towers W, Harty T, Jones RJA (2012) A review of the role of excess soil moisture conditions in constraining farm practices under Atlantic conditions. Soil Use Manage 4:580–589

Schulte RPO, Creamer RE, Donnellan T, Farrelly N, Fealy R, O'Donoghue C, O'hUallachain D (2014) Functional land management: a framework for managing soil-based ecosystem services for the sustainable intensification of agriculture. Environ Sci Policy 38:45–58

Sullivan DM, Hart JM, Christensen NW (1999) Nitrogen uptake and utilisation by Pacific Northwest crops. A Pacific Northwest Publication, Oregon State University, Corvallis

Turbé A, De Toni A, Benito P, Lavelle P, Ruiz N, Van der Putten WH, Labouze E, Mudgal S (2010) Soil biodiversity: functions, threats and tools for policy makers, Bio Intelligence Service, IRD, and NIOO, Report for European Commission (DG Environment)

United Nations Development Programme (UNDP) (2016) Sustainable development goals (website) http://www.undp.org/content/undp/en/home/sdgoverview/post-2015-development-agenda.html. Accessed 06/07/2016

van Eekeren N et al. (2008) Soil biological quality after 36 years of ley-arable cropping, permanent grassland and permanent arable cropping. applied soil ecol 40(3):432–446

Van Orshoven J, Terres J, Tóth T (2013) Definition and scientific justification for the common biophysical criteria—updated common bio-physical criteria to define natural constraints for agriculture in Europe. Contributors: Jones R, Le-Bas C, Nachtergale F, Rossiter D, Van Orshoven J, Schulte R, van Velthuizen H. European Commission Joint Research Centre and Institute for Environment and Sustainability

Soils and Productivity

14

Brian Tobin, Sara E. Vero, Owen Fenton, Lilian O'Sullivan, Pat Tuohy, Rochelle Fritch, and Rachel Creamer

Abstract

This chapter explores the predominant land-based production systems in Ireland, specifically describing grassland, arable and forestry production. Grass is the most important agricultural crop in Ireland representing the main feed source for the livestock sector. Arable production represents a relatively small area, largely due to excess soil moisture conditions that are typical of many Irish soils. Despite favourable conditions for forestry in Ireland, the level of forest cover is low by European standards and has historically been associated with sub-marginal and marginal agricultural land. The importance of soil structure in the provision of nutrients, water and air is emphasised along with best management practices in relation to the management cycle. The role of soil texture is outlined, as is the agroclimate and soil interaction in Irish soils. The agricultural drainage conditions of Irish soils fall into the temperate zone, where the main role of drainage is the removal of excess water in the root zone of crops from surplus rainfall. This follows with a historical account of land drainage schemes in Ireland. Finally, the issues of the main production systems on poorly drained soils and the cultural practices to maintain trafficability are described. Improving the overall trafficability of any soil first requires the identification of factors causing either excessive moisture or poor soil structure. Correctly ascertaining these issues allows implementation of appropriate preventative and remediation measures, and should be the initial step in any management strategy.

Keywords

Primary productivity • Drainage • Water table
Trafficability • Soil structure

B. Tobin (✉)
UCD Forestry, School of Agriculture and Food Science, University College Dublin, Dublin 4, Ireland
e-mail: brian.tobin@ucd.ie

S. E. Vero
Agricultural Catchments Programme, Teagasc, Johnstown Castle, Wexford, County Wexford, Ireland

O. Fenton · L. O'Sullivan · R. Creamer
Teagasc Agriculture and Food Development Authority, Johnstown Castle Environment Research Centre, Wexford, Ireland

L. O'Sullivan
Irish Environmental Protection Agency, Johnstown Castle, Wexford, County Wexford, Ireland

P. Tuohy
Teagasc Moorepark, Animal and Grassland Research and Innovation Centre, Fermoy, County Cork, Ireland

R. Fritch
UCD School of Agriculture and Food Science, University College Dublin, Dublin 4, Ireland

R. Creamer
Wageningen University and Research Atlas—Wageningen Campus, Wageningen, The Netherlands

14.1 Introduction

The land area of Ireland is 6.9 million hectares (M ha) of which approximately 65% is used for agriculture (DAFM 2015). Grasslands and rough grazing account for 80 and 11% of agricultural land use, respectively, while a relatively small portion of the agricultural area is dedicated to crops, fruits and horticulture (8%) (DAFM 2015). About 11% (737,904 ha) of the total area is used for forestry (DAFM 2015). Primary productivity through agriculture and forestry represent the main land use activities in Ireland and are critically important to the economic, social and cultural fabric of Irish society; being particularly important for the viability of rural areas. In Ireland, agriculture is dominated by grass-based livestock production. Grass utilisation across the country is largely controlled by differences in climate (especially rainfall) and soil type.

R. Creamer and L. O'Sullivan (eds.), *The Soils of Ireland*, World Soils Book Series,
https://doi.org/10.1007/978-3-319-71189-8_14

14.2 Land-Based Production Systems in Ireland

14.2.1 Grassland

Grass is the most important agricultural crop in Ireland representing the main feed source for the livestock sector. Historically, Irish agriculture has been pasture-based. The clearing of the forests started around 6000 years ago to make way for the increase of prehistoric farming (O'Connell and Molloy 2001). Over thousands of years, this clearing of forest culminated in grasslands dominating the Irish landscape. The 'making of plains', with the clearance of woodland for grazing, was considered a top achievement of the heroes of Irish mythology (Feehan 2003). The long history and culture around pastoral farming, the challenge of growing and harvesting arable crops on wetter soils, along with the competitiveness of grass-based dairy and beef production industries, means that grasslands are the prevalent land use in Ireland. Ireland's temperate maritime climate in north Atlantic Europe, associated with its long growing season, provides a natural advantage in relation to grass growth productivity (Rath and Peel 2005). This advantage provides a potential to grow between 12 and 16 tonnes of dry matter per hectare per annum on certain soil types with optimised soil fertility (O'Donovan et al. 2011). This is further complemented by the permanent nature of Irish grasslands. In Ireland, the percentage of permanent grassland (74%), versus temporary leys (17%), is higher compared to most other western European countries (Smit et al. 2008). Permanent grassland in Ireland contains a diverse range of plant communities, from highly intensive sown perennial ryegrass monocultures and white clover-ryegrass mixtures, to semi-improved and semi-natural grasslands. These different grassland types, with their differing levels of management and nutrient inputs, vary widely in their net primary production. Current intensive Irish grassland systems focus on perennial ryegrass monocultures, the productivity of which is dependent on high water and nutrient availability (Feehan et al. 2012). The sowing of sward mixtures, with the inclusion of legumes and deep rooting plants, may allow the production potential of grasslands to be maintained with a reduction of nutrient inputs (Finn et al. 2013). Temperatures above 6 °C allow for continuous pasture growth in Irish conditions (Connaughton 1973). Adequate rainfall and temperatures which permit grass growth for up to 10 months of the year in most areas along with sufficient radiation, allow for substantial production of pasture (O'Mara 2011). Utilisation of this grass growth is largely controlled by soil drainage capacity and climate (Tuohy et al. 2015). While most Irish soils do produce excellent grasslands, excess soil moisture does represent a threat to primary productivity and can increase the risk of negative environmental impacts (Schulte et al. 2012).

14.2.2 Arable Systems

Arable production represents a small portion of overall land use in Ireland accounting for about 380,000 ha (DAFM 2015; TTCSCG 2012). The relatively small land area dedicated to arable production is largely due to the excess soil moisture conditions that are typical in many Irish soils. Favourable conditions in the root zone are critical for crop production to support the adsorption of nutrients and water. Sufficient air-filled pore space is necessary to facilitate plant respiration and gaseous exchange (Moore 2001). However, conditions of excess water in the rooting zone can drastically reduce crop yields and plants may eventually die in cases of prolonged waterlogging due to such a lack of root zone oxygen (FAO 1997). A large proportion of Irish soils is characterised by poor internal drainage which is compounded by regionally high precipitation rates. In spring, poorly drained soils take more time and energy to warm up due to their higher water content, which in turn can delay crop/grass growth. Consequently, arable production is mostly concentrated on free-draining soils, such as the Brown Earths and Brown Podzolics in the south and east regions of the country, with their associated lower precipitation rates and deep water tables. In Ireland, these areas are mostly found stretching diagonally from east Cork in the south to Co. Louth in the north-east. Although the typical Calcareous Brown Earths found in the west of Ireland are free-draining, these soils tend to be shallow and therefore rooting depth limits their capacity for arable production. While the Luvisols that are widely found across the midlands of Ireland make good tillage soils and can be used for a variety of tillage crops and vegetable production, they exhibit a predisposition towards lodging of cereal crops. Lodging is the process by which small-grained cereals are displaced from their vertical stance, which can potentially lower yields and result in a deterioration of soil structure (Berry et al. 2004). Winter wheat, winter barley, winter oats, spring wheat, spring barley, spring oats, oilseed rape, spring beans and potatoes are the main crops produced. Spring barley is the most important arable crop grown accounting for between 50 and 60% of national cereal area (DAFM 2016). Irish cereal yields on well-suited soils can be among the highest in the world (TTCSCG 2012).

14.2.3 Forestry

The mild oceanic influence on its climate also provides Ireland with ideal conditions for growing many temperate

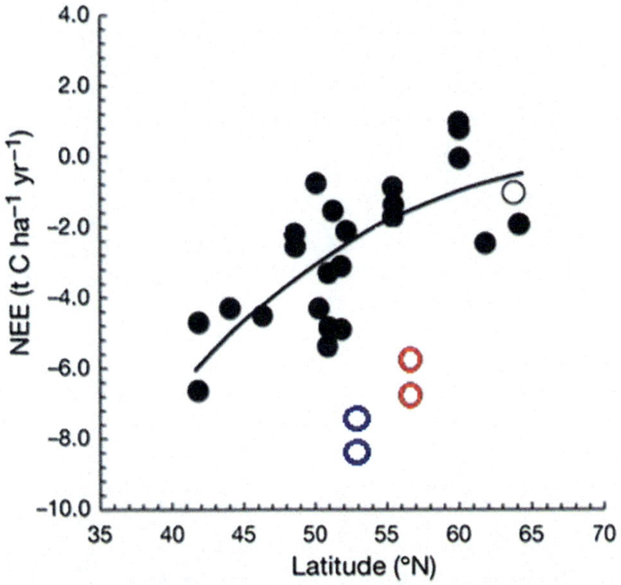

Fig. 14.1 Net ecosystem exchange (NEE) of the network of Euroflux forest sites plotted against latitude (adapted from Valentini et al. 2000). Note that the convention followed here is that a negative value indicates carbon absorbed by a forest and positive values a release to the atmosphere. The closed symbols represent natural or planted stands with traditional management and the open symbols identify intensively managed plantations. In red, the productivity of spruce stands in the UK (Clement et al. 2012), and in blue, spruce stands in Ireland (Black et al. 2007, 2009). The diagram demonstrates the relative climatic advantage Ireland enjoys for tree growth in Europe

forest species, leading to some of the highest productivity rates from conventionally managed forestry in Europe (Fig. 14.1). Despite favourable growing conditions, however, the level of forest cover in Ireland is low (11%) by European standards, where average forest cover among EU-27 countries stands at 37% (DAFM 2015). The mild, damp oceanic influences, however, also contribute their own difficulties in the management of soils to maintain their sustainability.

Historically, forestry in Ireland has been associated with sub-marginal and marginal agricultural land, largely because early land use policy developed by the Land Commission retained the best areas for food production (Gray 1963). As a result, the performance of tree crops on such land has been mixed. However, since the early 1990s when afforestation by private landowners began to overtake state afforestation, the quality of land in forestry improved substantially. Since that time, approximately 300,000 ha have been afforested (DAFM 2015). Though much of the early management of the forest estate was modelled on continental practice, such continued influence has been queried (Neeson 1991) in light of the substantially higher productivity rates of forests in Ireland. More recently, dynamic growth models parameterised specifically for Irish forestry (known as GROW-FOR), were developed by Broad and Lynch (2006) and allow growth and yield of a number of important species to be forecasted more readily than before.

Forest soil productivity assessment in Ireland initially followed the UK yield class system (Hamilton and Christie 1971). Since productivity on a particular soil type is species-specific, an indicator species was used, i.e. Sitka spruce (*Picea sitchensis* (Bong.) Carr.). This species is capable of adapting to a wide diversity of site types and has the widest range of yield classes of any conifer species grown commercially in the country. Since it occupies such a large proportion of the forest estate [52.5% according to the 2013 national forest inventory (NFI 2013)], it was an obvious choice as an indicator species (Bulfin 1987). Prior to afforestation, various systems were used to classify the potential of sites for tree growth. Classifications based on historic land use as indicated by Ordnance Survey maps (OCarroll 2012) were used to identify areas where improvements in fertility had taken place. Ground vegetation-based schemes (e.g. Anderson 1960) were used to prescribe suitable species in state afforestation in the 1960s. More recently, developments have built on this experience and the ecological site classification system used in the UK (Pyatt et al. 2001) was adapted for use in Ireland (CLIMADAPT) by Ray et al. (2009). Farrelly and Gallagher (2015a) developed a site classification system which covers the variation of site types commonly available in Ireland and provides a prescription for the potential species or forest types that might be grown in each class as well as the productivity that could be expected on such site types. This system was developed particularly to investigate the potential of identifying further site types suitable for afforestation in the face of increasing competition from agriculture and other land use enterprises (Farrelly and Gallagher 2015b). Government policy has placed a commitment on the expansion of productive forest area to 18% by 2046 (DAFM 2014). This is intended to sustain a viable processing sector as well as provide scope to mitigate increasing greenhouse gas (GHG) emissions from agricultural expansion and intensification. As a consequence, currently underutilised land (e.g. unenclosed areas) may have to provide the area needed to meet such targets.

14.3 Soil Structural Quality in Ireland

Good soil structure is essential for crop production, as this controls the provision of nutrients, water and air in soils. Soil structure refers to the shape, size and strength of structural units, 'peds' and to their spatial arrangement, often referred to as the 'building blocks' of a soil. The benefits of good soil structure from an agronomic and an environmental perspective include root support, provision of water and air for the growth of food and fibre, cycling of nutrients into plant

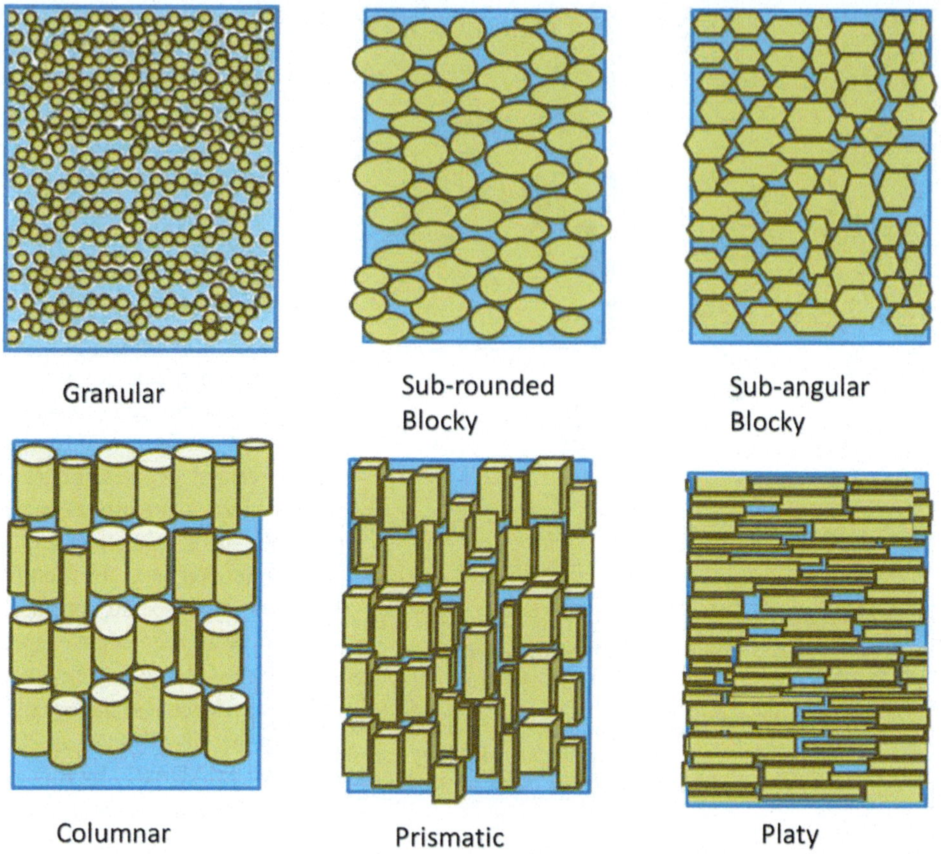

Fig. 14.2 Basic soil structure depends on the type of fundamental aggregate units, called peds, which can be used to describe the structural behaviour of soil types. The definitions above are based on the description provided by the FAO handbook (2006)

usable forms, purification of water through the percolation process, the storage and cycling of carbon, while also representing the largest biological habitat on earth. A review of how soil structure is affected under adverse climatic conditions was conducted by Guimarães et al. (2015) and included an erosion toolkit for tillage soils in Ireland.

The optimum soil structure is of a 'crumb' nature which acts like small building blocks which fit together reasonably well but have small spaces sufficient for air and water to pass through. The poorest soil structure has a platy composition (Fig. 14.2), in which layers of soil compress on top of each other, creating 'plates'. This is frequently caused by machinery compaction and found in the sub-soil, slowing the movement of water and drastically limiting aeration. Poor soil structure results in: limitations for root development, a reduction in the surface area for nutrient absorption, restricted drainage and a lack of oxygen for soil organisms.

Case Study 14.1: Soil Structure

There are several types of soil structural units (peds) (Fig. 14.2). A soil which has no structure at all is described as apedal, where no definite aggregates or peds can be defined (FAO 2006), this is usually present in extremely sandy soils or very heavy clay soils.

The crumb structure (described above) can be characterised as having small-sized (generally <1 cm) peds which have rounded edges. These rounded edges mean that the peds do not fit directly together and as a result create a network of pores between the peds that provides space for air and water. A blocky soil structure consists of peds which are generally >1 cm in size. Such peds can have rounded edges, similar to the crumb structure (sub-rounded blocky) or angular edges (blocky, sub-angular blocky). As the edges of a ped become angular, they fit together more closely in the soil matrix. However, this reduces the pore network and therefore the space for air and water within the soil. This has consequences for plant growth, drainage and nutrient availability in the soil. Soils with high clay content can often be found to have a prismatic structure in the sub-soil (below the surface horizon). A prismatic structure has round edges but an elongated column with flat sides. This can have consequences in sub-soil drainage as the flat sides can fit closely together, further reducing the pore network. Finally, a platy structure usually develops in the sub-soil (but can also be present in a surface horizon) due to a heavy load on the soil. A platy structure does not usually occur in nature, but is often the result of machinery loading or animal pressure on the soil in wet conditions. If a platy structure develops, this further reduces the drainage capacity of the soil and results in wetter soil, in which the soil structure is very weak and prone to further compaction. It is extremely difficult to remediate soils which have developed strong platy structures in the sub-soil and therefore careful management of soils in wet conditions is strongly recommended to avoid creating such problems.

Inappropriate management can result in the degradation of soil structure, which impairs soil quality and hinders productivity. The assessment of soil structural quality is therefore extremely useful for maximising sustainability and monitoring the effects of management. Assessments usually involve a combination of physical and biochemical indicators. For practicality, methodologies for quality assessment should be fast and accurate, and are frequently based on visual field indicators. Semi-quantitative visual soil evaluation techniques have been used successfully for some decades in agricultural systems to measure structural quality and these were critically reviewed by Emmet Booth et al. (2016), who concluded that texture influences, on-farm sampling procedures and more holistic assessments of soil structure all required further research. However, there has been little work carried out in assessing the structure of forest soils.

Soil structure integrity is often linked to land management. Due to the climatic regime in Ireland, the potential for poaching (penetration of the soil surface by animal hooves) is particularly prevalent in spring, autumn, winter and during high-intensity rainfall events. Under grassland management, the continuous presence of well-developed root mat provides some protection against soil structural damage, as the roots break up the soil peds into smaller units and contribute soil organic matter through roots exudates. Soil organic matter acts as a kind of glue which strengthens the individual peds. This constant provision of organic matter through root exudates throughout the year means that the soil structure in grassland systems is relatively strong. There are exceptions to this rule when grazing takes place in wet soil conditions the peds are weakened by the high soil moisture and repeated trampling by animals or trafficking by machinery can result in the peds weakening and smearing, either creating larger soil structural units or in the worst case scenario platy structure.

Good soil management and avoidance of animal grazing or trafficking (for slurry spreading, fertilisation or reseeding) in wet soil conditions will reduce any damage in grassland soils. Surface soil compaction often occurs due to repetitive trampling (poaching) leading to the reduction in soil strength, resulting in dense unstable peds at the surface of the soil, reducing soil infiltration and nutrient losses to water (Heathwaite and Jones 1996). This often results in patchy grass growth, resulting in areas of bare soil. Surface compaction while damaging in the short term, can be easily rectified through natural physical (wetting and drying cycles) and biological (earthworm burrows and root channels) amelioration. However, the relatively recent trend in Ireland has been to increase the length of grazing into the winter season (out-wintering) as a result of economic benefits (reducing the housing period and therefore associated feed and storage costs). This can be extremely detrimental to grass swards, especially on recently sown leys (Palmer et al. 2007).

Sub-surface damage occurs as a result of repetitive machinery operations during moist soil conditions. The application of fertilisers and slurry (following the closed period) in spring to enhance the first grass growth of the year can result in damage to soil structure and result in increases of soil bulk density.

Arable soils undergo a very different management cycle, there is repeated trafficking throughout the year, for the various crop types. The main machinery events associated with soil compaction in cultivated soils (Fig. 12.3), all have the potential to cause sub-soil compaction due to the increasing weight of machinery and number of passes. The use of power harrows and heavy machinery to produce a fine tilth suitable for precision seed drilling in the autumn also weaken topsoil structure, particularly in weakly structured fine sands and silt soils, resulting in pores becoming clogged with clay and silt after a rainfall event and development of a surface crust or cap (Palmer et al. 2007). Timing of machinery operations is often difficult to schedule with the increased use of contract machinery and labour, where farmer-based decisions on the condition of their field are not taken into account (Palmer et al. 2007; Creamer et al. 2010) (Fig. 14.3).

14.4 Managing Water Availability (Excess and Deficit)

Soil texture refers to the relative proportions of the particle sizes in the mineral fraction of the soil, specifically the proportion of sand, silt and clay <2 mm in diameter (Gardiner and Radford 1980) and can be measured in many ways each having an effect on the application in question (Fenton et al. 2015). Soils with equal amounts of sand, silt and clay provide a loam texture and represent the optimum soil texture. Sand particles are characterised by larger grain sizes often resulting in crumb or apedal soil structure which means

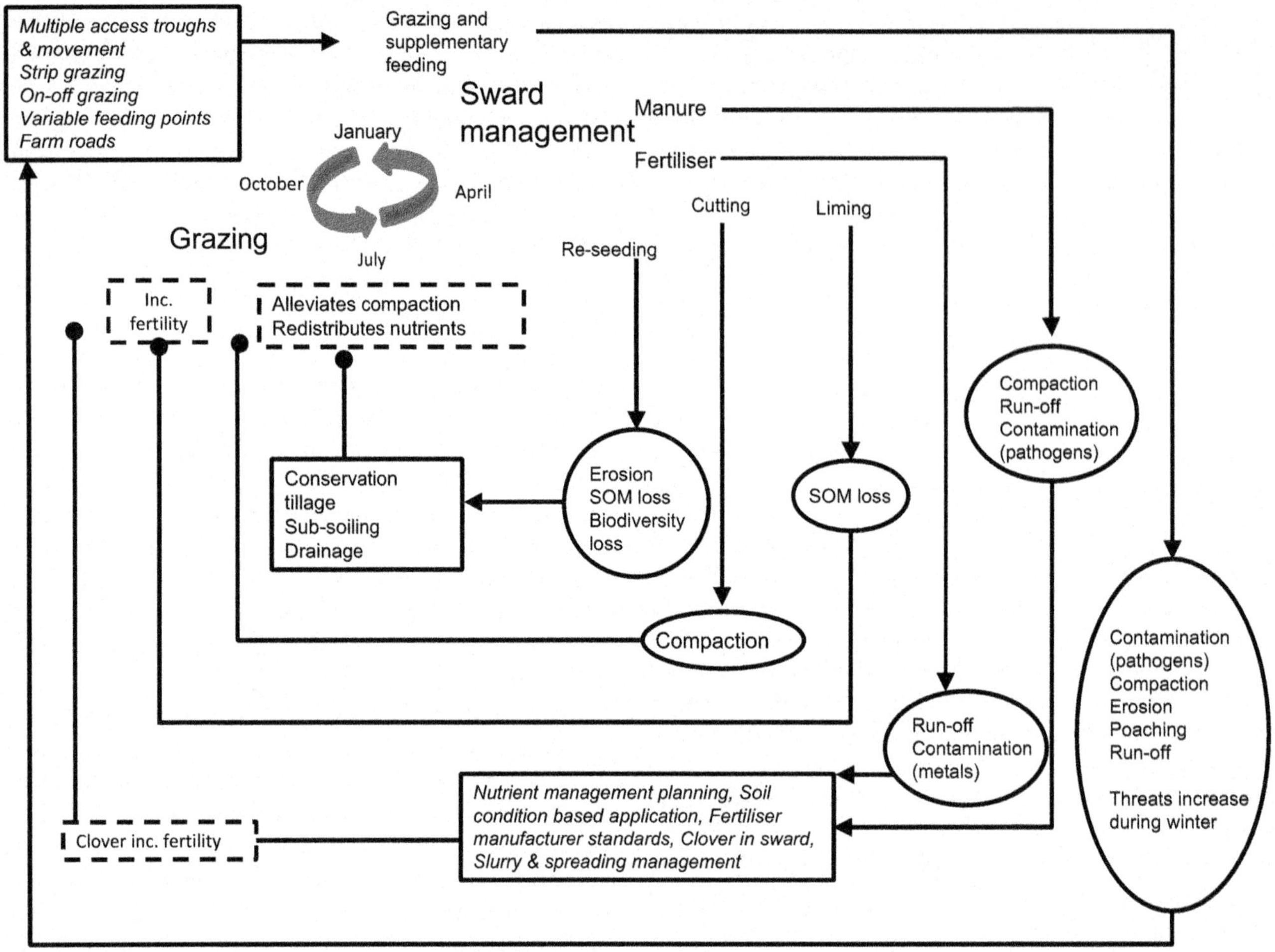

Fig. 14.3 The main operational events in the grassland cycle, based on Creamer et al. (2010), where grazing is shown as a continuous event in line with a shift towards extended year-round grazing in some countries. Figure reproduced with permission, originally published in Creamer et al. (2010). http://doi:10.1111/j.1475-2743.2010.00288.x

that the water holding capacity of sandy soils can be very low. As a result, water will infiltrate these soils rapidly (Fig. 14.4) and soils with high sand content, therefore, possess good aeration and drainage (Brady and Weil 1996). The speed of infiltration is the amount of water absorbed per hour and is measured in millimetres and can be measured using many different apparatus, e.g. an infiltrometer.

Clay particles are the finest mineral particles and while a certain amount of clay is desirable, excessively clayey soils will retain water resulting in cold sticky soils in wet weather that on drying can become hard and brittle. An extensive literature review by Coyle et al. (2016) explored the range of soil moisture conditions and deficits in relation to land use for the Atlantic pedo-climatic zone, which includes Ireland. With respect to soil function and soil management, Coyle et al. (2016) documented that primary production occurs across a range of soil moistures and deficits, e.g. herbage growth in pasture is found to be greatest at soil moisture conditions around field capacity (as found earlier by Ball and Douglas in 2003). This occurs in well to more often in moderately drained soils rather than on poorly drained equivalents. The latter will carry surplus water for prolonged periods (Shalloo et al. 2004) in contrast to well-drained soils that are more prone to drought conditions (Laidlaw 2009), even though trafficability and herbage utilisation are increased on well-drained soils (Schulte et al. 2012).

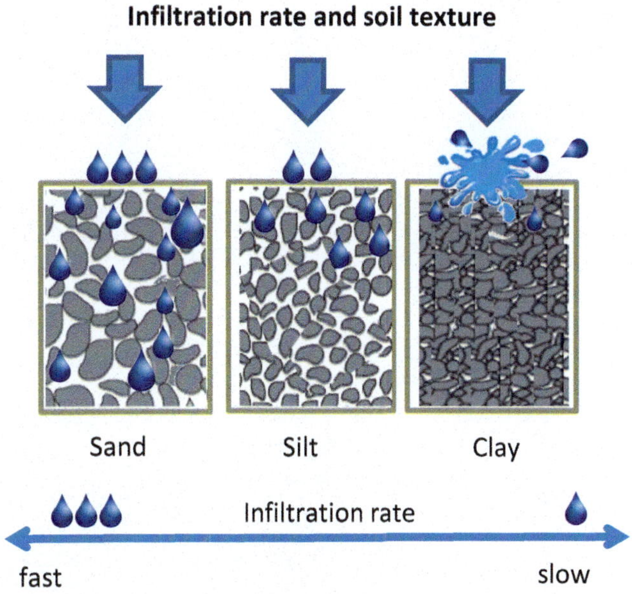

Fig. 14.4 Water infiltration rates are related to soil texture. Soils with higher sand content exhibit a higher infiltration rate compared to soils with high clay content

14.5 Agroclimate and Soil Interaction

In Ireland, precipitation is considered the meteorological variable that shows the greatest spatial and temporal variation (Hamilton et al. 1988) giving rise to high spatial variation in soil water dynamics (Schulte et al. 2006). Soil moisture deficit (SMD) is the precipitation required to bring the soil moisture content back to field capacity (see Chaps. 2 and 13). Field capacity is the amount of water the soil can hold against gravity, therefore a negative SMD indicates water surplus. This metric has been implemented within decision support frameworks for determining field conditions for various drainage classes, and hence, their behaviour as it relates to drainage, compaction and trafficability (Herbin et al. 2011; Piwowarczyk et al. 2011; Vero et al. 2013; Tuohy et al. 2016a). Soils will change their physical state with weather and season and in Ireland rainfall alters the cohesion between soil particles due to moisture content (Guimarães et al. 2015). This will inform management decision making in relation to timing of operations, animal trafficking and crop production as outlined above. Ultimately, good soil management relies on properly identifying soil characteristics and planning appropriate management having regard for current and future weather conditions (Collins et al. 2004).

A recent study utilising a high-resolution version of the EC-Earth model (Haarsma et al. 2013) suggests an increase in the frequency of extreme windstorms affecting Western Europe in future autumn seasons due to global warming. This will translate into an increased risk of storm damage, flooding and soil saturation with associated risk of damage to Irish agriculture and forestry. Drier summers along the east coast will lead to restrictions in primary production, due to water limitation, in both grassland and tillage crops, and could cause increased pressure on surface waters for irrigation. This will put more pressure on soils with impeded drainage characteristics and time required for recovery will be extended leading to knock-on effects for crop utilisation. In the case of forests, where stocks and age are increasing at a national level, this will result in substantial increases (at least double) in forest damage across Europe by the end of the century (Gardiner et al. 2010). For grass-based and tillage production systems, more concerted management effort will be required to maintain trafficability and fertility status.

14.6 Land Drainage

14.6.1 Drainage Status of Irish Soils

The agricultural drainage conditions around the world can be grouped according to three drainage zones: the temperate zone, the arid and semi-arid zone and the humid and semi-humid zone (Smedema et al. 2000; Schultz 2001). Under this classification, Ireland lies in the temperate zone, where the main role of drainage is the removal of excess water in the root zone of crops from surplus rainfall, while a secondary objective is to provide good trafficability for farm machinery and livestock (Schultz et al. 2007). Ireland is characterised by a cool temperate maritime climate where precipitation exceeds evapotranspiration (Lee 1986). Meteorological records show that excess rainfall and small moisture deficits are a dominant factor in Irish agriculture. Precipitation ranges from 750 to 1400 mm yr^{-1}, in low lying areas, while in mountainous areas, annual averages in excess

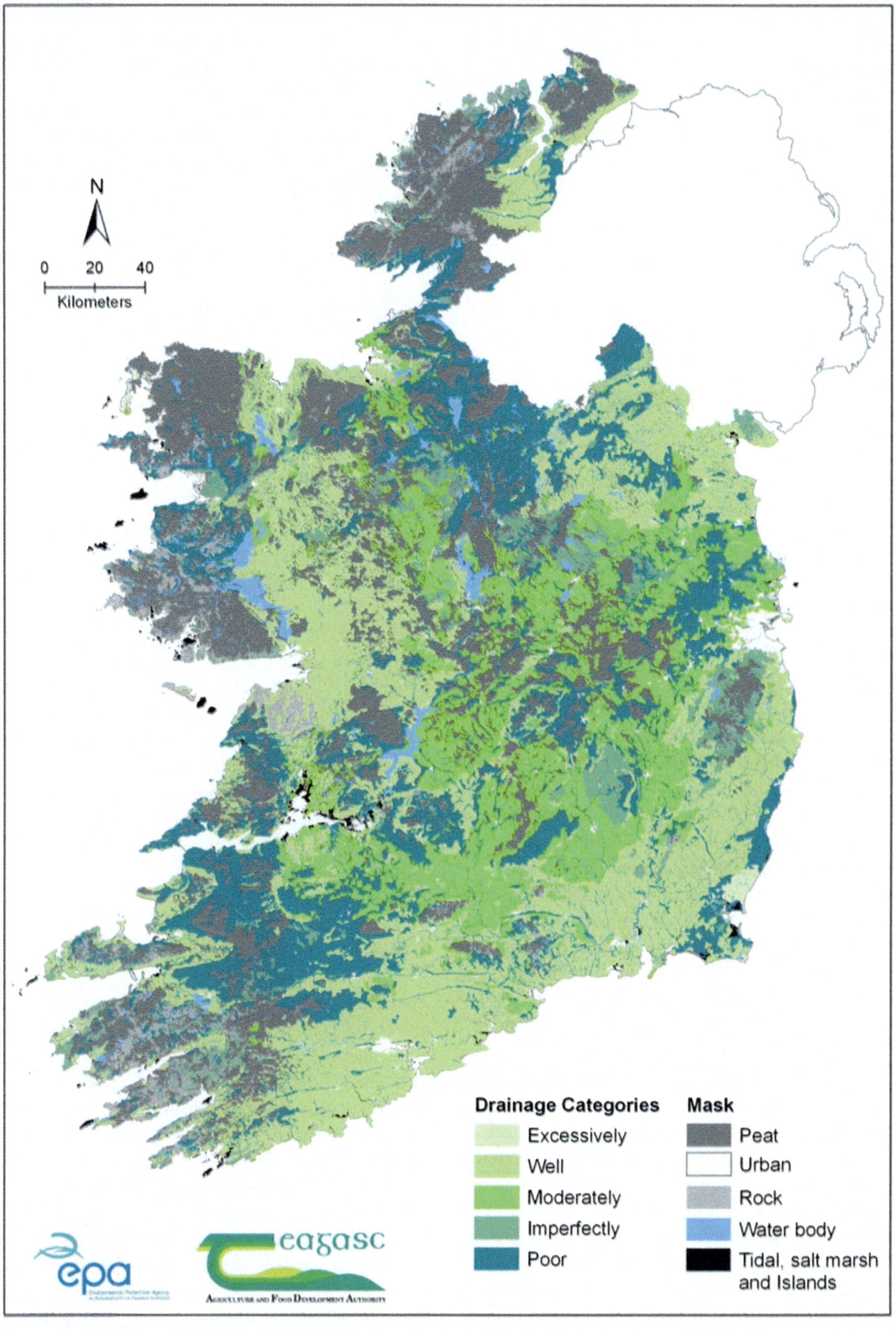

Fig. 14.5 Indicative soil drainage map from Creamer et al. (2016) showing the categorisation of taxonomic soil subgroups into drainage classes, based on a set of diagnostic criteria applied to the Irish Soil Information Systems by Creamer et al. (2014)

of 2000 mm are commonplace (http://www.met.ie/climate-ireland/rainfall). Evapotranspiration ranges from 390 to 570 mm (Mills 2000).

Approximately 49.5% (3.4 M ha) of the total land area of Ireland is classified as 'marginal land' which is affected by natural limitations related to its soil, topography, relief and climate. Such land is distributed throughout the country but is more widespread along the west coast (Gardiner and Radford 1980). With a large portion of the land area of Ireland classified as significantly wet with impeded drainage (Collins and Cummins 1996), or with poor natural drainage status (Fig. 14.5), excess soil moisture represents a major limitation in relation to primary productivity. The principal limitation of marginal land is its poor drainage status and much requires artificial drainage (Burdon 1986) if its agricultural productivity is to be maintained or improved. On poorly drained soils in wet growing seasons, the topsoil may never undergo significant drying, as persistent rainfall maintains high soil moisture contents (Mulqueen 1974). On grassland farms, this can inhibit soil trafficability, reduce the potential number of grazing days and hinder silage harvesting (Shalloo et al. 2004). The principal drainage problems in Ireland include seepage and springs (38%), soils with permeability issues (31%), and high water tables (24%) (Galvin 1971).

14.6.2 History of Land Drainage Schemes in Ireland

The state has been directly involved in field drainage from the 1940s to the 1980s when grants were available for this purpose. The first major scheme was set up by the Department of Agriculture in 1949. The Land Rehabilitation Project provided state aid for drainage work on farms and through the project, some 1.2 M ha of land were improved (Burdon 1986; Ryan 1986). Prior to this, a number of arterial drainage works (Arterial Drainage Act 1945) incorporating most major river catchments and important tributaries were carried out which facilitated efficient drainage of the surrounding farmland. This work commenced in 1842 under the control of the Office of Public Works. The total area drained under the various schemes, both arterial and field drainage, was approximately 2 million ha, accounting for nearly 30% of the total land area (Burdon 1986; Ryan 1986). Unfortunately, due to poor record keeping, the vast majority of the field drainage work carried out is now undocumented.

Approximately 20% of the utilised agricultural area in Ireland has undergone field drainage, compared with 65% in the UK and 74% in the Netherlands (Collins and Cummins 1996). In many instances, the results have not been satisfactory and additional drainage has still been required (Galvin 1969; Mulqueen and Hendriks 1986). The Land Rehabilitation Project, while providing a financial contribution to the cost of the works, imposed a very rigid design model which was rigorously enforced by the field officers responsible for authorising payment. The model, which was used across all drainage problems, typically involved installing clay pipes without any porous envelope material. The success of the technique was limited. No attempt was made to tailor specific solutions to particular drainage problems and unfortunately, these state-aided drainage works continued for many years without evaluating the serious limitations of the system.

Subsequently, a large body of research on various aspects of land drainage was carried out. Initial work focused on the drainage of peat (Burke 1967; Burke and McCormack 1969). Latterly more interest was taken in mineral soils and their inherently more varied and complex nature but which cover a larger proportion of Irish agricultural land. During the following decades, a much greater understanding of the complexities of typical drainage problems, brought about by the combination of an intricate geological and glacial history and relatively high rainfall, was established. Where permeable soils are affected by a high water table, these can be drained in a conventional manner by providing avenues for drainage flow and sufficient outfall (Galvin 1986; Tuohy et al. 2016b). For soils affected by artesian seepage or springs, considerable progress has been made in the development of investigative techniques and in the design of drainage systems for seepage problems (Mulqueen and Gleeson 1982). The solution usually requires relatively deep drains (1.5–2.5 m) to maximise the drainage potential of high transmissivity layers.

Impermeable soils, especially those of a plastic nature, give rise to major problems. Where infiltration and percolation are impeded, action is required to improve soil structure and facilitate sufficient movement of water through the soil profile (Tuohy et al. 2016a). Knowledge of the hydrological characteristics of Irish soils has advanced to the stage where with proper investigation and application of the known principles of water flow in soils, it is possible to design a site-specific drainage system (shallow or groundwater) where appropriate as part of a rural sustainable drainage approach (Teagasc 2013; Ibrahim et al. 2013). While land drainage is likely to increase productivity, this will also result in the aeration of wet soils resulting in the microbial oxidation of organic matter and the release of carbon dioxide to the atmosphere (O'Sullivan et al. 2015). The agricultural sector is faced with the challenge of meeting food security targets, while simultaneously reducing greenhouse gas emissions (GHG) in line with EU targets. The dual objective of increasing productivity while reducing sector emissions requires land to be managed in an integrated way to optimise the sectors response to GHG mitigation and C sequestration.

14.6.3 Grassland Production on Poorly Drained Soils in Ireland

National agricultural policy (Department of Agriculture, Fisheries and Food 2010) is placing pressure on Irish farms to increase productivity. The Food Harvest 2020 report set a target of increasing the value of primary agricultural output by 33% by 2020, relative to the average position in the 2007–2009 period. This target coupled with the abolition of EU milk quotas in 2015 (Läpple and Hennessy 2012) caused an expansion of farms and farming enterprises. Given the dominance of grassland systems in Irish agriculture (O'Mara 2008), expansion of agricultural production in Ireland will undoubtedly rely on increased pasture production (Patton et al. 2012) to facilitate economically sustainable growth.

Producing milk and beef from grazed grass has been an important aspect of the Irish agricultural economy for some decades (O'Donovan et al. 2011). The cost of grazed grass is significantly lower than the cost of both purchased concentrates and conserved forages (Finneran et al. 2012), so much so that for each extra tonne of grass dry matter utilised per hectare, profitability per hectare is increased by €160 (Dillon 2011). On poorly drained soils, however, complexities are added to such production systems, which are aggravated by inclement weather conditions. In a study of the profitability of dairy production on a heavy clay loam soil over a 10-year period, it has been shown that there was significant variation in both the cost and production performance on farms with poorly drained soils due to variations in meteorological conditions from year to year. As much as up to 50% of the variation in profitability from year to year could be attributed to annual rainfall (Humphreys et al. 2012). In wetter years, there was up to 25% lower herbage production for the same inputs, more difficult grazing conditions, shorter grazing season (225 versus 255 days), a lower proportion of grazed grass in the diet, higher supplementation with concentrates, lower body condition score at the end of the grazing season and lower annual milk sales. Similar results were reported by Shalloo et al. (2004).

Inevitably, land availability will become a limit to future expansion (O'Donnell et al. 2008). As a consequence, there is likely to be increasing pressures on existing pastures, as well as the potential for new enterprises emerging on the more marginal soil types (Patton et al. 2012). With these factors in mind, it is imperative that means of optimising soil management and improving conditions on poorly drained grassland soils are explored (Bilotta et al. 2007). Increasing both grass production and utilisation on poorly drained soils would result in increased farm profitability (Shalloo et al. 2004), thereby improving the competitiveness of agriculture, food and the wider bio-economy in the international arena (Dillon et al. 2005).

14.6.4 Crop Production on Poorly Drained Soils

All crops require an amenable root zone where the soil provides a suitable medium for the adsorption of water and nutrients and ample air present to facilitate the exchange of gases and plant respiration (Moore 2001). Air-filled porosity values of 0.10–0.12 m^3/m^3 are seen as the critical minimum value required to ensure adequate diffusion of air to plant roots (Carter 1988; Lipiec and Hatano 2003) to maintain plant health, while optimum air-filled porosity for pasture production should ideally be in the range of 0.15 and 0.20 m^3/m^3 (Drewry et al. 2008).

The water required by a plant is equal to the sum of the amount lost by evapotranspiration and that required to build plant tissue. This varies with crop species and variety (Critchley and Siegert 1991) and in time and space as the evapotranspirative demand varies with crop condition and local climate (Allen et al. 1998; Ali 2010). For optimum plant growth, soil water content should remain above this requirement level but below an excessive level, which would reduce the soil air content below a critical level (Brady and Weil 1996). Crop yields are frequently limited on poorly drained soils due to the adverse effect of excess water and a deficiency of air in soil pore space at rooting depth (Scott et al. 1989), which restricts respiration and growth. In cases of prolonged waterlogging, plants eventually die due to a lack of oxygen in the root zone (FAO 1997).

Poorly drained soils also tend to require a greater amount of energy and time to warm up than well-drained soils. The specific heat capacity of a dry mineral soil is only one-fifth of the specific heat capacity of water. As a result, significantly more energy is required to warm a poorly drained soil than an equivalent dry soil (Braun and Kruijne 1994). In addition, the cooling effect from the greater evaporation from a wet soil further delays a temperature rise. In temperate climates, both these effects cause a delay of growth in spring (Oosterbaan and Nijland 1994).

14.6.5 Land Drainage and Forestry

In comparison with annual production systems, the lower frequency of management interventions associated with forestry reduce substantially the difficulties of trafficability. Site access for harvesting is also improved by the ability to delay such operations until summer months. Nonetheless, drainage works have been an essential aspect of afforestation activities in Ireland throughout the twentieth century as much of the land available for afforestation was of a poorly draining nature. Surface-water Gleys, Groundwater Gleys and peaty soils are particularly prone to waterlogging and anaerobic conditions can occur over prolonged periods inhibiting root growth. Trees planted directly into such soils experience extremely poor growth and stability, and may be held in check for many years. Such forest sites are also difficult to access for mechanical harvesting without further damaging the fine-textured soils. The ideal soil preparation method should increase the aeration by draining excess water [thereby increasing fertility (Schaible and Dickson 1990)], breaking indurated layers (e.g. iron pans) and reducing effects of competing ground vegetation (Wills et al. 2001). In common with agriculture, forestry has experienced severe limitation in making use of peatlands, largely due to waterlogging and infertility. Following experiences in Wales and Scotland particularly, much pre-establishment peatland drainage in Ireland in the 1950s and 1960s was achieved with single and double (more common) mouldboard ploughs. These produced a raised ribbon with a deep furrow to one or both sides and became the standard technique for establishment on Peat soils (OCarroll et al. 1981). Though it proved excellent for establishment and growth, restrictions to vertical and horizontal root growth led to very unstable crops and the resulting terrain impeded harvesting operations. The technique was found by OCarroll et al. (1981) to be inferior to tunnel ploughing. Hendrick et al. (1984) also found it produced poorer stem form than the sub-surface drainage channels from the tunnel ploughs. The direction of open furrow ploughs was also important and for best stability had to be oriented in the same direction as the prevailing wind. The presence of collector drains at sufficient depth and frequency in peat soils is also important in reducing the occurrence of bog flows following periods of particularly high rainfall (Hendrick 1990).

In investigating the effects of a number of cultivation methods on drainage in mineral Gley soils, Wills et al. (2001) found that the influence of mounding drains on site drainage was limited to an area up to 3 m on either side. The same study also found that sub-surface mole drains, on a suitable site, lowered the water table more uniformly and effectively. Hendrick (1989) also found mole-drained Gley soils allowed deeper and more evenly distributed rooting compared to double mouldboard ploughing. In a tree pulling experiment, the trees on mole-drained sites had an average maximum turning moment almost double that of trees established on double mouldboard plough ribbons.

Despite early failures, some forests have been found to have benefited from nineteenth century drainage schemes. Conry (1973) investigated the effect that stone drains, inserted into a Blanket Peat area in Co. Laois, had on spruce forest growth. Though installed as part of a famine relief scheme in c. 1880, they were too widely spaced to be very effective, but did have good localised effects on tree growth. Investigations into the soil surrounding the drains found substantial improvements in the air/moisture ratio. Though the drains still carried water, the productivity improvements reduced substantially with distance from the drains.

14.6.6 Cultural Practices to Maintain Trafficability

Maintaining trafficability on fine-textured and wet soils during forest harvesting operations is frequently difficult. This is particularly the case in forestry systems where extremely heavy machinery is often required to cope with challenging terrain and the large dimensions of a timber crop. This problem is addressed and ameliorated through the use of 'brash mats'. Harvesting machines are required to access sites along specific paths and to cut and process timber lengths in front of them, leaving the residue 'brash' material (tree tops, off-cuts, branches and foliage etc.) to form a mat on top of which the machinery travels. This practice serves the dual purpose of maintaining site fertility by retaining material of high nutrient status distributed across sites, but also protects against rutting damage by cushioning the wheels and tracks (and by limiting the effect spatially if the cushioning is insufficient to completely avoid compaction).

Recent research (Vero et al. 2013; Piwowarczyk et al. 2011) has ascertained the utility of using SMD as a predictor when modelling grassland soil trafficability. This work considered the vulnerability of the soil to significant or prolonged compaction to be the primary indicator of reductions in trafficability. For slurry spreading operations, Vero et al. (2013) determined that SMDs above 10 mm minimised compaction and yield loss, while slurry spreading at <10 mm SMD caused damage which significantly impaired yield and compacted the upper soil layer for greater than 2 months. Piwowarczyk et al. (2011) considered the trafficability of the soil from a livestock perspective and ascertained that decreasing SMD rendered the soil less suitable for grazing. The results of these studies indicated threshold values for the prevention of soil compaction which closely resembled those suggested by Earl (1996). SMD can be predicted for various drainage classes with acceptable accuracy up to 48 h in advance, and may provide a simple and readily implementable tool for supporting farm decision making.

Improving the overall trafficability of any soil first requires the identification of factors causing either excessive moisture or poor soil structure. Correctly ascertaining these issues allows implementation of appropriate preventative and remediation measures, and should be the initial step in any management strategy.

14.7 Conclusions

Changes in future climate in the form of increasing storm intensity accompanied by heavier rainfall are expected, in the west of Ireland; while drier summers are predicted for the east. Increased incidence and quantity of rainfall outside the growing season is also predicted. This will put more pressure on soils with impeded drainage characteristics and time to recovery will be extended with knock-on effects for crop utilisation. In the case of forests (combined with an increasing and ageing forest stock), this will result in substantial increases (at least double) in forest damage across Europe by the end of the century. The inherent productivity of a soil (i.e. its capacity to support plant growth) is dictated by the interplay of several factors—nutrient and water availability, structural quality and drainage status. The latter factor, in Ireland, exerts a major control over the other aspects and is the primary influence on trafficability and nutrient transport. It further represents the single greatest opportunity to improve days-at-grass for the dairy production systems dominating Irish agriculture, and hence, potentially offers substantial financial benefits to both individual farmers and national output. While a relatively small proportion of Ireland's total land area has been drained compared to other EU member states, various strategies to address the challenge of prolonged soil saturation have been employed over the last several decades. Novel new approaches promise to maintain and improve the sustainable delivery of other environmentally significant aspects of soil quality, including carbon sequestration, pollution mitigation as well as the economic viability of production.

References

Ali MH (2010) Crop water requirement and irrigation scheduling. In: Ali MH (ed) Fundamentals of irrigation and on-farm water management: volume 1. Springer, New York, pp 399–452

Allen RG, Pereira LS, Raes D, Smith M (1998) Crop evapotranspiration: guidelines for computing crop requirements irrigation and drainage paper no 56. FAO, Rome

Anderson ML (1960). The selection of tree species. An ecological basis of site classification for conditions in Great Britain and Ireland. Oliver and Boyd, Edinburg, London

Arterial Drainage Act (1945) Government Publications Office, Dublin, Ireland

Ball BC, Douglas JT (2003) A simple procedure for assessing soil structural, rooting, and surface conditions. Soil Use Manag 19: 50–56

Berry PM, Sterling M, Spink JH, Baker CJ, Sylvester-Bradley R, Mooney SJ, Tams AR, Ennos AR (2004) Understanding and reducing lodging in cereals. Adv Agron 84:217–271

Bilotta GS, Brazier RE, Haygarth PM (2007) The impacts of grazing animals on the quality of soils, vegetation, and surface waters in intensively managed grasslands. Adv Agron 94:237–280

Black K, Bolger T, Davis P, Nieuwenhuis M, Reidy B, Saiz G, Tobin B, Osborne B (2007) Inventory and eddy covariance-based estimates of annual carbon sequestration in a Sitka spruce (*Picea sitchensis* (Bong.) Carr.) forest ecosystem. Eur J For Res 126:167–178

Black K, Byrne KA, Mencuccini M, Tobin B, Nieuwenhuis M, Reidy B, Bolger T, Saiz G, Green C, Farrell EP, Osborne B (2009) Carbon stock and stock changes across a Sitka spruce chronosequence on surface-water gley soils. Forestry 82:255–272

Brady NC, Weil RR (1996) The nature and properties of soils, 11th edn. Prentice Hall Publishers, New Jersey

Braun HMH, Kruijne R (1994) Soil conditions. In: Ritzema HP (ed) Drainage principles and applications. International Institute for Land Reclamation & Improvement, Wageningen, pp 77–110

Broad L, Lynch T (2006) Growth models for Sitka spruce in Ireland. Ir For 63:53–79

Bulfin M (1987) Availability of land for forestry in Ireland and its suitability for Sitka spruce. Ir For 44:18–31

Burdon D (1986) Hydrogeological aspects of agricultural drainage in Ireland. Environ Geol 9:41–65

Burke W (1967) Principles of drainage with special reference to peat. Ir For 24:8–11

Burke W, McCormack P (1969) Gravel drains for peat. Ir J Agric Res 8:285–287

Carter MR (1988) Temporal variability of soil macroporosity in a fine sandy loam under mouldboard ploughing and direct drilling. Soil Tillage Res 12:37–51

Clement R, Jarvis PG, Moncrieff JB (2012) Carbon exchange of a Sitka spruce plantation in Scotland over five years. Agric For Meteorol 153:106–123

Collins JF, Cummins T (1996) Agroclimatic atlas of Ireland. Agricultural Meteorology, Dublin

Collins JF, Larney FJ, Morgan MA (2004) Climate and soil management. In: Keane T, Collins JF (eds) Climate, weather and Irish agriculture, 2nd edn. Joint Working Group on Applied Agricultural Meteorology, Dublin

Connaughton MJ (1973) The grass growing season in Ireland. Agrometeorological Memorandum no 5 Dublin, Met Éireann

Conry MJ (1973) The effect of 19th century stone drains on the growth of Sitka spruce. Ir For 30:79–87

Coyle C, Creamer RE, Schulte RPO, O'Sullivan L, Jordan P (2016) A functional land management conceptual framework under soil drainage and land use scenarios. Environ Sci Policy 56:39–48

Creamer RE, Brennan F, Fenton O, Healy MG, Lalor STJ, Lanigan GJ, Regan JT, Griffiths BS (2010) Implications of the proposed soil framework directive on agricultural systems in Atlantic Europe—a review. Soil Use Manage 26:198–211

Creamer RE, Simo I, Reidy B, Carvalho J, Fealy R, Hallett S, Jones R, Holden A, Holden N, Hannam J, Massey P, Mayr T, McDonald E, O'Rourke S, Sills P, Truckell I, Zawadzka J, Schulte RPO (2014) Irish soil information systems. Synthesis report (2007-S-CD-1-S1). EPA STRIVE Programme, Wexford. Available for download from http://erc.epa.ie/safer/reports

Creamer RE, Simo I, O'Sullivan L, Reidy B, Schulte RPO, Fealy RM (2016) Irish soil information systems: soil properties map, final technical report 18. EPA STRIVE Programme, Wexford. Available for download from http://erc.epa.ie/safer/reports

Critchley W, Siegert K (1991) Water harvesting—a manual for the design and construction of water harvesting schemes for plant production. FAO, Rome

DAFM (2014) Forests, products and people Ireland's Forest Policy—a renewed vision. Department of Agriculture Food and the Marine, Dublin

DAFM (2015) Information on the LULUCF actions to limit or reduce emissions and maintain or increase removals from activities defined

under Decision 529/2013/EU. Department of Agriculture, Food and the Marine, Dublin
DAFM (2016) Draft adaptation plan for the agriculture and forest sector. Department of Agriculture, Food and the Marine, Dublin
Department of Agriculture, Fisheries and Food (2010) Food harvest 2020. A vision for Irish agri-food and fisheries. Available online www.agriculture.gov.ie/media/migration/agrifoodindustry/foodharvest2020/2020FoodHarvestEng240810.pdf
Dillon P (2011) The dairy industry—planning for 2020. In: Proceedings of the Teagasc national dairy conference, Teagasc, Carlow, pp 1–25
Dillon P, Roche JR, Shalloo L, Horan B (2005) Optimising financial return from grazing in temperate pastures In: Murphy JJ (ed) Utilisation of grazed grass in temporal animal systems. Academic Publishers, Wageningen, pp 131–147
Drewry JJ, Cameron KC, Buchan GD (2008) Pasture yield and soil physical property responses to soil compaction from treading and grazing—a review. Aust J Soil Res 46:237–256
Earl R (1996) Prediction of trafficability and workability from soil moisture deficit. Soil Tillage Res 40:155–168
Emmet Booth JP, Forristal PD, Fenton O, Ball BC, Holden NM (2016) A review of visual soil evaluation techniques for soil structure. Soil Use Manage 32:623–634
FAO (1997) Management of agricultural drainage water quality. FAO water report no 13. FAO, Rome
FAO (2006) Guidelines for soil description, 4th edn. FAO, Rome
Farrelly N, Gallagher G (2015a) The development of a site classification for Irish forestry. Ir For 72:166–188
Farrelly N, Gallagher G (2015b) The potential availability of land for afforestation in the Republic of Ireland. Ir For 72:120–138
Feehan J (2003) Farming in Ireland—history heritage and environment. Faculty of Agriculture, University College Dublin, Dublin
Feehan J, Sheridan H, Egan D (2012) The grasses of Ireland. Teagasc
Fenton O, Vero S, Ibrahim TG, Murphy PNC, Sheriff SC, Ó hUallacháin D (2015) Consequences of using different soil texture determination methodologies for soil physical quality and unsaturated zone time lag estimates. J Contam Hydrol 182:16–24
Finn JA, Kirwan L, Connolly J, Teresa Sebastia M, Helgadottir A, Baadshaug OH, Belanger G, Black A, Brophy C, Collins RP, Cop J, Dalmannsdottir S, Delgado I, Elgersma A, Fothergill M, Frankow-Lindberg BE, Ghesquiere A, Golinska B, Golinski P, Grieu P, Gus-tavsson A-M, Hoglind M, Huguenin-Elie O, Jorgensen M, Kadziuliene Z, Kurki P, Llurba R, Lunnan T, Porqueddu C, Suter M, Thumm U, Luescher A (2013) Ecosystem function enhanced by combining four functional types of plant species in intensively managed grassland mixtures: a 3-year continental-scale field experiment. J Appl Ecol 50:365–375
Finneran E, Crosson P, O'Kiely P, Shalloo L, Forristal D, Wallace M (2012) Stochastic simulation of the cost of home-produced feeds for ruminant livestock systems. J Agric Sci 150:123–139
Galvin LF (1969) Land drainage survey—II. Ir J Agric Res 8:1–18
Galvin LF (1971) Land drainage survey III: maps of Ireland showing the general distribution of drainage problems and of drainage schemes. Ir J Agric Res 10:213–221
Galvin LF (1986) Aspects of land drainage development in Ireland in the last twenty-five years. In: J Vos (ed) Symposium on 25th international course on land drainage, twenty-five years of drainage experience, The Netherlands International Institute for Land Reclamation and Improvement, Wageningen, pp 131–140
Gardiner MJ, Radford T (1980) Soil associations of Ireland and their land use potential explanatory bulletin to soil map of Ireland. National Soil Survey of Ireland, An Foras Talúntais, Dublin
Gardiner B, Blennow K, Carnus JM, Fleischner P, Ingemarson F, Landmann G, Lindner M, Marzano M, Nicoll B, Orazio C, Peyron JL, Reviron MP, Schelhaas M, Schuck A, Spielmann M, Usbeck T (2010) Destructive storms in European forests: past and forthcoming impacts final report to European Commission—DG environment. European Forest Institute, Joensuu, Finland, 138 pp
Gray HJ (1963) The economics of Irish forestry. J Stat Soc Enq Soc Ir 21:18–24
Guimarães RML, Fenton O, Murphy BW, Tormena CA (2015) Soil structure under adverse weather/climate conditions. In: Ball BC, Munkholm LJ (eds) Visual soil evaluation: realizing potential crop production with minimum environmental impact. CABI
Haarsma RJ, Hazeleger W, Severijns C, de Vries H, Sterl A, Bintanja B, van Oldenborgh BJ, van den Brink HW (2013) More hurricanes to hit western Europe due to global warming. Geophys Res Lett 40 (9):1783–1788
Hamilton GJ, Christie JM (1971) Forest management tables. Forestry Commission Booklet No 34. HMSO, London
Hamilton J, Lennon P, O'Donnell B (1988) Objective analysis of monthly climatological fields of temperature, sunshine, rainfall percentage and rainfall amount. J Climatol 8:109–124
Heathwaite AL, Jones PJ (1996) Contribution of nitrogen species and phosphorus fractions to stream water quality in agricultural catchments. Hydrol Process 10:971–983
Hendrick E (1989) The effect of cultivation method on the growth and root anchorage of Sitka spruce. Ir For 46:19–28
Hendrick E (1990) A bog flow at Bellacorrick forest, Co. Mayo. Ir For 47:32–44
Hendrick E, OCarroll N, Pfeifer AR (1984) Effect of ploughing direction and method on the stem form of south coastal lodgepole pine. Ir For 41:66–76
Herbin T, Hennessy D, Richards KG, Piwawarcyzk A, Murphy JJ, Holden NM (2011) The effects of dairy cow weight on selected soil physical properties indicative of compaction. Soil Use Manag 27:36–44
Humphreys J, Mihailescu E, Casey IA (2012) An economic comparison of systems of dairy production based on N fertilized grass and grass-white clover grassland in a moist maritime environment. Grass Forage Sci 67:519–525
Ibrahim TG, Fenton O, Richards KG, Fealy RM, Healy MG (2013) Spatial and temporal variations of nutrient loads in overland flow and subsurface drainage from a marginal land site in south east Ireland. Biol Environ (Proc R Ir Acad Ser B) 113B(2):169–186
Laidlaw AS (2009) The effect of soil moisture content on leaf extension rate and yield of perennial ryegrass. Ir J Agric Food Res 48:1–20
Läpple D, Hennessy T (2012) The capacity to expand milk production in Ireland following the removal of milk quotas. Ir J Agric Food Res 51:1–11
Lee J (1986) Aspects of agricultural land use in Ireland. Environ Geol 9:23–30
Lipiec J, Hatano R (2003) Quantification of compaction effects on soil physical properties and crop growth. Geoderma 116:107–136
Mills G (2000) Modelling the water budget of Ireland-evapotranspiration and soil moisture. Ir Geogr 33:99–116
Moore G (2001) Soil guide: a handbook for understanding and managing agricultural soils. Perth, Western Australia
Mulqueen J (1974) Drainage of impeded soils. Ballinamore, An Foras Taluntais, Co Leitrim
Mulqueen J, Gleeson TN (1982) Some relationships of drainage problems in Ireland to solid and glacial geology, geomorphology and soil types. In: Gardiner MJ (ed) Land drainage, Proceedings of the CEC seminar. A.A Balkema, Rotterdam, pp 11–31
Mulqueen J, Hendriks R (1986) The use of groundwater surveys in the diagnosis and solution of a drainage problem. Ir J Agric Res 25:285–298
Neeson E (1991) A history of Irish forestry. The Lilliput Press, Dublin
NFI (2013) National Forest Inventory, Republic of Ireland—Results. Forest Service, Johnstown Castle, Co. Wexford

O'Connell K, Molloy M (2001) Farming and woodland dynamics in Ireland during the Neolithic. Biol Environ: Proc Royal Ir Acad 101B(1–2):99–128

O'Donnell S, Shalloo L, Butler AM, Horan B (2008) A survey analysis of opportunities and limitations of Irish dairy farmers. J Farm Manage 13:419–434

O'Donovan M, Lewis E, O'Kiely P (2011) Requirements of future grass-based ruminant production systems in Ireland. Ir J Agric Food Res 50:1–21

O'Mara F (2008) Country pasture/forage resource profile. Available at http://www.fao.org/ag/AGP/agpc/doc/Counprof/Ireland/Ireland.htm . Accessed September 2014

O'Mara F (2011) Country pasture/forage resource profiles Ireland. FAO, Rome (online). http://www.fao.org/ag/agp/agpc/doc/counprof/PDF%20files/Ireland.pdf. Accessed on 15 Sept 2016

O'Sullivan L, Creamer RE, Fealy R, Lanigan G, Simo I, Fenton O, Carfrae J, Schulte RPO (2015) Functional land management for managing soil functions: a case-study of the trade-off between primary productivity and carbon storage in response to the intervention of drainage systems in Ireland. Land Use Policy 47:42–54

OCarroll N (2012) Two simple site classification systems. Ir For 69:238–241

OCarroll N, Carey ML, Hendrick E, Dillon J (1981) The tunnel plough in peatland afforestation. Ir For 38:27–40

Oosterbaan RJ, Nijland HJ (1994) Determining the saturated hydraulic conductivity. In: Ritzema EP (ed) Drainage principles and applications. International Institute for Land Reclamation & Improvement, Wageningen, pp 435–476

Palmer RC, Burton RGO, Hannam J, Creamer R (2007) Comparison of soil structural conditions in Tone and Parrett catchments during winter periods 2002–03 and 2005–06. Environment Agency Contract Report

Patton D, Shalloo L, Pierce KM, Horan B (2012) A biological and economic comparison of 2 pasture-based production systems on a wetland drumlin soil in the northern region of Ireland. J Dairy Sci 95:484–495

Piwowarczyk A, Giuliani G, Holden NM (2011) Can soil moisture deficit be used to forecast when soils are at high risk of damage owing to grazing animals? Soil Use Manag 27:255–263

Pyatt G, Ray D, Fletcher J (2001) An ecological site classification for forestry in Great Britain Bulletin 124 Forestry Commission Publications, P.O. Box 501, Leicester, LE94 OAA

Rath M, Peel S (2005) Grassland in Ireland and the UK in Grassland: a global resource. In: McGilloway (ed). Wageningen Academic Publishers, The Netherlands

Ray D, Xenakis G, Tene A, Black K (2009) Developing a site classification system to assess the impact of climate change on species selection in Ireland. Ir For 66:101–123

Ryan T (1986) Agricultural drainage practices in Ireland. Environ Geol 9:31–40

Schaible R, Dickson DA (1990) Effects of drainage intensity and planting position on the growth and nutrition of second rotation Sitka spruce on shallow peat. Ir For 47:19–27

Schulte RPO, Richards K, Daly K, Kurz I, McDonald EJ, Holden NM (2006) Agriculture, meteorology and water quality in Ireland: a regional evaluation of pressures and pathways of nutrient loss to water. Biol Environ: Proc Royal Ir Acad 106B(2):117–133

Schulte RPO, Fealy R, Creamer RE, Towers W, Harty T, Jones RJA (2012) A review of the role of excess soil moisture conditions in constraining farm practices under Atlantic conditions. Soil Use Manage 4:580–589

Schultz B (2001) Irrigation, drainage and flood protection in a rapidly changing world. Irrig Drain 50:261–277

Schultz B, Zimmer D, Vlotman WF (2007) Drainage under increasing and changing requirements. Irrig Drain 56:S3–S22

Scott HD, DeAngulo J, Daniels MB, Wood LS (1989) Flood duration effects on soybean growth and yield. Agron J 81:631–636

Shalloo L, Dillon P, O'Loughlin J, Rath M, Wallace M (2004) Comparison of a pasture-based system of milk production on a high rainfall, heavy-clay soil with that on a lower rainfall, free-draining soil. Grass Forage Sci 59:157–168

Smedema LK, Abdel-Dayem S, Ochs WJ (2000) Drainage and agricultural development. Irrigat Drain Syst 14:223–235

Smit HJ, Metzger MJ, Ewert F (2008) Spatial distribution of grassland productivity and land use. Eur Agric Syst 98:208–219

Teagasc (2013) Teagasc manual on drainage and soil management: a best practice manual for Ireland's farmers. In: Moore M, Fenton O, Tuohy P, Ibrahim T (eds). Teagasc Carlow, Ireland

TTCSCG (Teagasc Tillage Crop Stakeholder Consultative Group) (2012) Tillage sector development plan 2012. A report compiled by Teagasc Tillage Crop Stakeholder Consultative Group Teagasc, Ireland

Tuohy P, Fenton O, Holden NM, Humphreys J (2015) The effects of treading by two breeds of dairy cow with different live weights on soil physical properties, poaching damage and herbage production on a poorly drained clay-loam soil. J Agric Sci 153:1424–1436

Tuohy P, Humphreys J, Holden NM, Fenton O (2016a) Runoff and subsurface drain response from mole and gravel mole drainage across episodic rainfall events. Agric Water Manage 169:129–139

Tuohy P, Humphreys J, Holden NM, O' Loughlin J, Reidy B, Fenton O (2016b) Visual drainage assessment: a visual soil assessment method of land drainage design. Ir J Agric Food Res (in press). https://doi.org/10.1515/ijafr-2016-0003

Valentini R, Matteucci G, Dolman AJ, Schulze E-D, Rebmann C, Moors EJ, Granier A, Gross P, Jensen NO, Pilegaard K, Lindroth A, Grelle A, Bernhofer C, Grünwald T, Aubinet M, Ceulemans R, Kowalski AS, Vesala T, Rannik Ü, BerbigierP Loustau D, Guethmundsson J, Thorgeirsson H, Ibrom A, Morgenstern K, Clement R, Moncrieff Montagnani L, Minerbi S, Jarvis PG (2000) Respiration as the main determinant of carbon balance in European forests. Nature 404:861–865

Vero SE, Antille DL, Lalor STJ, Holden NM (2013) Field evaluation of soil moisture deficit thresholds for limits to trafficability with slurry spreading equipment on grassland. Soil Use Manag 30(1):69–77

Wills J, Sundström E, Nieuwenhuis M, Keane M (2001) The impact of soil preparation method on water-table depth in Irish forest plantations on wet mineral soils. Ir For 57:46–58

Soil Fertility and Nutrient Cycling

15

David P. Wall, Lilian O'Sullivan, Rachel Creamer, and Michael J. McLaughlin

Abstract

Plant production and nutrient cycling are two of the key functions that intensively farmed soils must perform. Nutrient cycling is the capacity of the soil to provide nutrition for food, fuel and fibre crops that are grown across the landscape. This chapter explores the availability of macronutrients and micronutrients in Irish soils. Overall, the nutrient status of the 11 Great Soil Groups identified in Ireland will have been influenced by agricultural production to some degree. The more intensively farmed and productive soils such as Brown Earths, Brown Podzolics, Luvisols, Alluvial soils, Surface water and Groundwater Gleys are more likely to receive active management by farmers to influence their soil fertility status. Weathering and soil formation processes that place soils in each Great Soil Group category have a major influence on their response to fertiliser and manure additions and also how they maintain their long-term soil fertility status. Of the 16 essential plant nutrients, the primary nutrients—N, P, K and secondary nutrients, sulphur (S), Magnesium (Mg) and calcium (Ca) are required in largest quantities by forages and crops. Total fertiliser use on Irish farms has declined in recent years mainly due to increasing fertiliser costs and regulation of nitrogen and phosphorus use. Crucially, a balance between maintaining soil fertility at optimum levels while minimising nutrient losses to the water and the atmosphere must be achieved.

Keywords

Soil fertility • pH • Lime • Cation exchange capacity Macronutrients • Micronutrients

D. P. Wall (✉) · L. O'Sullivan · R. Creamer
Teagasc Agriculture and Food Development Authority, Johnstown Castle Environment Research Centre, Wexford, Ireland
e-mail: David.Wall@teagasc.ie

L. O'Sullivan
Irish Environmental Protection Agency, Johnstown Castle, County Wexford, Ireland

R. Creamer
Wageningen University and Research Atlas—Wageningen Campus, Wageningen, The Netherlands

M. J. McLaughlin
Soil Science, School of Agriculture, Food and Wine, The University of Adelaide, CSIRO Land and Water, Glen Osmond, SA 5064, Australia

15.1 Soil Fertility Levels on Irish Farms

There has been a sharp decrease in total fertiliser (N, P and K) use on Irish farms in recent years (Fig. 15.1). This is mainly due to the combined effects of increasing fertiliser prices, and the regulation of N and P fertiliser use. This decrease has been most severe on grazed grassland, with fertiliser P use decreasing by 63% between 2003 and 2008. In 2009, national P and K fertiliser use was at its lowest level for more than two decades. Since then, fertiliser P and K use has remained at approximately half of what it was two decades ago. In total, Irish farmers currently spend approximately €613 million on fertilisers, and fertilisers represent one of the biggest single input costs (15–20% of total variable costs) on dairy farms (Wall et al. 2015). When used efficiently, fertiliser inputs help to attain target crop yields and represent good value for money.

Soil testing every four years is a compulsory requirement for Irish farms applying for EU nitrates derogation as these soils are farmed more intensively. However, even within these regulations, it is important to remember that the primary function of soil testing on the farm should be to inform a farmer of the soil fertility status and to plan fertiliser applications. This is particularly true in the case of lime and K, which are not included under Irish national regulations. Soil tests submitted to Teagasc indicate that soil fertility levels have declined dramatically in recent years as a result of reduced fertiliser use. Soil test data up to 2016 indicates that 90% of grassland soils have less than optimum balance of P, K and pH status (Fig. 15.2). These very low levels of

R. Creamer and L. O'Sullivan (eds.), *The Soils of Ireland*, World Soils Book Series,
https://doi.org/10.1007/978-3-319-71189-8_15

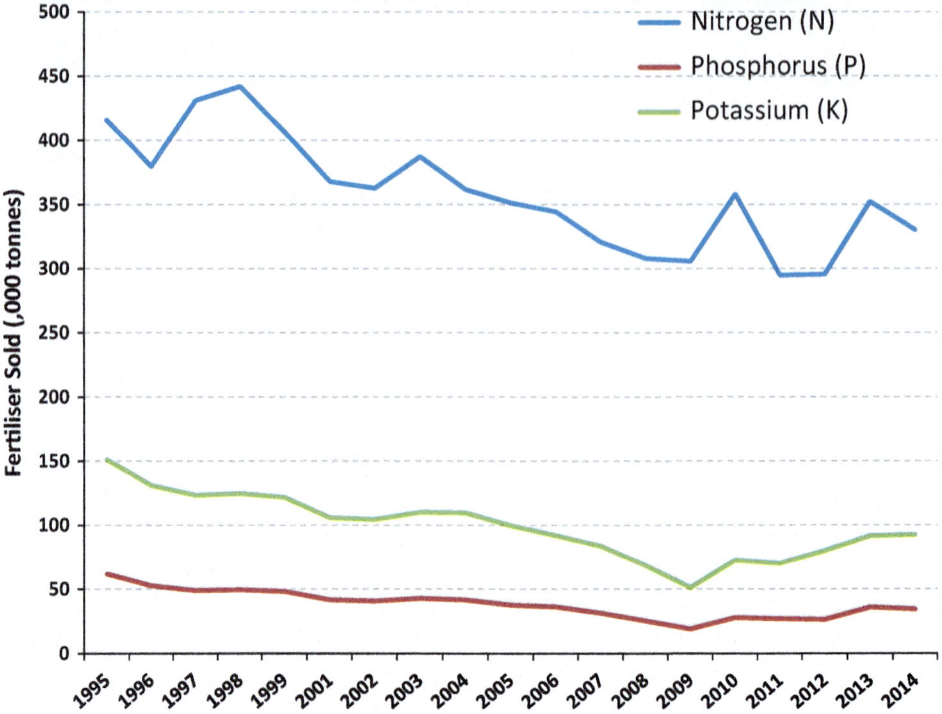

Fig. 15.1 Fertiliser N, P and K sales in Ireland between 1995 and 2014. *Source* DAFM

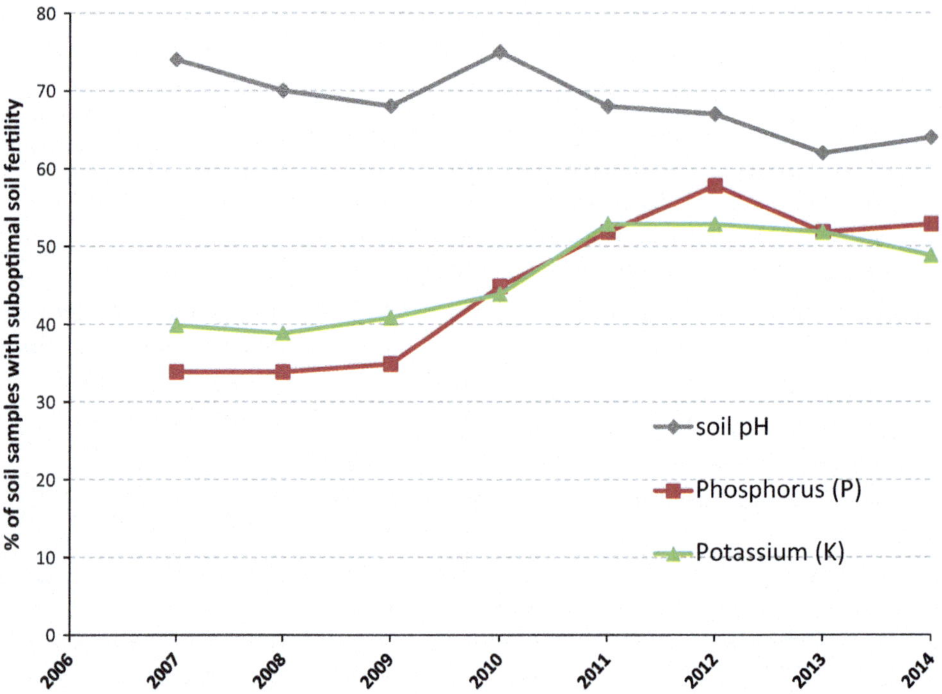

Fig. 15.2 The % of soil samples from dairy farms analysed through Teagasc with low soil pH, P and K fertility levels (Teagasc 2016)

soil fertility pose a significant threat to achieving increased productivity and profitability on dairy farms. Investing in soil fertility will pay dividends in terms of increasing the carrying capacity of the farm through increased grass production and offsetting more expensive feedstuffs that would otherwise need to be imported. Organic fertilisers are a valuable source of N, P and K. Cattle slurry is by far the most common form of organic fertiliser applied in Ireland. Over the last number of years, improving the utilisation of slurry and reducing fertiliser costs has been a major ambition for Irish farms. This has been achieved on farm by better facilities for the management of slurry during the winter and more slurry applied on grassland in early spring when N utilisation is highest.

The nutrient status of the 11 Great Soil Groups identified in Ireland will have been influenced by agricultural production to some degree. However, four of these Great Soil Groups are usually found in either upland or mountainous areas, and as a result these soils will exhibit; (1) a shorter growing season (Podzols and Lithosols), (2) be too shallow (Lithosols), (3) be dominated by rock outcrops (Rendzina), or (4) be too wet (Ombrotrophic peat soils consisting of raised and blanket bogs) to yield a return on investment in soil fertility. These characteristics limit the use of these soils from an agricultural perspective and so these soils are, therefore, more likely to be used for extensive rough grazing by drystock. The more intensively farmed and productive soils such as Brown Earths, Brown Podzolics, Luvisols, Alluvial soils, Surface water and Groundwater Gleys are more likely to receive active management by farmers to influence their soil fertility status. However, the weathering and soil formation processes under which these have been created and the distinct soil properties that places these soils in each Great Soil Group category has a major influence on their response to fertiliser and manure additions and also how they maintain their long-term soil fertility status. Of the 16 essential plant nutrients, the primary nutrients—N, P, K and secondary nutrients, sulphur (S), Magnesium (Mg) and calcium (Ca) are required in largest quantities by forages and crops.

15.2 Acidity, Soil PH and Lime

Soil pH, measured on a pH scale, is the measure of acidity in the soil. The natural pH of the soil is primarily dependent upon soil parent material and rainfall. For example, soils that are formed from granite parent material, such as that found in the Wicklow mountains, will be more acidic while in contrast, those derived from limestone parent material weather to produce more alkaline soils with a naturally higher pH. The pH status of the soil is a key factor which regulates nutrient availability in soils and is optimum in a pH range of 6.3–7.0. In Ireland, soils formed from limestone parent material such as the Calcareous Brown Earths found in the west of Ireland and most Luvisols which are widely found across the midlands, may naturally have high pH status (>7.0) due to higher levels of calcium carbonate ($CaCO_3$) and, therefore, do not require lime applications. In contrast, Brown Podzolics, usually found over sandstone (Fig. 15.3) naturally have inherently low nutrient and lime status (pH < 5.5) due to the leaching of elements, in particular iron (Fe) and aluminium (Al), from the top soil layer.

15.2.1 Cation Exchange Capacity (CEC)

CEC represents the total quantity of negative charges to attract cations and is expressed in milliequivalents of negative charge per 100 g of dry soil (meq/100 g soil). As a result of negative charges, or clay minerals and organic matter, positively charged nutrient cations are retained from leaching through soils (Fig. 15.4). In clay minerals, these negative charges occur as a result of internal substitution of structural elements in the clay mineral lattice e.g. Mg substituting for Al, or Al substituting for Si. This creates a permanent negative charge (i.e. does not vary with pH) at the surface of the clay mineral which retains cations to balance the charge. Hence the amount of negative charge on the soil is called the cation-exchange capacity.

In addition, a second charge is present both in organic matter and the surface (external) faces of clay minerals, which is a pH-dependent negative charge through the presence of oxygen-containing chemical groups. This surface charge gives rise to pH-dependent negative charge, with CEC increasing as pH increases. CEC is a very important property of soils, as soils with low CEC are said to be poorly buffered, can acidify easily, and cannot hold nutrient cations against leaching. Soils with high CEC are well buffered and do not acidify easily and tend to be more fertile as they retain nutrient cations against leaching.

Acidification is a naturally occurring process but can be accelerated by plant, animal or human activity. Acidification is usually associated with leaching of base cations, but base cations can leach without acidification occurring. Acids donate and bases accept hydrogen ions. In neutral soils, base cations including Ca, K, Mg and sodium (Na) dominate the exchange capacity. Acidity occurs if base cations such as these, become replaced with hydrogen (H^+) or aluminium (Al^{3+}) ions. Accordingly, soil acidity is driven by substances or processes that increase the H^+ ion concentration in the soil solution. The balance between hydrogen (H^+) ions and hydroxyl (OH^-) ions, underpins whether a soil is acidic or alkaline. In Ireland, regionally high rainfall rates are likely to promote acidity. This is due to the production of H^+ ions and the leaching of base cations by excess water moving through the soil profile. Rainwater combines with carbon dioxide (CO_2) to form carbonic acid (H_2CO_3) (Brady and Weil 2010) (Eq. 15.1):

$$CO_2 + H_2O \rightarrow H_2CO_3 \rightarrow HCO_3^- + H^+ \quad (15.1)$$

When H^+ ions are adsorbed on a clay surface, they attack the mineral structure releasing Al^{3+} which, in turn, becomes adsorbed on the colloid cation exchange sites.

$$Al - Clay \rightarrow Al^{3+} + H_2O \rightarrow AlOH^{2+} + H^+ \quad (15.2)$$

Anthropogenic activity including agriculture, such as the applications of fertiliser and organic matter can contribute to soil acidity as their oxidation results in the production of H^+ ions (Sumner and Yamada 2002). Besides rainfall, high N use is considered a major driver of soil acidity in Ireland

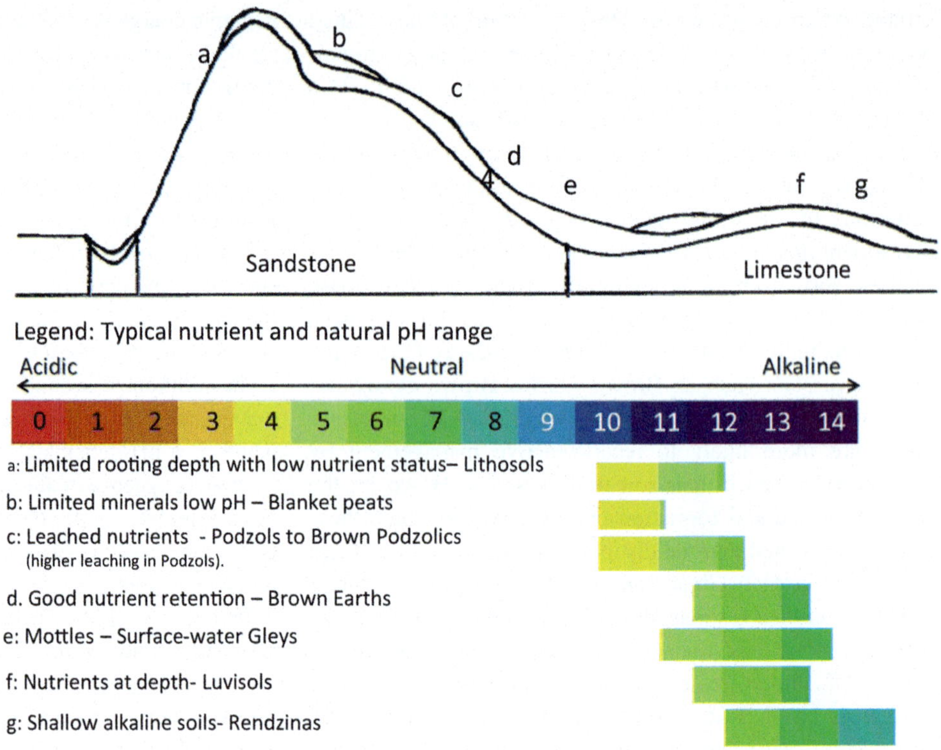

Fig. 15.3 Typical nutrient and pH range of soils that are widely found across the Irish landscape

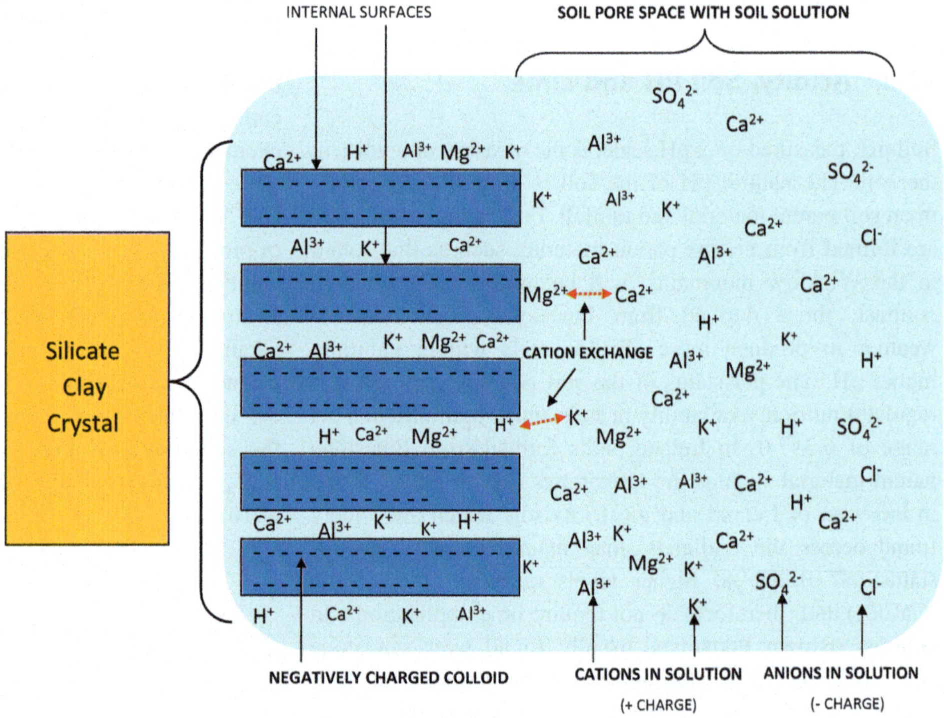

Fig. 15.4 Cations held on to a negatively charged clay particle. Clay particles, the smallest particles, are flat electrically negatively charged colloids (internal structure shown). Ions with positive charges are cations, and the negatively charged anions attract and hold on to cations. Nutrients such as ammonium, potassium, magnesium and calcium are attracted and held on to the clay particle

(Fig. 15.5) (Fertiliser Association of Ireland 2016). For example, Eq. 15.3 shows the production of H^+ ions associated with Calcium Ammonium Nitrate (CAN) fertiliser:

$$NH_4NO_3 + 2O_2 \rightarrow 2NO_3^- + 2H^+ + H_2O \quad (15.3)$$

Low pH can result in Al and manganese (Mn) toxicity, which is one common reason for crop failure in acid soils.

However, suboptimal soil pH (acidity) is easily overcome by regular applications of lime (Fig. 15.6).

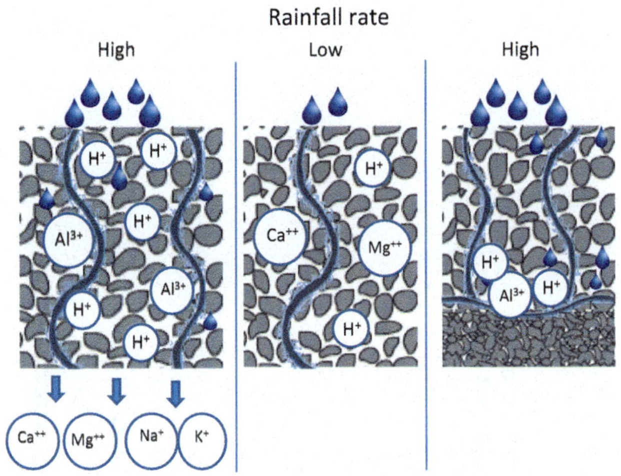

Fig. 15.5 Leaching of soil nutrients as affected by soil drainage characteristics and rainfall quantities. Highest leaching of base cations occurs in free draining soils under high rainfall compared to more poorly drained soils where acidity and other cations may be concentrated within impeded soil layers

As with all nutrients, lime applications must be managed so that leaching is reduced. Lime is constantly being lost from the soil and should, therefore, be managed as part of the nutrient management plan. Drainage water is estimated to remove approximately 250–625 kg/ha of lime equivalent each year, depending on soil type. Free draining soils lose lime more quickly than heavier soils. Extra attention is required in 'light' (sandier textured) land (with low CEC), especially where there is no limestone in the soil parent material or bedrock. Lime required to neutralise acidity produced by fertiliser N use will vary from 180 to 220 kg/ha/year per 100 kg/ha of N applied as CAN or urea, respectively (Fertiliser Association of Ireland 2016). On more poorly drained soils such as Surface-water Gleys, once corrected the pH status will likely be maintained for longer as leaching of alkalinity from the profile is limited. Although Ca can cause soil particles to disaggregate and increase aeration and drainage, in low Ca clay soils the Mg will have a more dominant role on soil structure making particles stick together (Sinclair et al. 2014). Hence a balance in the Ca:Mg ratio is important: where the soil is already high in Mg no additional Mg will be necessary (Sinclair et al. 2014). Buffering capacity refers to the amount of lime that is required to adjust the soil pH and is dependent on soil CEC. Soils vary in their buffering capacity which means that even if soils have the same pH the amount of lime required to change the soil pH will vary. Heavier textured soils with higher clay content or higher organic matter levels (higher CEC) tend to have a higher buffering capacity which results in a higher lime requirement. While these soils have a higher lime requirement, the higher buffering capacity means that these soils should retain lime better into the future.

Lime usage in Ireland has declined in the period 1970–2014 (Fig. 15.6) from an average of 1.7 million tonnes in the 1980s to an average of 725,000 tonnes in the 2000s (FAI and Teagasc 2016). In 2012, of the approximately half million tonnes of lime applied nationally, about half was applied in the south of the country. This regional relationship is due to several factors including the acidic nature of the soils, (such as the old red sandstones of the Munster Basin) and the high concentration of tillage cropping in these counties. Different types of lime can be applied to correct pH status. In Ireland, Ca ground limestone is the most commonly used due to its fast-acting properties that can produce a rapid pH adjustment. In general, the smaller the particle size of the lime, the faster it will react in the soil. Dolomitic Mg ground limestone can be used, and although slower to react it has a higher liming value and represents a good source of Mg for soils that might require this nutrient (Wall and Plunkett 2016). Alternatively, granulated limes, such as fine-grained lime with a particle size of less than 0.1 mm diameter are very reactive and are best-suited where pH is in an optimum range that needs to be maintained.

The target soil pH for grass production on peat soils is lower (5.5) as these soils have low mineral content and, therefore, low exchangeable acidity (acidity that originates form Al- and Fe-rich clays) (Fig. 15.7). However, applications of Ca, Mg and micronutrients are usually required when peat soils are drained for agricultural use. Another important consideration when liming, is that the breakdown of the grass sod may be increased due to the elevated biological activity which may increase the risk of poaching. This increased poaching risk may be offset by way of a little and often approach to lime applications. A soil pH of between 6.3 and 6.6 is recommended to support the biological processes in soils, in particular for N fixation by clover. While this is pertinent for all farming systems, this is particularly emphasised in organic farming systems where nutrient management is focused on feeding the soil, rather than feeding the plants to maintain the longer term nutrient cycling (Table 15.1).

15.3 Macronutrients

15.3.1 Nitrogen (N)

Nitrogen is the main driver of plant growth and is the cornerstone of all protein production. Organic matter (OM) is the main store of N in soils (can be in excess of 5000 kg N/ha), therefore, soils with higher soil OM contents

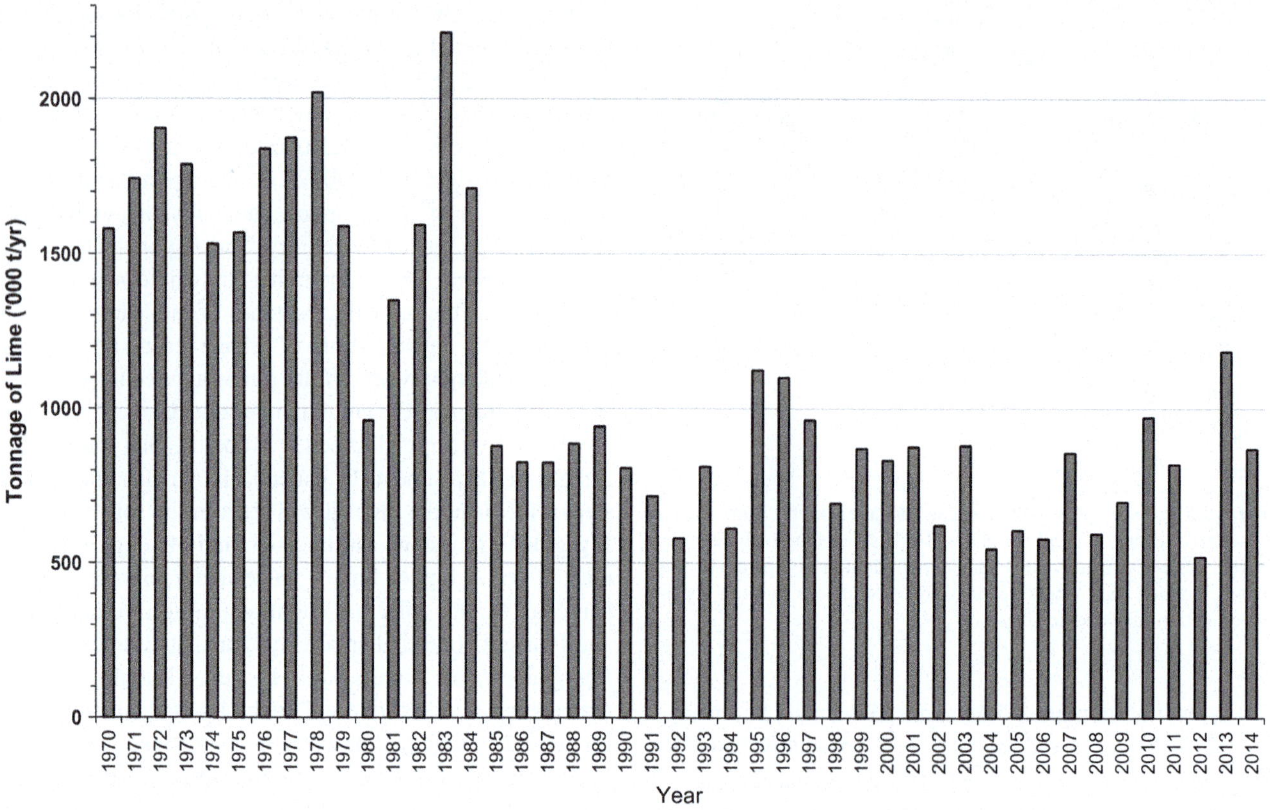

Fig. 15.6 Lime Usage from 1970–2014 based on Fertiliser Association of Ireland (FAI) and Teagasc technical bulletin series no. 2 (2016)

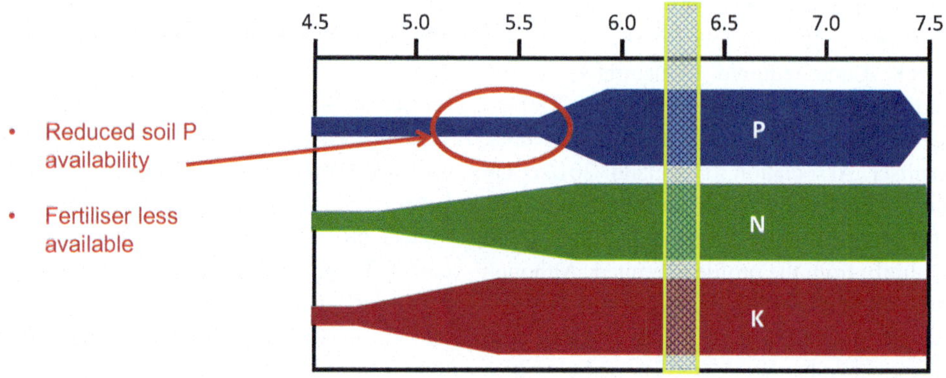

Fig. 15.7 Soil pH has a large influence on soil nutrient availability. Aim to maintain mineral soils in the pH range 6.3–7.0 and peat soil in the pH range 5.5–5.8 to maximise nutrient supply. A pH range between 6.3 and 7.0 for mineral soils is optimum

Table 15.1 Optimum soil pH recommendations for a variety of crops

Crop	Optimum pH
Beet, Beans, Peas and Oilseeds	7.0
Cereals and Maize	6.5
Grassland	6.3
Grassland (High Molybdenum)	<6.2
Potatoes	6.0

such as Peats and soils with histic or humic horizons have the potential to release more N to the growing crop. The rate at which N is released is dependent upon the C:N ratio (10–12:1 is optimum) and soil conditions such as temperature, moisture and oxygen. This means that OM added to the soil can either increase the rate of decomposition, releasing N (mineralisation), where the added material is rich in N, or where there is a lot of C being added, immobilization of soil N takes place, locking the N away. Therefore, knowledge of the C:N ratio through soil tests is important and if the ratio is high, amend with N appropriately to encourage a balance of 10–12:1. Manures and composts are an efficient way to ensure cycling of nutrients within the farm management system.

Generally, the highest N release from soils occurs in late spring and early autumn when soil temperature and moisture levels are favourable. However, N is a volatile nutrient, which moves freely through soils, plants, water and air. It exists in many different compound forms, some of which are available for uptake by plants (e.g. ammonium and nitrate), some of which are greenhouse gases (e.g. nitrous oxide (N_2O) and some of which can be lost in drainage water (e.g. nitrate (NO_3^-). Grassland and crops grown on lighter (sandier) and well-drained soils such as Brown Earths and Brown Podzolics are likely to be very responsive to N fertiliser inputs. However, care must be taken when applying N, especially during wetter months as there is a higher risk of the N being leached via draining water. In contrast, Gley soils are more poorly drained due to either impermeable clay-rich layers which stop water and nutrients from draining through (Surface-water Gleys) or from a high or fluctuating water table (Groundwater Gleys). When conditions are favourable, these Gley soils can release N from their OM reserves, such as during periods when they are drying out or beginning to wet-up. During prolonged wetter periods when Surface water and Groundwater Gleys and Alluvial soils (which are prone to flooding) become waterlogged there is a risk of denitrification where N will be lost to the atmosphere as N_2O and dinitrogen gas. Many Gley soils have artificial drainage systems which help these soils to dry out more quickly in spring increasing the grass growing season and reducing the risk of N loss.

Different soils vary in capacity to produce grass and crops due to their intrinsic physical and chemical properties, their biology and the climatic environment in which they occur. It is considered that Irish grassland soils have the capacity to supply large quantities of N (23–114 kg N ha^{-1} over a 5-week growing period) in the absence of fertiliser inputs (McDonald et al. 2014) through biological N mineralisation processes (Fig. 15.8). In 2011, it was indicated that the N use efficiency of Ireland's grass-based dairy production systems to be between 27 and 37% (Ryan et al. 2011) compared to the worldwide N use efficiency across all agricultural systems of 37% (Van der Hoek 1998). This highlights the importance of customised nutrient management plans that can increase efficiencies such as a reduced requirement for external inputs (Wall and McDonald 2014).

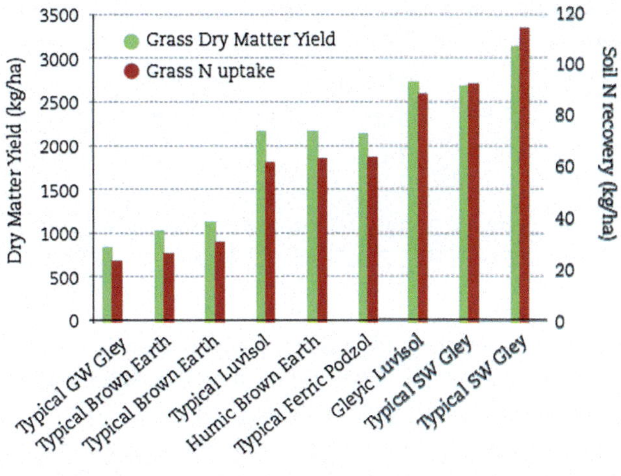

Fig. 15.8 Average soil N recovery and grass dry matter yield for a range of Irish grassland soils over a 5-week grass growth interval under optimised growth conditions (Wall and McDonald 2014)

Important too is the large contribution of mineral N that can be obtained from the atmosphere. This is done through the sowing of legumes, which 'fix' the atmospheric N through their symbiosis with Rhizobium bacteria in the legume roots. The most popular legume used is white clover which potentially can transfer 80–100 kg N/ha/year to growing grass. Red clover can reach higher rates of fixation (200–300 kg N/ha/year), but does not respond well to defoliation. Therefore, red clovers should only be used on permanent or arable silage ground. In arable rotations, legumes can be included as a green manure or as an inter-crop, ensuring a replenishment of N into the soil. In organic systems, the N mineralised from the soil can only be replenished by additions of manures or composts whereas in conventional agriculture the use of chemical fertilisers or a combination of chemical and organic N inputs is typical.

15.3.2 Phosphorus (P)

In the soil, the vast majority of P is bound so tightly that it is unavailable to plants. In organic systems, with no input of water-soluble P fertilisers, many plants rely on their symbiosis with specialised fungi called mycorrhiza to better access P from the soil particles and, in return, the plants supply the fungi with sugars. Phosphorus plays a central role in the energy regulation of all organisms and its availability is likely to be most limited early in the season when soils are cold and wet. However, the 'behaviour' of P is completely opposite to that of N, as it is strongly bound to the soil or in soil OM. When mining soil P for plant growth, we need to

Table 15.2 Soil index system for P for grassland and arable crops (Morgan's extractable P test 1941)

P index	Grassland	Arable
1	0–3.0	0–3.0
2	3.1–5.0	3.1–6.0
3	5.1–8.0	6.1–10.0
4	>8.0	>10.0

keep an eye on the total P reserves that remain in the soils. There are dramatic variations in the total P reserves between soils and this can simply be the result of the geology of the soil (limestone versus acidic rock) and of past nutrient management. Plant available P in the soil is a small proportion of these total P reserves (usually less than 1%) and is measured in Ireland, using the Morgan's P soil test. Soil type can influence how soil test P levels respond to P fertiliser applications and more heavy textured soils such as Luvisols or Surface-water Gleys have the potential to 'lock up' fertiliser P. Also, soils based on shale bedrock have increased Fe and Al, and also have increased potential to lock up fertiliser P. Therefore, farmers need to adopt different fertiliser management strategies to overcome this issue such as applying the crop P allowance using a 'little and often' approach rather than all at once. Care should also be taken when applying P fertilisers and slurries on more poorly drained Gley soils as they pose a higher risk for P loss after heavy rainfall if water flows off the surface (see Chap. 16), representing a loss on investment and environmental impact.

In Ireland, agriculture is dominated by grass-based dairy production systems. In these systems, P is a particularly important input for profitable intensive grass-based dairy farming as it is an essential nutrient for production of economically viable herbage yields (Ruane et al. 2014). Phosphorus reserves must be further monitored in dairy systems due to the dairy cow investment of P into milk production which will require larger inputs to offset the loss in milk. Soil P is measured using an index system which is calculated on the basis of the Morgan's P soil test (Table 15.2).

In conventional farming, a Morgan's P of 'Index 3' (5–8 mg per litre for grassland, or 6–10 for tillage/arable) is recommended. However, there is anecdotal evidence that 'Index 2' (3–5 mg per litre for grassland, or 3–6 for tillage/arable) may be sufficient. Where soil P levels fall in Index 1 (0–3), this will result in P reserves being exhausted resulting ultimately in P-deficiency in grass and other crop production. Soluble fertiliser P or rock phosphate may be applied, along with manures or compost for a number of years, until a repeat soil test shows more favourable results. Also, where a soil tests positive for Index 4, it is important to discontinue P applications for 2 to 3 years and retest the soil at that stage to monitor soil P levels after that time.

15.3.3 Potassium (K)

Potassium is another essential nutrient required in large quantities by plants for plant health including reproduction and growth. It is necessary to regulate the opening and closing of stomata during photosynthesis and is, therefore, essential to regulate CO_2 uptake. The behaviour of K lies somewhere in between that of N and P: it can be bound, but also move quite freely through soil, water and plants. Brown Earths and Brown Podzolic soils may have high K reserves and supply for crop production, derived from their parent material, especially if formed over shale and mudstones. However, significant quantities of K can be removed from lighter textured, sandier soils such as sandy loams. In contrast, Luvisols and Surface water Gleys are the most prominent K fixing soils where K fixation is more prominent when OM levels are low following long-term arable production. To mitigate against K fixation and maintain adequate soil K supply for crop uptake more frequent applications of K fertilisers or manures are warranted on these soils. On Peat soils both K and Mg are easily released which can result in low concentrations. Where K concentrations in soils are suboptimal for grass or crop production, it is often called the 'hidden' deficiency with the main effect of a K-deficiency in grassland associated with a change in the botanical composition of grasslands—changing from productive species, such as ryegrass, to unproductive or unpalatable species such as bentgrass or cocksfoot.

15.3.4 Sulphur (S)

Sulphur is required by plants to convert N into proteins and plant S levels should be greater than 0.2% of dry matter (DM), and the N:S ratio should be less than 15:1. Plant S content and plant N:S ratios are useful guides in assessing the likelihood of a response to S fertilisers. In Ireland, S levels in the soil tend to be lower than in many parts of Europe, because of the lack of heavy industry and the associated S depositions from the atmosphere (typically 1–2 kg/ha/year). Sulphur behaves much like N in soils where a considerable amount of the available S is held in organic matter and must be mineralised to release S (as sulphate) for crop uptake. In sandier textured and well-drained soils (Brown Podzolics, Brown Earths) crops can be very responsive to S applications as S can be leached as sulphate as water percolates through the topsoil. Sulphur deficiencies may be more apparent on these soil types where high levels of N fertiliser is being used, or where high yielding tillage and silage crops are removing large quantities of S. Therefore, applications of N + S fertilisers are required to meet crop demand. More poorly drained, heavier textured and high OM soils (such as Gley soils or Peats) have greater potential to release S from OM reserves, making it available to meet the demands of crop uptake. It is important not to oversupply S where it is not required, as excess S can depress the uptake of selenium (Se) in herbage and reduce the absorption of copper (Cu) by animals.

Murphy (2000) studied the response to sulphur across 140 grassland sites in Ireland. Seventy of them showed a DM yield response to S fertiliser. The responses were higher in mid-summer and autumn than in spring. On the 70 sites that showed a response to S, the application of S fertiliser increased the total annual DM yield by 15% on average (Fig. 15.9).

15.3.5 Magnesium (Mg)

Magnesium is an essential plant nutrient, particularly in the photosynthesis process as a building block of chlorophyll. Magnesium is involved in the transport of sugars through plants and root growth may thus be limited under Mg deficiency. Often growth reductions from Mg deficiency occur before any deficiency symptoms are seen on the plant. As a result, Mg deficiency can be a major limitation in crop production. Where the soil test for Mg is low and liming is necessary, magnesic limestone should be used if possible. However, a herbage test is more reliable than a soil test to check the Mg that is available to the animals. It is often necessary to supply a source of Mg (e.g. calcined magnesite), to animals, particularly dairy cows, in spring to prevent hypomagnesaemia (grass tetany). This is mainly because there is a low absorption of Mg from the lush spring grass as it passes rapidly through the animal's digestive tract. Magnesium concentration in herbage is lowest in the spring.

15.4 Micronutrients

Micronutrients are only required in small quantities by crops and pastures (generally only grams per hectare are taken up each year), but when deficiencies occur large reductions in crop yield or pasture growth can occur. The main micronutrients used in agriculture are boron (B), cobalt (Co), copper (Cu), iodine (I), iron (Fe), manganese (Mn), molybdenum (Mo), selenium (Se) and zinc (Zn). The role of micronutrients in plants varies according to the nutrient in question. Boron is one of the most important micronutrients affecting cell membrane stability and is essential for flowering in plants. Copper, Fe, Mn and Zn are essential for many plant enzyme systems, and Fe and Mn are particularly important for photosynthesis. Cobalt and Mo are essential in legumes for efficient symbiotic atmospheric N fixation by *Rhizobium* bacteria in root nodules. Iodine and Se are not essential micronutrients for plants, but are essential for healthy growth of animals and humans.

Cobalt, Cu, Fe, Mn and Zn have a similar behaviour in that in soil they are present as positively charged cations. Hence, availability is lowest in alkaline soils due to strong retention of the micronutrient cations by the soil, and precipitation with carbonates and hydroxides is common. Hence these micronutrients are extremely immobile in soil and efficient uptake by plants depends on good root architecture (extensive fine root systems) or the presence of mycorrhizal fungi which assist many plants to acquire these nutrients from soil. Liming of soils may also induce deficiencies of these micronutrients as they are 'tied up' in the soil by the increased soil pH and the presence of lime particles. Boron, on the other hand, is present as the uncharged boric acid molecule in most soils and hence is extremely mobile, and can easily leach through soils, especially in high rainfall areas. Iodine, Mo and Se are present in soil as negatively charged ions, with Mo and Se present in combination with oxygen (as selenite/selenate for Se and as molybdate for Mo), and these micronutrients are also relatively mobile in soil, compared to the micronutrient cations. Iodine, Mo and Se are generally least available in acidic soils. Molybdenum (Mo) is essential for many plant functions such as the conversion of nitrates into amino acids in the plant or the conversion of inorganic P into organic P forms. However, Mo interacts with soil pH differently to other micronutrients such as Cu and Zn in that it becomes more bioavailable as the soil pH increases. On soil with high Mo concentrations and high pH levels, high levels of Mo uptake in grass and overconsumption by ruminant livestock can lead to toxicity and result in deficiencies in other nutrients, typically Cu. Copper is an element that is essential (in small quantities) for animal health and vigour. The occurrence of Cu deficiencies can be exacerbated where soils are over limed, due to its interaction with greater Mo availability at high pH. Grass with high Mo concentrations can antagonise Cu absorption by the animal, leading to deficiency symptoms. To prevent this occurring, on soils with high Mo concentrations it is recommended to maintain soil pH within a slightly lower range, around pH 6.0–6.2, thus controlling Mo availability. The following describes micronutrients in Irish soils:

15.4.1 Copper (Cu)

Soils derived from sandstones, granites and peats are likely to have low Cu levels. The amount of Cu normally absorbed from food by the ruminant is less than 10%. Soil Cu levels of below 5 mg/l in mineral soils are rare north of a line from

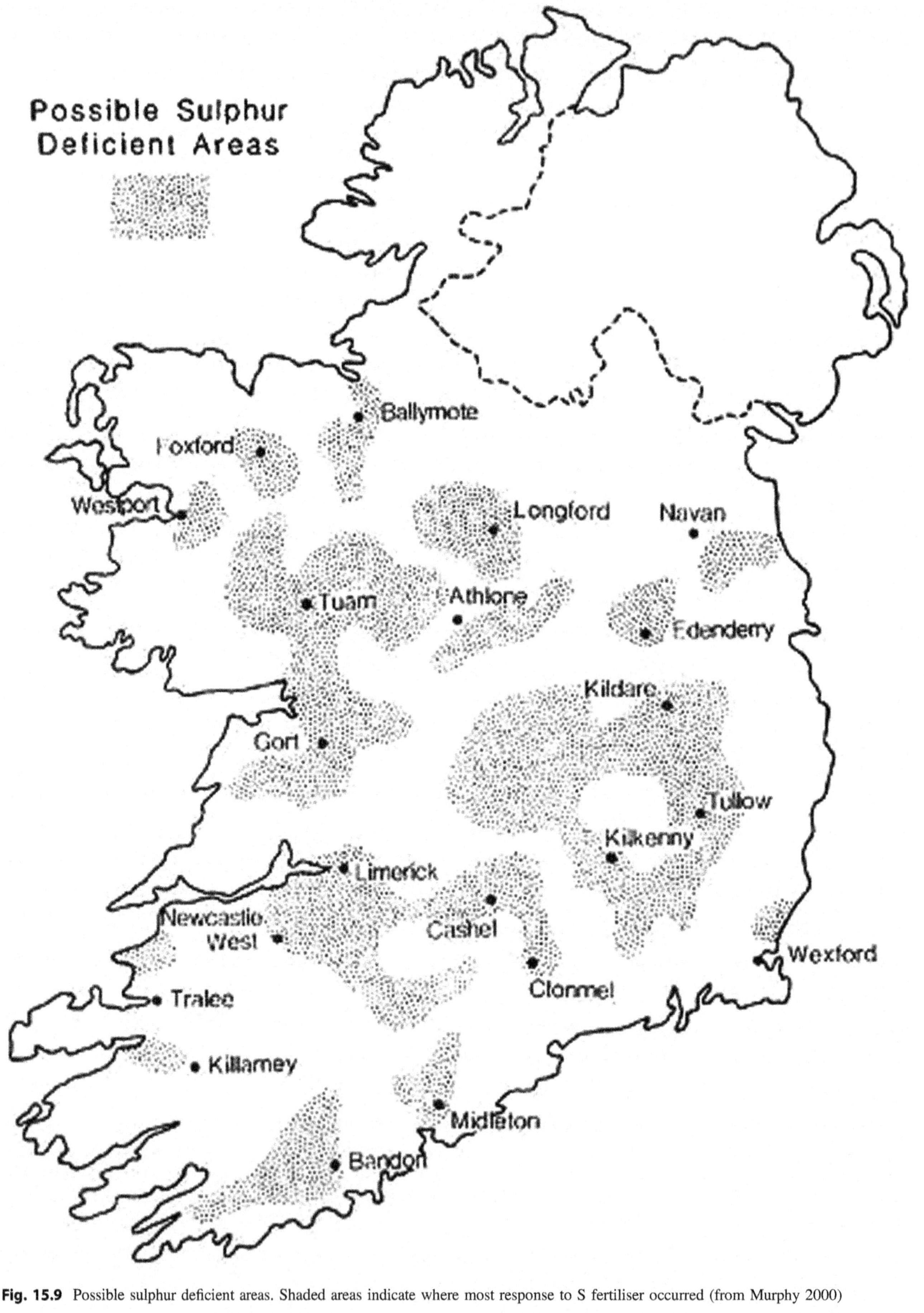

Fig. 15.9 Possible sulphur deficient areas. Shaded areas indicate where most response to S fertiliser occurred (from Murphy 2000)

Galway to Dublin. Low soil concentrations are common on peat land and in the light sandy soils which occur in counties Wexford, Carlow, Waterford, east Wicklow, north Clare and Kerry. Soil Cu tests are not useful as definitive diagnostic tests owing to the complexity of the problem of Cu availability and utilisation by the ruminant. It is better to use herbage Cu levels together with Mo and S values.

15.4.2 Zinc (Zn)

Zn deficiency is most likely on sandy soils with high pH. Approximately 23% of Irish soils sampled have soil concentrations of Zn below 2.0 mg/l. The occurrence of soils with Zn levels below 2 mg/l is highest in counties Donegal (65%), Louth (47%), Carlow (29%), Tipperary (25%), Cork (24%) and Meath (24%). Zinc deficiency in grassland occurs at ethylenediaminetetraacetic acid (EDTA) extractable soil concentrations below 0.6 mg/l. The main deficiency symptom is reduced grass emergence and stunted pasture growth. Soil concentrations below 0.6 mg/l are rare and the occurrence of pasture affected by Zn deficiency is unusual.

Adequate Zn in the herbage is important for animal health. Low Zn in herbage affects the immune system. Severe deficiency affects hoof and hair growth in cattle and wool growth in sheep. At soil concentrations of less than 2.0 mg/l EDTA extractable Zn, the concentration in grass will normally range from 11.3 to 22.0 mg/l. Herbage concentrations of less than 20 mg/l are regarded as low for livestock.

15.4.3 Cobalt (Co)

Soils giving rise to low Co in herbage are typically formed from granite and sandstones. However, low herbage Co levels can occur even on soils rich in Co, if the total Mn content of the soil is high. Hence, when determining the level of Co in a soil, it is also necessary to analyse for total Mn in order to determine whether Co deficiency can occur. It also determines whether Co can be applied to the soil, or if it must be fed directly to the animals.

Most of the soils which respond to pasture treatments are located in counties Mayo and Kerry; and to a lesser extent in counties Limerick, Carlow, Waterford and Galway. Cobalt deficiency induces symptoms known as pine and poor thrift in sheep. A concentration of 0.1 mg/kg Co in herbage is required to supply sufficient Co for animal needs. Total Mn concentrations in soil greater than 600 mg/kg can reduce the plant uptake of Co by pasture species to below 0.1 mg/kg. Investigations have shown that up to 25% of the stomach contents of sheep may be ingested soil. Cobalt deficiency rarely occurs in pastures with soil Co greater than 6 mg/kg and total Mn above 600 mg/kg, due to Co intake by soil ingestion.

15.4.4 Selenium (Se)

Selenium, like Co and I, is not required by plants but it is required by animals. The Se content of Irish pastures can vary widely. Some soils, mainly in counties Meath, Limerick and Tipperary have excess Se, which can cause Se toxicity in livestock. Selenium deficiency occurs most frequently in the sandstone soils of south Munster and the clear-water-limestones of west Munster and east Galway.

Selenium toxicity tends to occur at herbage concentrations above 3 mg/kg Se. Soil testing for Se can be used to diagnose Se toxicity; Rogers et al. (1989) found soil Se levels ranging from 3 to 132 mg/kg in Se-toxic fields. Se toxicity causes loss of hair and extension and sloughing off of the hooves of livestock. On soils with high Se, there is no certain method of avoiding Se toxicity in livestock. The affected areas may be fenced off and used to grow cereals or hay as the Se content of hay is lower than its content in pasture.

Soil testing for Se is not of great value in diagnosing Se deficiency. Analysis of forage or blood is usually used instead. McGrath (2000) found that Se in pasture was very variable and that one could not rely on a single determination to diagnose the Se status of grassland. Selenium deficiency occurs at herbage concentration below 0.1 mg/kg. It causes stillbirth and retention of afterbirth in livestock. Poor conception rates and embryonic mortality are also common.

15.5 Future Nutrient Management

Farmers regularly manage the fertility of the soils on their farms by applying fertilisers and organic manures to build-up or maintain the supply of nutrients required for the grass or crop types they produce. Organic fertilisers are a valuable source of N, P and K. Cattle slurry is by far the most common form of organic fertiliser applied in Ireland. Over the last number of years, there has been a major drive on Irish farms to improve the utilisation of slurry and reduce fertiliser costs.

However, experienced farmers will know that not all soils (or fields) have the same production potential (or suitability for certain crop types) or response in terms of their soil fertility status to the nutrients that are applied. While several factors will determine production potential and fertility status, one major denominator is soil type and most farms will have at least two different soil types, if not more, dispersed within the farm boundary. Given the nature of the Irish landscape and its origins (glaciation) (Chap. 2) there is often large variability in soils over relatively small spatial areas, even within town lands or farms. This poses a challenge for individual farmers and their advisors when planning nutrient and fertiliser management strategies for their farms.

A blanket approach, where all fields, receive and 'are perceived to respond the same' to similar nutrient application rates may not happen in reality. This is because different soil types possess different characteristics including the parent material (rock type, glacial till or mineral deposit) that soil is derived from and its nutrient composition, soil texture (i.e. proportions of sand, silt and clay present), soil organic matter level, water holding capacity and drainage class (i.e. free draining vs. poorly draining).

As soils vary naturally in their nutrient status, it is appropriate that nutrient applications need to be tailored to the particular soil type in order to achieve the correct nutrient balance required for the growing crop. To implement soil fertility advice for individual soil types at farm level, knowledge of the soils present is required. The Irish Soil Information Systems has mapped our soils at a regional/catchment scale which is a major step towards identifying the typical soil types a farmer could expect to find on his farm. The future challenge is to achieve improved nutrient management and increased profitability and sustainability of farming systems by linking soil fertility knowledge to these different soil types and to train and support professional advisors and farmers to identify and manage soil appropriately. Organic fertilisers are a valuable source of N, P and K. Cattle slurry is by far the most common form of organic fertiliser applied in Ireland. Over the last number of years, there has been a major drive on Irish farms to improve the utilisation of slurry and reduce fertiliser costs. This has been achieved on farm by better facilities for the management of slurry during the winter and more slurry been applied on grassland in early spring when N utilisation is highest.

References

Brady N, Weil R (2010) Elements of the nature and properties of soils, 3rd edn. Prentice Hall, Upper Saddle River New Jersey

Fertiliser Association of Ireland and Teagasc (2016) Soil pH & lime, technical bulletin series no. 2, Fertiliser Association of Ireland in association with Teagasc

Morgan MF (1941) Chemical soil diagnosis by the universal soil testing system. Connecticut Agricultural Experimental Station Bulletin 450, Connecticut

Murphy M (2000) Beneficial effects of sulphur on agricultural crops and some implications for livestock nutrition, Teagasc. Johnstown Castle, Co., Wexford

McDonald NT, Watson CJ, Laughlin RJ, Grant J, Lalor STJ, Wall DP (2014) Soil tests for predicting nitrogen supply for grassland under controlled environmental conditions. J Agric Sci, 1–14. https://doi.org/10.1017/S0021859614000264

McGrath D (2000) Soil and herbage heavy metal/trace element variability and relationships at farm and regional level. End of project report for Armis 4610. Teagasc, Johnstown Castle, Wexford

Rogers PAM, Fleming GA, Gately TF (1989) Mineral imbalances and mineral supplementation. Teagasc, Dublin

Ruane EM, Treacy M, McNamara K, Humphreys J (2014) Farm-gate phosphorus balances and soil phosphorus concentrations on intensive dairy farms in the south-west of Ireland. Ir J Agric Food Res 53:105–119

Ryan W, Hennessy D, Murphy JJ, Boland TM, Shalloo L (2011) A model of nitrogen efficiency in contrasting grass-based dairy systems. J Dairy Sci 94:1032–1044

Sinclair A, Crooks B, Coull M (2014) Soils information, texture and liming recommendations, Technical Note TN656, Scotland's Rural College (SRUC)

Sumner ME, Yamada T (2002) Farming with acidity in communications in soil science and plant analysis 33(15–18):2467–2496. Available at: http://www.tandfonline.com.ezproxy.is.ed.ac.uk/doi/pdf/10.1081/CSS-120014461. Accessed: 02/11/2016

Teagasc (2016) Soil analysis and trends Available at: https://www.teagasc.ie/media/website/crops/soil-amp-soil-fertility/All-Soil-Results-Enterprises.pdf. Accessed: 14/12/2016

Van Der Hoek KW (1998) Nitrogen efficiency in global animal production. Environmental Pollution 102(suppl 1):127–132

Wall DP, McDonald N (2014) Soils—the nutrient reservoir, vol 9: no 3. Autumn 2014 ISSN 1649-8917, Teagasc, Carlow

Wall DP, Forrestal P, Plunkett M (2015) Sub-optimal soil fertility costing dairy farmers dearly. Sustainable Dairy Expansion, TResearch, vol 10, no 2, Summer 2015. ISSN 1649-8917, pp 42–43

Wall DP, Plunkett M (2016) Major and micro nutrient advice for productive agricultural crops. Teagasc 2016. ISBN no. 978-1-84170-632-0

Soils and Water Quality

16

Karen Daly, Karl Richards, Per-Erik Mellander, Phil Jordan, Daire Ó hUallacháin, Sophie Sheriff, Sara E. Vero, and Owen Fenton

Abstract

Ireland has a diverse physical landscape of upland and lowland inland waters represented by 13,200 km of river and 12,000 lakes framed by an estimated 3171 km of coastline providing a range of freshwater ecosystem services. The quality and quantity of water flowing through Ireland's inland and coastal waters are determined by the complex interaction of anthropogenic activities with soil and climate. It is the mosaic of soils that characterise the catchment draining a waterbody that regulates the flow of water through a catchment and the movement of nutrients and contaminants. In this chapter, we describe the role of soil as the matrix through which, water and contaminants can be transported and retained within the Irish landscape, and the impacts on water quality in Ireland. Here we reference current water quality policies relating to diffuse of pollution from soils and describe processes and functions identified in Irish soils that influence nutrient and sediment transport to affect water quality in Ireland.

Keywords

Agriculture • Nutrients • Eutrophication • Water quality Soil processes

K. Daly (✉) · K. Richards · P.-E. Mellander · D. Ó hUallacháin · O. Fenton
Teagasc Environment Research Centre, Johnstown Castle, Wexford, County Wexford, Ireland
e-mail: Karen.daly@teagasc.ie

S. E. Vero
Agricultural Catchments Programme, Teagasc, Wexford, County Wexford, Ireland

P. Jordan
Geography and Environmental Sciences, Ulster University, Coleraine, Northern Ireland, UK

S. Sheriff
School of the Environment, University of Dundee, Dundee, DD1 4HN, Scotland

16.1 Introduction

Of the total land area in Ireland (6.9 million ha), almost 65% is used for agricultural production with approximately 11% used for forestry (DAFM 2015). Agricultural land use in Ireland is comprised of 91% grassland and 9% arable crop production (DAFM 2015). The management of animals, nutrients and pesticides on agricultural soils can affect water quality due to their direct interaction, primarily with soils. This interaction is mediated to a large extent by soil pore water interactions at the pedo-scale, however, at larger catchment scales, these interactions are defined by hydrological pathways that transport nutrients, pesticides and sediment to stream networks. In this chapter we reference the importance of agricultural nutrients such as phosphorus (P) and nitrogen (N) for productivity (see also Chaps. 14 and 15) but expand on their impacts on water quality via interactions within the soil matrix. Finally, a model of soil-water interaction at small and large scales is described.

16.2 Provision of Pathways for Water Flow and Transfer Mechanisms for Nutrients

A catchment is a basin-shaped area of land that is comprised of a system of streams that catches precipitation and transports water, sediment and other material in a drainage network. The Irish agricultural landscape is a heterogeneous mix of topography, soil type and drainage, soil depth and bedrock. Headwaters are streams that occur in the upper catchment and may be intermittent, flowing during wetter periods, while others may be more permanent. In steep areas, the stream will be fast flowing and will have a lot of energy to transport materials such as sediment, rocks or gravel. Small headwater catchments are likely to vary in land use which typically consists of mosaics of small-sized agricultural fields. These fields are separated by ditches and

R. Creamer and L. O'Sullivan (eds.), *The Soils of Ireland*, World Soils Book Series,
https://doi.org/10.1007/978-3-319-71189-8_16

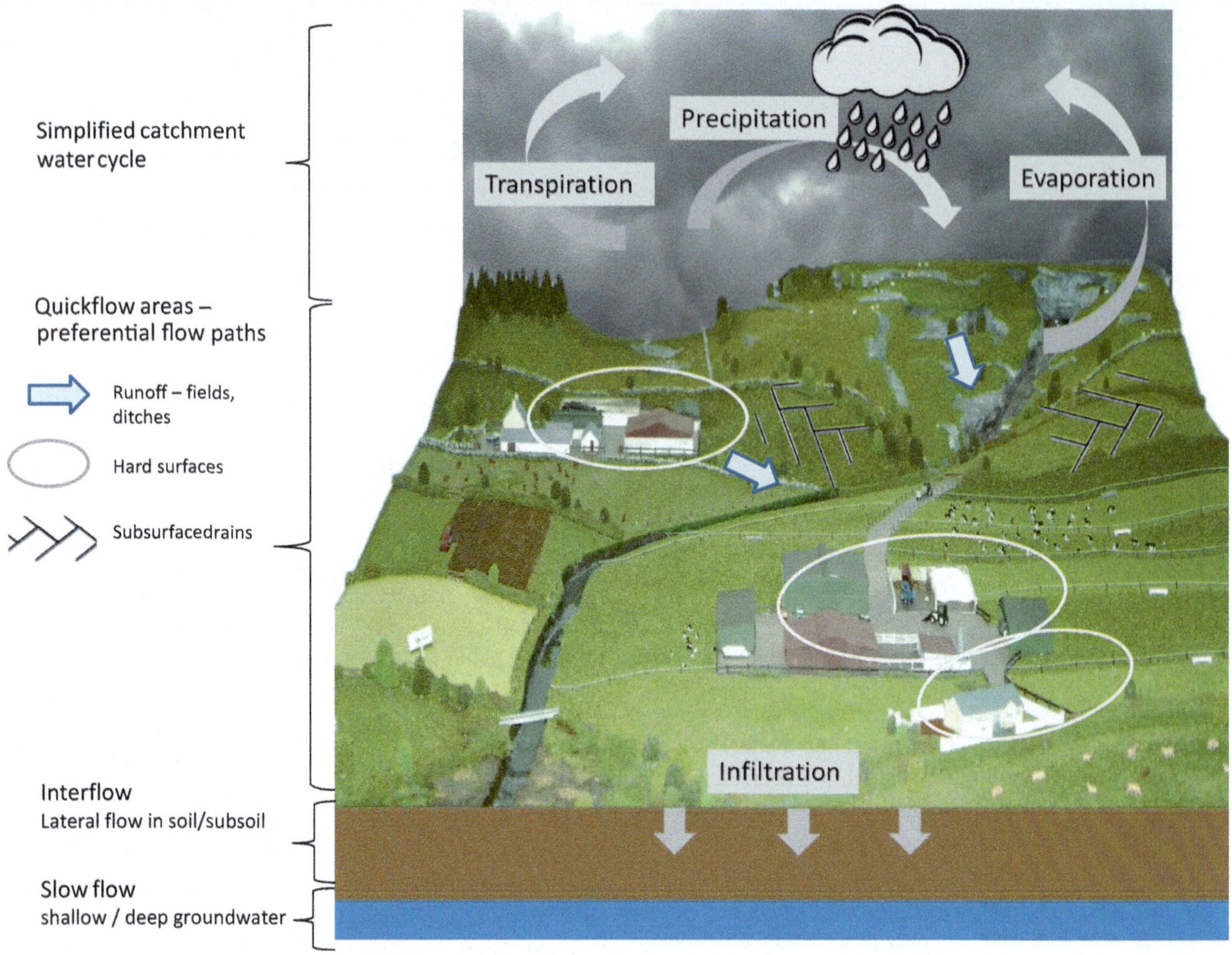

Fig. 16.1 Simplified catchment water cycle and the flow types or hydrological pathways that can be found

hedgerows giving rise to large variability in dominating flow paths, water storage and hydrological connectivity. These factors, together with the weather, dictate the pathways and catchment river flow characteristics. While such flow controls give rise to spatial variability, rainfall and evapotranspiration influence temporal hydrological variability, with higher rainfall typical in winter and higher evapotranspiration demand in summer. December and January are the wettest months while April or June tending to be the driest (Met Éireann—The Irish Meteorological Service). This results in more frequent high magnitude river flow peaks with increased base flow in wintertime, and fewer river flow peaks and lower base flow in summer. Base flow refers to the portion of the stream that has seeped into the channel slowly over time from deep and shallow subsurface flow (water movement which takes place through the soil). Therefore, base flow is the main source of stream water during dry weather. There is also a spatial difference with a gradient of rainfall and evapotranspiration across Ireland from east to west. Typically, the annual rainfall of the eastern half of the country is between 750 and 1000 mm and in the western half between 1000 and 1400 mm while in the mountains rainfall exceeds 2000 mm).

At catchment scale, hydrological pathways can be divided into three general flow types (Fig. 16.1). The first is *quickflow* which can be a result of run off from fields, ditches, hard surfaces, preferential flow paths and subsurface drains. The second is *interflow,* which is the lateral flow in the soil and/or subsoil, commonly in a soil layer more permeable than the material it rests on. The third flow type, *slowflow,* is shallow or deep groundwater (Fig. 16.2). Since water is a vehicle for nutrient transfer (quantitatively and qualitatively), the dominating pathway will to some degree determines nutrient loss to water in terms of nitrogen (N) and phosphorus (P) form (*e.g.* dissolved or particulate) and quantity. Generally, it is perceived that excessive N is leached to groundwater and will be transferred to streams via belowground pathways such as *quickflow* in subsurface drains, or via *interflow* or *slowflow* pathways where transformations may occur, depending on travel time and the

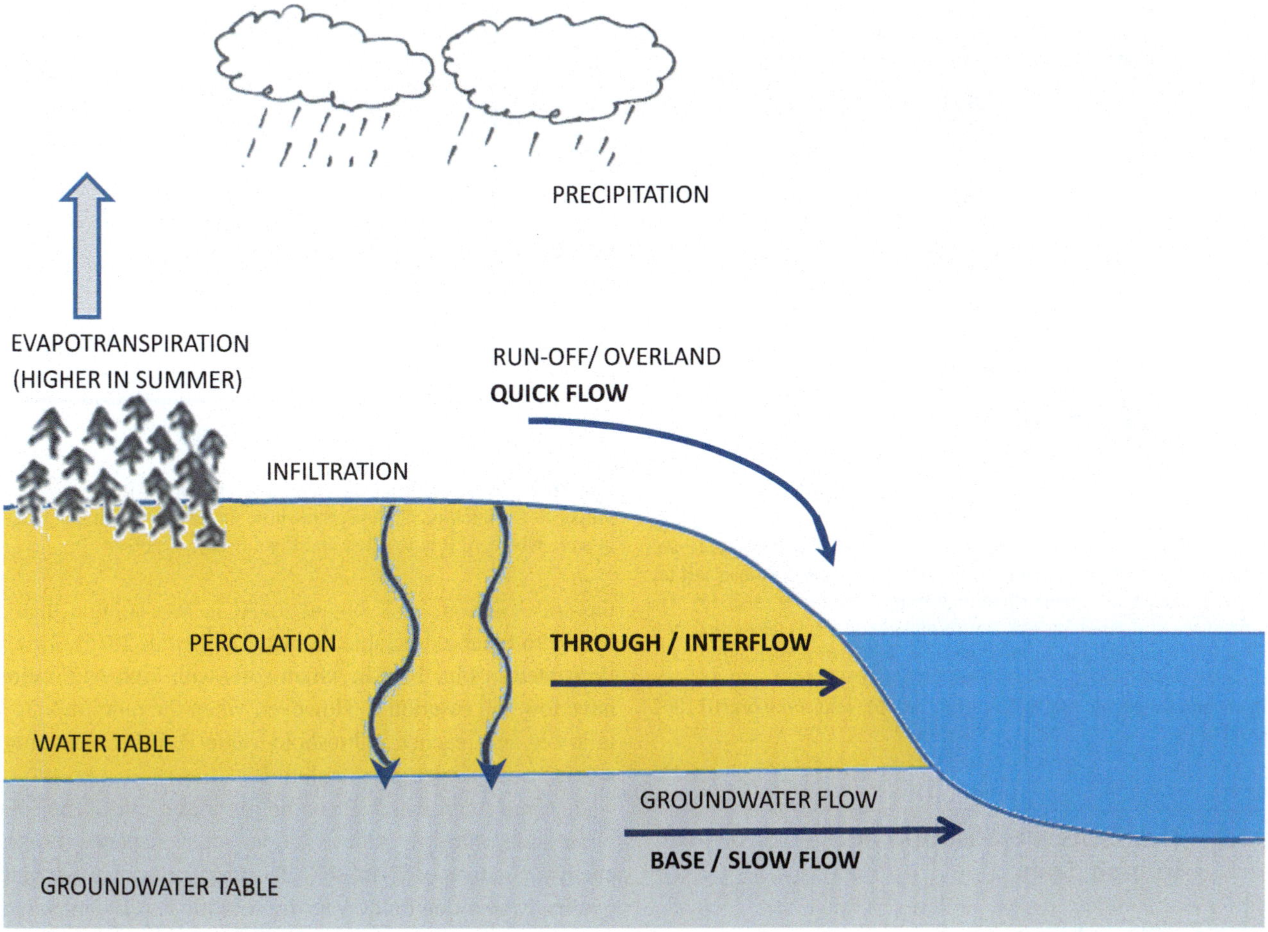

Fig. 16.2 Flow pathways: quickflow, interflow and slowflow. When temperatures are higher in summer, evapotranspiration increases and at this time, base flow can become the main source of water to streams and rivers

biogeochemical environment. Particulate P is predominantly transported via *quickflow* pathways. Similarly, dissolved P can be transported in *quickflow* but also via subsurface macropores and preferential flow (*e.g.* Doody et al. 2006) or *slowflow* pathways (Mellander et al. 2016).

The different structure, texture, mineral content or drainage status of the soil will determine how much water can run off the land and also the potential for erosion losses. Poorly drained soils, such as Surface-water Gleys, commonly give rise to a flashy hydrology whereby streams are loaded with a high volume of water over a short span of time. This is because these soils are characterised by a fine texture and a gleyed subsurface horizon (argillic B horizon (Btg) that impedes the downward movement of water thereby promoting this flashy hydrology, particularly in wetter periods. This results in high flow peaks from surface *quickflow* pathways. In such cases, there is a need to manage the *quickflow* components of run off to moderate nutrient transfers, specifically P. An essential nutrient for plant nutrition and animal health, this element is removed in farm produce (milk, meat, hay and silage) and needs to be replaced by added fertiliser P to maintain an optimum supply in soil for grass and crop production (see Chap. 15). Fertiliser P applications that exceed crop requirements can elevate soil P levels causing an excess of this nutrient in an available and soluble form that can be transported via hydrological pathways to receiving waters. Enrichment of freshwaters with P causes the overproduction of algae and cyanobacteria, leading to eutrophic conditions. Therefore, P as a key nutrient in productive agricultural systems needs to be in balance with crop demands and managed on soils so as to avoid losses to water. However, soil hydrology can exert a greater influence on nutrient losses, specifically P, whereby poorly drained soils can generate larger losses of P to streams compared to catchments defined by well-drained soils, (Fig. 16.3).

In well-drained soils such as Brown Earths, water will percolate to groundwater and river flow will be composed of a large proportion of groundwater from *slowflow* pathways. If the rainfall exceeds the soils infiltration capacity, these soils will generate *quickflow* and may result in particulate loss from surface soil erosion.

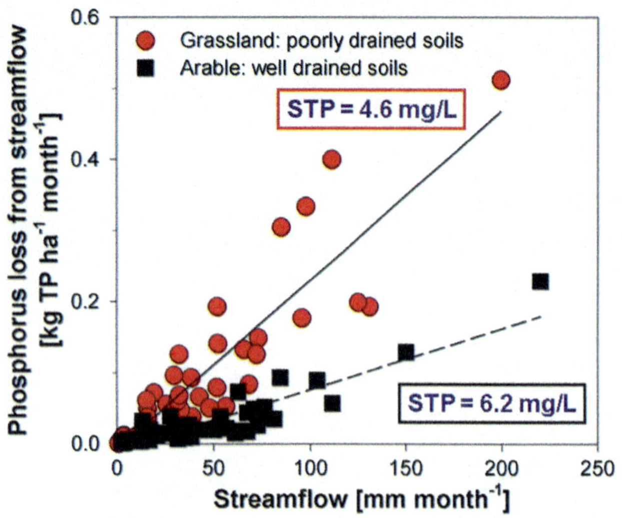

Fig. 16.3 Phosphorus losses in stream flow from two catchments with contrasting soil drainage. Soil test P (STP) represents the mean soil test P levels in each catchment (modified from Mellander et al. 2015). The differences in slope show how source pressure (STP) is decoupled from overall P transport due to differences in soil hydrological flow paths (flashier, more *quickflow* on the poorly drained grassland soil). Original figure published by Mellander et al. (2015), https://doi.org/10.1002/hyp.10415

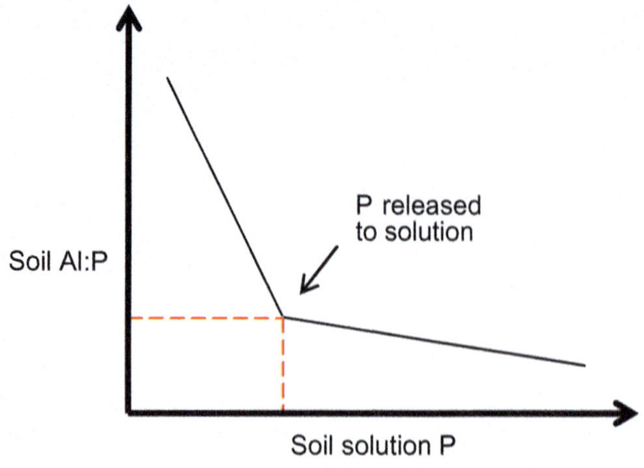

Fig. 16.4 The relationship between the ratio of Al to P in soil can influence P solubility. The diagram show the change-point that occurs in soils where Al:P is lowered and P solubility increases

16.3 Nutrients Mobilisation and Attenuation in Irish Soils

16.3.1 Phosphorus

At farm scale, agricultural P is managed on Irish soils by balancing inputs with off-takes (see Chap. 15) and to some degree, by soil type. Mineral and organic soils react differently to added P, such that high organic matter soils such as peats and mineral soils with organic surface horizons cannot retain P and are at risk of nutrient loss to water if surface quickflow pathways are present (Daly et al. 2001). These differences in soil types are delineated by the amount of organic matter in soil, which causes low nutrient absorption. The poor P retention capacity of peat and surface horizons with organic matter >20% (>11.6% organic carbon) associate these soil types with high risk in terms of water quality.

In mineral soils, P bound in soil is associated with aluminium (Al) and iron (Fe) in acidic and neutral soils and calcium (Ca) in calcareous soils and soils derived from limestone parent material (Daly et al. 2015). Phosphorus added to soil is assimilated into the solid phase by a series of sorption reactions that vary according to soil type. High P sorbing soils in Ireland typically contain high amounts of Al and Fe, which react with added P by providing sites for P sorption and retention. Soils with high Al contents are highly buffered in that they have a large capacity for P held with strong binding energies and can control the release of P to the soil solution, as P bound to Al is less soluble than P bound to other extractable metals (Daly et al. 2015). This is illustrated in Fig. 16.4 In which soils with high Al:P ratios, have low soil solution P. However, when the ratio of Al to P is lower and reaches a threshold value P, it becomes more soluble and is released to solution.

The ease with which P can be mobilised out of the solid phase and into the soil solution often depends on the chemical nature of P in soil. This has relevance to water quality largely due to the varying solubility of P compounds and the predominant forms of soil P. Mobilisation of inorganic P is largely controlled by the reverse sorption and desorption reactions that take place between the solid phase of soil and the soil solution phase. In Irish soils, P bound to Al is often less soluble than P bound to Fe (Daly et al. 2015) and these physicochemical forms can influence the concentrations of soluble P that are mobilised from the soil matrix and into a hydrological pathway.

16.3.2 Phosphorus Attenuation in Irish Soils

Once mobilised into a transport pathway, the movement of P across the landscape can be intercepted by re-adsorption back into the soil matrix or precipitation out of the soil solution into a solid form. The chemical attenuation of mobile and soluble P from hydrological pathways as they reach parts of the landscape that have the capacity to re-adsorb nutrients such as P is an important factor in neutralising their effects on water quality. In Irish soils high Ca in soil is associated with precipitation of P out of solution and onto the solid phase (Daly et al. 2015). Water quality research from a karst catchment in the west of Ireland illustrated high concentrations of available P at surface soil horizons but low groundwater P concentrations. P transport

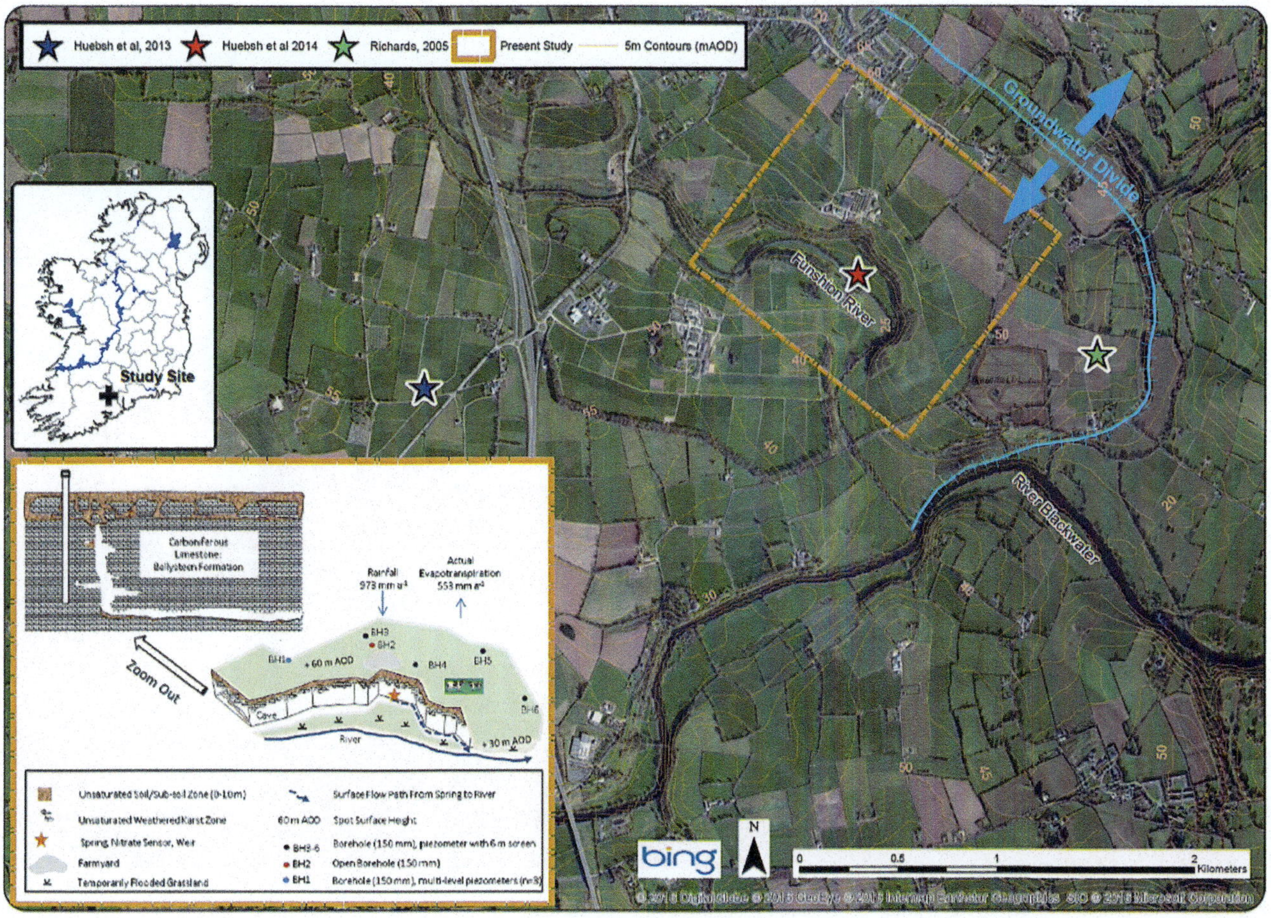

Fig. 16.5 Site locations of recent Karst studies in Ireland, delineation of present site, groundwater divide and major surface-water bodies in the area (Fenton et al. 2017). Original figure published by Fenton et al. (2017), http://dx.doi.org/10.1016/j.agee.2017.01.014

through the groundwater-fed system may have been intercepted by karst material potentially precipitating P, out of solution, resulting in low groundwater P concentrations, despite high surface soil P concentrations (Mellander et al. 2012). Other sites in Ireland have reported similar scenarios of high soil P concentrations at the surface horizons on well-drained soils, with low orthophosphate concentrations in the underlying groundwater. In Fig. 16.5, the presence of a limestone parent material layer has intercepted soluble P fractions by precipitating P out of solution, resulting in the low orthophosphate concentrations in the underlying groundwater (Fenton et al. 2017).

16.3.3 Nitrogen

Nitrogen (N) is one of the key limiting nutrients for plant growth globally and modern farming practices rely heavily on applying both organic and inorganic fertilisers to increase crop growth. Crop growth is limited by the availability of N due to the rapid loss of applied N to soil, which can be due to immobilisation in the soil, increasing the organic soil N reserves or through loss to the environment. Introduction of fertiliser to soil starts a cascade of reactions that are controlled by micro-organisms. These include the microbial transformations that control the immobilisation (lock up) or mineralisation (release) of soil organic N, and transformation of organic N to nitrate (NO_3) via ammonium (NH_4) (Fig. 16.6). In terms of water quality, the main N form of concern is NO_3 as it is highly susceptible to leaching through soil to water. In addition, NO_3 can be lost to the atmosphere under anaerobic conditions through denitrification in soil or water, resulting in N loss as N_2O, nitrogen oxides (NO_x) and nitrogen gas (N_2). Loss of N to the environment has major environmental impacts which include eutrophication of inland rivers/lakes and coastal waters resulting in algal blooms and changes to the aquatic ecosystem. Contamination of drinking water by NO_3 has been linked to both negative health impacts such as blue baby syndrome (methaemoglobinemia), cancer (Non-Hodgkin Lymphoma) and to positive health benefits in terms of controlling pathogens in the human gut. The loss of N to the atmosphere has a number of environmental impacts:

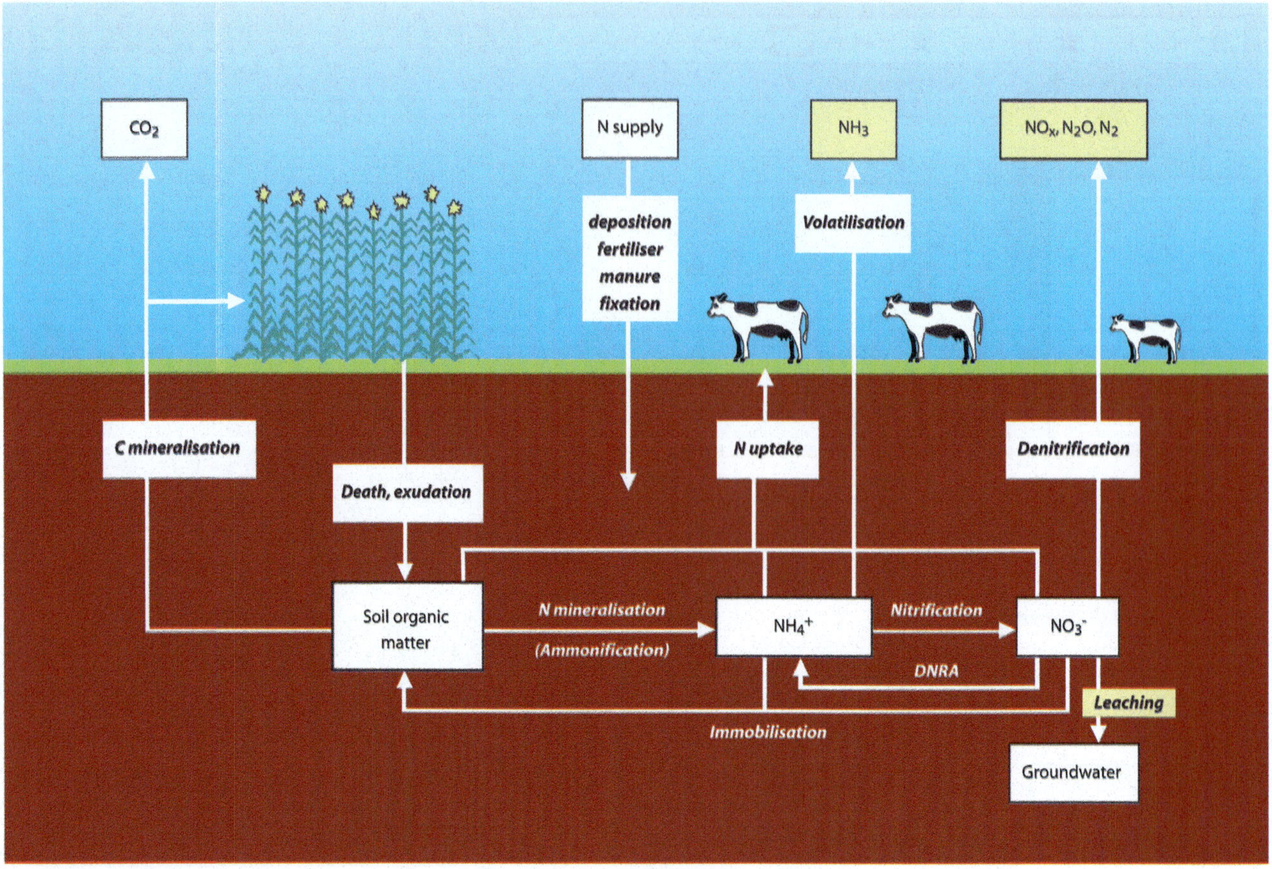

Fig. 16.6 Simplified nitrogen cycle in grazed grassland soils. Nitrogen added to soil undergoes a large number of microbial transformations converting non-crop available organic forms to crop available inorganic forms (NH_4 and NO_3). Soil nitrate (NO_3) is susceptible to loss to water or the atmosphere. Figure originally published by Karl Richards, Diana Selbie, Enda Cahalan, Samuel Dennis, Maria Ernfors, Eddy Minet, Gary Lanigan, Stan Lalor, John Murphy. in TResearch, Volume 6: Number 2. Summer 2011, Teagasc

- Soil eutrophication and acidification (NH_3)
- Greenhouse gas emissions (N_2O)
- Formation of particulate matter and destruction of tropospheric ozone (NO_x)

The soil NO_3 pool is the most susceptible form of N for loss to the environment. The proportion of NO_3 lost to water or the atmosphere is mainly controlled by soil hydrology or soil wetness. Soil texture is a very important factor influencing the fate and transport of N in the environment. Soil texture influences the speed of water movement through soil, the proportion of rainfall infiltration or overland flow, the depth to which water infiltrates, soil oxygen concentration and other chemical properties such as soil C (see Chap. 17). Dry aerobic soils (dominated by sandy textures) are most susceptible to NO_3 leaching whereas wet anaerobic soils (dominated by clay and silt textures) are most susceptible to N_2O losses via denitrification. Other soil factors such as bulk density, which influence soil hydrology, also have an important role in determining the proportion of NO_3 leached or denitrified.

16.4 Sediment Transport, Retention and Agricultural Impacts

The interaction between soil and water can also cause erosion and transport of particulates carrying nutrients such as N and P. The erosion, transport and deposition of fine sediment are hydro-geomorphic processes, and although a natural phenomenon, excessive transfer of sediment into rivers may have environmental and economic impacts.

Agricultural activities including overgrazing, river bank erosion via animal poaching or management resulting in bare ground, have all been implicated in causing conditions that can lead to sediment erosion. This loss of soil may be increased when storm runoff events coincide with management phases of bare soil. Excessive loss of soil from agricultural landscapes preferentially depletes fine sediment, sediment-associated nutrients such as P and organic matter, all of which are crucial soil components required to sustain agricultural production (Creamer et al. 2010). Unsurprisingly, loss of high-value top soil from agricultural land surfaces can have agronomic and economic implications. From an environmental point of view, soil erosion and associated elevated sediment inputs, have been identified as a significant threat to freshwater ecosystems (Richter et al. 1997; Izagirre et al. 2009; Lawler et al. 2016).

Arable land is typically more susceptible to soil erosion than grassland (Van Oost et al. 2009; Regan et al. 2012). On arable land, periods of low groundcover between cropping cycles, or adjacent to row crops such as potatoes and maize are 'windows of opportunity' for soil erosion (Boardman and Favis-Mortlock 2014, Boardman 2015). Intensively managed grassland is, however, increasingly being identified as a source of sediment (Bilotta et al. 2010) for example, the increased likelihood of compaction and bare ground due to overgrazing associated with high stocking densities (Bilotta et al. 2007), increased runoff and associated transport of particle-bound pollutants, such as P (Haygarth et al. 2006).

The mass transfer of sediment fines in most Irish catchments can be associated with a high particulate P load which has been lost in high energy runoff events due to erosion of soil surfaces or the enrichment of fines from other sources (such as banks) from soluble P that has been lost in overland flow due to dissolution (Jordan et al. 2005)—or both mechanisms.

However, Sherriff et al. (2015) found that suspended sediment yields (SSY) in five Irish intensively managed agricultural catchments were low in comparison to similar catchment and landscape settings elsewhere in Europe. Suspended sediment yield was higher in catchments dominated by poorly drained soils than those with well-drained soils. Furthermore, on poorly drained soils, catchments with a greater proportion of arable land use have reported the highest annual average SSY. Well-drained soils dominated by arable crops did, however, show the potential to supply significant quantities of sediment. The authors concluded that the complexity of landscape features (a high density of field boundaries including, hedgerows, drainage ditches and small irregular field sizes) resulting in low connectivity may provide resilience to hillslope soil erosion and/or sediment transport, despite spatial dominance and intensity of agriculture. The diffuse transfer controls of both sediments and P from Irish catchments can, therefore, be linked in some or several ways and linked to the flashiness of the river system (Douglas et al. 2007).

16.5 Modelling Soil Water

Soil water is a critical component of the three-phase system, (solid, liquid and gaseous components which can be present in varying amounts) and influences soil strength, plant and microbial growth and nutrient transport. As water content fluctuates over time and due to the limitations of field instrumentation (cost and time), numerical models that simulate water and nutrient movement within soil have become popular alternatives. Such models mathematically simulate water movement using the Richards equation (Richards 1931), and may also incorporate equations for evapotranspiration, plant growth and chemical transport. Models can be either one-dimensional (representing a vertical soil profile), two-dimensional (representing a hill slope transect) or three-dimensional (representing both vertical and lateral water movement in a complex landscape). Models require input data including soil and solute properties and meteorological records or projections. The data requirements, complexity of the simulations and uncertainty of the results increase as the model user moves from one to three-dimensional approaches. Wagener et al. (2001) and Konikow (2011) discussed the importance of selecting a model which best reflects the processes and problems in question, while also taking data availability into consideration. A more complex or 'data hungry' model (e.g. a three-dimensional model) is not necessarily more accurate or suitable for a particular scenario than a relatively simple one-dimensional model. All models simplify soil characteristics (Fig. 16.7), and hence, should be considered a useful tool for understanding the environment, rather than being an exact representation of reality (Konikow 2011). This is particularly true given the popularity of soil databases (e.g. ROSETTA (Schaap et al. 2001) as a means of supplying input data.

Of particular interest in Ireland, is the modelling of nutrients, from sources such as surface applied fertilisers or septic tanks, through the soil to receptors such as groundwater, surface-waters and wells. Vero et al. (2014) used a numerical model (Šimůnek et al. 2013) supplied with soil data from the Waterford National Soil Survey to estimate the time lag or delay of nitrate transport to groundwater. The study reported that trends in response to water quality mitigation measures would first be observed in <1 year, for soil groups ranging from Surface-water Gleys to Podzols.

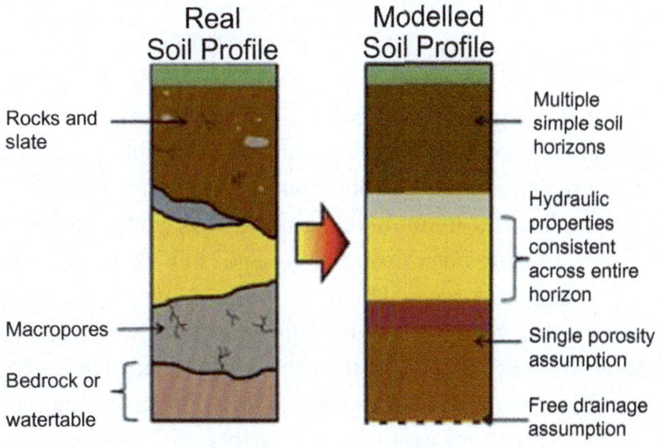

Fig. 16.7 Comparison of an actual soil profile and the simplification employed in numerical models

However, long-term time lags could approach three years for profiles up to 1.5 m in depth and would exceed this value for deeper profiles and heavier soils. It was also revealed that the quality of model input data becomes of greater importance where long-term time lags are of interest. Deep and multi-layered soil profiles have similarly increased requirements. Time lag ranges determined by Vero et al. (2014), particularly when coupled with groundwater time lags, could preclude certain agricultural catchments from the attainment of EU Water Framework Directive objectives within the set deadlines (2015, 2021 and 2027). This is an example of the utility of numerical models, which can aid policymakers and land managers in the effective design of environmental legislation based on scenario testing. Other applications include assessment of plant-water availability, flood behaviour, irrigation needs and temperature dynamics.

16.6 Conclusions

Soils in Ireland play an important role in nutrient retention and mobilisation that serves functions such as soil fertility in the supply of nutrients for primary productivity. The interaction between soil fertility and soil-water processes can direct nutrients into a hydrological pathway that is connected to a waterbody. This interaction can lead to surface or groundwater nutrient losses, which depends to a large extent on the mosaic of soil types that define the hydrological pathways present for nutrient transfer, at catchment scale. However, it is worth considering the influence that soil chemical attributes such as organic matter, Al, Fe and Ca can exert on nutrient mobilisation from the soil matrix into solution at the smaller process scales.

The Water Framework Directive (WFD) (Directive 2000/60/EC of the European Parliament and of the Council) sets specific requirements regarding the protection of existing satisfactory water resources and the restoration of unsatisfactory water resources to either Good, or High Ecological Status objectives. All water bodies are required to meet their WFD objectives by 2027. Under the WFD, agriculture is recognised as a pressure on water quality, specifically nutrient enrichment of surface and groundwaters from agricultural sources. N and P from animal and chemical sources when used in excess of crop demands can be transported to water bodies via hydrological pathways in soils; however, the role of soil-water interactions and soil chemistry that defines Irish soil types which has been shown to influence these losses, and are not considered in current water quality policy. Current statutory instruments aim to control the use of agricultural nutrients in isolation of these drivers of nutrient transfer. Mitigating against these factors will require better use of soil information and land use at catchment scale, and targeting measures to areas where soil type is driving nutrient losses. Furthermore, this will also require input from and working with the farming community to ensure that proposed measures are acceptable and cost-effective.

References

Boardman J (2015) Extreme rainfall and its impact on cultivated landscapes with particular reference to Britain. Earth Surf Proc Land 40:2121–2130

Boardman J, Favis-Mortlock DT (2014) The significance of drilling date and crop cover with reference to soil erosion by water, with implications for mitigating erosion on agricultural land in South East England. Soil Use Manag 30:40–47

Bilotta GS, Brazier RE, Haygarth PM (2007) The impacts of grazing animals on the quality of soils, vegetation, and surface waters in intensively managed grasslands. In: Donald LS (ed) Advances in agronomy, Academic Press, pp 237–280

Bilotta GS, Krueger T, Brazier RE, Butler P, Freer J, Hawkins JMB, Haygarth PM, Macleod CJA, Quinton JN (2010) Assessing catchment-scale erosion and yields of suspended solids from improved temperate grassland. J Environ Monit 12:731–739

Creamer RE, Brennan F, Fenton O, Healy MG, Lalor STJ, Lanigan GJ, Regan JT, Griffiths BS (2010) Implications of the proposed soil framework directive on agricultural systems in Atlantic Europe—a review. Soil Use Manag 26:198–211

Department of Agriculture, Food and the Marine (DAFM) (2015) Information on the LULUCF actions to limit or reduce emissions and maintain or increase removals from activities defined under Decision 529/2013/EU. Department of Agriculture, Food and the Marine, Dublin, Ireland

Daly K, Styles D, Lalor S, Wall D (2015) Phosphorus sorption, supply capacity and availability in soils with contrasting parent material and chemical properties. Eur J Soil Sci 66:792–801

Daly K, Jeffrey D, Tunney H (2001) The effect of soil type on phosphorus sorption capacity and desorption dynamics in Irish grassland soils. Soil Use Manag 17:12–20

Douglas RW, Menary W, Jordan P (2007) Phosphorus and sediment transfers in a grassland river catchment. Nutr Cycl Agroecosyst 77(199–212):3

Doody D, Moles R, Tunney H, Kurz I, Bourke D, Daly K, O'Regan B (2006) Impact of flow path length and flow rate on phosphorus loss in simulated overland flow from a humic gleysol grassland soil. Sci Total Environ 372:247–255

Fenton O, Mellander P-E, Daly K, Wall DP, Jahangir MMR, Jordan P, Hennessey D, Huebsch M, Blum P, Vero S, Richards KG (2017) Integrated assessment of agricultural nutrient pressures and legacies in karst landscapes. Agr Ecosyst Environ 239:246–256

Haygarth PM, Bilotta GS, Bol R, Brazier RE, Butler PJ, Freer J, Gimbert LJ, Granger SJ, Krueger T, Macleod CJA, Naden P, Old G, Quinton JN, Smith B, Worsfold P (2006) Processes affecting transfer of sediment and colloids, with associated phosphorus, from intensively farmed grasslands: an overview of key issues. Hydrol Process 20:4407–4413

Izagirre O, Serra A, Guasch H, Elosegi A (2009) Effects of sediment deposition on periphytic biomass, photosynthetic activity and algal community structure. Sci Total Environ 407(21):5694–5700

Jordan P, Menary W, Daly K, Kiely G, Morgan G, Byrne P, Moles R (2005) Patterns and processes of phosphorus transfer from Irish grassland soils to rivers—integration of laboratory and catchment studies. J Hydrol 304:20–34

Konikow LF (2011) The secret to successful solute-transport modelling. Ground Water 49(2):144–159

Lawler D, Rymszewicz A, Conroy L, O'Sullivan JJ, Bruen M, Turner J, Kelly-Quinn M (2016) SILTFLUX literature review. EPA, Dublin

Mellander P-E, Jordan P, Wall DP, Melland AR, Meehan R, Kelly C et al (2012) Delivery and impact bypass in a karst aquifer with high phosphorus source and pathway potential. Water Res 46:2225–2236

Mellander P-E, Jordan P, Shore M, Melland AR, Shortle G (2015) Flow paths and phosphorus transfer pathways in two agricultural streams with contrasting flow controls. Hydrol Process 29:3504–3518

Mellander P-E, Jordan P, Shore M, McDonald N, Wall DP, Shortle G, Daly K (2016) Identifying contrasting controls and surface water signals from groundwater phosphorus flux. Sci Total Environ 541:292–302

Regan JT, Fenton O, Healy MG (2012) A review of phosphorus and sediment release from Irish tillage soils, the methods used to quanitfy losses and the current state of mitigation practice. Biol Environ Proc R Irish Acad 112(B):157–183

Richards LA (1931) Capillary conduction of liquids through porous mediums. Physics 1(5):318–333

Richter BD, Braun DP, Mendelson MA, Master LL (1997) Threats to imperiled freshwater fauna. Conserv Biol 11(5):1081–1093

Schaap MG, Leij FJ, van Genuchten MTh (2001) ROSETTA: a computer program for estimating soil hydraulic parameters with hierarchical pedotransfer functions. J Hydrol 251:163–176

Sherriff SC, Rowan JS, Melland AR, Jordan P, Fenton O, Ó hUallacháin D (2015) Investigating suspended sediment dynamics in contrasting agricultural catchments using ex situ turbidity-based suspended sediment monitoring. Hydrol Earth Syst Sci 19:3349–3363

Šimůnek J, Šejna M, Saito H, Sakai M, van Genuchten M Th (2013) The HYDRUS 1D software package for simulating the movement of water, heat and multiple solutes in variably saturated media. Version 4.16 HYDRUS software series 3, Department of Environmental Science, University of California Riverside, Riverside, California, USA, p 340

Van Oost K, Cerdan O, Quine TA (2009) Accelerated sediment fluxes by water and tillage erosion on European agricultural land. Earth Surf Proc Land 34:1625–1634

Vero SE, Ibrahim TG, Creamer RE, Grant J, Healy MG, Henry T, Kramers G, Richards KG, Fenton O (2014) Consequences of varied soil hydraulic and meteorological complexity on unsaturated zone time lag estimates. J Contam Hydrol 170:53–67

Wagener T, Boyle DP, Lees MJ, Wheater HS, Gupta HV, Sorooshian S (2001) A framework of development and application of hydrological models. Hydrol Earth Syst Sci 5(1):13–26

Soils and Carbon Storage

17

Kenneth A. Byrne, Gary Lanigan, Rachel Creamer, and Florence Renou-Wilson

Abstract

Soils act as both sources and sinks of atmospheric C and as such there is great interest in investigating the impact of land use and land use change on C stocks in soils. Peatlands occupy ~20% of the irish landscape and store 75% of all C stocks. While pristine peatlands are long term C sinks, drainage and land use may change these systems to sources of soil C. Forests are generally recognised to be C sinks although the role of soil differs between soil types. Studies to date suggest that afforested Gley soils are C sinks while afforested Brown Earths may lose C. Grassland is the dominant land use and site based studies suggest that it could be a C sink of 0.5 t C ha^{-1} $year^{-1}$. Cropland is a net C source with cultivation being the principal driver of this loss. Options to maintain or enhance C stocks in tillage systems include the use of cover crops, straw and manure incorporation and a move to minimum tillage. Despite the progress that has been made in quantifying C stocks in Irish soils and in understanding the impact of land use on soil C stocks many information gaps remain.

Keywords

Carbon cycle • Soil carbon • Land use

K. A. Byrne (✉)
Department of Biological Sciences, School of Natural Sciences, University of Limerick, Limerick, Ireland
e-mail: Ken.Byrne@ul.ie

G. Lanigan · R. Creamer
Teagasc Environment Research Centre, Johnstown Castle, Wexford, County Wexford, Ireland

R. Creamer
Wageningen University and Research Atlas—Wageningen Campus, Wageningen, The Netherlands

F. Renou-Wilson
School of Biology and Environmental Science, University College Dublin, Belfield, Dublin 4, Ireland

R. Creamer and L. O'Sullivan (eds.), *The Soils of Ireland*, World Soils Book Series,
https://doi.org/10.1007/978-3-319-71189-8_17

17.1 Introduction

Since the industrial revolution, anthropogenic activities have caused a rapid increase in atmospheric carbon dioxide (CO_2) concentrations. This has stimulated scientific interest in Earth's carbon (C) stocks and flows. Despite the progress that has been made in both quantifying and understanding these C stocks and flows (Denman et al. 2007) further research is still required regarding the amount of C exchanged between terrestrial ecosystems and the atmosphere (van der Werf et al. 2009; Le Quéré et al. 2012). It is widely agreed that after fossil fuel emissions, land use and land use change is the next largest contributor of anthropogenic emissions to the atmosphere (IPCC 2007) and is the most uncertain component of the global C cycle (Houghton et al. 2012). Based on a review of published estimates Scharlemann et al. (2014) suggest that the global soil C stock is approximately 1500 pentagrams of C, but cautions that there are considerable differences between studies. Differences in estimates (which could also apply to national scale studies) can be due to a range of factors such as sampling period, sampling intensity, spatial resolution of soil data, differences in sampling methods (e.g. by depth or horizon), availability of bulk density data, as well as laboratory methods for determination of soil C content and differences in approaches to calculations. In addition, some studies may include inorganic C or varying amounts of stone content (Batjes 1996) (Fig. 17.1).

17.2 Carbon Stocks in Irish Soils

Ireland's temperate maritime climate with relatively high rainfall and moderate temperatures is particularly suited to the accumulation of soil organic matter (Collins et al. 2004) and, therefore, C. Despite the widespread recognition of the importance of the topic among soil scientists (e.g. Batjes 1996) and of the many services provided by soil organic

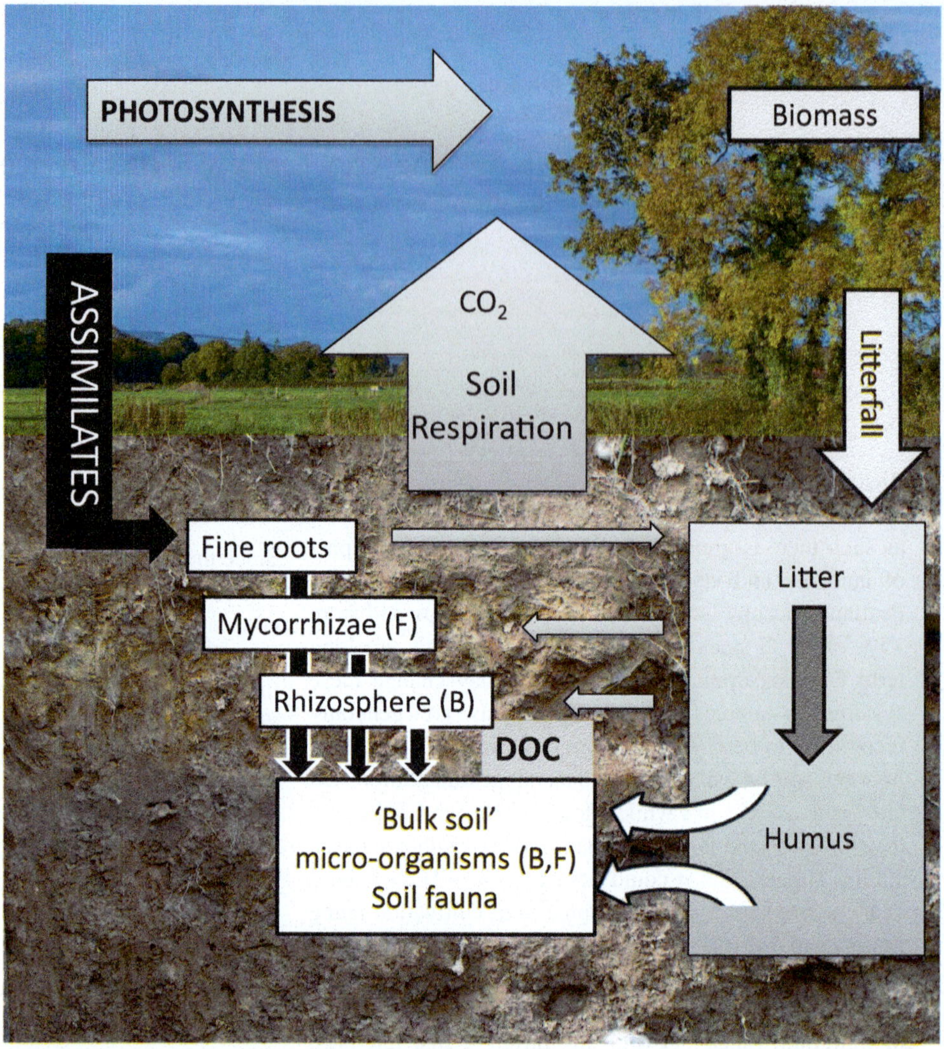

Fig. 17.1 Basic concept model of soil carbon dynamics in mineral soils adapted with permission, based on Kutsch et al. (2009) originally published by Cambridge University Press

matter, prior to the early 2000s, studies on C in Irish soils were limited (Brogan 1966; McGrath and Zhang 2003). The first national assessment of C stocks in Irish soils was carried out by Tomlinson (2005) who mapped total soil C stocks at 2 × 2 km resolution using soil maps (Gardiner and Radford 1980) and published soils data. Tomlinson (2005) estimated that total soil C stocks in 1990 and 2000 were 2048 and 2021 Mt, respectively. This study was limited by the lack of bulk density data for Irish soils but did establish a baseline for soil C stocks and underlined the importance of peat soils which were estimated to contain 53% of C stocks on 17% of the land area. Eaton et al. (2008) combined land cover data from CORINE with soil C density (a term which denotes a mass of C per unit area) values from England and Wales (Bradley et al. 2005) and N Ireland (Cruickshank et al. 1998) and estimated that soil C stocks to 1 m depth in 2000 were 1469 Mt with peatlands accounting for 36% of this. While these studies were valuable they lacked specific data on C stocks in Irish soils. Xu et al. (2011) addressed this by combining two databases (Fay et al. 2007; Kiely et al. 2010) and estimated national soil C stocks to 50 cm depth at 1474 ± 181 Mt. Within the last national soil survey, the Irish Soil Information Systems (SIS) project estimated the soil organic carbon (SOC) stocks for mineral soils in Ireland (Simo et al. 2016). Apart from the work of Hammond (1981), the full distribution, depth and associated bulk density data of peat soils have not yet been recorded and the production of a single integrated map that includes both peat soils and the Irish SIS data is recommended for future research (Simo et al. 2016). Reporting of SOC stocks relies upon suitable data on SOC content and soil bulk density, along with methods to upscale point data to comprehensive spatial estimates. Calculations of SOC stocks for the total profile of the Irish SIS data were based upon the classical method of calculating SOC stock for a given depth and summing the SOC stock of successive horizons in the profile. In the absence of measured data, the use of pedotransfer functions (PTF) described by Reidy et al. (2016) were used with Irish SIS data to calculate

SOC stock at soil horizon level to 1 m depth. Simo et al. (2016) applied the equation by Schrumpf et al. (2011) to estimate the SOC in each horizon (OCh):

$$\mathrm{OCh}\left(\mathrm{t\,ha^{-1}}\right) = \text{soil bulk density}\left(\mathrm{g\,cm^{-3}}\right) * \text{soil C concentration}(\%) * \text{soil thickness}(\mathrm{cm}) \quad (17.1)$$

where horizon depths did not align with the depth intervals selected to calculate SOC, the following equation was applied to calculate SOC stock to a certain depth:

$$\mathrm{SOC}(100\,\mathrm{cm}) = \mathrm{OCh_1} + \mathrm{OCh_2} + \left(\frac{\mathrm{OCh_3}}{\mathrm{Dh_3}}(100 - \mathrm{UBh_3})\right) \quad (17.2)$$

where OCh_1, Och_2, Och_3 = Organic C (mg ha^{-1}) calculated with Eq. 17.2 for individual horizons, Dh = depth and UB = Upper boundary in this horizon (h) (cm).

With this data, a number of methods can be applied to produce a SOC stock map. The maps presented in Figs. 17.2 and 17.3, were developed using a scored approach based upon the relative proportion of each subgroup within the soil association. The constituent subgroup SOC proportions were summed together to generate the soil association SOC stock (t ha^{-1}) value. Once validated, the values were plotted in the soil subgroup map of Ireland.

17.3 Irish Soils and Carbon Storage Potential

Figure 17.4 presents a conceptual model for the soil carbon (C) sequestration function. This refers to the rate of a 'permanent' increase in SOC and comprises the transfer and retention of atmospheric carbon dioxide (CO_2) into the SOC pool (Lal 2006). The rate of soil C sequestration is highly variable and is influenced by a range of factors including climate, soil type, land use and management practices (Ostle et al. 2009).

While peat soils provide the largest stocks of SOC, the sequestration rate is slow compared to other land uses. Degradation of peat can very quickly result in a substantial loss of SOC through implementation of drainage systems. The plantation or natural occurrence of forestry can result in a large SOC store, however, management practices associated with afforestation can have short to mid-term impacts on the long-term storage of carbon. Permanent grassland systems are considered an essential store of SOC, due to the extensive area of grassland land use in Ireland. The wetter grassland soils (with higher clay content) tend to store greater amounts of SOC compared to the drier, more freely draining soils, which tend to be more aerated resulting in greater decomposition of SOC. Finally, arable soils are generally low in SOC content, due to aeration of the surface horizons following tillage.

17.3.1 Peatlands

Peatlands have accumulated and stored C in the peat over thousands of years. Irish peatlands developed in the post-glacial period (Holocene) after the recession of the ice sheets some 10,000 years ago and have accumulated peat and thus C, since then, although at variable rates. In peatlands, the decay of the organic matter occurs at a faster rate within the aerobic zone, above the water level than in the anaerobic layer below the water level. However, because of continuous, albeit very slow decomposition of peat (unless frozen), this C store will diminish if no additional organic material is formed above. Therefore, active peat formation in peatlands and thus a positive C balance is a prerequisite for the long-term maintenance of the peat C store.

Peatlands store C in different parts of their ecosystem (biomass, litter, peat layer, mineral subsoil layer and pore water), with each pool having its own dynamics and turnover. The peat layer is a main long-term store of C as peat largely consists of organic material (by definition more than 30% of the dry mass), with a C content ranging from 48 to 63% (Heathwaite and Gottlich 2003)—an average of 51% can be calculated from various Irish peat analyses (Hammond 1981; Tomlinson 2005; Renou-Wilson et al. 2008). Recent estimates show that worldwide, peatlands hold on average 1125 t C ha^{-1} (total 450 Gt) which is the largest C density of any terrestrial ecosystem (Joosten et al. 2016). Estimates in C pools vary widely depending on assumptions of the extent and depth of peatlands, as well as lack of adequate data on the bulk density of different peat deposits (Charman 2002; Turunen et al. 2002). However, despite those uncertainties, the global peatland C store represents approximately a third of the global soil C pool (Carter and Scholes 2000). The total C content in the peat worldwide is also equivalent to 75% of all atmospheric C (mainly in the form of CO_2 and CH_4)—estimated to be 750 billion tonnes in 1990 (Houghton 1997).

The extent of the soil C store in Irish peatlands is also a source of large uncertainties due to lack of relevant data. Using the General Soil Map (Gardiner and Radford 1980) and a series of county maps to calculate the extent of peatlands and relying on disparate data sources to estimate peat depth, Tomlinson (2005) estimated that Irish peatlands had a C stock of 1071 Mt C. Eaton et al. (2007) estimated the peat C stock to be almost 50% higher at 1503 Mt, using the peat bog category from CORINE 2000, C density from (Cruickshank et al. 1995) and the average depth values from Hammond (1981). Of relevance, these two estimates account

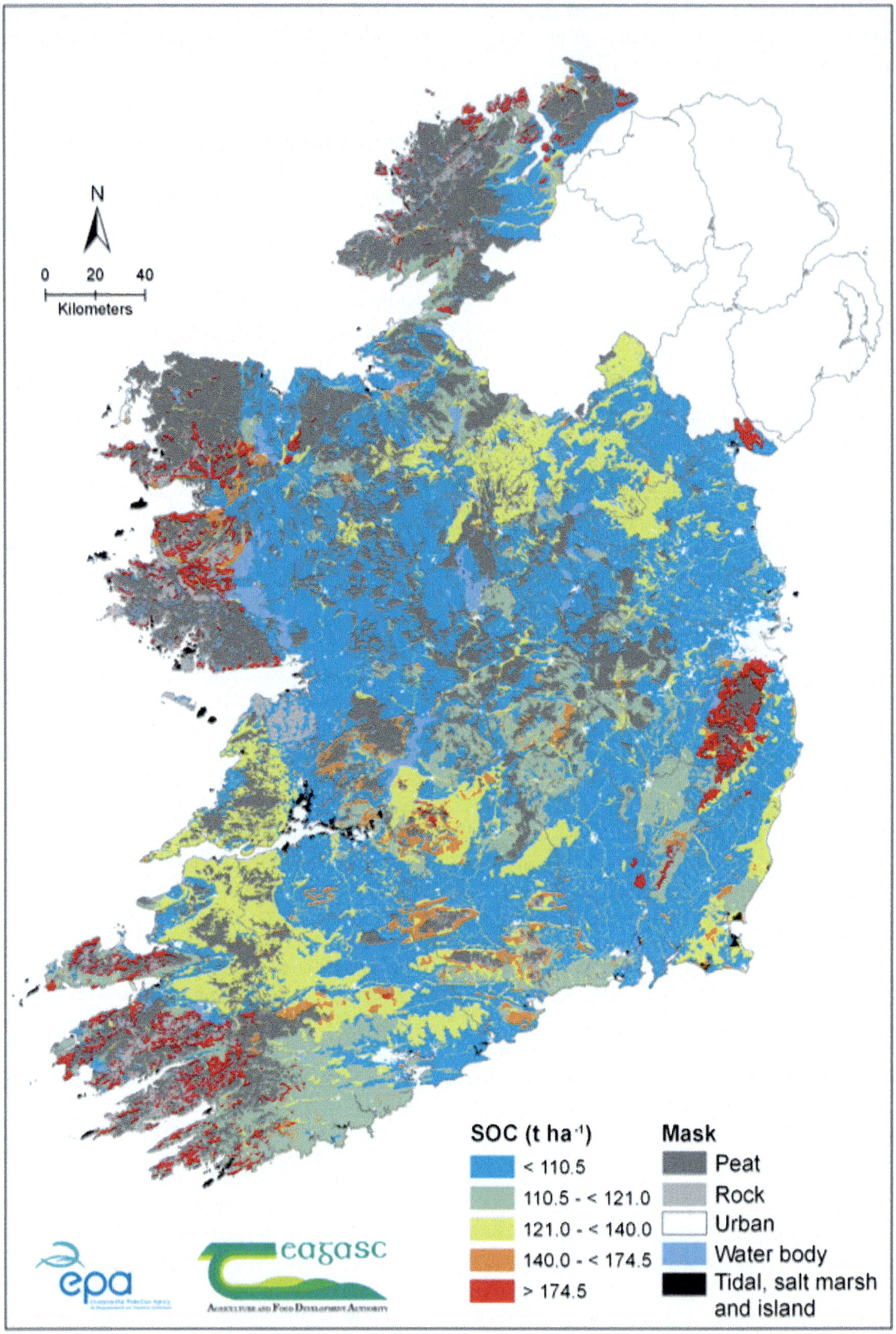

Fig. 17.2 Soil organic carbon stock map of Ireland for the top 50 cm. Map from Simo et al. (2016) based upon a weighting of the subgroup constituents of the Irish Soil Information Systems data (EPA STRIVE programme), summed together for soil associations and developed within ArcGIS software

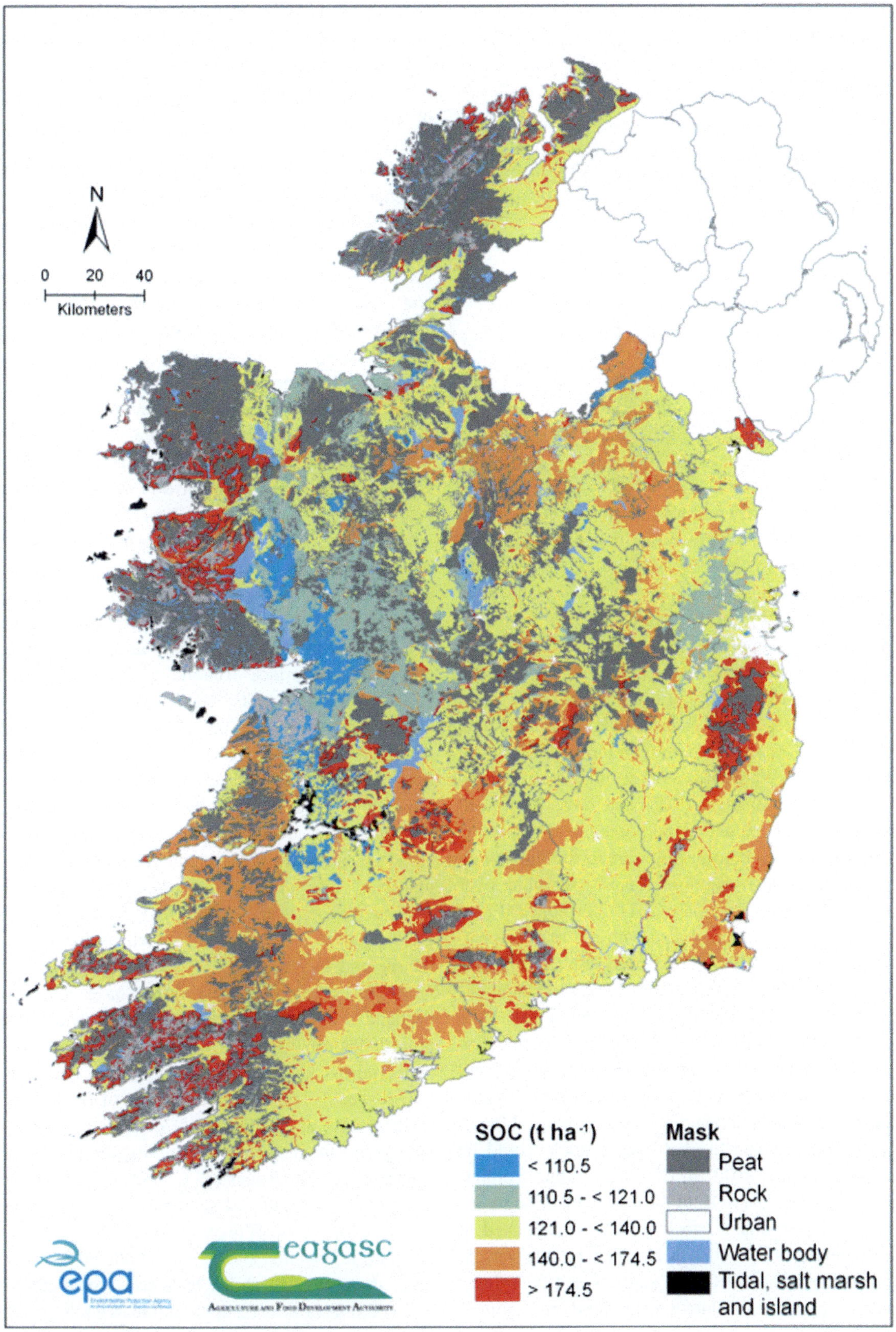

Fig. 17.3 Soil organic carbon stock map of Ireland for the top 1 m. Map from Simo et al. (2016) based upon a weighting of the subgroup constituents of the Irish Soil Information Systems data (EPA STRIVE programme), summed together for soil associations and developed within ArcGIS software

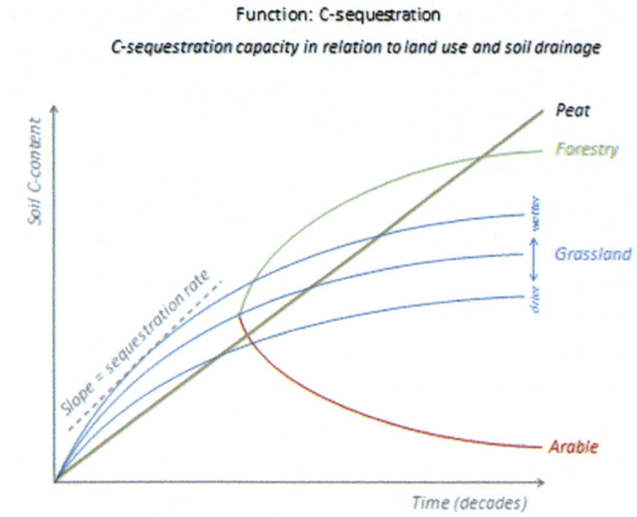

Fig. 17.4 Conceptual model of soil carbon sequestration from Coyle et al. (2016)

for 53% (Tomlinson 2005) and 62% (Eaton et al. 2008) of the total soil organic C stock present in Ireland. Cruickshank et al. (1998) calculated the peat SOC stock for Northern Ireland to be 162.1 Mt C, accounting for 42% of the total soil C stock on only 15% of the land area.

Given these high proportions, it is critical to improve the accuracy of these estimates by firstly, improving the peat soil map and secondly, by developing a model to predict peat depths as the data is usually inconsistent in quality and coverage, an issue raised by both Tomlinson (2005) and Eaton et al. (2008). While intact raised bog depth can be reasonably modelled (Ingram 1982) and have been well investigated for peat extraction purposes, peat depth on blanket peatlands varies considerably due to the undulating base mineral surface and, therefore, requires sampling over large areas. Holden and Connolly (2011) developed a Peat Depth Model (PDM) and, combined with the map of peatland spatial extent, the DIPMV2 (Connolly et al. 2009), they estimated that blanket peatland in the Wicklow Mountains contained 2.3 Mt C. This compares to the previously published values ranging from 0.45 Mt C (Eaton et al. 2008) to 2.18 Mt C (Tomlinson 2005). This suggests that the peatland SOC stock for Ireland to date has been well underestimated. If the modelled data used in this study are extrapolated for all blanket bogs, the national blanket bog C stock would be estimated to be around 1073 Mt C. When adding Tomlinson's figure for basin peat SOC to the new blanket bog SOC figure, the total peat soil C stock would be estimated at 1566 Mt C (Renou-Wilson et al. 2011). New remote sensing platform together with new C data analysis and peat depth will be refined in the near future which would lead to more accurate estimates of Ireland's peat SOC stocks.

Without a doubt, Irish peatlands are a huge C store, likely containing more than 75% of all the soil organic C in Ireland. Such resource should, therefore, be carefully managed as the size and functioning as well as disturbance of any large C pool is of importance to future climate change.

In order for the C stock to continue to increase, the rate of C sequestration in peatlands should be larger than the total rate of C losses. Large variations regarding both rates have been recorded spatially and temporally. There has been a great endeavour to understand processes of C sequestration and losses in Irish peat soils, the most important of which concerns the amounts and controls of gas emission, mainly CO_2 and CH_4.

Natural Irish peatlands (those with little human impact) act as long-term CO_2 sinks (Glencar Atlantic blanket bog C budget (gaseous and fluvial) averaged nearly 30 g C m^{-2} $year^{-1}$) although the annual increment may be relatively small and may shift between a C sinks and sources from year-to-year (Koehler et al. 2011). Overall, natural peatlands act as long-term C stores primarily as a result of a persistently high water table, which creates conditions within the peat, whereby the amount of C fixed by the peatland vegetation during photosynthesis is greater than that released through ecosystem respiration, CH_4 emissions and leaching/surface runoff of dissolved organic C (DOC) and fire events. GHG dynamics are significantly altered when a peatland undergoes a change in land use; all of which generally involve the same fundamental ecosystem changes (i.e. drainage and lowered water table levels) thus producing essentially similar negative effects. Increased emissions of carbon dioxide (CO_2) and N_2O, and reduced methane (CH_4) emissions have been widely reported for drained grasslands on organic soils (Renou-Wilson et al. 2012, 2014), for industrial peatlands (Wilson et al. 2007, 2012) as well as forested peat soils (Byrne and Farrell 2005).

Recent studies of GHG fluxes have also shown that peatland rewetting/restoration can bring back some of the biogeochemical functions of natural peatlands. A rewetted industrial blanket bog in Bellacorick (Co. Mayo) was monitored over 5 years and was a sink of 95 g C m^{-2} $year^{-1}$ (Wilson et al. 2016). Of particular significance in rewetted/restored sites is the ability for certain vegetation assemblages to sequester C. Since certain vegetation assemblages are better at sequestering C than others (Dunn et al. 2016; Renou-Wilson et al. 2016), plant community structure in rewetted sites could be altered in order to maximise the peatland's ability to store C. Such ecosystem service could be prioritised in the future as a mitigation measure against climate change (Ward et al. 2013) (Fig. 17.5).

Fig. 17.5 Rewetted industrial blanket bog at Bellacorick Co. Mayo. Photograph courtesy of Florence Renou-Wilson

17.3.2 Forestry

Since the beginning of the twentieth century, the area of forest in Ireland has grown steadily from less than 1% of the land area to 11.1% or 731,652 ha (John Redmond, personal communication). This is dominated by plantations, 57% of which are in public ownership (Redmond et al. 2007). Exotic conifer species account for 74% of the total area and broadleaf species for the balance. Peat is the dominant soil type with blanket peat accounting for 30.25% of the total area with a further 12.2% on basin peat and 1% on cutaway peat. Gleys, Podzols, Brown Earths and Brown Podzolics account for 25.3, 10.4, 8.0 and 5.3%, respectively with the balance being represented by small areas on other soil types (Forest Service 2013). Along with undisturbed peat soils, the maintenance and establishment of forestry ecosystems (natural or plantation), especially hardwood species, offer the greatest C sequestration potential in the long term (Coyle et al. 2016; Wigley and Schimel 2000; Kirk 2004; Xiong et al. 2014). This is further enhanced on poorly drained, fine-textured soils (Coyle et al. 2016).

Afforestation is generally considered to create a net C sink through the sequestration of C in biomass and soil organic matter (Byrne 2010). However, the net C balance will change across temporal and spatial scales depending on soil type, species and management. Site preparation for afforestation or reforestation involving drainage and cultivation such as mounding can alter the moisture status of the soil. Wellock et al. (2014) compared soil C stocks in a chronosequence of ash (*Fraxinus excelsior* L.) to a reference grassland site. While the grassland has a soil C stock to 30 cm depth of 90.2 t C ha^{-1} afforestation caused a decline to 66.7 t C ha^{-1} 27 years after which these losses were reversed with a return to 71.3 t C ha^{-1} at 47 years. Site preparation at these sites may have accelerated organic matter decomposition and soil C stocks continued to decline until these losses were compensated by organic matter inputs from litterfall and root turnover. Other studies to date provide contrasting results. Reidy and Bolger (2013) measured soil C stocks across a chronosequence of Sitka spruce (*Picea sitchensis* (Bong.) Carr.) afforestation on Gley soils in central Ireland and estimated that on average 1.83 t C ha^{-1} was sequestered in the soil. Working in the same sites and using a combination of forest inventory, C gas flux measurement and modelling assessments, Black et al. (2009) estimated that soil C stocks increased by 2.2–2.5 t C ha^{-1} $year^{-1}$ across the chronosequence. Selecting sites for such a study with identical soil properties and similar pre-afforestation land use is very difficult and suggests that differences in site topography can influence soil C accumulation. Drainage and afforestation of peat soils shift the balance of organic matter accumulation towards increased decomposition and C loss (Minkkinen et al. 2008). Despite the significance of peat soils in Ireland relatively little information is available about the C balance of peatland forests. Byrne and Farrell (2005) measured total soil respiration in a range of afforested sites on blanket peat and found a large range in values from 1 to 2.6 t C ha^{-1} $year^{-1}$. While heterotrophic respiration was not measured the authors suggest that it could be similar to total respiration of 1.4–1.6 t C ha^{-1} $year^{-1}$ measured at clearfell sites in the same study area. Studies on other types of peatland are limited but suggest that soil CO_2 emissions could be higher on more fertile sites. Byrne et al. (2007) measured soil respiration in an afforested cutaway peatland of residual woody fen/*Phragmites peat* overlying a sub-peat mineral soil consisting of glacial till and clay. The rate of C loss due to decomposition of the residual peat, root biomass and forest floor was estimated to be 7.20 t C ha^{-1} $year^{-1}$. Despite these high losses the authors suggested that the site was a net C sink due to the high productivity (and therefore high C uptake) of the forest stand. More recently, Lane (2016) assessed soil respiration in first rotation Sitka spruce on Peaty Gley sites in south-west Ireland and estimated that CO_2 emissions due to soil organic matter decomposition were 2.8–5.2 t C ha^{-1} $year^{-1}$. Based on studies to date there is good evidence that organic soils have the potential to be C sources while recognising the need for further studies to assess above- and below-ground organic matter inputs through litterfall, root turnover and harvest residues. Mineral soils have the potential to be short-term sources following site preparation and afforestation but further work is needed across a range of soil types and forest sites to assess C stocks and organic matter properties. Other areas in need of attention are soil C in reforested sites and the impact of the removal of harvest residues (for energy generation) on soil C (Fig. 17.6).

17.3.3 Grassland

A large number of global studies have demonstrated that grasslands act as a C sink with the rate of C sequestration strongly dependent upon the management of these ecosystems. European temperate grasslands have been shown to be a C sink, with sequestration rates of circa 1 t C ha^{-1} $year^{-1}$ (Soussana et al. 2004; Gottschalk et al. 2007). Research indicates that SOC concentrations of Irish mineral grassland soils range from 3.2 to 6.3% (Kiely et al. 2010; Xu et al. 2011). Using combined empirical and GIS approaches, Khalil et al. (2013) estimated the SOC density from 0 to 100 cm depth for Irish grasslands on mineral soils with an organic surface layer to be 213.8 t C ha^{-1}. In addition, almost half of the SOC stocks are in physically or chemically recalcitrant forms with evidence that there may also be a vertical movement of C down through the soil profile (Torres-Sallan et al. in press).

Annual net C sequestration, either calculated from flux data as the net biome productivity (NBP) of the system, or from direct measurements of soil organic C have indicated a range of values from −1 t C ha^{-1} $year^{-1}$ to +2 t C ha^{-1} $year^{-1}$, with Irish pastures circa −0.5 t C ha^{-1} $year^{-1}$ (where negative values indicate net CO_2 uptake, Soussana et al. 2004). As Irish managed grasslands constitute 90% of utilisable agricultural area, this is clearly a considerable sink. However, under UNFCCC rules only net-net sequestration is allowable (i.e. additional sequestration relative to a baseline year, for example, 2005). This makes verification of the impact of management especially challenging as there are small C inputs (0.25 t C ha^{-1}) entering into a very large background sink (200 t C ha^{-1}). Long-term trials, such as the Park Grass experiment at Rothamsted, England and Hillsborough, Northern Ireland, indicate that optimal fertilisation (nitrogen, phosphorus, potassium) soil pH as well as slurry additions can increase SOC in grassland. However, once stocks reach equilibrium, new C additions may only serve to prime the microbial community, and any C inputs to the system are respired over a short period of time. Low-intensity grazing (as opposed to silage cuts) has also been shown to enhance field-scale C balance (Klumpp et al. 2011; Ni Chonchubhair et al. 2016). However, at higher stocking rates (>2 LU ha^{-1}) this effect is negated. Drainage of both histic (>20% OC) and humic (>3.5% OC corrected by clay content—see Creamer et al. 2014) soils can also result in a net SOC loss (O'Sullivan et al. 2015) due to slaking of aggregates, which in turn, results in increased mineralisation of passive, slow-turnover C pools. However, in the case of peat soils, net CO_2 loss associated with

Fig. 17.6 Peaty Podzol, Devil's Bit, Co. Tipperary. Photo courtesy of Michael Clancy

drainage may be counterbalanced by a reduction in nitrous oxide (N_2O) emissions, Renou-Wilson et al. (2014) reported that grassland over nutrient-rich peat was a CO_2 source of 233 g C m^{-2} $year^{-1}$ compared to nearly neutral in a nutrient-poor shallow drained grassland over peat.

17.3.4 Tillage

Cropland soils generally store less SOC than grazing land because cropland has greater disturbance from cultivation, a lack of organic manure being returned to the system, a winter fallow period and as a consequence, has less root and shoot material returned to the soil. Across the European Union (EU), croplands are a small net C source of 0.17 t C ha^{-1} $year^{-1}$ (Ciais et al. 2010). In Ireland, continuous spring barley crops have been observed to be a net C source of circa 1 t C ha^{-1} $year^{-1}$ (Ceschia et al. 2010). The mechanism for SOC loss from ploughing is hypothesised to be due to a combination of increased decomposition of SOC due to soil aeration and soil aggregate destruction and a reduction in aggregate formation (Oades 1984; Six et al. 1998, 1999). The following options are proposed to maintain or enhance SOC levels in tillage systems.

- Cover crops/rotations: Crop rotations that include cover crops, perennial grasses and legumes maximise soil C inputs and maintain a high proportion of active C. Because the growth of cover crops represents an additional input of organic C into the farming system compared to leaving the land fallow, they present the opportunity to increase or slow the decline of SOC in tillage situations, especially as there is a strong relationship between length of the fallow period and C loss from the system. Cover crops are slow to increase the total SOM/SOC levels, but increase the active pools quicker.
- Straw and manure incorporation: Straw incorporation increases SOC as organic matter is directly input back into the soil. In the UK, measured SOC increases were

equivalent to c. 22% of the crop residue/straw incorporation (up to 32 t OC ha^{-1} over 23 years) (Bhogal et al. 2009). Manure inputs have also been shown to build SOC stocks, depending on the source, with both farmyard manure (FYM) and green compost considered to have been particularly effective (Bourdin et al. 2014).

- Reduced/minimum tillage: The concept of no-till or reduced tillage is that aggregates are disrupted less, leading to reduced SOC loss. However, while studies in Oak Park, Carlow have demonstrated that net C fluxes and SOC levels in the top 30 cm are increased there is evidence that ploughing may simply redistribute SOC over a greater depth profile and when looked at over a full soil profile, to say 1 m depth, often no difference is observed between ploughing and reduced tillage (Abdalla et al. 2014).

17.3.5 Bioenergy Crops

Perennial energy crops have the potential to sequester additional C in soil biomass if established on former cropland (0.44 Mg soil C ha^{-1} $year^{-1}$ for poplar and willow and 0.66 Mg soil C ha^{-1} yr^{-1} for miscanthus). However, there were no positive or even negative effects on the C balance if energy crops are established on former grassland (Don et al. 2012). As miscanthus is a C4 crop, it has been possible to utilise ^{13}C stable isotope tracing to examine the rate of input of C derived directly from miscanthus into the soil. Given that the net C flux of miscanthus is very high (over 6 t C ha^{-1} $year^{-1}$, Ni Chonchubhair et al. 2016), there were no observed differences in total SOC between grassland and bioenergy plots. However, 79% of labile particulate organic matter was derived from miscanthus, indicating that C from the crop was entering the high turnover C pools and being utilised relatively rapidly. There were no differences in recalcitrant C pools 10 years post-conversion, even when tillage and perennial bioenergy was compared (Zimmerman et al. 2012).

17.3.6 European and Global Goals Relating to Soil Carbon Sequestration

The EU is committed to tackling climate change and more than ever before, the role of soil in climate change mitigation is under consideration. Key climate and energy targets within the EU are set in the 2020 climate and energy package and the 2030 climate and energy framework. By 2030, the overall EU target is a reduction of 40% on 1990 greenhouse gas emissions. Ireland will have to reduce its emissions by 30% by that time, relative to 2005 emissions. In October 2014, the European Council recognised the lower mitigation potential of agriculture and the land use sector along with the need to ensure coherence between the EU's food security and climate change objectives. Under the Paris Agreement and EU 2030 Climate Package, C sinks can contribute to greenhouse gas emissions reduction, thereby broadening the options for greenhouse gas emissions reductions. Ireland will have 5.6% (of the 2005 base year) flexibility from land use, larger than any other member state except Latvia. There is, however, a challenge in terms of MRV (measurement, reporting and verification) of these stocks, and in particular, the impact of management on these C stocks. The change in C inputs relative to the large C stocks already present in the soil require high spatial resolution in terms of sampling or a long time period (decades) between sampling. This, in turn, requires the establishment of long-term experiments in order to deconvolute the impacts of altered management on SOC/soil organic matter.

References

Abdalla M, Hastings A, Helmy M, Prescher A, Osborne B, Lanigan G, Forristal D, Killi D, Maratha P, Williams M, Rueangritsarakul K, Smith P, Nolan P, Jones MB (2014) Assessing the combined use of reduced tillage and cover crops for mitigating greenhouse gas emissions from arable ecosystem. Geoderma 223–225:9–20

Batjes NH (1996) Total carbon and nitrogen in the soils of the world. Eur J Soil Sci 47:151–163

Bhogal A, Nicholson FA, Chambers BJ (2009) Organic carbon additions: effects on soil bio-physical and physico-chemical properties. Eur J Soil Sci 60:276–286

Black K, Byrne KA, Mencuccini M, Tobin B, Nieuwenhuis M, Reidy B, Bolger T, Saiz G, Green C, Farrell EP, Osborne B (2009) Carbon stocks and stock changes across a Sitka spruce chronosequence on surface water gley soils. Forestry 82(3):255–272

Bourdin F, Sakrabani R, Kibblewhite MG, Lanigan GJ (2014) Effect of slurry dry matter content, application technique and timing on emissions of ammonia and greenhouse gas from cattle slurry applied to grassland soils in Ireland. Agr Ecosyst Environ 188:122–133

Bradley RI, Milne R, Bell J, Lilly A, Jordan C, Higgins A (2005) A soil carbon and land use database for the United Kingdom. Soil Use Manag 21:363–369

Brogan JC (1966) Organic carbon in Irish pasture soils. Irish J Agric Res 5(2):169–176

Byrne KA (2010) The role of plantation forestry in Ireland in the mitigation of greenhouse gas emissions. Irish Forestry 67(1–2):86–96

Byrne KA, Farrell EP (2005) The effect of afforestation on soil carbon dioxide emissions in blanket peatland in Ireland. Forestry 78 (3):217–227

Byrne KA, Cabral R, Farrell EP (2007) *Commercial afforestation*. In: Wilson D, Farrell EP (eds) CARBAL. Carbon gas balances in industrial cutaway peatlands in Ireland (Final report). Bord na Móna, Newbridge, County Kildare

Carter AJ, Scholes RJ (2000) Spatial global database of soil properties. IGBP global soil data task CD-ROM. International Geosphere-Biosphere Programme (IGBP) Data Information Systems, Toulouse, France

Ceschia E, Béziat P, Dejoux JF, Aubinet M, Bernhofer Ch, Bodson B, Buchmann N, Carrara A, Cellier P, Di Tommasi P, Elbers JA, Eugster W, Grünwald T, Jacobs CMJ, Jans WWP, Jones M, Kutsch W, Lanigan G, Magliulo E, Marloie O, Moors EJ, Moureaux C, Olioso A, Osborne B, Sanz MJ, Saunders M, Smith P, Soegaard H, Wattenbach M (2010) Management effects on net ecosystem carbon and GHG budgets at European crop sites. Agric Ecosyst Environ 139:363–383

Charman D (2002) Peatlands and environmental change. Wiley, Chichester

Ciais P, Wattenbach M, Vuichard N, Smith P, Piao SL, Don A, Luyssaert S, Janssens IA, Bondeau A, Dechow R, Leip A, Smith PC, Beer C, Van Der Werf GR, Gervois S, Van Oost K, Tomelleri E, Freibauer A, Schulze ED, Carboeurope Synthesis Team (2010) The European carbon balance. Part 2: croplands. Glob Change Biol 16(5):1409–1428

Collins JF, Larney FJ, Morgan MA (2004) Chapter 6: climate and soil management. In: Keane T, Collins JF (eds) Climate, Weather and Irish Agriculture. Agmet, Dublin, pp 119–160

Connolly J, Holden NM, Ward SM (2009) Mapping peat soils in Ireland: updating the derived Irish peat map. Irish Geogr 42(3):343–352

Coyle C, Creamer RE, Schule RPO, O'Sullivan L, Jordan P (2016) A functional land management conceptual framework under soil drainage and land use scenarios. Environ Sci Policy 56:36–48

Creamer R, Simo I, Reidy B, Carvalho J, Fealy R, Hallett S, Jones R, Holden A, Holden N, Hannam J, Massey P, Mayr T, McDonald E, O'Rourke S, Sills P, Truckell I, Zawadzka J. Schulte R (2014) Irish soil information systems synthesis report. EPA Strive Programme, Wexford. Available for download from http://www.epa.ie/pubs/reports/research/land/EPA%20RR130%20WEB.pdf

Cruickshank MM, Tomlinson RW, Bond D, Devine PM, Edwards CJW (1995) Peat extraction, conservation and the rural economy in Northern Ireland. Appl Geogr 15(4):365–383

Cruickshank MM, Tomlinson RW, Devine PM, Milne RM (1998) Carbon in the vegetation and soils of Northern Ireland. Biology and environment. Proc R Ir Acad 98:9–21

Denman KL, Brasseur G, Chidthaisong A et al (2007) Couplings between changes in the climate system and biogeochemistry. In: Solomon S, Qin D, Manning M et al (eds) Climate change 2007: the physical science basis. Contribution of working group I to the fourth assessment report of the intergovernmental panel on climate change. Cambridge University Press, Cambridge, UK, NY, USA, pp 499–587

Don A, Osborne BA, Hastings A, Skiba U, Carter MS, Drewer J, Flessa H, Freibauer A, Jones MB, Lanigan GJ, Mander U, Monti A, Valentine J, Walter T, Zenone T (2012) Land-use change to bioenergy production in Europe: implications for the greenhouse gas balance and soil carbon. Glob Change Biol Bioenergy 4:372–391

Dunn C, Graham J, Roberts TS, Freeman C (2016) Plant Species effects on the carbon storage capabilities of a blanket bog complex. Wetlands 36:47–48

Eaton JM, McGoff NM, Byrne KA, Leahy P, Kiely G (2007) The impact of agricultural land cover change on soil organic carbon stocks in Ireland. Cost action 639. In: Jandl R, Olsson M Greenhouse-gas budget of soils under changing climate and land use (burnout). Federal Research and Training Centre for Forests, Natural Hazards and Landscape (BFW), Vienna, pp 75–78

Eaton JM, McGoff NM, Byrne KA, Leahy P, Kiely G (2008) Land cover change and soil organic carbon stocks in the Republic of Ireland 1851–2000. Clim Change 91(3):317–334

Fay D, McGrath D, Zhang C, Carrigg C, O'Flaherty V, Carton OT, Grennan E (2007) Towards a national soil database. Final report. Environmental Protection Agency, Johnstown Castle, Ireland

Forest Service (2013) The second national forest inventory—Republic of Ireland—results. Forest Service, Department of Agriculture, Food and the Marine, Johnstown Castle Estate, Co. Wexford, Ireland

Gardiner MJ, Radford T, (1980). Soil associations of Ireland and their land use potential: explanatory bulletin to soil map of Ireland. Soil survey bulletin no. 36. An Foras Talúntais, Dublin

Gottschalk P, Wattenbach M, Neftel A, Fuhrer J, Jones M, Lanigan G, Davis P, Campbell C, Soussana J-F, Smith P (2007) The role of measurement uncertainties for the simulation of grassland net ecosystem exchange (NEE) in Europe. Agr Ecosyst Environ 121(1–2):175–185

Hammond RF (1981) The peatlands of Ireland. Soil survey bulletin no. 35. An Foras Talúntais, Dublin, Ireland

Heathwaite AL, Gottlich K (eds) (2003) Mires—process, exploitation and conservation. Chichester, Wiley

Holden NM, Connolly J (2011) Estimating the carbon stock of a blanket peat region using a peat depth inference model. Catena: doi: https://doi.org/10.1016/j.catena.2011.1002.1002

Houghton J (1997) Global Warming. The complete briefing. Cambridge University Press, Cambridge, U.K.

Houghton RA, House JI, Pongratz J et al (2012) Carbon emissions from land use and landcover change. Biogeoscience 9:5125–5142

Ingram HAP (1982) Size and shape in raised mire ecosystems: a geophysical model. Nature 297(5864):300–303

IPCC (2007) Climate change 2007: synthesis report. contribution of Working Groups I, II and III to the fourth assessment report of the intergovernmental panel on climate change. IPCC, Geneva, Switzerland

Joosten H, Sirin A, Couwenberg J, Laine J, Smith P (2016) The role of peatlands in climate regulation. In: Bonn A, Allott T, Evans M, Joosten H, Stoneman R (eds) Peatland restoration and ecosystem services—science, policy and practice. Cambridge University Press, pp 66–79

Khalil MI, Kiely G, O'Brien P, Müller C (2013) Organic carbon stocks in agricultural soils in Ireland using combined empirical and GIS approaches. Geoderma 193–194:222–235

Kiely G, Eaton J, McGoff N, Xu X, Leahy P (2010) Soil carbon measurement and modelling of soil carbon stocks and stock changes in Irish soils (Final report). Environmental Protection Agency, Johnstown Castle, Ireland

Kirk G (2004) The biogeochemistry of submerged soils. Wiley, NJ

Klumpp K, Tallec T, Guix N, Soussana JF (2011) Long-term impacts of agricultural practices and climatic variability on carbon storage in a permanent pasture. Glob Change Biol 17(3534–3545):2011

Koehler A-K, Sottocornola M, Kiely G (2011) How strong is the current carbon sequestration of an Atlantic blanket bog? Glob Change Biol 17:309–319

Kutsch WL, Bahn M, Heinemeyer A (2009) Soil carbon relations: an overview. In: Kutsch WL, Bahn M, Heinemeyer A (eds) Soil carbon dynamics: an integrated methodology. Cambridge University Press, pp 1–15

Lal R (2006) In: Bhatti JS, Lal R, Apps MJ, Price MA (eds) Climate change and managed ecosystems. CRC, Taylor and Francis, UK, USA

Lane RJ (2016) Soil carbon dynamics in afforested organo-mineral soils. PhD thesis, University of Limerick, Ireland

Le Quéré C, Andres RJ, Boden T et al (2012) The global carbon budget 1959–2011. Earth Syst Sci Data Discuss 5:1107–1157

McGrath D, Zhang CS (2003) Spatial distribution of soil organic carbon concentrations in grassland of Ireland. Appl Geochem 18:1629–1639

Minkkinen K, Byrne KA, Trettin CC (2008) Climate impacts of peatland forestry. In: Strack M (ed) Peatlands and climate change. International Peat Society, Jyväskylä, pp 98–122

Ni Chonchubhair O, Osborne B, Finnan J, Lanigan G (2016) Comparative assessment of ecosystem C exchange in Miscanthus and reed canary grass during early establishment. GCB Bioenergy. https://doi.org/10.1111/gcbb.12343

O'Sullivan L, Creamer RE, Fealy R, Lanigan G, Simo I, Fenton O, Carfrae J, Schulte RPO (2015) Functional land management for managing soil functions: a case-study of the trade-off between primary productivity and carbon storage in response to the intervention of drainage systems in Ireland. Land Use Policy 47:42–54

Oades JM (1984) Soil organic matter and structural stability: mechanisms and implications for management. Plant Soil 76:319–337

Ostle NJ, Levy PE, Evans CD, Smith P (2009) UK land use and soil carbon sequestration. Land Use Policy 47:42–54

Redmond J, Gallagher G, Černý M, Russ R (2007) Ireland's nation forest inventory—results. In: Nieuwenhuis M, Redmond J, O'Donovan C (eds) Proceedings of NFI conference, National Forest Inventory Republic of Ireland. Forest Service, Johnstown Castle Estate, Co. Wexford, Ireland

Reidy B, Bolger T (2013) Soil carbon stocks in a Sitka spruce chronosequence following afforestation. Irish Forestry 70(1–2)

Reidy B, Simo I, Sills P, Creamer RE (2016) Pedotransfer functions for Irish soils—estimation of bulk density (ρb) per horizon type. SOIL 2:25–39

Renou-Wilson F, Keane M, McNally G, O'Sullivan J, Farrell EP (2008) BOGFOR programme—final report: a research programme to develop a forest resource on industrial cutaway peatlands in the Irish midlands. Coford, Dublin

Renou-Wilson F, Bolger T, Bullock C, Convery F, Curry JP, Ward S, Wilson D, Müller C (2011) BOGLAND—sustainable management of peatlands in Ireland. Johnstown Castle, Co. Wexford, STRIVE Report No 75 prepared for the Environmental Protection Agency (EPA): 157

Renou-Wilson F, Wilson D, Müller C (2012) Methane emissions from peat soils under grassland: impact of rewetting. In: 14th International Peat congress. International Peat Society, Stockholm, Sweden

Renou-Wilson F et al (2014) The impacts of drainage, nutrient status and management practice on the full carbon balance of grasslands on organic soils in a maritime temperate zone. Biogeosciences 11 (16):4361–4379

Renou-Wilson F, Müller C, Moser G, Wilson D (2016) To graze or not to graze? Four years GHG balances and vegetation composition from a drained and a rewetted organic soil under grassland. Agric Ecosyst Environ 222:156–170

Scharlemann JPW, Tanner EVJ, Hiederer R, Kapos V (2014) Global soil carbon: understanding and managing the largest terrestrial carbon pool. Carbon Manag 5(1):81–91. https://doi.org/10.4155/cmt.13.77

Schrumpf M, Schulze ED, Kaiser K, Schumacher J (2011) How accurately can soil organic carbon stocks and stock changes be quantified by soil inventories? Biogeosciences 8:1193–1212. https://doi.org/10.5194/bg-8-1193-2011

Simo I, Creamer RE, O'Sullivan L, Reidy B, Schulte RPO, Fealy RM (2016) Irish soil information systems: soil properties map final technical report 18. EPA STRIVE programme, Wexford. Available for download from http://erc.epa.ie/safer/reports

Six J, Elliott ET, Paustian K, Doran JW (1998) Aggregation and soil organic matter accumulation in cultivated and native grassland soils. Soil Sci Soc Am J 62:1367–1377

Six J, Elliott T, Paustian K (1999) Aggregate and soil organic matter dynamics under conventional and no-tillage systems. Soil Sci Soc Am J 63:1350–1358

Soussana J-F, Loiseau P, Vuichard N, Ceschia E, Balesdent J, Chevallier T (2004) Carbon cycling and sequestration opportunities in temperate grasslands. Soil Use Manag 20:219–230

Tomlinson RW (2005) Soil carbon stocks and changes in the Republic of Ireland. J Environ Manage 76:77–93

Torres-Sallan G, Schulte RPO, Lanigan G, Byrne KA, Reidy B, Simo I, Six J, Creamer RE (in press) Carbon distribution in subsoil aggregates differs between soil types. Scientific Reports

Turunen J, Tomppo E, Tolonen K, Reinikainen A (2002). Estimating carbon accumulation rates of undrained mires in Finland—application to boreal and subarctic regions. Holocene (12):69–80

van der Werf GR, Morton DC, DeFries RS et al (2009) CO_2 emissions from forest loss. Nat Geosci 2:737–738

Ward S, Ostle N, Oakley CJ, Quirk H, Henrys PA, Bardgett RD (2013) Warming effects on greenhouse gas fluxes in peatlands are modulated by vegetation composition. Ecol Lett. doi:https://doi.org/10.1111/ele.12167 (18 Aug 2013)

Wellock ML, Rafique R, LaPerle CM, Peichl M, Kiely G (2014) Changes in ecosystem carbon stocks in a grassland ash (*Fraxinus excelsior*) afforestation chronosequence in Ireland. J Plant Ecol 7 (5):429–438

Wigley TML, Schimel DS (2000) The carbon cycle. Cambridge University Press

Wilson D, Tuittila E-S, Alm J, Laine J, Farrell EP, Byrne KA (2007) Carbon dioxide dynamics of a restored maritime peatland. Ecoscience 14(1):71–80

Wilson D, Renou-Wilson F, Farrell C, Bullock C, Müller C (2012) Carbon restore—the potential of Irish peatlands for carbon uptake and storage. Climate change research programme. Report series no. 17 prepared for the Environmental Protection Agency, Johnstown Castle, Co. Wexford, Ireland by University College Dublin: 32

Wilson D, Farrell C, Fallon D, Moser G, Muller C, Renou-Wilson F (2016) Multi-year greenhouse gas balances at a rewetted temperate peatland. Glob Change Biol. https://doi.org/10.1111/gcb.13325

Xiong X, Grunwald S, Myers DB, Ross CW, Harris WG, Comerford NB (2014) Interaction effects of climate and land use/land cover change on soil organic carbon sequestration. Sci Total Environ 493:974–982

Xu X, Liu W, Zhang C, Kiely G (2011) Estimation of soil organic carbon stock and its spatial distribution in the Republic of Ireland. Soil Use Manag 27(2):156–162

Zimmermann J, Dauber J, Jones MB (2012) Soil carbon sequestration during the establishment phase of Miscanthus × giganteus: a regional-scale study on commercial farms using 13C natural abundance. Glob Change Biol Bioenergy 4:453–461

18 The Living Soil: Biodiversity and Functions

Olaf Schmidt, Tom Bolger, Rachel Creamer, Fiona Brennan, and Alan D. W. Dobson

Abstract

Soil biodiversity encompasses an enormous array of life on the planet. Soil organisms are essential for most processes and functions in the soil. Soil biological knowledge is critical for understanding functions such as nutrient supply to plants, carbon sequestration and greenhouse gas emissions, all of which are key to meeting the global challenges of food security and climate change mitigation. Soil microorganisms include a wide array of bacterial, archaeal and eukaryotic taxa. Microbes are extremely diverse and abundant, with up to 10 billion microorganisms predicted in a single gram of soil. Nematodes provide a good case study of a soil faunal group that encompasses all feeding habits within their own taxonomic group. Their different feeding habits mean that nematodes provide many and different connections in soil food webs. Microarthropods are discussed, as representing a well-studied soil animal group in Ireland in terms of their community ecology and biogeography, with several species of the Mesostigmata identified. Knowledge of larger animals such as earthworms is more complete, with 27 species recorded from Ireland.

O. Schmidt (✉)
UCD School of Agriculture and Food Science, University College Dublin, Belfield, Dublin 4, Ireland
e-mail: olaf.schmidt@ucd.ie

T. Bolger
UCD School of Biology and Environmental Science, University College Dublin, Belfield, Dublin 4, Ireland

R. Creamer
Wageningen University and Research Atlas—Wageningen Campus, Wageningen, The Netherlands

F. Brennan
Teagasc Environment, Soils and Land-Use Department, Johnstown Castle, County Wexford, Ireland

A. D. W. Dobson
School of Microbiology and Environmental Research Institute, University College Cork, Lee Road, Cork, Ireland

Keyword

Soil biology • Soil ecology • Soil organisms
Soil biodiversity

18.1 Introduction

Soil biodiversity encompasses a huge array of life on the planet. The variety of soil biodiversity is astounding, ranging from bacteria, which are from 1–100 μm in size (i.e. completely invisible to the eye) through to the macrofauna, such as earthworms, ants, woodlice and centipedes, which are 2 mm or larger in size and can be easily seen (Jeffery et al. 2010; Orgiazzi et al. 2016). Soil organisms are essential for most soil processes and functions including nutrient cycling, organic matter breakdown, maintenance of soil structure, support of plant growth, and pest and disease suppression (Curry 1994; Bardgett 2005).

In terms of agricultural production, the intensification of management systems has led, in some cases, to reduced soil biodiversity due to increased mechanisation, addition of nutrients and the use of monoculture systems. Examples from Ireland are presented by Curry et al. (2002) and Schmidt et al. (2001) who studied earthworm populations in intensively managed potato and wheat production systems. Sustainable farming approaches acknowledge the key role of soil biodiversity for the production of food and fibre, as well as the delivery of ecosystem services (Coleman et al. 2004). Whilst much of the soil biota remains unexplored, there are some key groups of soil organisms that are generally well studied and known to be important in the delivery of these soil functions. These groups include bacteria and fungi (due to the large number present in soil and their roles especially in nutrient cycling); nematodes (they include root feeders but also antagonists of plant pests as well as decomposer species that regulate nutrient cycling); and earthworms which, being the physical engineers of the soil, are responsible for the movement of mineral particles, organic matter and other

R. Creamer and L. O'Sullivan (eds.), *The Soils of Ireland*, World Soils Book Series,
https://doi.org/10.1007/978-3-319-71189-8_18

organisms in the soil, thereby exerting a strong influence on the soil structure.

The soil habitat is a highly structured, in many respects unique environment, where interactions between the soil biological communities are based upon access to food resources, air and water (Coleman et al. 2004). Different species live in different parts of the soil, some in water films, others in air spaces, and their feeding strategies (which together form food webs) can be very complex. The food webs that exist below ground are made up of organisms that eat either living plants or dead organic material and the organisms that in turn eat them. Food webs are used to describe the feeding preferences of all animals within the soil and how they interact, but this not only tells us about who eats who but also about the carbon and nitrogen flows taking place in the soil (Dix 2017).

A food web with a more complex structure, is believed to be more efficient and resilient to stresses like drought or disturbance, with many pathways for energy and nutrients to move along as they are transformed from organic material to a form that plants can use (Creamer et al. 2016; Morriën et al. 2017). Thus, sustainable soil management aims to achieve as diverse a below-ground community as possible, in terms of species, feeding groups, genetic diversity and representation of functional 'traits' that define ecological roles.

Within soil food webs, certain species are so important that they are often called 'keystone species' or 'ecosystem engineers' that modify—through activities such as feeding and burrowing—the soil environment and resource availability to other soil organisms and plants (Jones et al. 1994). These species tend to be the most studied taxa and they can be useful as 'bioindicators' reflecting the overall biological status of soils (Bouffaud et al. 2016; Dirilgen et al. 2016).

There is a great need to further our knowledge on soil biology, biodiversity and ecological functions, and to understand how the diversity found in different soil types or under land managements can alter the soil functions we require as a society for the delivery of 'ecosystem services'. Some soil types are of national importance in Ireland but are underrepresented in the international literature, requiring a greater national research effort. Pertinent examples include; oligotrophic and mineratrophic peat soils (raised bogs, fens) and reclaimed cut-over peatlands (Dickinson and Dooley 1969; Bolger et al. 2002; Curry and Schmidt 2006).

18.2 Soil Microbiome

Microorganisms underpin the majority of functions provided by soil, and thus research investigating the soil microbiome is increasingly recognised as essential to the sustainable management and use of soils. Microbes are the main drivers of nutrient cycling in soils, playing an active role in nutrient supply to plants. They are the main decomposers of organic matter, they can transform gaseous nitrogen into plant-available forms (but also plant-available forms into gaseous forms), they increase the availability of certain nutrients for plants, and they produce essential vitamins and plant hormones (Swift et al. 1979; Coleman et al. 2004). Consequently, knowledge of the activity of important microbial functional groups in soil is critical to understanding nutrient supply to plants, carbon sequestration and greenhouse gas emissions, all of which are key to meeting global challenges of food security and mitigating climate change. The soil microbiome is intrinsically important for plant establishment and productivity, even determining which plants can establish in which soil. It is also important for the resistance and resilience of the soil system to abiotic stresses, for stimulating plant defences, and for preventing pathogen invasions. The latter two functions are especially important for economic and food security reasons as it is estimated that between 20 and 40% of total crop production globally is lost due to a combination of disease pathogens and pests (Savary et al. 2012). In addition to plant pathogens, the soil environment plays an important role in the transmission of human and animal pathogens, and insights into their fate within soil can offer opportunities for the development of mitigation options to break the cycle of infection (Keith et al. 2016).

In Ireland, as elsewhere, an understanding of the soil microbiome was largely hampered until recently by technological limitations; insights from culture-based techniques into the composition and functioning of the soil microbiome as a whole have been restricted. However, with the advent of new molecular and sequencing methodologies (see next section), microbial ecology research has been revolutionised and major scientific advances have been made globally within the last 10–20 years. These include a recent complete revision of the tree of life (Hug et al. 2016) and the identification of a number of pathways and new organisms involved in major nutrient cycles (Strous et al. 1999; Treusch et al. 2005; Daims et al. 2015; van Kessel et al. 2015).

From a microbiological perspective, soils represent the most complex environment on the planet. Soils are highly dynamic, and both temporally and spatially heterogenous. Determining factors in the microbial environment, including gradients of environmental variables such as aerobic status, nutrient availability and pH, can change rapidly over small distances (within microns) and within very short time spans. Soil particles also have large surface areas that provide a habitat to soil microbes, for instance, 1 g of clay soil can contain up to 24 km^2 of surface area within the pore space (Jeffery et al. 2010). The huge surface areas present in soil combined with the inherent spatial and temporal variability enables soil to support an enormous number and diversity of ecological niches, and consequently of microbial communities. Microbial populations do not exist in spatially segregated niches, rather soil is composed of complex interdependent communities with a cascade of lifestyles.

Major research efforts in soil microbiology are currently underway in Ireland. Examples of particularly active research areas are studies investigating the fate of enteric pathogens in soils (Brennan et al. 2010a, b; 2014); studies quantifying the impacts on the soil microbiome of management, vegetation, and fertiliser, lime and biochar applications (Fox et al. 2016; Massey et al. 2016; Moynihan et al. 2013, 2015; Randall 2016); as well as studies identifying the microbes and pathways that underpin nutrient cycling, carbon sequestration and greenhouse gas losses (Ernfors et al. 2014; Fox et al. 2014; Gahan and Schmalenberger 2014, 2015; Jones et al. 2014; Samad et al. 2016). Another strand of research has focused on the exploitation of soil microorganisms in the degradation and detoxification of pollutants (Lillis et al. 2009; Courtney et al. 2014; Sawulski et al. 2015).

18.3 Measuring Soil Microbial Diversity and Functions

Microbial communities within soil are very diverse, and this diversity is reflected in the crucial roles they play in regulating soil fertility, plant health and in the cycling of carbon, nitrogen and other important nutrients. Each gram of soil contains up to 10 billion microorganisms of wide taxonomic diversity including bacterial, archaeal and eukaryotic taxa (Knietch et al. 2003). This diversity is mirrored by the diversity of their protein-encoded functions, encompassing a vast array of physiologies and potential life history strategies. However, less than about 1% of this microbial diversity is cultivable by traditional techniques. There are different methods which can be applied to assess the abundance, diversity and function of the microbial communities in soils. Genetic methods describe the DNA make-up of the microbial community, through the extraction and amplification of DNA and RNA material. Phenotypic methods identify specific signatures on the cell walls called phospholipid fatty acids, which can be used to identify different components of the microbial community, such as Gram-positive and Gram-negative bacteria, actinomycetes, ectomycorrhizal fungi, etc. Finally, the third kind of method that can be applied to the microbial community is the measurement of the activity of the microbes, this includes measures such as microbial biomass, respiration and enzyme activity (Coleman et al. 2004).

More recently, metagenomics-based approaches (Schloss and Handlesman 2003) have been applied to study the genetic material recovered directly from the soil. This has allowed microbial ecologists to characterise the taxonomic, phylogenetic and functional diversity of soil microbial communities to an extent that would have been unthinkable up until very recently (Delmont et al. 2011).

This increased capacity to explore soil microbial communities has allowed many important scientific questions to be addressed in relation to soil microbial ecology, including: What is the full extent of soil microbial diversity, both functionally and phylogenetically? How do microbial communities interact with and influence one another either synergistically or antagonistically? Which microorganisms are linked to which soil habitats? How do microbial abundances change with changing physical and chemical conditions? Do highly diverse microbial communities confer resilience in soils? Does horizontal gene transfer play a role in the stability of microbial communities? And the topical issue of how sensitive microbial communities are likely to be to climate change?

Function-based metagenomic approaches involve generating a soil metagenomic library and the subsequent screening of this library for the presence and expression of specific functions and activities, such as novel biocatalysts, natural products and bioactive small molecules with potential biopharmaceutical applications (Fig. 18.1). Two examples from Ireland are the novel phytases isolated from a metagenomic library derived from a soil cultivated with winter wheat (*Triticum aestivum*) at the Teagasc Oak Park research centre in Co. Carlow (Tan et al. 2014) and an antibacterial protein isolated from a metagenomic library from an organic field trial site; again from Teagasc Oak Park (O'Mahony et al. 2015).

18.4 Soil Fauna

Our knowledge of the diversity of animals that live below ground is, generally, much less complete than that of above-ground animals. This is not surprising since most soil animals are very small, they are hard to observe, difficult to study in their natural environment and superficially do not have the aesthetic appeal of above-ground animal groups that have been studied by natural historians and ecologists for a long time. Soil invertebrate animals include many taxa, from microscopic protozoa (single celled organisms, now classified as their own kingdom Protista) and nematodes, to microarthropods, to the so-called macrofauna including woodlice, millipedes, insects (often larval stages) and earthworms (Curry 1994; Orgiazzi et al. 2016). To pick one taxon from this list, it has been estimated that globally there could be between 40,000 and 10,000,000 species of nematodes (Yeates et al. 2009), that occur in all terrestrial soils with abundances of up to a million individuals per m^2 in some soils (Vervoort et al. 2012).

Fig. 18.1 General scheme for functional metagenomic-based approaches to exploit the soil microbiome

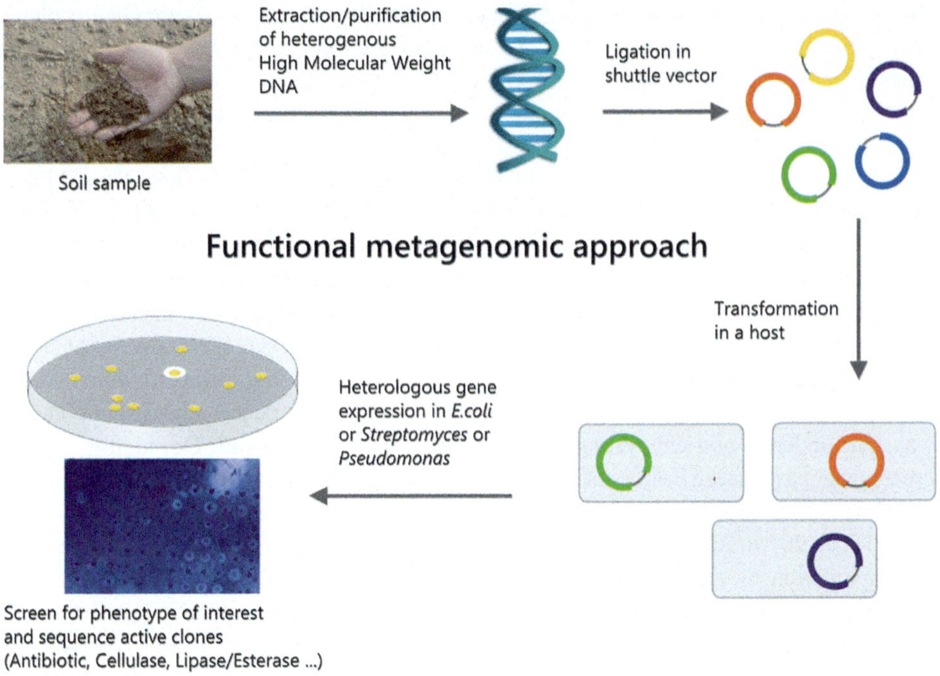

Soil animals have varied feeding strategies, from herbivores feeding on roots, to decomposers feeding on decaying organic matter, to omnivores and predators of other animals. Some animal groups even encompass all possible feeding habits within their own taxonomic grouping. Nematodes (phylum Nematoda) are a fine example of this because they feed on a wide variety of different food types within the soil, with each feeding type having specialised mouthparts (Fig. 18.2). Some nematodes eat bacteria; others feed on fungi, whilst some species feed on living plant roots, carnivorous nematodes predate on protozoa and other nematodes, whilst others can eat all of these things (Neher 2010; Yeates et al. 2009). These feeding habits mean that nematodes provide many and different connections in soil food webs, contributing to the complexity and functions of food webs in the cycling of carbon, nitrogen and other nutrients. A recent study conducted in Ireland (on samples from Scotland) used stable isotope ratio analysis of carbon and nitrogen to assign free-living soil nematodes to feeding groups (Melody et al. 2016).

Our knowledge on soil fauna is often a result of studying pest dynamics rather than the beneficial role of the soil fauna. For instance, soil nematodes have been studied for many years as plant parasites, which are damaging to our crops, such as the root-feeding 'eelworms' (Moore 1977). However, there is such a variety of nematodes found in the soil that they also provide many beneficial services such as insect parasitic species that we now use as agents in the biological control of various pests (Griffin et al. 1991), and free-living predatory soil nematodes (Keith et al. 2009) that are important members of virtually all soil food webs, to name just a few examples. The latter study highlighted how little we know about predatory soil nematodes by adding 11 new species (out of a total of 13 recorded) for the nematode order Mononchida for the island of Ireland.

Our knowledge of soil biodiversity is probably more complete for larger animals such as earthworms (phylum Annelida) because they are easier to see, they are easily recognisable animals for land users and the general public, and they are known widely to be beneficial for soil structure, water and air movement, decomposition and nutrient cycling (Curry 1994). For all of these reasons, it is to be expected that our knowledge about these animals, their distribution and abundance, response to land management practices and also their functions, is more advanced compared to the knowledge about other soil animal taxa. This is borne out by the wealth of published studies on earthworms from Ireland, represented by 27 species in the family Lumbricidae. In particular, many studies have assessed earthworm populations and species composition in Irish forest, grassland as well as arable soils under various cropping systems (Bolger et al. 2002; Curry et al. 2002, 2008; Schmidt and Curry 1999; Schmidt et al. 2001; Keith et al. 2012). In spite of this interest, new species records are still being added to the Irish list (Muldowney and Schmidt 2002; Melody and Schmidt 2012). This information has been utilised as part of a wider European study on the abundance of earthworms across different countries in Europe (Rutgers et al. 2016, Fig. 18.3).

The proper application and implementation of biodiversity monitoring and data recording activities to soil animals are in their infancy (Orgiazzi et al. 2016). The power and usefulness of good biodiversity data records was illustrated

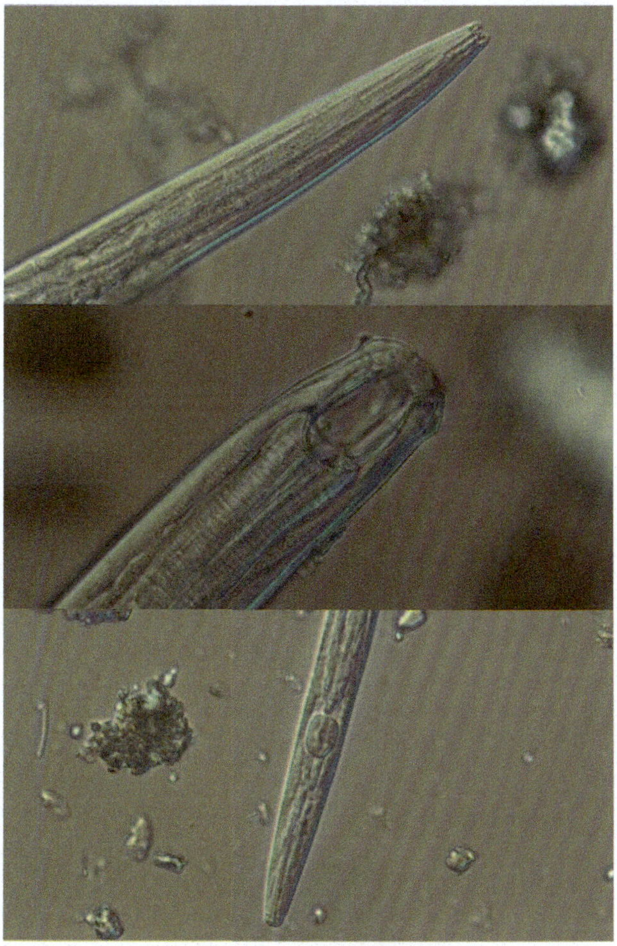

Fig. 18.2 Nematodes with specialised feeding parts from Irish grassland soils: top *Panagrolaimus* (bacterivorous), centre *Mononchus* (predatory) and bottom *Aphelenchoides* (fungivorous). Photos by: Tamas Salinki, Wageningen University and Research

compellingly by the recent publication of the first continental mapping exercise for any soil animal taxon, mapping European earthworms (Rutgers et al. 2016). Participation in this exercise by Ireland was largely made possible by the CréBeo project that assessed soil biodiversity at 60 sites in the Republic of Ireland (Keith et al. 2012), augmented by other published population studies. A similar effort was conducted in Northern Ireland on the invasive 'New Zealand flatworm' (*Arthurdendyus triangulatus*), that feeds on earthworms. The published country maps (Fig. 18.3) and accompanying paper offer new insights into the size, composition and environmental drivers of earthworm communities in Ireland as compared to that in other European countries.

The fact that most studies cited in this chapter were on single soil animal species or on specific taxa on their own belies the reality of soil food webs and the complexity and multi-species interactions they entail. Few studies in Ireland have attempted to study direct interactions between soil animals and soil microbiota, again earthworms feature amongst these. For instance, Thakuria et al. (2010) described the bacteria that live on the gut walls of earthworms, Stromberger et al. (2012) tried to identify all microbial taxa living in the walls of earthworm burrows and Schmidt et al. (2016) quantified the carbon flow from photosynthetic microorganisms on the soil surface to earthworms (and springtails). Linking the above and below ground, Muldowney et al. (2003) investigated possible correlations between the abundance of badgers and earthworms in grassland-dominated habitats in County Kilkenny.

Microarthropods are an example of a relatively well-studied soil animal group in Ireland in terms of abundances, habitat preferences and species records. Microarthropods are the most abundant arthropods in temperate soils and pores and do not burrow (Swift et al. 1979). They comprise taxa such as Pseudoscorpiones, Protura, Diplura and Symphyla which, whilst being relatively widespread, are never abundant, and Collembola and Acarina which are very abundant and diverse. Works by Legg and O'Connor (1997) and Cawley (2009) have provided an overview of the biogeography of the pseudoscorpions in Ireland but no comprehensive work has been done on their ecology. Even less has been published on the Protura, Diplura and Symphyla in Ireland (e.g. Blackith and Good 1991). However, there is an extensive literature on the two most abundant groups, the Acarina (mites) and Collembola. In fact, several species of soil mite (Mesostigmata) new to science have been described based on Irish material (e.g. Halbert 1915; Evans 1982).

In Ireland, mites and Collembola are common and abundant in virtually all soils. In pioneering population studies in Ireland, Curry (1969) recorded abundances of up to 106,000 m^{-2} mites and 105,360 m^{-2} Collembola in an old grassland at Celbridge, Co Kildare. These are amongst the highest abundances ever recorded in grassland and may not be typical of grasslands in Ireland. For example, the numbers found in a similar, albeit more intensely managed, site at Grange, Co. Meath were only 22,000 and 28,313 m^{-2} for mites and Collembola respectively (Bolger and Curry 1980). There is also significant seasonal variation in abundance. In a grassland at Kilmore, Co. Wexford, numbers of mites varied between 88,000, 126,000 and 68,000 m^{-2} in January, May and September of the same year and the abundances of Collembola were similarly variable at 62,200, 59,640 and 16,200 m^{-2} (Bolger and Curry 1984).

Despite these variations, abundances differ between habitats. Mites are often more abundant in forests and peatland. For instance, abundances as high as 112,590 m^{-2} were found in a Norway Spruce stand in Kilkenny (Heneghan and Bolger 1996) and of 59,303 m^{-2} from a birch stand on peat (Table 18.1). However, because they are so interrelated, it is difficult to separate the effects of habitat and soil type, but habitat does appear to have a more pronounced

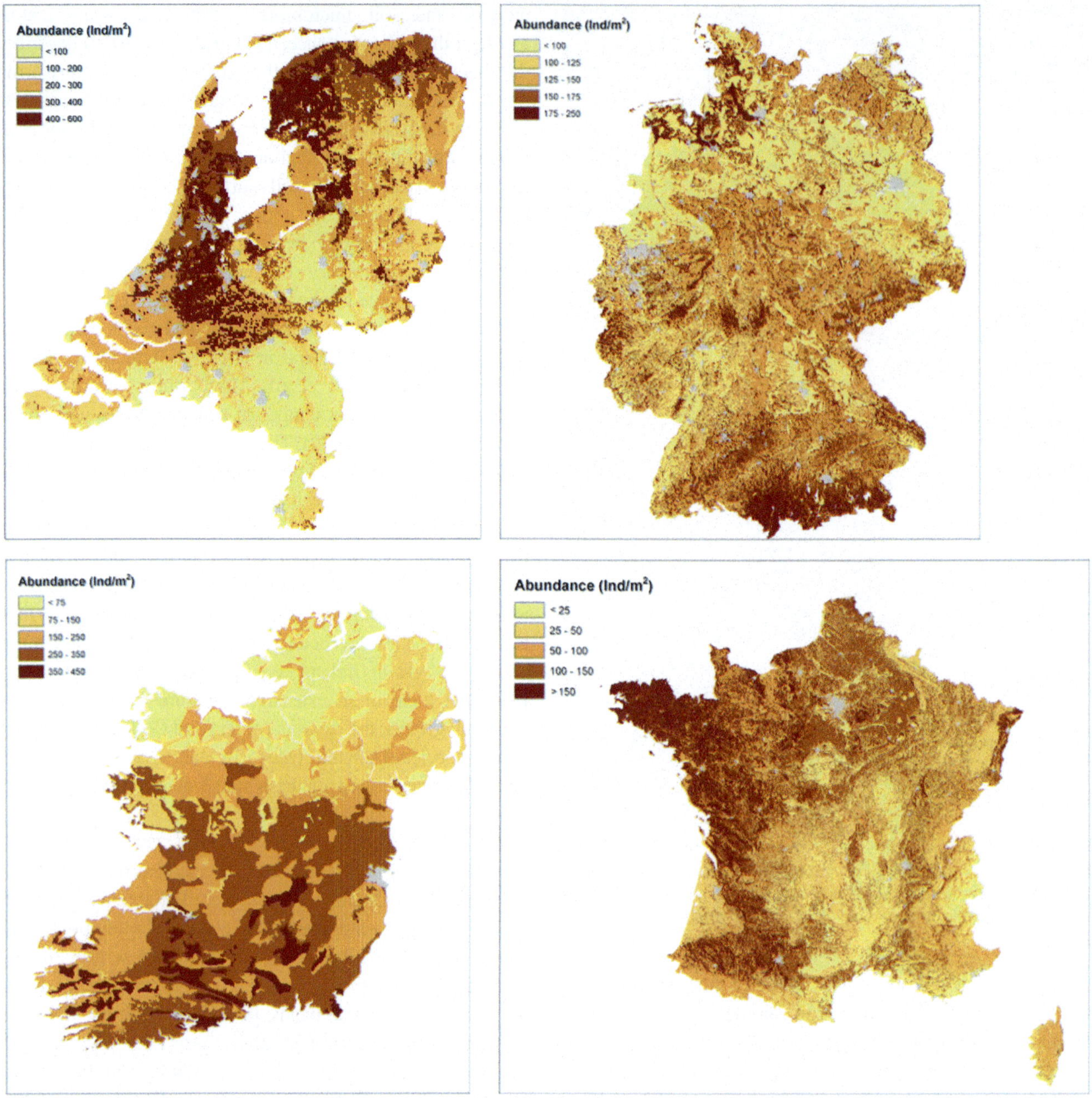

Fig. 18.3 Earthworm abundance maps created by country-specific models for Ireland, France, Germany and the Netherlands. The ranges in the legends for these maps are chosen to depict existing gradients optimally. Reprinted with permission from Elsevier from Rutgers et al. (2016) Copyright © 2015 Elsevier B.V.

effect than soil type per se. For example, within single sites where soil type did not vary, abundances varied significantly under different vegetation cover (Table 18.1). At the J.F.K. Arboretum in Co. Wexford, the trees were planted in blocks on the same soil type and at Killaune the areas sampled were essentially part of the same bog.

Despite the fact that mites and Collembola have been extensively studied in Ireland, it still appears that many species which occur remain unrecorded. Bolger et al. (2002) showed that the rate at which new species of Collembola are being added to the Irish records was not declining, suggesting that many species remain uncovered, and a recent

Table 18.1 Abundances of Mites and Collembola recovered from adjacent stands of differing vegetation at two sites

	Mites	Collembola
J.F.K. Arboretum, Co. Wexford		
Cupressocyparis leylandii	12,800	5000
Tsuga heterophylla	40,100	46,700
Chamaecyparis lawsoniana	8600	5600
Pinus radiata	19,300	14,300
Picea sitchensis	29,200	26,000
Grassland	32,800	38,000
Killaun Bog, Co. Offaly		
Calluna vulgaris	56,110	6767
Pinus contorta	51,872	1927
Betula pubescens	59,303	6130

The J. F. Kennedy Arboretum (North American Plots) is on mineral soil, Killaun Bog is a raised bog

compilation (Bolger, unpublished; Arroyo et al. 2017) of mites from two of the best known orders (Mestigmata and Oribatida) showed that over 170 species have been added to the 394 species listed by Luxton in 1998.

18.5 Conclusions

Soil organisms are essential for most soil processes and functions such as nutrient cycling, organic matter breakdown, maintenance of soil structure and many more. So far in this book, the importance of soil structure for plant productivity has been emphasised (Chap. 14), as has the importance of soil fertility and nutrient cycling (Chap. 15), the importance of microbial transformations in relation to water quality (Chap. 16) and indeed the importance of carbon storage for climate change mitigation (Chap. 17). Soil biodiversity is sometimes described as the engine room of the soil and justifiably so, as it is a key driver of many of the processes described in the aforementioned chapters. Despite this, there is still a great need to further knowledge on soil biology, biodiversity and ecological functions. A biodiversity baseline is required for different soils under different land use types to understand the delivery of soil functions required by society. Also, the complexity of soil food webs demands that research must extend beyond single soil animal species or taxa towards capturing multi-species interactions more realistic of the complexity of soil food webs. Finally, a case has been made (Keith et al. 2016) for the need to integrate soil biodiversity research into larger socio-ecological contexts.

References

Arroyo J, O'Connell T, Bolger T (2017) Oribatid mites (Acari: Oribatida) recorded from Ireland: Catalogue, historical records, species habitats and geographical distribution, combinations, variations and synonyms. Zootaxa 4328:1 https://doi.org/10.11646/zootaxa.4328.1

Bardgett R (2005) The biology of soil: a community and ecosystem approach. Oxford University Press, Oxford, p 242

Blackith RE, Good JA (1991) Protura in Ireland. Bull Irish Biogeograph Soc 14:84–89

Bolger T, Curry JP (1980) Effects of cattle slurry on soil arthropods in grassland. Pedobiologia 20:246–253

Bolger T, Curry JP (1984) Influences of pig slurry on soil microarthropods in grassland. Rev Écol Biol Sol 21:269–281

Bolger T, Schmidt O, Purvis G, Curry JP (2002) The biodiversity, function and management of soil invertebrate populations. In: Convery F, Feehan J (eds) Achievement and Challenge: Rio + 10 and Ireland. University College Dublin, The Environmental Institute, pp 2–10

Bouffaud ML, Creamer RE, Stone D, Plassart P, van Tuinen D, Lemanceau P, Wipf D, Redecker D (2016) Indicator species and co-occurrence in communities of arbuscular mycorrhizal fungi at the European scale. Soil Biol Biochem 103:464–470

Brennan FP, Abram F, Chinalia FA, Richards KG, O'Flaherty V (2010a) Characterization of environmentally persistent *Escherichia coli* isolates leached from an Irish soil. Appl Environ Microbiol 76:2175–2180

Brennan FP, O'Flaherty V, Kramers G, Grant J, Richards KG (2010b) Long-term persistence and leaching of *Escherichia coli* in temperate maritime soils. Appl Environ Microbiol 76:1449–1455

Brennan FP, Moynihan E, Griffiths BS, Hillier S, Owen J, Pendlowski H, Avery LM (2014) Clay mineral type effect on bacterial enteropathogen survival in soil. Sci Total Env 468–469:302–305

Cawley M (2009) New records for Irish false-scorpions (Arachnida: Pseudoscorpiones), also incorporating a county checklist. Bull Irish Biogeograph Soc 33:99–114

Coleman DC, Crossley DA Jr, Hendrix PF (2004) Fundamentals of soil ecology, 2nd edn. Elsevier, Amsterdam, p 386

Courtney R, Feeney E, O'Grady A (2014) An ecological assessment of rehabilitated bauxite residue. Ecol Engineer 73:373–379

Creamer RE, Hannula SE, Leeuwen JPV, Stone D, Rutgers M, Schmelz RM, de Ruiter PC, Hendriksen NB, Bolger T, Bouffaud ML, Buee M, Carvalho F, Costa D, Dirilgen T, Francisco R, Griffiths BS, Griffiths R, Martin F, da Silva PM, Mendes S, Morais PV, Pereira C, Philippot L, Plassart P, Redecker D, Römbke J, Sousa JP, Wouterse M, Lemanceau P (2016) Ecological network analysis reveals the inter-connection between soil biodiversity and ecosystem function as affected by land use across Europe. Appl Soil Ecol 97:112–124

Curry JP (1969) The qualitative and quantitative composition of the fauna of an old grassland site at Celbridge, Co. Kildare. Soil Biol Biochem 1:219–227

Curry JP (1994) Grassland invertebrates: ecology, influence on soil fertility and effects on plant growth. Chapman & Hall, London. 437 pp

Curry JP, Schmidt O (2006) Long-term establishment of earthworm populations in grassland on reclaimed industrial cutaway peatland in Ireland. Suo: Mires Peat 57:65–70

Curry JP, Byrne D, Schmidt O (2002) Intensive cultivation can drastically reduce earthworm populations in arable land. Eur J Soil Biol 38:127–130

Curry JP, Doherty P, Purvis P, Schmidt O (2008) Relationships between earthworm populations and management intensity in cattle-grazed pastures in Ireland. Appl Soil Ecol 39:58–64

Daims H, Lebedeva EV, Pjevac P, Han P, Herbold C, Albertsen M, Jehmlich N, Palatinszky M, Vierheilig J, Bulaev A, Kirkegaard RH, von Bergen M, Rattei T, Bendinger B, Nielsen PH, Wagner M (2015) Complete nitrification by *Nitrospira* bacteria. Nature 528:504–509

Delmont TO, Robe P, Cecillon S, Clark IM, Constancias F, Simonet P, Hirch PR, Vogel TM (2011) Accessing the soil metagenome for studies of microbial diversity. Appl Environ Microbiol 77: 1315–1324

Dickinson CH, Dooley M (1969) Fungi associated with Irish peat bogs. Proc R Irish Acad B: Biol Geol Chem Sci 68:109–135

Dirilgen T, Arroyo J, Dimmers WJ, Faber J, Stone D, Martins da Silva P, Cavalho F, Schmelz R, Griffiths BS, Francisco R, Creamer RE, Sousa JP, Bolger T (2016) Mite community composition across a European transect and its relationships to variation in other components of soil biodiversity. Appl Soil Ecol 97:86–97

Dix E (2017) The role of microarthropods in carbon and nitrogen dynamics in grassland soils. Unpublished Ph.D. thesis, University College Dublin, 266 pp

Ernfors M, Brennan F, Richards K, McGeough K, Griffiths BS, Laughlin RJ, Watson CJ, Philippot L, Grant J, Minet E, Moynihan E, Mueller C (2014) The nitrification inhibitor dicyandiamide increases mineralization-immobilization turnover in slurry amended grassland soil. J Agr Sci 152:137–149

Evans GO (1982) Observations of the genus *Protogamasellus* with a description of a new species. Acarologia 23:303–313

Fox A, Kwapinski W, Griffiths BS, Schmalenberger A (2014) The role of sulfur—and phosphorus-mobilizing bacteria in biochar-induced growth promotion of *Lolium perenne*. FEMS Microbiol Ecol 90: 78–91

Fox A, Gahan J, Ikoyi I, Kwapinski W, O'Sullivan O, Cotter PD, Schmalenberger A (2016) Miscanthus biochar promotes growth of spring barley and shifts bacterial community structures including phosphorus and sulfur mobilizing bacteria. Pedobiologia 59: 195–202

Gahan J, Schmalenberger A (2014) The role of bacteria and mycorrhiza in plant sulfur supply. Front Plant Sci 5:723. https://doi.org/10.3389/fpls.201400723

Gahan J, Schmalenberger A (2015) Arbuscular mycorrhizal hyphae in grassland select for a diverse and abundant hyphospheric bacterial community involved in sulfonate desulfurization. Appl Soil Ecol 89:113–121

Griffin CT, Moore JF, Downes MJ (1991) Occurrence of insect parasitic nematodes (Steinernematidae, Heterorhabditidae) in the Republic of Ireland. Nematologica 37:92–100

Halbert JN (1915) Clare Island survey. 39. Acaridina ii. Terrestrial and marine Acarina. Proc R Irish Acad 31(39): 45–136

Heneghan L, Bolger T (1996) Effects of acid rain components on soil microarthropods: a field manipulation. Pedobiologia 40:413–438

Hug LA, Baker BJ, Anantharaman K, Brown CT, Probst AJ, Castelle CJ, Butterfield CN, Hernsdorf AW, Amano Y, Ise K, Suzuki Y, Dudek N, Relman DA, Finstad KM, Amundson R, Thomas BC, Banfield JF (2016) A new view of the tree of life. Nature Microbiol 1:16048. https://doi.org/10.1038/nmicrobiol.2016.48

Jeffery S, Gardi C, Jones A, Montanarella L, Marmo L, Miko L, Ritz K, Pérès G, Römbke J, van der Putten WH (eds) (2010) European atlas of soil biodiversity. European Commission, Publications Office of the European Union, Luxembourg, 128 pp

Jones CG, Lawton JH, Shachak M (1994) Organisms as ecosystem engineers. Oikos 69:373–386

Jones CM, Spor A, Brennan FP, Breuil MC, Bru D, Lemanceau P, Griffiths B, Hallin S, Philippot L (2014) Recently identified microbial guild mediates soil N_2O sink capacity. Nature Clim Change 4:801–805

Keith AM, Griffin CT, Schmidt O (2009) Predatory soil nematodes (Mononchida) in major land-use types across Ireland. J Natural Hist 43:2571–2577

Keith AM, Boots B, Hazard C, Niechoj R, Arroyo J, Bending GD, Bolger T, Breen J, Clipson N, Doohan FM, Griffin CT, Schmidt O (2012) Cross-taxa congruence, indicators and environmental drivers in soils under agricultural and extensive land management. Eur J Soil Biol 49:55–62

Keith AM, Schmidt O, McMahon BJ (2016) Soil stewardship as a nexus between Ecosystem Services and One Health. Ecosyst Serv 17:40–42

Knietch A, Waschkowitz S, Bowien A, Henne A, Daniel R (2003) Metagenomes of complex microbial consortia derived from different soils as sources for novel genes conferring formation of carbonyls from short-chain polyols on *E. coli*. J Microbiol Biotechnol 5:46–56

Legg G, O'Connor JP (1997) A review of the Irish pseudoscorpions (Arachnida: Pseudoscorpiones). Bull Irish Biogeograph Soc 20:105–117

Lillis L, Doyle E, Clipson N (2009) Comparison of DNA- and RNA-based bacterial community structures in soil exposed to 2,4-dichlorophenol. J Appl Microbiol 107:1883–1893

Luxton M (1998) The oribatid and parasitiform mites of Ireland, with particular reference to the work of J. N. Halbert (1872–1948). Bull Irish Biogeograph Soc 22:1–72

Massey PA, Creamer RE, Whelan MJ, Ritz K (2016) Insensitivity of soil biological communities to phosphorus fertilization in intensively managed grassland systems. Grass Forage Sci 71:139–152

Melody C, Schmidt O (2012) Northward range extension of an endemic soil decomposer with a distinct trophic position. Biol Lett 8:956–959

Melody C, Griffiths BS, Dyckmans J, Schmidt O (2016) Stable isotope analysis ($\delta^{13}C$ and $\delta^{15}N$) of soil nematodes from four different feeding groups. PeerJ 4:e2372. https://doi.org/10.7717/peerj.2372

Moore JF (1977) Studies on occurrence and control of *Longidorus* and *Xiphinema* nematodes. Irish J Agr Res 16:301–310

Morriën E, Hannula SE, Snoek LB, Helmsing NR, Zweers H, de Hollander M, Soto RL, Bouffaud ML, Buée M, Dimmers W, Duyts H, Geisen S, Girlanda M, Griffiths RI, Jørgensen HB, Jensen J, Plassart P, Redecker D, Schmelz RM, Schmidt O, Thomson BC, Tisserant E, Uroz S, Winding A, Bailey MJ, Bonkowski M, Faber JH, Martin F, Lemanceau P, de Boer W, van Veen JA, van der Putten WH (2017) Soil networks become more connected and take up more carbon as nature restoration progresses. Nature Commun 8:14349. https://doi.org/10.1038/NCOMMS14349

Moynihan E, Richards K, Ritz K, Tyrrel S, Brennan F (2013) The impact of soil type, biology and temperature on the environmental persistence of non-toxigenic *E. coli* O157. Biol Env Proc R Irish Acad 113B:41–46

Moynihan EL, Richards KG, Brennan FP, Tyrrel SF, Ritz K (2015) Enteropathogen survival in soil from different land-uses is predominantly regulated by microbial community composition. Appl Soil Ecol 89:76–84

Muldowney J, Schmidt O (2002) *Allolobophora cupulifera* Tétry (Oligochaeta: Lumbricidae) in Ireland: first records for the British Isles. Megadrilogica 9:29–32

Muldowney J, Curry JP, O'Keeffe J, Schmidt O (2003) Relationships between earthworm populations, grassland management and badger density in County Kilkenny, Ireland. Pedobiologia 47:913–919

Neher DA (2010) Ecology of plant and free-living nematodes in natural and agricultural soil. Annu Rev Rev Phytopathol 48:371–394

O'Mahony M, Henneberger R, Selvin J, Kennedy J, Doohan F, Marchesi JR, Dobson ADW (2015) Inhibition of the growth of *Bacillus subtilis* DSM10 by a newly discovered antibacterial protein from the soil metagenome. Bioengineered 6:89–98

Orgiazzi A, Bardgett RD, Barrios E, Behan-Pelletier V, Briones MJI, Chotte JL, De Deyn GB, Eggleton P, Fierer N, Fraser T, Hedlund K, Jeffery S, Johnson NC, Jones A, Kandeler E, Kaneko N, Lavelle P, Lemanceau P, Miko L, Montanarella L, Moreira FMS, Ramirez KS, Scheu S, Singh BK, Six J, van der Putten WH, Wall DH (eds) (2016) Global soil biodiversity atlas. European Commission, Publications Office of the European Union, Luxembourg. 176 pp

Randall K (2016) The microbial ecology of the soil–plant interface. Unpublished Ph.D. thesis, University College Dublin

Rutgers M, Orgiazzi A, Gardi C, Römbke J, Jänsch S, Keith AM, Neilson R, Boag B, Schmidt O, Murchie AK, Blackshaw RP, Pérès G, Cluzeau D, Guernion M, Briones MJI, Rodeiro J, Piñeiro R, Díaz-Cosin DJ, Sousa JP, Suhadolc M, Kos I, Krogh PH, Faber JH, Mulder C, Bogte JJ, van Wijnen HJ, Schouten AJ, de Zwart D (2016) Mapping earthworm communities in Europe. Appl Soil Ecol 97:98–111

Samad MS, Bakken LR, Nadeem S, Clough TJ, de Klein CAM, Richards KG, Lanigan GJ, Morales SE (2016) High-resolution denitrification kinetics in pasture soils link N_2O emissions to pH, and denitrification to C mineralization. PLoS ONE 11:e0151713. https://doi.org/10.1371/journal.pone.0151713

Savary S, Ficke A, Aubertot JN, Hollier C (2012) Crop losses due to diseases and their implications for global food production losses and food security. Food Secur 4:519–537

Sawulski P, Boots B, Clipson N, Doyle E (2015) Differential degradation of polycyclic aromatic hydrocarbon mixtures by indigenous microbial assemblages in soil. Lett Appl Microbiol 61:199–207

Schloss PD, Handlesman J (2003) Biotechnological prospects from metagenomics. Curr Opinion Biotechnol 26:1135–1145

Schmidt O, Curry JP (1999) Effects of earthworms on biomass production, nitrogen allocation and nitrogen transfer in wheat–clover intercropping model systems. Plant Soil 214:187–198

Schmidt O, Curry JP, Hackett RA, Purvis G, Clements RO (2001) Earthworm communities in conventional wheat monocropping and low-input wheat–clover intercropping systems. Ann Appl Biol 138:377–388

Schmidt O, Dyckmans J, Schrader S (2016) Photoautotrophic microorganisms as a carbon source for temperate soil invertebrates. Biol Lett 12:20150646. https://doi.org/10.1098/rsbl.2015.0646

Stromberger ME, Keith AM, Schmidt O (2012) Distinct microbial and faunal communities and translocated carbon in *Lumbricus terrestris* drilospheres. Soil Biol Biochem 46:155–162

Strous M, Fuerts JA, Kramer EHM, Logemann S, Muyzer G, van de Pas-Schoonen KT, Webb R, Kuenen JG, Jetten MSM (1999) Missing lithotroph identified as new planctomycete. Nature 400:446–449

Swift MJ, Heal OW, Anderson JM (1979) Decomposition in terrestrial ecosystems. University of California Press, Berkeley and Los Angeles, p 372

Tan H, Mooij MJ, Barret M, Hegarty PM, Dobson ADW, O'Gara F (2014) Identification of novel phytase genes from an agricultural soil-derived metagenome. J Microbiol Biotechnol 24:113–118

Thakuria D, Schmidt O, Finan D, Egan D, Doohan FM (2010) Gut wall bacteria of earthworms: a natural selection process. ISME J 4: 357–366

Treusch AH, Leininger S, Kletzin A, Schuster SC, Klenk HP, Schleper C (2005) Novel genes for nitrite reductase and Amo-related proteins indicate a role of uncultivated mesophilic crenarchaeota in nitrogen cycling. Env Microbiol 7:1985–1995

van Kessel MAHJ, Speth DR, Albertsen M, Nielsen PH, op den Camp HJM, Kartal B, Jetten MSM, Lücker S (2015). Complete nitrification by a single microorganism. Nature 528:555–559

Vervoort MTW, Vonk JA, Mooijman PJW, van den Elsen SJJ, van Megen HHB, Veenhuizen P, Landeweert R, Bakker J, Mulder C, Helder J (2012) SSU ribosomal DNA-based monitoring of nematode assemblages reveals distinct seasonal fluctuations within evolutionary heterogeneous feeding guilds. PLoS ONE 7:e47555. https://doi.org/10.1371/journal.pone.0047555

Yeates GW, Ferris H, Moens T, van der Putten WH (2009) The role of nematodes in ecosystems. In: Wilson JW, Kakouli-Duarte T (eds) Nematodes as Environmental Indicators. CAB International, London, pp 1–44

Soils and Archaeology

19

Thomas Cummins, Helen Lewis, Clíodhna Ní Lionáin, and Stephen Davis

Abstract
Archaeology is the study of the human past from its material remains, most of which are made of or found within soils and sediments. Past human actions impact the soil record, as seen through relics of changes in soil characteristics and qualities, changes to sedimentation, and the presence of archaeological features and artefacts preserved within modern soils. Soil and sediment conditions control what survives in the burial environment, what decomposes, and consequently influence all archaeological sites, artefacts, and ecological remains. The study of these remains, through survey, excavation, and post-excavation analyses, informs our understanding of past cultures and environments, providing insight into how people have interacted with the soil, both directly, through settlement, land use, and monument construction, and indirectly, by altering local ecosystems over time. Soils can be considered repositories of traces of human action, and in turn the soils of Ireland have formed under the continuous influence of people, up to the present day, when most land in Ireland is actively managed for agriculture, forestry, extraction or construction. Consequently, all land managers are stewards of soil-bound heritage, and have the opportunity and responsibility to recognize the archaeological heritage value of land in their care, and to participate in conserving this value as a public good. This chapter reviews some of the soil evidence for Irish landscape history, the heritage content of soils, archaeological work that has helped discover that heritage, and issues surrounding the management of the cultural heritage in soils.

Keywords
Soil heritage • Archaeological heritage • Cultural soils Sediments • Geoarchaeology • Archaeological prospection • Taphonomy • Dowth

19.1 Introduction

Archaeology is the study of the human past from its material remains, most of which are made of or found within soils and sediments. Past human actions impact the soil record, as seen through relics of changes in soil characteristics and qualities, changes to sedimentation, and the presence of archaeological features and artefacts preserved within modern soils. Soil and sediment conditions control what survives in the burial environment, what decomposes, and consequently influence all archaeological sites, artefacts, and ecological remains. The study of these remains, through survey, excavation, and post-excavation analyses, informs our understanding of past cultures and environments, providing insight into how people have interacted with the soil, both directly, through settlement, land use, and monument construction, and indirectly, by altering local ecosystems over time. Soils can be considered repositories of traces of human action, and in turn the soils of Ireland have formed under the continuous influence of people, up to the present day, when most land in Ireland is actively managed for agriculture, forestry, extraction or construction. Consequently, all land managers are stewards of soil-bound heritage, and have the opportunity and responsibility to recognize the archaeological heritage value of land in their care, and to participate in conserving this value as a public good. This chapter reviews some of the soil evidence for Irish landscape history, the heritage content of soils, archaeological work that has helped discover that heritage,

T. Cummins (✉)
UCD School of Agriculture and Food Science, University College Dublin, Belfield, Dublin 4, Ireland
e-mail: thomas.cummins@ucd.ie

H. Lewis · S. Davis
UCD School of Archaeology, University College Dublin, Belfield, Dublin 4, Ireland

C. Ní Lionáin
Devenish Nutrition, Belfast, Ireland

R. Creamer and L. O'Sullivan (eds.), *The Soils of Ireland*, World Soils Book Series,
https://doi.org/10.1007/978-3-319-71189-8_19

and issues surrounding the management of the cultural heritage in soils.

19.2 Irish Landscape History from the Soil and Sediment Record

The soil record in Ireland extends from the end of Pleistocene glaciation to the present day. During most of this time the island has been occupied by humans, and thus all Irish soils have developed with human influence along with natural soil formation factors. From the end of the last ice age, parent materials such as rock and glacial till have been constantly weathering and transforming into soil, affected over time by climate, topography and organisms (Jenny 1941; for studies in Ireland see e.g. Dimbleby 1965; Moles et al. 1995; Moles and Moles 2002). Since the arrival of humans into the Irish landscape at least 10,000 years ago (cf. Dowd and Carden 2016), their activities have influenced the factors of soil formation, especially those related to biota. The activities of early settlers and farmers, who felled trees for land clearance and introduced tillage crops and grazing domesticated animals, converted previously forested landscapes into long-term grasslands by at latest the end of the Neolithic period (O'Connell and Molloy 2001). Agricultural landscapes have been largely maintained to the present day, with varying intensity, focus and technology of farming practices. Expansion and contraction of settlement areas and alterations to drainage have caused soils to change over time, against a backdrop of larger-scale climate shifts. Effects of cultural activity on soil development are widespread in the archaeological and environmental record across northern Europe, highlighting the impact of human cultural practices on the environment over time (e.g. Roberts 2014).

Irish evidence suggests similar trajectories over time to those of the rest of northern Europe (e.g. Weir et al. 1971; Macphail 1986; Macphail and Robertson 1989). Soil changes related to erosion (soil loss and slope movement), altered drainage, intensifying tillage practices, the impact of settlement construction and of industry are among the most evident. Vegetation clearance, followed by centuries without forest canopy, has increased the impact of rainfall and erosion, leading to waterlogging of soils, reduced profile thickness, podzolisation, and, in some locations, the initial steps in peat formation (Clayton 1973–4; Cunningham et al. 1999). Through centuries of agricultural management, liming agents and nutrient-rich manures have been applied to fields (Collins 2008) to maintain a relatively high pH, active rooting and soil faunal activity, sustaining the productive capacity of the soil.

Soil exposure through deforestation, crop rotation and tillage practices may all increase soil erosion by wind and water action, particularly on slopes. Given the climate and topography of Ireland, erosion is seen today (e.g. O'Sullivan et al. 2014; McCormick et al. 2014; Drew 1983; Huang and O'Connell 2000) and in the archaeological record (e.g. Moles et al. 1999). A typical 'stable' grassland today would have seen at least one phase (but usually many) of past forest/scrub clearance, and often also several phases of past tilling, some prehistoric and some historic. The types of impact expected are surface (e.g. deflation, colluviation, crusting) and sub-surface (e.g. nutrient leaching, clay redistribution, size-sorting of inclusions). Identifying land clearance archaeologically requires rigorous modelling of soil profile changes over time, in addition to proxy environmental indicators such as pollen or charcoal records. Pollen records exist in ancient bogs and in lake and stream sediments found across Ireland (e.g. Jessen 1949; Mitchell 1951; Smith 1961; Thompson 1997; Lomas-Clarke and Barber 2004; Edwards and Whittington 2001; O'Connell and Molloy 2001; Turner et al. 2015; Mitchell et al. 2013), but little attention has been paid as yet to linking local soil records with these regional proxies of vegetation change over time (but see Verrill and Tipping 2010). Many archaeological soils in Ireland show localised indicators of past vegetation and turf clearance, breaking up of the ground, agricultural interventions, and the impact of amendments related to ancient tilled fields (e.g. Case et al. 1969). These are seen through physical disturbance indicators ranging from ridged field systems and lynchets to 'buried' spademarks to microscopic clay accumulation features typical of cultivation (e.g. 'agricutans') identified through soil micromorphology. Amendment indicators may include certain artefacts, manures, ash, seaweed inclusions, chemical alterations, and related morphological evidence such as of soil faunal activity (e.g. Barker 1985; Case et al. 1969; Courty et al. 1989; Jongerius 1983; MacPhail 1990; Romans and Robertson 1983).

In addition to the soil record, about one-fifth of Ireland has or had peat cover (cf. Hammond 1981). Peat consists of surface organic sediments created when plant litter production exceeds decomposition; the cool, moist Irish climate and high water retention, provide good conditions for plant growth, and poor conditions for decomposer organisms. Peat has formed in large shallow basins (former post-glacial lakes) in the midlands. In upland areas from the mid-Holocene Epoch on, blanket peats have formed, sometimes covering prehistoric fields and settlements (e.g. O'Sullivan and Sheehan 1996). The question has long been asked whether early agriculture or agricultural changes somehow caused blanket peat to start forming in the past (Mitchell 1972), or if peat formation was triggered by climatic change with an increased moisture surplus (either increased rainfall or decreased evapotranspiration—see Thorp and Glanville 2003). Caulfield et al. (1998) dated discontinuous peat formation to before the occupation of

Neolithic field systems at Belderrig, Co. Mayo, and matched the later date-range of pine trees rooted in peat across west Mayo with a synchronous event in Scotland. These lines of evidence suggest that peat development was due to climate change, and not agriculture.

19.3 Soil and Sediments as a Repository of Archaeological Heritage

In addition to providing information on landscape history, soils and sediments are the preservation environment for most archaeological artefacts and features. All of these elements of the archaeological record can be studied through survey, excavation and post-excavation approaches adapted to the specific soil, sedimentary and geological environment.

The archaeological heritage of Ireland is afforded statutory protection through the National Monuments Act (1930) and its subsequent amendments (1954, 1987, 1994, 2004), which provide for the establishment and maintenance of statutory inventories, including the Record of Monuments and Places (RMP), based on the Sites and Monuments Record (SMR). Known monuments and zones of notification for operations around monuments are shown on the maps of the National Monuments Service (NMS), and can be viewed on the NMS websites (www.archaeology.ie; www.heritagemaps.ie). Where sub-surface features are known, they may be similarly protected, and require consideration within planning processes. However, at present in Ireland these are not systematically included on the Sites and Monuments Record. These features may be found through various survey methods, and their states of preservation and potential significance should be assessed and then revised on a regular basis through auger and/or bulk sample analyses. Guidelines for reporting and protecting archaeological sites are available at www.archaeology.ie.

There are currently no guidelines available for soils in Irish archaeology. However, by understanding the characteristics of local soils, broad predictions can be made about the likelihood of preservation of aspects of archaeological heritage. For instance, an acidic, freely-draining soil such as a Podzol on granite is unlikely to preserve bones or unburnt organic remains, and may not respond well to certain geophysical survey techniques. A Cambisol soil over limestone and under tillage is likely to have disturbed artefacts, preserved but fragmented bones, no wood, and can be expected to deliver excellent magnetometry results. A Stagnic Groundwater Gley soil in a floodplain is likely to contain organic remains, with surface features masked by alluvial deposits. Technologies including aerial photographs, satellite imagery or LiDAR (Light Detection And Ranging) may allow larger features to be seen, appearing as anomalies based on variations in topography, soil moisture and plant growth. The predictive qualities of the soil and sedimentary record allow archaeologists to tailor assessment, conservation, and research activities. Soils are very locally diverse, however, and broad models do not allow for the variation often seen across just a few meters, thus restricting accuracy of zero-impact intervention approaches. Soil maps (e.g. Creamer et al. 2014), aerial photographs, LiDAR, geophysics and other remote-sensing approaches, while all useful for archaeology, also have limitations. Assessment of conservation environments is best achieved through localised methods such as soil auger survey, bulk soil analyses of surface samples, or by test-pitting. Around Ireland, there are tens of thousands of ancient remains preserved as features in the soil, a proportion of which also have above-ground remains, built of earth and/or stone. Earthen monuments vary from large and clearly visible structures, such as passage tombs, ringforts, hillforts or henge monuments (ditch and bank, sometimes with stone or timber circles), to remains with no surface expression. Earthworks and buried features survive from times with no written records, from diverse societies from whom little else remains.

The depth-limit of an archaeological excavation is usually set at the depth of construction disturbance, or at the level below which no further traces of human activity can be found. In Ireland, this is normally at the level of glacial deposits or rock. An artefact-free depth of alluvium or sand does not necessarily indicate the lack of earlier activity at the site, as these Holocene deposits can cover earlier activity. As such, archaeologists must also be aware of the sedimentary environment and history of a site, and tailor assessment strategies accordingly. Assessment and monitoring may also go further, in the case where a development will affect drainage, and thereby survival of deeper buried remains on and around the site (e.g. French 2004).

When survey and/or excavation are completed, finds are documented, conserved and stored, and the record of excavated contexts and features forms a permanent site description and stratigraphic record. The excavated soil is not preserved, although bulk samples may be taken for extracting biological remains, and undisturbed monoliths may be prepared for thin-section preparation. This process of "preservation by record" creates an intellectual resource that offsets the fact that interventions, including archaeological interventions, irreversibly disturb the deposits and soils studied. Excavation records are accessible through government libraries and the online services noted above, but the amount and detail of the soil and sediment information recorded is highly variable both in original recording and in publication. Soil and sediment characteristics underpin the archaeological record and so wider access to primary records is greatly needed (Fig. 19.1).

Artefacts and environmental remains decompose more rapidly in aerobic soil environments, leaving a biased

Fig. 19.1 Artefacts in their soil context at Belderrig, Co. Mayo. Left: Unconsolidated surficial deposits exposed by coastal erosion, post-glacially transformed at the surface into mineral soil. Blanket peat developed over the mineral soil, and was later cut away for domestic fuel, with the truncated peat remnant reclaimed as grassland by cultivation and liming. Archaeological excavation is seen inside the yellow fence. Right: A close-up of the upper soil profile shows stone artefacts associated with Mesolithic activity in two layers. At this site, evidence of Mesolithic activity is overlain by Neolithic field walls enclosing land over a large area, all below the former peat (Warren 2008). This specific cliff profile was not excavated, as it was unstable and unsafe for work; the status, identity, and heritage value of these particular objects, and the objects themselves, are now lost to the sea

residue of the most-durable materials, such as worked stone, charcoal, ceramics and glass. The survival of metals and bone is more variable, being very dependent on soil pH and redox potential, related to soil minerals and oxygen availability. Soil disturbance from fauna, flora and human land use also has a major impact on the condition of archaeological materials. The anaerobic conditions in bogs and other wetlands have led to a high level of preservation of organic remains, including 'bog bodies' (e.g. Kelly 2013; Bermingham and Delaney 2006), timber trackways (e.g. Raftery 1996; Moloney et al. 1995; Gowen et al. 2005), crannog-type lake settlements (e.g. Fredengren et al. 2010; Bermingham and Moore 2015), and objects like the Faddan More Psalter (Gillis 2012). However, in these environments objects such as low-fired pottery, bones and shells see variable or zero preservation.

The survival of human remains illustrates the different preservation potential of varying soil and sedimentary environments (Fig. 19.2). In the anaerobic, strongly-acid conditions of Irish peat bogs even soft tissues such as hair, skin, and internal organs may be preserved, including the gut contents of 'bog bodies'. The rigid mineral material of bone, calcium hydroxylapatite, dissolves in the acidic solution of peat bogs, leaving behind a bone-ghost of collagen protein. In acidic but aerobic soil environments, complete decay of bone and soft tissues may occur, leaving no surviving human remains (Fig. 19.2, right). Bones in all soils are subject to decay, including in alkaline conditions such as over limestone, as rainwater, organic acids, and biota have an impact in all soil types. The decay process takes time, and more soluble materials decompose or alter first, e.g. organic components such as collagen often disappear before bone mineral, and porous bones decompose more quickly than denser bones. Animal bones and shells provide information regarding ancient human lifestyles and environments, and are subject to the same decomposition processes as human bones. Suitable preservation conditions with appropriate soil chemistry and little disturbance allow investigations on demography, diet, genetics, origins, health and lifestyles from the bone record (e.g. McKenzie et al. 2015; Knudson et al. 2012).

Other soil processes that greatly affect the archaeological record include burial by surface-casting earthworms, mineral weathering, leaching by rainwater, size-sorting by biota and water, and processes involved in the cultural practices of cultivation, especially tilling and amendment. For example, anecic earthworms generate a surface layer of granular soil

Fig. 19.2 Soil conditions lead to varying preservation. Left: Human remains preserved in the acid, anaerobic environment of a waterlogged peatland, a Dystric Histosol: fingers of Oldcroghan Man (reproduced courtesy of the National Museum of Ireland). Centre: Bones, consisting of mineral remains with the protein fraction—all soft tissues and a wooden coffin decomposed—in the alkaline aerobic burial environment of a probable Cambisol at St. Brigid's Hospital, Ballinasloe (Rogers et al. 2006). Right: Excavated grave cut with no human remains, following complete decomposition of organic remains and solution of calcium phosphate bone minerals in a strongly acid Podzol; Templeteenaun, Co. Wicklow (O'Sullivan et al. 2009)

crumbs by digesting and excreting soil particles creating surface casts. As they do not ingest larger objects, the latter travel downwards over time, forming a lower layer often recognized as a stone line in the soil profile, but which can also include pottery, bones—anything above the size of a sand grain. This process is frequently seen in Irish soils and affects the stratigraphy of the archaeological record, as do biomechanical processes more widely (Johnson 2002). Earthworms are typically most active in calcareous soils under stable grassland conditions. Tillage of such soils brings the size-sorted artefacts and stones into the upper part of the profile, meaning they are doubly disturbed in the vertical dimension. Tillage also moves objects horizontally, and this impact is especially strong on slopes, where objects also move due to gravity, and objects with less mass are moved further downslope by gravity and slope erosion than those with greater mass (Lewis 1998). As such, artefacts in alternating grassland and tillage, typical of much enclosed agricultural land in Ireland in recent centuries, can show substantial displacement and size-sorting vertically and horizontally. Understanding these dynamics, the relative locations of objects within the soil can also be used to identify ancient sites. Immediately after cultivation, organised surveys of ploughed soil by trained fieldwalkers can be used as a systematic method of recording artefacts to provide information on site types and chronology. In fields with little slope, or in combination with other survey approaches, this method can also be used to estimate the potential location of remains enclosed within the soil.

Ancient land use and soils can be modelled by understanding these processes. For example, where a pebble-free topsoil horizon is found underneath a monument, it can be suggested that it represents pre-monument earthworm sorting undisturbed by ploughing, at least in the immediate term before monument construction. This indicates that the monument was built in a grassland clearing or a pasture field that was grazed for some time (probably decades) prior to the construction of the monument. Earthworms and other soil fauna can also influence soil drainage and erosion, thus affecting the burial environment, and may cause organic remains such as charcoal to be broken up (Goldberg and Macphail 2006; Courty et al. 1989).

19.4 Geoarchaeology

Geoarchaeology combines earth science disciplines with the cultural approaches of archaeology to interpret ancient use of space, site construction and materials, landscape history and environmental changes. It makes use of many specialized physical, chemical and morphological approaches to identify deposited materials and the formation processes they represent. Few geoarchaeological studies have been undertaken in Ireland, compared to some other parts of Europe (e.g. Cornwall 1958; Keeley and Macphail 1981; Fédoroff and Courty 1987; Revel 1991; French 2003; French et al. 2007). Soils 'buried' under monuments and used as construction materials in earthen monuments were, however, a focus of Irish research in the early to mid-twentieth century. These early studies demonstrated the potential of Irish soil and sediment remains to reflect past landscape conditions and site histories. Buried soils and soils used in monument construction often reveal substantial landscape change over time, with monument soils indicating landscape characteristics different to modern conditions at the same location. Modelling how these changes occurred over time suggests that substantial soil erosion and alteration have occurred since at least the Neolithic Period across Ireland (e.g. Common 1940; O'Kelly 1951). For instance, a great deal of discussion focused on how variations in soil color can inform on monument and landscape history (e.g. O'Kelly 1951; Brade-Birks 1951; Proudfoot and Simmons 1958; Proudfoot 1960). Modern approaches to archaeological soils have become increasingly sophisticated: there are many more methods for studying ancient soils, and increased knowledge available from international research that allows a better interpretation of ancient soil history, human creation of sediments, and past land-use impact. The potential of past climate change to be related to these alterations in Ireland has been noted above, but the record of soil history is not well understood or dated on dryland soils, even where substantial monuments have been excavated. There are also many issues relating to pre-monument land use, the building of monuments, continuing and altered soil development, and the impact of a monument on the underlying soil profile (for some recent Irish examples see, e.g. Ostericher 2014; Ostericher and Lewis, in press).

19.5 Archaeological Prospection

Archaeological prospection enables certain aspects of the archaeological record to be characterised without damage. The preference of archaeologists is for heritage objects to remain buried and for non-destructive methods to be employed where possible, with excavation always a last resort (DoAHGI 1999). Non-destructive methods include the analysis of documentary sources, maps, aerial photographs, folklore and local knowledge, as well as non-invasive or low-impact fieldwork, such as walk-over survey and remote sensing (e.g. LiDAR, geophysical and geochemical survey). More invasive investigation is required in advance of development (e.g. buildings, roads), where at least the upper soil will be removed entirely, destroying any surviving remains. Archaeologists would also recommend investigation by field-walking for tilled fields, as these agricultural practices are constantly eroding, disturbing, and exposing new and deeper archaeological deposits. Even where excavation is required, archaeologists prefer to leave at least part of each site for future study. For example, the Viking settlement of Woodstown, which led to the establishment of Waterford city, was identified on the planned route of the N25 roadway (since re-routed) during a program of archaeologically-monitored test-trenching. As important finds were retrieved, further investigation was undertaken in small trenches, but most of the site was left in place (Russell and Hurley 2014).

LiDAR works on the principle of transmitting laser beam pulses and calculating distance by measuring the time taken for these beams to return, and record very low-profile surface topography, which may reveal archaeological features (Opitz and Cowley 2013; Crutchley 2010). Geophysical survey describes a range of remote sensing techniques (e.g. magnetometry, electrical resistivity, ground penetrating radar) used to map sub-surface features (Gaffney and Gaffney 2011; Gaffney and Gater 2003). Such approaches use soil, sediment and material properties to identify anomalies, which occur in patterns and types that may indicate archaeological features. While geophysical survey may be airborne (e.g. Gallagher et al. 2016) for archaeology it is usually ground-based owing to the spatial resolution required.

Geochemical surveys involve minor intervention, usually by sampling from the topsoil. Such geochemical survey can be multi-elemental; however, the main approach used in Ireland has been phosphate analysis (e.g. Ullrich 2010, 2013). Phosphate analysis is also of particular interest in agronomic nutrient management. In recent years, there has been an increasing degree of overlap between agronomic soil survey and archaeological prospection (e.g. Hoefer et al. 2009, 2012). For instance, the UK DART Project ('Detection of Archaeological Residues using remote sensing Techniques') aimed to quantify some of the variables responsible for the appearance of archaeological sites through remote sensing (cf. Donoghue et al. 2006; Stott et al. 2015). Other projects have explored the possibility of using

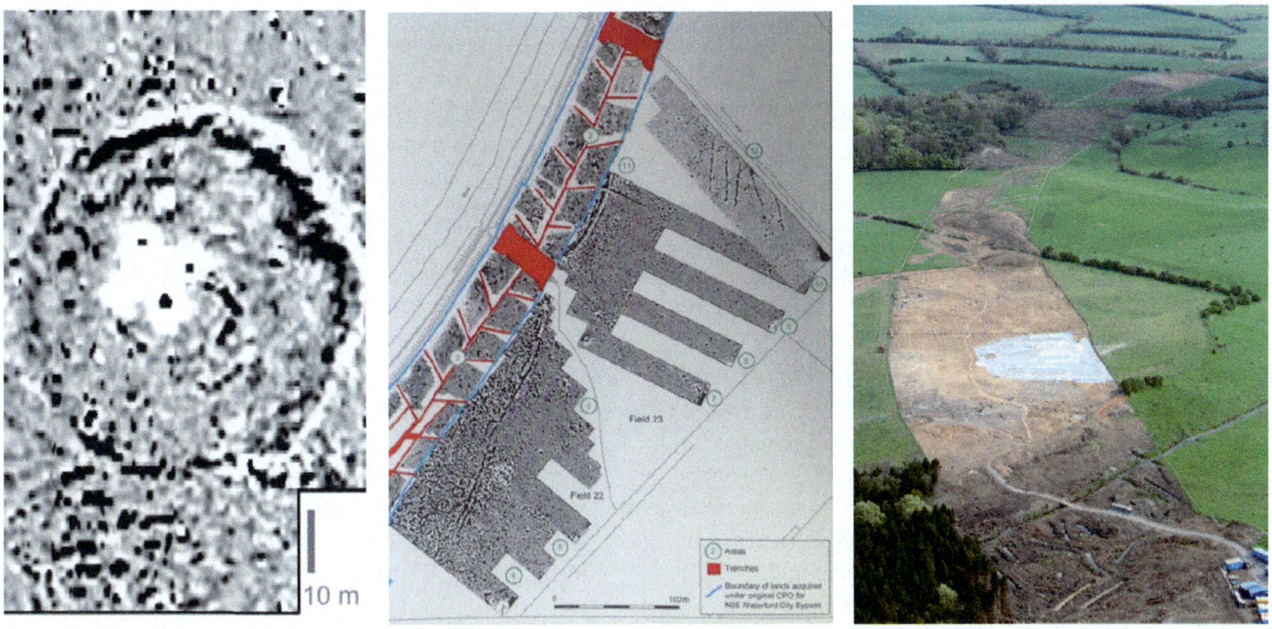

Fig. 19.3 Methods of archaeological prospection. Left: Geophysical prospection image from a fluxgate gradiometer survey of the top of Lyons Hill, Co. Dublin (Leigh 2010). Variations in the signal represent fluctuations in the local magnetic field, responding to soil materials such as burning and ditch fills. Centre: Test trenching at Woodstown, Co. Waterford (red) led to the discovery of an early Viking-Age settlement. Subsequent fluxgate gradiometer survey (stippled grey areas) revealed information about the layout of the site (Russell and Hurley 2014, Fig. 3.2). Right: Site of the Lismullen henge, showing the road-take land with the discovery site covered by plastic, stripped topsoil, test trenches, and also revealing subsoil colours with increasing organic-matter content from light to dark. Image Muireann NíBhrolcháin

multispectral satellite remote sensing data to identify archaeological features in grasslands as a function of vegetation health (Bennett et al. 2013), by employing methods derived from forestry applications (e.g. Tewari et al. 2003; Tuominen 2009). The use of hyperspectral imaging, especially in tandem with increasingly advanced UAV (Unmanned Airborne Vehicles, or 'drones') technology is likely to be significant in the future of precision agriculture, with the data also having significant and as yet untapped archaeological value (Doneus et al. 2014).

Surveys and excavations on construction projects during Ireland's economic boom greatly increased the number of recorded sites, and expanded our knowledge of the cultural materials contained in the soil (e.g. Grogan et al. 2007; McKenzie et al. 2015; O'Connell 2013; Russell and Hurley 2014). Even systematic investigation does not, however, always capture all sites. For instance, in Ireland it is standard practice to excavate trial pits and test-trenches, but pre-construction environmental assessment does not regularly include topsoil sieving or fieldwalking, and monitoring of topsoil stripping is variable. It is likely that many artefacts once held in the topsoil have been lost to the record because of this, despite otherwise rigorous assessment processes (Fig. 19.3).

Case Study 19.1 Farming Landscapes Through Time—Examples from County Meath

The Archaeological Ensemble of the Bend in the Boyne, more widely known as the Brú na Bóinne World Heritage Site (WHS), is one of only two UNESCO archaeological WHS in Ireland. It was inscribed in 1991 owing to the scale and extent of its passage tomb cemeteries, its long history of settlement and burial, and its exceptional megalithic art. Brú na Bóinne occupies c. 3331 ha, divided into a core area of 770 ha, with buffer zones to the north (798 ha) and south (1763 ha) (UNESCO 1993). Brú na Bóinne is known worldwide for its Neolithic monuments, in particular its concentration of more than 41 passage tombs dating to the Middle Neolithic (c. 3600–3100 BC), including the 'megatombs' of Knowth, Dowth and Newgrange, marking the pinnacle of the passage tomb tradition in Ireland (Eogan and Doyle 2010). The area is also agriculturally rich, with Meath having the second highest proportion of dry, lowland mineral soils in Ireland (Lee 1974, 152–3) and one of the lowest proportions of 'marginal' land (<20%—ibid., 154). Modern land use is predominantly pastoral, with only 41 ha of the core area under arable agriculture (Lydon and Smyth 2014). However, agriculture in Brú na Bóinne is of much greater antiquity, extending back to the early Neolithic (c. 4000–3600 BC), associated with the construction of large rectangular houses, the remains of which underlie the great tomb of Knowth (cf. Smyth 2011; Eogan and Roche 1997). The first farmers practiced arable and pastoral agriculture (McClatchie et al. 2014) and flourished for a brief time (c. 3700–3600 BC) (McSparron 2008). It has been suggested that this early Neolithic 'boom' was followed by a middle Neolithic 'bust', with a decrease in agricultural activity and a change in domestic architecture to less substantial structures (Whitehouse et al. 2014). However, this argument should be balanced against the enormous increase in monument construction at this time (Smyth 2014).

In the late Neolithic (c. 3100–2500 BC), the focus moved away from tombs towards the building of large ritual spaces, possibly for seasonal gatherings. Brú na Bóinne continued to be a major focus of monument construction, incorporating a high density of monuments, including at least seven embanked enclosures and four post-circles (Stout 1991). Evidence for later prehistoric activity within the area is scant. The Bronze Age (c. 2500–500 BC) has been interpreted as a time of reduced monument construction within Brú na Bóinne, although in the early part of this period significant settlement activity has been identified (Mount 1994; O'Kelly et al. 1981). Iron Age (c. 500 BC to AD 400) activity is less elusive, with burials at Knowth (McGarry 2012) and suggestions of Iron Age activity at Newgrange (Ó Néill 2013).

Besides the surveys and excavations in the area, there has been a little work on using soils and sediments to better understand the prehistoric landscape of the area. For instance, preliminary studies of the geomorphology and environmental records of the region have revealed a post-glacial landscape history that fits with a regional model of Neolithic landscape clearance, and have suggested potential relationships between monument siting and aspects of the river system (such as one round barrow possibly having been built on an island at Newgrange—Foster and Turner 2009). A recent soil micromorphology study of soils at Knowth (Ostericher 2014; Ostericher and Lewis, in press) has developed a model of land use and monument construction at the site, and explored issues of soil change over time. Despite the demonstrated strengths of soil micromorphology for interpretation of earthen monuments and archaeological landscapes (e.g. Courty et al. 1989; French 2003; French et al. 2005, 2007; Goldberg and Macphail 2006; Macphail 2007), the approach has been applied only rarely at Irish sites (e.g. Ellis 2002; Lewis 2003; Verrill 2006; Verrill and Tipping 2010).

At Knowth the buried soil showed a history of stable vegetation, probably Mesolithic forest, which was disturbed locally, probably related to pre-monument settlement activity (e.g. Eogan and Roche 1997), and monument construction (Ostericher 2014). Variations in turves used to build the monument showed wetter grassland soils as well as tilled soils, reflecting local land use at the time of the monument's construction. Evidence of scarping and turf stripping were found in the soil profiles, while pre-passage tomb surfaces and use of space were also discussed (Ostericher 2014; Ostericher and Lewis, in press). The findings augment our understanding of the site and its landscape history, demonstrating the value of advanced analytical soil approaches for archaeological interpretation in Ireland.

In the Early Medieval period (c. AD 400–1000) a number of large, complex sites are known at Brú na Bóinne. These include at least five multivallate (more than one rampart) enclosures, likely to have been of high status and potentially under the control of Knowth, the seat of kings of Northern Brega from the 7–10th centuries AD (Swift 2008; Mac Shamhráin 2004). While there is little direct evidence for agricultural practice in Brú na Bóinne at this time, studies of Early Medieval agriculture in Ireland show an intense focus on cattle, especially in the earlier part of the Early Medieval period, with cultivation of barley, oats and wheat intensifying from the 8th century AD onwards (e.g. O'Sullivan et al. 2014; McCormick et al. 2014). Field systems appear to have clustered around settlement enclosures (ringforts), but the land use in these fields has yet to be systematically investigated. Later in the medieval period, field systems with a thoroughfare and associated house structures or enclosures appear. These 'Deserted Medieval

Fig. 19.4 Left, Dowth Henge. Image Ken Williams. Right, magnetic Gradiometry Survey showing field system on the Lands of Dowth Hall. (Data courtesy of Dr. Knut Rassman and the Römisch-Germanische Kommission)

Villages' (DMVs) are widely recorded across Ireland (Glasscock 1971; O'Connor 1998; Barry 2002), with some examples recently identified by lidar survey in the Boyne Valley (e.g. Dowth; Wardstown—Davis 2011).

A Palimpsest of Farming Landscapes on the Lands of Dowth Hall

Located around 7 km east of Slane village, the lands at Dowth Hall represent one of the largest land-holdings (c. 174 ha) in the Brú na Bóinne WHS. These represent a remarkably intact demesne landscape, which has retained its integrity since at least the 13th century AD, mostly under the ownership of the Netterville family (early 14th–early 19th century; Byrne et al. 2008). The land is currently under pasture, and has been home to a variety of farming landscapes from the Neolithic on. The Neolithic presence is most apparent from two small passage tombs and Dowth Henge (Site Q), located on a prominent east-west shale ridge. It is thought that most Irish henges were made by scarping the enclosed area for the construction of the earthen banks (Stout 1991), an interpretation corroborated by recent soil micromorphological study at Rath Maeve near Tara (Ostericher 2014), but Dowth Henge has both internal and external ditches, which may have provided much of the earthen material used in its bank (Fig. 19.4).

Present-day land management strategies, focused primarily on restoring the fertility of the land, have helped in the discovery of medieval farming landscapes at Dowth Hall. Systematic estate-wide soil analysis for agronomic nutrient management identified unusually high phosphate levels in three areas. Enhanced phosphate levels relative to the surrounding area can reflect past human activity, with potential cultural sources including human and animal waste, refuse, burials, and wood ash (Holliday and Gartner 2007, 301–5). Geophysical survey was subsequently carried out over the Dowth Hall phosphate anomalies (Davis 2013), leading to the identification of two concentric, multivallate enclosures (early medieval ringforts), one of which has an associated field system, and a later medieval settlement (DMV).

The DMV is located immediately south of Dowth Castle and church, and consists of a series of plots aligned either side of a natural break of slope that probably functioned as a thoroughfare. The Dowth DMV has been entirely levelled, but it persists as anomalies in soil characteristics, both physical (e.g. soil depth owing to ditch features) and chemical (e.g. variations in soil magnetic properties; variations in organic matter content). Soils of DMV sites preserve residual traces of past agricultural regimes and land management practices, which can be revealed through geophysical survey, geochemical survey, and LiDAR.

However, chronological resolution can only be provided by more invasive techniques, such as excavation or augering (Eogan 2008).

In the 18th century AD, another farming landscape emerged at Dowth Hall, which deliberately referenced both its medieval and prehistoric past. The demesne landscape surrounding the neoclassical villa of Dowth Hall incorporated features of an earlier medieval manor, including fish ponds and possibly the deer park, and referenced monuments created by prehistoric farmers, most notably Dowth Henge. The main drive has deliberately exploited the henge by skirting its edge to incorporate visible elements from the prehistoric, medieval, and 18th century landscapes. The desire to incorporate the historical narrative into later landscapes continues with the vision of the current landowners to create a living, sustainable, heritage and farming landscape, and to raise awareness of the farmers of Dowth who came before them.

19.6 Managing Soil Heritage

Managing the cultural heritage of Irish soils and sediments requires an understanding of the palimpsest-like nature of our landscape. Recognising heritage value in earth-bound materials must develop through an understanding of and engagement with people and their values in the lived landscape. One example where this has been achieved is the Burren Programme, which has established cooperation in managing cultural and biotic heritage in an Irish living agricultural landscape (Dunford 2016). Country-wide agri-environmental payment schemes (REPS, AEOS and GLAS) have aimed to protect visible archaeological monument heritage. The GLAS scheme at the time of publication pays €120 annually per monument in grassland, or €146 in tillage, for one to twenty monuments under the heading "protection and maintenance of monuments". Since eligible monuments must be "visible in the opinion of the [agricultural] advisor", no soil-bound or ploughed monuments are recognised or protected from cultivation, scrub removal, grazing pressure or other disturbances during agriculture (DAFM 2015) (Fig. 19.5).

Land managers involved in farming, forestry, recreation and ecosystem services should be aware of the need to identify and protect upstanding monuments, and their obligations under the relevant statutory legislation. Between the known monuments, and throughout the soil profile, there are discontinuous phases of small features, dispersed artefacts, and undiscovered monuments, from a landscape of continuous occupation. In Ireland, the National Monuments Acts 1930–2004 clearly protect all artefacts, ecofacts, soils, sediments and constructions of archaeological significance, and land managers are required to be aware of this legislation. However, there are major gaps in protection. For instance soil disturbance is not controlled where it is related to agricultural activity outside designated protected zones. In such areas, field-walking programmes could be important.

One of the goals of archaeological heritage management is to preserve the archaeological record in situ. This is difficult in the face of changing climate and land use practices, and particularly where drainage and tillage regimes are altered. Soil environments change naturally over time, and it is not possible to preserve soil and sediment contexts as they were when they were first deposited. However, we can endeavour to prevent major changes that would lead to accelerated decay and loss. It is known that building construction, quarrying, mining, and road development will greatly change nearby soil conditions, especially drainage, resulting in increased decay of many types of archaeological objects and deposits (e.g. French et al. 1999; French 2004). It is also known that continued practices, some traditional, can lead to accelerated erosion and loss of the archaeological record. For instance, repeated ploughing leads to increased erosion and destruction of buried archaeology by deflation and impact of soil disturbing implements. Where assessment suggests degradation, many approaches to conservation can be proposed. One approach to retaining waterlogged conditions, where desired, is to prevent lateral drainage by using impermeable rigid sheeting inserted vertically around a site, leaving a wet island inside. This has been done at Corlea, Co. Offaly, where an 80 m stretch of a 1 km Iron Age trackway has been preserved for a length of 80 m. Timber from the remaining 920 m of known trackway was destroyed during peat extraction and presumably continues to degrade elsewhere in drained sections of the bog where the water table fluctuates.

Monitoring using geochemical sensing, as well as repeated sampling of biotic remains, has been the basis of management in many locations (e.g. PARIS1–5), and preservation in situ can offer visitors the possibility to connect with a site in ways that are not provided by dislocated museum exhibits. Dublin's Viking city is at least partly preserved by record (Wallace 2016), while parts of the Viking city of York in England are preserved in situ. Current work on a Roman boat preserved under Guy's Hospital in London, UK, demonstrates the state of the art (Van Walt 2016), with monitoring of redox potential and other parameters for five to eight years, and access kept open, to be followed by a backup plan of excavation and preservation by record if the materials deteriorate.

One of the most important ways of protecting our soil heritage is through education. From an early age, people should be made aware of the potential of soil as an

Fig. 19.5 Fieldwalking to look systematically for artefacts turned up by cultivation at Tlachtga, Hill of Ward, Co. Meath, 2015. This method, with plotting of individual artefact locations, can be used to characterize the spatial arrangement of dispersed artefacts in topsoil. It may indicate hot-spots for further investigation, but also reveal subtler distributions that reflect wider processes in the ancient cultural landscape. Image Katherine McCormack

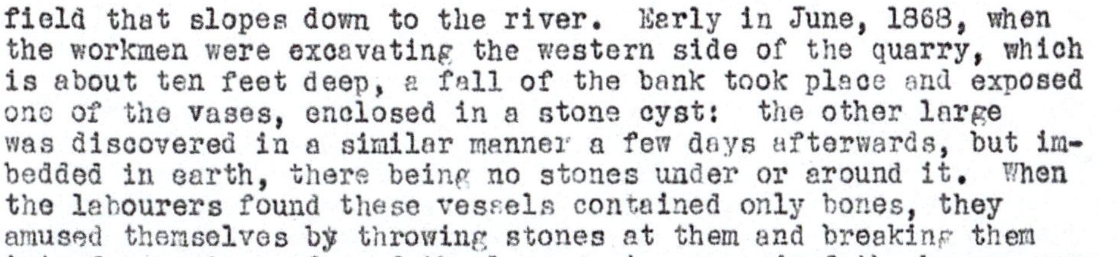

field that slopes down to the river. Early in June, 1868, when
the workmen were excavating the western side of the quarry, which
is about ten feet deep, a fall of the bank took place and exposed
one of the vases, enclosed in a stone cyst: the other large
was discovered in a similar manner a few days afterwards, but im-
bedded in earth, there being no stones under or around it. When
the labourers found these vessels contained only bones, they
amused themselves by throwing stones at them and breaking them
into fragments; afew of the larger pieces, and of the bones, were
preserved and brought to Richard A. Gray, Esq., County Surveyor, who
kindly placed them at my disposal; I visited the locality, and got
the particulars of their discovery from the workmen, who likewise
gathered for me all the pieces they could collect of the broken
vessels: in arranging them I detected the third or smaller vase,

Fig. 19.6 Extract from a file of the National Museum of Ireland relating the account of discovery of a Bronze-Age cist with ceramic vessel containing human bone material, during quarry operations at Palmerstown in west Dublin. It is important that everyone understands about soil-bound archaeological heritage, since it may turn up unexpectedly, as recorded here

archaeological and environmental store of heritage information, and have a sense that 'strange' objects found in the ground can potentially be important (Fig. 19.6). In the nineteenth century, soil heritage consisted of treasure in most people's minds; in the twentieth, cultural materials and stratigraphic sequences became generally understood; now we are increasingly recognising the landscape and post-deposition processes too.

References

Barker G (1985) Prehistoric farming in Europe. Cambridge University Press, Cambridge

Bennett R, Welham K, Hill RA, Ford A (2013) Airborne spectral imagery for archaeological prospection in grassland environments—an evaluation of performance. Antiquity 87:220–236

Bermingham N, Moore C (2015) The Drumclay crannog-dwellers: revealing 1000 years of lakeside living. Current Archaeol 299:30–37

Bermingham N, Delaney M (2006) The bog body from Tumbeagh. Wordwell, Bray

Brade-Birks SG (1951) Addendum. In: O'Kelly MJ (ed) (1951) Some soil problems in archaeological excavation. J Cork Hist Archaeol Soc LVI: 40–42

Byrne FJ, Jenkins W, Kenny G, Swift C (2008) Excavations at Knowth, vol 4: Historical Knowth and its hinterland. Royal Irish Academy, Dublin

Case HJ, Dimbleby GW, Mitchell GF, Morrison MES, Proudfoot VB (1969) Land use in Goodland Townland, Co Antrim, from Neolithic times until today. J R Soc Antiqu Irel 99:39–53

Caulfield S, O'Donnell RG, Mitchell PI (1998) 14C dating of a Neolithic field system at Céide Fields, County Mayo, Ireland. Radiocarbon 40(2):629–640

Clayton B (1973–4) Palaeoecological investigations at Crossnacreevy, Co. Down. Ulst J Archaeol 36/37:42–51

Collins JF (2008) Quickening the Earth: soil minding and mending in Ireland. UCD School of Biology and Environmental Science, Dublin

Common RH (1940) The blue layer. In: Herring IJ, May AmcL, Cloghnagalla Cairn, Boviel, Co Londonderry. Ulst J Archaeol 3rd Ser I:48–49 http://www.jstor.org/stable/20627310

Cornwall IW (1958) Soils for the archaeologist. London

Courty M-A, Goldberg P, Macphail RI (1989) Soils and micromorphology in archaeology. Cambridge University Press

Creamer R, Simo I, Reidy B, Carvalho J, Fealy R, Hallett S, Jones R, Holden A, Holden N, Hannam J, Massey P, Mayr T, McDonald E, O'Rourke S, Sills P, Truckell I, Zawadzka J, Schulte R (2014) Irish soil information systems: integrated synthesis report. Environmental Protection Agency, Wexford. http://erc.epa.ie/safer/resource?id=1bfe6109-3b92-11e4-b233-005056ae0019, with survey mapping at http://soils.teagasc.ie/

Crutchley S (2010) The light fantastic. Using airborne lidar in archaeological survey. Historic England

Cunningham D, Farrell EP, Collins JF (1999) Soil responses to land-use change ± a study in south-west Ireland. For Ecol Manag 119:63–76

DAFM (2015) Green, Low-carbon, Agri-environment Scheme (GLAS) Specification for Tranche 2 Participants 31st December 2015. Department of Agriculture Food and the Marine. Available at http://agriculture.gov.ie/farmerschemespayments/glas/glastranche2/, accessed 11 Sept 2016

Davis S (2011) The unseen Hill of Ward: new insights from LIDAR data. Archaeol Irel 25(4): 36–40. http://www.jstor.org.ucd.idm.oclc.org/stable/41550248

Davis S (2013) New geophysical surveys at Dowth Henge, Brú na Bóinne. PAST Newsl Prehist Soc 75:2–3. http://www.prehistoricsociety.org/files/PAST_75_for_website.pdf

Davis S, Lewis H, Turner J, Foster G, O'Donnell L, Collins B, McClatchie M, Allen MJ, McCormick F (in press) Environmental evidence from Knowth and its landscape setting. In: Eogan G, Cleary K (eds) Excavations at Knowth, vol 6. Royal Irish Academy, Dublin

Dimbleby GW (1965) Post-glacial changes in soil profiles. Proc R Soc B(161):355–362

DoAHGI (1999) Framework and Principles for the Protection of the Archaeological Heritage. Department of Arts, Heritage, Gaeltacht and the Islands. Dublin: the Stationery Office

Doneus M, Verhoeven G, Atzberger C, Wess M, Rus M (2014) New ways to extract archaeological information from hyperspectral pixels. J Archaeol Sci 52:84–96

Donoghue DNM, Beck AR, Galiatsatos N, McManus KB, Philip G (2006) The use of remote sensing data for visualising and interpreting archaeological landscapes. In: Baltsavias E, Gruen A, van Gool L, Pateraki M (eds) Recording, modeling and visualization of cultural landscape. Taylor & Francis, London, UK

Dowd M, Carden RF (2016) First evidence of a Late Upper Palaeolithic human presence in Ireland. Quat Sci Rev 139:158–163. https://doi.org/10.1016/j.quascirev.2016.02.029

Drew DP (1983) Accelerated soil erosion in a karst area: the Burren, western Ireland. J Hydrol 61:113–124

Dunford B (2016) The Burren Life Programme (BLP) 2016 Progress Report. Burren Life Programme, Carran. http://www.burrenprogramme.com/wp-content/uploads/2015/12/Burren-Life-Progress-Report-2016.pdf

Edwards KJ, Whittington G (2001) Lake sediments, erosion and landscape change during the Holocene in Britain and Ireland. CATENA 42:143–173

Ellis C (2002) Ráith na Ríg, Tara: soil micromorphological and bulk sample analysis. Discovery Programme Reports: 6. Royal Irish Academy/Discovery Programme, Dublin, 127–165

Eogan G (2008) Excavations at Knowth 4. Royal Irish Academy, Dublin

Eogan G, Roche H (1997) Excavations at Knowth 2. Dublin, Royal Irish Academy

Eogan G, Doyle P (2010) Guide to the Passage Tombs at Brú na Bóinne. Wordwell, Bray

Fédoroff N, Courty MA (1987) Paléosols. In: Miskowsky J-C (ed) Géologie de la préhistoire. Masson, Paris, pp 251–280

Foster G, Turner J (2009) The fluvial geomorphology of the lower Boyne valley. In: Lewis H, Gallagher C, Davis S, Turner J, Foster G, Courtney L, Goucher K, Scaife R, Langdon C, Holmes N, Mulroony G, Allen MJ, Brady C, Guinan L, Brown A, Cummins T, Meehan R (eds) An integrated, comprehensive GIS model of landscape evolution and land use history in the River Boyne vally. Phase II report 2009, Appendix XII. Published online at: http://www.heritagecouncil.ie/fileadmin/user_upload/INSTAR_Database/GIS_Landscape_and_Landuse_History_in_Boyne_River_Valley_Progress_Report_09.pdf

Fredengren C, Kilfeather, Stuijts I (2010) Lough Kinale: studies of an Irish lake (The Discovery Programme Lake Settlement Project: Discovery Programme Monograph No. 8). Wordwell, Bray

French C (2003) Geoarchaeology in action: studies in soil micromorphology and landscape evolution. Routledge, London

French C (2004) Hydrological monitoring of an alluviated landscape in the lower Great Ouse valley at Over, Cambridgeshire: results of the gravel extraction phase. Environ Archaeol 9:1–12

French C, Davis M, Heathcote J (1999) Hydrological monitoring of an alluviated landscape in the lower Great Ouse valley, Cambridgeshire: interim results of the first three years. Environ Archaeol 4:41–56

French C, Lewis H, Scaife R, Allen M (2005) New perspectives on Holocene landscape development in the southern English chalklands: the upper Allen valley, Cranborne Chase, Dorset. Geoarchaeol Int J 20(2):109–134

French C, Lewis H, Allen MJ, Green M, Scaife R, Gardiner J (2007) Prehistoric landscape development and human impact in the upper Allen valley, Cranborne Chase, Dorset. McDonald Institute for Archaeological Research, Cambridge

Gaffney C, Gaffney V, (2011). Through an imperfect filter: geophysical techniques and the management of archaeological heritage. In D Cowley (ed.) Remote sensing for archaeological heritage management in the 21st century. Europae Archaeologiae Consilium. 117–128

Gaffney CF, Gater JA (2003) Revealing the Buried Past: geophysics for archaeologists. Tempus, Stroud

Gallagher V, Knights K, Carey S, Glennon M, Scanlon R, (2016) Topsoil Ebook—Atlas of Topsoil Geochemistry of the Northern Counties of Ireland Data from the Tellus and Tellus Border Project. Geological

Survey of Ireland, 2016. https://jetstream.gsi.ie/iwdds/delivery/GSI_Transfer/Tellus/EBook_Topsoils_Final_03Feb2016.pdf

Gillis J (2012) The methodology used for dewatering the vellum from the Faddan More Psalter. J Inst Conserv 35(2):137–151. https://doi.org/10.1080/19455224.2012.744678

Goldberg P, MacPhail RI (2006) Practical and theoretical geoarchaeology. Blackwell, Oxford

Gowen M, Ó Néill J, Philips m (eds) (2005) The Lisheen Mine archaeological project 1996—8. Wordwell, Bray, p 5

Grogan E, O'Donnell L, Johnston P (2007) The Bronze Age landscapes of the Pipeline to the West: an integrated archaeological and environmental assessment. Dublin, Wordwell

Hammond RF (1981) The Peatlands of Ireland. An Foras Taluntais, Dublin

Hoefer G, Lueck E, Ruehlmann J, Gundelach V Bachmann J (2012) Comparison of non-destructive geophysical methods to predict the heterogeneity of subsoil compaction. In: 2nd international conference on hydropedology, Leipzig 22–27 July 2012

Hoefer G, Gundelach V, Ruehlmann J, Bachmann J (2009) The capability of non-destructive geophysical methods in precision agriculture to capture subsoil mechanical strength. In: Precision agriculture '09, 7th European conference on precision agriculture, Wageningan, 6–8 July 2009

Holliday VT, Gartner WG (2007) Methods of soil P analysis in archaeology. J Archaeol Sci 34:301–333

Huang CC, O'Connell M (2000) Recent land-use and soil erosion history within a small catchment in Connemara, western Ireland: evidence from lake sediments and documentary sources. CATENA 41:293–335

Jessen K (1949) Studies in the Late Quaternary deposits and flora-history of Ireland. Proc R Irish Acad 52B:85–290

Leigh J (2010) Lyons Hill County Kildare Licence No. 10-R-119 15/10/2010, Client: Kill History Group. J.M. Leigh Surveys, unpublished

Jenny H (1941) Factors of soil formation. Soil Sci 52(5):415

Johnson DL (2002) Darwin would be proud: bioturbation, dynamic denudation, and the power of theory in science. Geoarchaeology 17 (1):7–40

Jongerius A (1983) The role of micromorphology in agricultural research. In: Bullock P, Murphy CP (eds) Soil micromorphology, volume 1: techniques and application. In: Proceedings of the 6th international working meeting on soil micromorphology, London 1981, 111–38. AB Academic Publishers, Berkhamsted

Keeley HCM, Macphail RI (1981) A soil handbook for archaeologists. Bulletin No. 18, 225–241. Institute of Archaeology, London

Kelly EP (2013) An archaeological interpretation of Irish Iron Age bog bodies. Archaeol Violence Interdiscip Approaches 2:232

Knudson KJ, O'Donnabhain B, Carver C, Cleland R, Price TD (2012) Migration and Viking Dublin: paleomobility and paleodiet through isotopic analyses. J Archaeol Sci 39:308–320. https://doi.org/10.1016/j.jas.2011.09.014

Lee J (1974) An analysis of land potential for grazing in Ireland with particular reference to farm size relationships. J Stat Soc Inq Soc Ireland 23:149-182

Lewis H (1998) The characterisation and interpretation of ancient tillage practices through soil micromorphology: a methodological study. Unpublished Ph.D. thesis, University of Cambridge

Lewis H (2003) Soil micromorphology and geoarchaeology at Parknabinnia Court Tomb (Clare Megalith 153), Co. Clare, Ireland. Unpublished manuscript. Available at: http://hdl.handle.net/10197/4533

Lomas-Clarke SH, Barber KE (2004) Palaeoecology of human impact during the historic period: palynology and geochemistry of a peat deposit at Abbeyknockmoy, Co., Galway, Ireland. The Holocene 14 (5):721–731

Lydon K, Smith G (2014) CORINE landcover 2012 Ireland final report. Environ Prot Agen. Available from http://www.epa.ie/pubs/data/corinedata/CLC2012_IE_Final_Report.pdf

Macphail RI (1986) Paleosols in archaeology: their role in understanding Flandrian pedogenesis. In: Wright VP (ed) Paleosols: their recognition and interpretation. Blackwell, Oxford

Macphail RI, Robertson L (1989) The application of micromorphology to the understanding of Holocene soil development in the British Isles; with special reference to early cultivation. In: Federoff N, Bresson, LM, Courty MA (eds) Soil micromorphology: INRA, pp 647–656

Macphail RI (2007) A Pre-Barrow Thin Section. In: Benson D, Whittle A (eds) Building memories: The Neolithic Cotswold Long Barrow at Ascott-under-Wychwood, Oxfordshire. Oxbow, Oxford

Macphail RI (1990) Soil micromorphological evidence of the impact of ancient agriculture. In: 14th international congress of soil science transactions, vol 7. International Society of Soil Science, Kyoto, pp 264–69

Mac Shamhráin A (2004) The island of St Patrick: church and ruling dynasties in fingal and meath, pp 400–1148. Four Courts, Dublin

McClatchie M, Bogaard A, Colledge S, Whitehouse NJ, Schulting RJ, Barratt P, McLaughlin TR (2014) Neolithic farming in north-western Europe: archaeobotanical evidence from Ireland. J Archaeol Sci 51:206–215

McCormick F, Kerr TR, McClatchie M, O'Sullivan A (2014) Early medieval agriculture, livestock and cereal production in Ireland, AD 400–1100. British archaeological reports international series 2647, Archaeopress, Oxford

McGarry T (2012) Late Iron Age burials at Knowth. In: Eogan G (ed) The archaeology of Knowth in the first and second millennia AD. Excavations at Knowth 5. Royal Irish Academy, Dublin, pp 687–693

McKenzie CJ, Murphy EM, Donnelly CJ (2015) The science of a Lost Medieval Gaelic Graveyard. The Ballyhanna research project. Transport Infrastructure Ireland, Dublin

McSparron C (2008) Have you no homes to go to? Neolithic housing. Archaeol Irel 22(3):18–21

Mitchell GF (1951) Studies in Irish Quaternary deposits: No. 7. Proc R Irish Acad 53B:111–206

Mitchell GF (1972) Soil deterioration associated with prehistoric agriculture in Ireland. In: Proceedings of the twenty fourth international geographical congress (Montreal), Symposium 1. Ottawa, Geological Survey of Canada, pp 59–69

Mitchell FJG, Stefanini BS, Marchant R (2013) A catalogue of Irish pollen diagrams. Proc R Irish Acad 113B:1–31

Moles NR, Moles RT (2002) Influence of geology, glacial processes and land use on soil composition and Quaternary landscape evolution in The Burren National Park, Ireland. Catena 47(4):291–321

Moles NR, Moles RT, O'Donovan G (1995) Evidence for the presence of Quaternary loess-derived soils in the Burren karstic area. Ir Geogr 28:48–63

Moles RT, Moles NR, Leahy JJ (1999) Radiocarbon-dated episode of Bronze Age slope instability in the southeastern Burren, County Clare. Ir Geogr 32:52–57

Moloney A, Bermingham N, Jennings D, Keane M, McDermott C, Ocarroll E (1995) Blackwater survey & excavations atrefact deterioration in Peatlands Loughmore Co. Mayo. Irish Archaeological Wetland Unit: Transactions Volume 4. Crannóg Publications, Dublin

Mount C (1994) Aspects of ritual deposition in the Late Neolithic and Beaker Periods at Newgrange, Co. Meath. Proc Prehist Soc 60: 433–443

O'Connell A (2013) Harvesting the stars. A Pagan Temple at Lismullin Co. Meath. Wordwell, Bray

O'Connell M, Molloy K (2001) Farming and woodland dynamics in Ireland during the Neolithic. Biol Environ Proc R Irish Acad 101B:98–128

O'Kelly MJ (1951) Some soil problems in archaeological excavation. J Cork Hist Archaeol Soc LVI 29–39:44

O'Kelly MJ, Cleary RM, Lehane D (1981) Newgrange, Co. Meath, Ireland. The Late Neolithic/Beaker period settlement. British Archaeological Reports 190, Oxford

Ó Néill J (2013) Being prehistoric in the Irish Iron Age. In: O'Sullivan M, Scarre C, Doyle M (eds) Tara—from the past to the future. Wordwell, Dublin, pp 256–266

Opitz R, Cowley D (2013) Interpreting archaeological topography. Oxbow, Oxford

Ostericher I (2014) Soil micromorphology of Rath Maeve and Knowth: landscape and monumentality in prehistoric Ireland. Unpublished MA Thesis, University College Dublin

Ostericher I, Lewis H (in press) Chapter 8. Soil micromorphology of Knowth. In: Eogan G, Cleary K (eds) Excavations at Knowth, vol 6. Royal Irish Academy, Dublin

O'Sullivan A, Sheehan J (1996) The Iveragh Peninsula: an archaeological survey of South Kerry. Cork University Press, Cork

O'Sullivan A, McCormick F, Kerr TR, Harney L, Kinsella J (2014) Early medieval dwellings and settlements in Ireland, AD 400–1100. British archaeological reports international series 2604, Archaeopress, Oxford

O'Sullivan A, Warren G, McDermott C, Lyne E (2009) Archaeological excavations of a medieval church and graveyard at Templeteenaun Ballinagee townland, Co. Wicklow 2004–2008, UCD School of Archaeology, University College Dublin

PARIS1–5, conferences on Preserving Archaeological Remains in Situ, held in 1996, 2001 (Museum of London Archaeology Service), 2006 (VU of Amsterdam, 2008), 2011 (Conservation and Management of Archaeological Sites, 2012 vol 14(1–4)), and 2015 (Conservation and Management of Archaeological Sites, 2016 vol 8(1–3)

Proudfoot B (1960) Appendix I: Report on soils and stratigraphy at Castle Skreen. In: Dickinson CW, Waterman DM, Proudfoot B, Mitchell GF Excavations at Castle Screen, Co. Down, Ulst J Archaeol 23:63–77

Proudfoot B, Simmons IG (1958) Appendix 3. Soil report. In: Waterman DM, Excavations at Ballyfounder Rath, Co.Down. Ulst J Archaeol 21:54–61

Raftery B (1996) Trackway excavations in the Mountdillon Bogs, Co. Longford. Irish Archaeological Wetland Unit Transactions 3. Dublin. Crannóg Publications

Revel JC (1991) Pédologie et archéologie. In: Guilaine J (ed) Pour une archéologie agraire. Armand Colin, Paris, pp 323–343

Russell I, Hurley MF (2014) Woodstown a Viking-Age settlement in Co. Waterford. Four Courts Press, Dublin

Roberts N (2014) The Holocene. Blackwell

Rogers T, Fibiger L, Lynch LG, Moore M (2006) Two glimples of nineteenth-century burial practice in Ireland: a report on the excavation of burials from Manorhamilton Workhouse, Co. Leitrim, and St. Brigid's Hospital, Ballinasloe, Co. Galway. J Ir Archaeol 15:93–104. http://www.jstor.org/stable/20650852

Romans JCC, Robertson L (1983) The general effects of early agriculture on the soil profile. In: Maxwell GS (ed) The impact of aerial reconnaissance on archaeology. Council for British Archaeology research report No. 49. Council for British Archaeology, London, pp 136–41

Smith AG (1961) Cannons Lough, Kilrea, Co., Derry: stratigraphy and pollen analysis. Proc R Ir Acad 61B:369–383

Smyth J (2011) The house and group identity in the Irish Neolithic. Proc R Irish Acad 111C:1–31

Smyth J (2014). Settlement in the Irish neolithic: new discoveries at the edge of Europe. Oxbow Books

Stott D, Boyd DS, Beck A, Cohn AG (2015) Airborne LiDAR for the detection of archaeological vegetation marks using biomass as a proxy. Remote Sens 7(2):1594–1618. https://doi.org/10.3390/rs70201594

Stout G (1991) Embanked enclosures of the Boyne region. Proc R Irish Acad 91C:245–284

Swift C (2008) The early history of Knowth. In: Byrne FJ, Jenkins W, Kenny G, Swift C, Excavations at Knowth: historical Knowth and its Hinterland, vol. 4. Royal Irish Academy, Dublin, pp 5–42

Tewari S, Kulhavý J, Rock BN, Hadaš P (2003) Remote monitoring of forest response to changed soil moisture regime due to river regulation. J For Sci 49:429–438

Thompson AD (1997) A lake sediment record of late glacial and Holocene environmental change from Molly's Lough, Co. Clare, Ireland. Ph.D. thesis, University of Wales, Aberystwyth

Thorp M, Glanville P (2003) Mid-Holocene sub-blanket-peat alluvia and sediment sources in the upper Liffey valley, Co., Wicklow, Ireland. Earth Surf Proc Land 28(9):1013–1024. https://doi.org/10.1002/esp.517

Tuominen J, Lipping T, Kuosmanen V, Haapanen R (2009) Remote sensing of forest health. In Ho PGP (ed.) Geoscience and remote sensing. InTech, Rijeka

Turner JN, Holmes N, Davis SR, Leng MJ, Langdon C, Scaife RG (2015) A multiproxy (micro-XRF, pollen, chironomid and stable isotope) lake sediment record for the Lateglacial to Holocene transition from Thomastown Bog, Ireland. J Quaternary Sci 30 (6):514–528. https://doi.org/10.1002/jqs.2796

Ullrich JM (2010) Geochemical contributions to the archaeological examination of promontory forts: the use of phosphate analysis techniques on Achillbeg and Achill Islands, Co. Mayo. Ph.D. Thesis, School of Archaeology, University College Dublin

Ullrich JM (2013) Assessing interior and exterior divisions of space using phosphate analysis spot test methods. Assem Sheff Grad J Archaeol 12:43–56. http://www.assemblage.group.shef.ac.uk/journal/issue-12

UNESCO (1993) Brú na Bóinne—Archaeological ensemble of the bend of the Boyne. Inscriptions on the World Heritage List, 17th session of the Committee Decision code CONF 002 XI. http://whc.unesco.org/en/decisions/3335

Van Walt (2016) Roman boat at Guy's Hospital. Van Walt Environmental Connect. http://www.vanwalt.com/blog/2016/05/van-walt-putting-your-name-in-lights-saving-your-money/

Verrill L (2006) Later prehistoric environmental marginality in Western Ireland: Multi-proxy investigations. Unpublished Ph.D. thesis, University of Edinburgh

Verrill L, Tipping R (2010) A palynological and geoarchaeological investigation into Bronze Age farming at Belderg Beg, Co., Mayo, Ireland. J Archaeol Sci 37(6):1214–1225

Wallace P (2016) Viking Dublin. The wood quay excavations. Irish Academic Press Limited

Warren G (2008) Fieldwork in Belderrig, Co. Mayo 2004–2008 04E0893. UCD School of Archaeology. http://hdl.handle.net/10197/3524

Weir AH, Catt JA, Madgett PA (1971) Postglacial soil formation in the loess of Pegwell Bay, Kent, England. Geoderma 5:131–149

Whitehouse NJ, Schulting RJ, McClatchie M, Barratt P, McLaughlin TR, Bogaard A, Marchant R, Gaffrey J, Bunting MJ (2014) Neolithic agriculture on the European western frontier: the boom and bust of early farming in Ireland. J Archaeol Sci 51:181–205

Soil: Construction, Planning and Extraction 20

Hendrik van der Kamp, Maryann Harris, and Vincent Gallagher

Abstract

The sealing of soils can result in reduced biodiversity, increased flooding of built areas and loss of productivity. Ireland has a sealed surface of 292 m^2 per person, much higher than the EU average owing to the extremely dense road network for a very sparsely populated country. To minimise soil sealing, identifying previously developed sites or brownfield sites, for development is increasingly being used. Risk assessment, environmental assessment, detailed risk management and mitigation strategies applied during construction followed by a post-construction monitoring phase are necessary for these developments due to potential contamination issues associated with historical uses. In an Environmental Impact Statements, construction projects must include a list of alternative uses for the receiving environment. One potential value is the agricultural value of the soil. Increasingly, urban agriculture is being recognised in urban planning which is viewed as having a role beyond food production. Flooding events are common in Ireland, and although implementation has been slow, sustainable urban drainage systems (SuDS) are being promoted in Ireland since 2005. Soil in the urban environment also link green areas together and can support the maintenance of biodiversity. Some construction projects require soil to be stored for reuse to protect the genetic seedbank of the site prior to construction. The emergence of green infrastructure and the concept of urban metabolism are also described in this chapter. Finally, the impacts of extraction, including mining, on soils are described.

Keywords

Brownfield • Planning • Green infrastructure

20.1 Introduction

Construction projects greatly impact soil and therefore, soil management should be integral in the planning and construction process in a number of ways. While at site level a good standard of management in relation to soil is implemented, often the broader impacts and wider considerations can be lacking. As such, decisions on where to zone land for new urban developments are more likely to depend upon considerations such as transport infrastructure rather than on the basis of soil quality. However, in its simplest form, most construction entails the movement of the earth. Depending upon the nature of the development proposal and the local topography, soil will need to be imported or exported from the site. In order to minimise the cost and traffic movements, and for soil conservation purposes, 'cut and fill calculations' are determined whereby the amount of material from cuts should roughly match the amount of fill needed for the construction project. It is a normal requirement that topsoil that is stripped to facilitate construction has to be stored on-site or nearby in order to be reused within the development. This is done to minimise the loss of organic material and avoid compaction.

The quality of urban soils for a given site can be degraded as a result of previous land use and management practices over time. In the World Reference Base for Soil Resources (United Nations 2015), urban soils can thus be referred to as a separate category (Technosols) depending on the extent to

H. van der Kamp (✉)
School of Transport Engineering, Environment and Planning, Dublin Institute of Technology, Dublin, Ireland
e-mail: henk.vanderkamp@dit.ie

M. Harris
School of Biology and Environmental Science, University College Dublin, Dublin, Ireland

M. Harris
Parks and Landscape Services Division, Dublin City Council, Dublin, Ireland

V. Gallagher
Department of Communications, Climate Action and Environment, Geological Survey of Ireland, Beggars Bush, Haddington Road, Dublin 4, Ireland

R. Creamer and L. O'Sullivan (eds.), *The Soils of Ireland*, World Soils Book Series,
https://doi.org/10.1007/978-3-319-71189-8_20

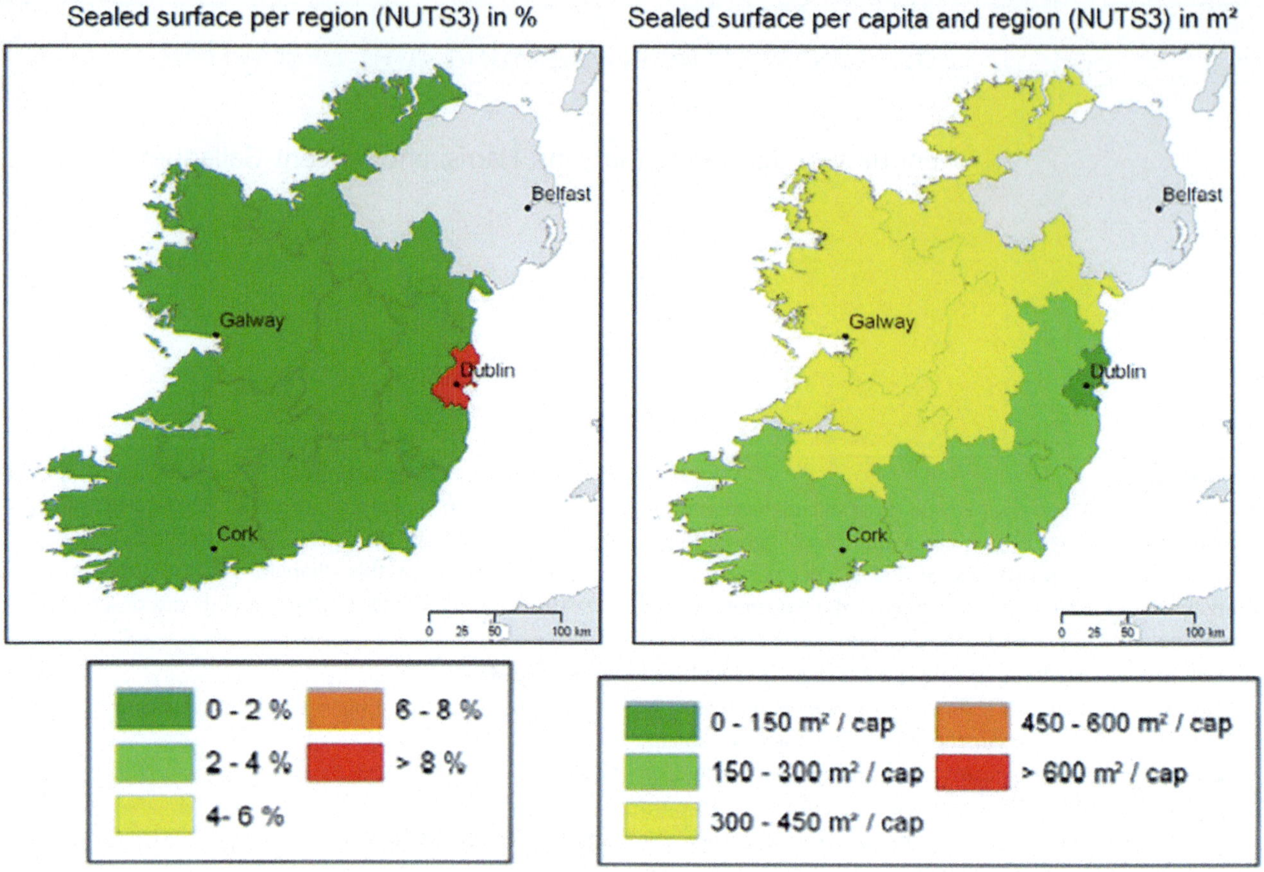

Fig. 20.1 Sealed surface by region and per person by region (Prokop et al. 2011) based on EEA-FTSP-Sealing-Enhancement European Mosaic: Pop EUSTAT, Environment Agency Austria Umweltbundesamt. As shown in the figure on the right, higher population density in urban Dublin decreases the sealed surface per capita, even though this area contains the highest sealed surface density (as shown by figure on the left)

which urban use has modified the soil structure. Technosols are described as soils that contain a significant amount of artefacts, are sealed by technic hard material or contain a geomembrane. They are also referred to as 'urban soils' (see Chap. 12). Geomembranes are often used for road construction, filtration of soil media and for archaeological preservation. In this chapter, we consider the role of soil as a platform for construction, options for sustainable spatial planning and finally, the impacts of extraction such as mining on the soil resource are presented.

20.2 Soil as a Platform for Construction

Urban development and infrastructure construction that covers the soil surface with impervious materials are known as soil sealing (EC 2016a). The sealing of soils as a result of construction activity is a matter of concern as it can result in reduced biodiversity, increased flooding of built areas and loss of productivity. Ireland has a sealed surface of 292 m^2 per person—about 30% above the European Union (EU) average, largely due to an extremely dense road network for a very sparsely populated country. There are 23 km of road per km, 1000 persons—one of the highest rates in the whole of the EU (Prokop et al. 2011). The sealed surface per capita rate (2010) is sharply divided on a regional basis with the rate in the Border, West, Mid-West and Midland regions of Ireland being double (300–450 m^2 per capita) that of the Dublin region (under 150 m^2 per capita) (Fig. 20.1). Potential impacts that construction projects may have on soils are land sealing, compaction (leading to loss of structure and soil drainage), increase/decrease in vulnerability status and changes in drainage patterns (EPA 2015). Compaction, while not sealing the soil as such, may lead to a reduction in the ability of these soils to store water and thus result in an increased run-off. The impact of subsoil removal is generally not significant in terms of altering subsoil geology as similar material to that being excavated is generally widespread throughout the area.

A strategy for minimising soil sealing in new developments is to identify lands that were previously used for urban development but which have become redundant. This is known as 'brownfield development'. In contrast, 'greenfield' development is urban development where agricultural land is typically converted into urban use. Although brownfield development is preferred over greenfield development, it creates its own challenges. Frequently, the historical uses and disturbances (see Chap. 12), including airborne deposition of substances, of brownfield lands can result in contamination of soil by the accumulation of toxic levels of substances. Due to potential contamination issues when redeveloping brownfield sites, there is likely to be a greater need for risk assessment during the planning and design phase, environmental assessment during planning phases and detailed risk management and mitigation strategies during construction. Also, a post-construction monitoring phase is required.

In Ireland, the Environmental Protection Agency (EPA) is the statutory body responsible for protecting the environment. Any unearthed contaminants will require varying degrees of remediation measures, which may require a licence from the EPA. The preparation of an Environmental Impact Statement requires that the receiving environment must always be described taking soils into account. Guidelines for the preparation of an Environmental Impact Statement have been issued by the relevant Government Department (DoECLG 2013a). Other guidelines are also used (e.g. IGI 2002) while the EPA has issued detailed advice notes (EPA 2015). Information on the soils in a study area is generally available from the EPA online mapping. The database indicates the types of soils in different parts of the country. This can form a basis for a description of the receiving environment in the case of a specific construction project. Trial pits are normally also done to provide additional information. Typically, these are done during the initial screening for an Environmental Impact Assessment to inform the nature of the assessment and not only the design and are used towards the development of the Environmental Impact Statement. While such trial pits are necessary for foundation design, they also provide useful information about the receiving environment. Where an Environmental Impact Statement is prepared for an urban development project and contaminated soil is suspected, soil sampling, geotechnical testing and chemical analysis for contaminants of potential concern are relevant measures to be undertaken (EPA 2015, p. 22).

During the construction stage, there is potential for contamination of soil which should be identified through prior risk assessment. These include soil contamination from spillages of fuel, lubricants, concrete and cement from construction plants or storage areas on-site. The risk of pollution may be increased in areas where excavation takes place or where the excavations or culverts may form preferential pathways for groundwater flow. To minimise impacts through compaction or spillage, construction management plans become an important feature of large construction projects. Such construction management plans typically specify routes for construction traffic and designate bunded areas where storage of liquids is done in a controlled manner and on an impervious surface. Where soil needs to be removed because of its contaminated nature, such removal of material from a site may require a licence under the Waste Management Act, 1996, prior to the undertaking of such works. The cleansing of contaminated soil is a necessary element of construction in those circumstances and may involve the removal of the soil and replacing it with clean soil. To cleanse contaminated soil different methods can be used. It can also be done 'in situ' either using chemical methods or natural processes (Swartjes 2011).

Radon is another important consideration related to soils and construction. Radon is a naturally occurring radioactive gas that comes from the natural breakdown (radioactive decay) of uranium. Radon accounts for more than half of the total radiation dose received by the Irish population (EPA 2016). It is found that, in rock and soil, Ireland has relatively high levels of radon gas. Concerns relate to the health implications as it is a known carcinogen. In 2006, a national action plan to deal with the relatively high-level concentration of Radon in houses in certain regions of Ireland was published by the Radiological Protection Institute of Ireland (RPII 2006). This report concluded that Ireland has an average value that is over four times higher than that of the United Kingdom and approximately twice that of Canada and the United States. The level of radon is measured in Becquerels per cubic metre—expressed as Bq/m^3—and the national 'reference level' is 200 Bq/m^3. In order to address the situation, Building Regulations (S.I. No. 497/1997) were amended in 1997 requiring that reasonable precautions are taken to avoid danger to health and safety caused by substances in the ground. Radon gas is such a substance and the Department of Environment, Heritage and Local Government has published guidance through Technical Guidance Document-C which requires that all homes built since 1st July 1998 have a potential means to extract radon from the substructure, such as a standby radon sump. In addition, all new houses in High Radon Areas require the fitting of an approved sealed membrane (radon barrier) of low radon permeability over the footprint of the building (RPII 2006, p. 19). For existing homes, the EPA recommends that one should consider taking action to reduce the level of radon in the home if the measurement is above this national reference level. Steps that are taken to reduce the radon level are generally known as 'radon remediation'. There are two main methods of radon remediation—either by preventing radon entering the home from the ground underneath or by removing the radon after it has entered the home. Testing is repeated post-remediation to ensure success.

20.3 Soils and Sustainable Spatial Planning

Soil is also treated as a material asset of natural origin that must be considered as such. Soils of the receiving environment of a construction project are described in an Environmental Impact Statement to enable assessment of the impacts that the construction project may have on that environment and also assess the value of the receiving environment in terms of potential alternative uses. A list of potential alternative benefits that soils may provide is included in the relevant EPA guidelines document (EPA 2015). This document states that soils and bedrock can be assessed in terms of their:

- Intrinsic scientific value/geological heritage
- Value as a material asset
- Agricultural value
- Habitat value
- Intrinsic part of the hydrological cycle (in qualitative and quantitative terms)
- Potential to interact with groundwater by facilitating movement or by chemically altering it
- Potential for erosion, removal or extraction
- Drainage characteristics
- Capacity as an organic filter and carbon sink
- Potential for negative environmental impact if contaminated material is present
- Potential for waste generation.

20.3.1 Urban Agriculture

In the past, the value of soil for food production was normally associated exclusively with greenfield development where a new development would convert previous agricultural use into urban use. Increasingly, however, agriculture also has an urban component. Unlike in industrial agriculture, urban agriculture sees its aims as more than food production and includes many other non-production oriented goals (Lohrberg et al. 2015). One such goal is the inclusion of agriculture in the metabolism of the city. …'*Urban agriculture may play a crucial role in the handling and circulation of water and nutrients and contribute to solutions for waste disposal, by acting as a sink for organic waste or slightly polluted water*'. (Lohrberg et al. 2015, p. 19). Soil can thus play a crucial role in achieving more closed nutrient and water cycles within the city as an ecological system.

Through encouragement and recognition in planning policies, urban agriculture can perform a much wider role in urban development. The following list is suggested:

- Social and community benefits by creating stronger social and community networks in local areas
- Recreation and leisure demands from the urban population (e.g. allotments)
- Education (e.g. educational gardens, school visits to children's farms)
- Opportunity for short-term land use on otherwise derelict and vacant land in cities
- Achievement of green infrastructure (see below).

Case Study 20.1: Urban Agriculture—An Alternative to Greenfield Expansion

For all of the goals listed above, soil forms the fundamental building block. One important aspect in this is the quality of the soils in our urban areas. It is not accidental perhaps that many important European cities were founded in the most fertile territories (Hillman 2006). Underlying reasons include the need to provide food at times of limited options to transport food over long distances and also the fact that cities often were formed at locations near the main transport corridors of the day: i.e. rivers. The trend of continued urban expansion into fertile agricultural lands should be questioned. Most cities grow outwards and policies of spatial planning typically designate agricultural land as zones for urban expansion. This means a conversion of agricultural greenfield land into urban land and in this process, the loss of capacity for food production is seldom taken into account. More often the decisions where to zone land for new urban development depend on transport infrastructure, land contours or physical convenience (e.g. avoiding the need to cross physical barriers such as rivers or canals). The agricultural land itself does not always have a positive zoning objective in spatial planning policies requiring its protection from a food production perspective and is rather regarded as a 'white zoning' category where no designated use is indicated. This is often the case for lands that are outside urban areas—periurban zones.

20.3.2 Urban Drainage Systems

Drainage systems for urban areas have undergone a radical change over time. From the open sewers of the old days to the twentieth century combined sewers, to separate systems where non-polluted surface water run-off is collected in a system that is separate to polluted (foul) water. In the past decade, attention has been placed on reducing the amounts of surface water run-off that are generated by the growing area of sealed surfaces in combination with the increased occurrence of floods. In Europe, a series of major damaging floods between 1998 and 2009 prompted the EU Floods Directive (2007/60/EC) that was entered into force in 2007 (EC 2016c). Since then, Member States must assess the risk from flooding and take measures to reduce flood risk while facilitating public say in the planning process (EC 2016c). In relation to new construction projects, the current standard is that it should not result in a significant increase in the amount of surface water run-off that is disposed of elsewhere. In order to meet this requirement, new developments should not result in an increase of surface water outflow from the site that was the case prior to development. In flood-prone areas, local authorities may not permit any surface water discharge. Holding facilities need to be provided where necessary but increasingly issues around the ongoing maintenance of such facilities mean that other methods of storm water retention must be integrated into site planning and architectural designs. In response to the challenge of minimising surface water outflow, the development of Sustainable Urban Drainage Systems (SuDS) is being promoted in Ireland since 2005 (Dublin Drainage Consultancy 2005). SuDS have been described as offering a 'total' solution to rainwater management while traditional drainage can be considered as only providing a 'collection and disposal' approach. The integrated approach of SuDS considers land use planning, water quality and quantity, amenity and habitat enhancements and is therefore considered a means to help achieve the objectives of the Water Framework Directive. In the Greater Dublin Area, the relevant policies in this area are contained in the Greater Dublin Strategic Drainage Study (DCC 2009).

The implementation of SuDS has been slow due to: lack of technical expertise in Ireland, no national standards or guidance, challenges of multidisciplinary approaches for delivery, limited applicability on some sites with high water tables/contamination, economic recession and failures at national planning level to integrate the principles of SuDS with development planning policies and decisions. For example, the Residential Density Guidelines (1999) seek to avoid excessive suburbanisation and to promote higher densities in appropriate locations but this is likely, for example, to reduce greenspace and as these are only guidelines they are not subject to environmental assessment.

Although there are policies to promote sustainable drainage systems, such as the Design Manual for Urban Streets these apply only to new developments which require planning consent. There are currently no incentives for retrofitting existing developments to reduce storm water discharge or soil sealing (Harris et al. 2016). Due to the economic downturn, the construction industry had contracted severely, and so the impact of recent policy changes has been delayed. In spatial planning terms, the need to provide attenuation capacity on a site for development is also looked at as an opportunity to provide environmental and urban design benefits. The creation of wetlands, for example, can be integrated with parklands and through 'raingardens' in residential and commercial developments in a highly urbanised context. An example of the latter can be found at Dublin Docklands which is designated as a Strategic Development Zone. This results in many examples where the responsibility for drainage design is combined with a landscape design to promote habitat creation and sustain biodiversity.

The use of exposed (unsealed) soil and the minimisation of sealing are also desirable from a microclimate point of view. Increasingly there is growing concern about the heat island effect that is experienced in urban centres where there are large amounts of hard and reflective surfaces with limited capacity for absorption of heat. Surface water and planting can assist in providing a cooling effect during summertime and perhaps provide shelter from wind during other seasons. The multiple benefits of SuDS are summarised in the Dublin city development plan as follows:

> …SuDS reduces flood risk, improves water quality and provides amenity through the use of permeable paving, swales, green roofs, rainwater harvesting, detention basins, ponds and wetlands. SuDS can achieve multiple objectives such as removing pollutants from urban run-off at source, controlling surface water run-off from developments and ensuring flood risk does not increase further downstream. Furthermore, SuDS offers the opportunity to combine water management with green space, which can increase amenity and biodiversity. Dublin City Council will carry out ongoing maintenance and monitoring of the sustainable drainage systems within the public domain (DCC 2016, p. 131).

Another important consideration for paying attention to soils in the urban environment is the need to maintain biodiversity. Soil improves the capacity to link green areas together. This function can be performed by not sealing soil but also by the creation of green areas on top of hard surfaces. For example, green roofs in combination with green walls can provide linkages between green areas thus improving the ecological quality of an area that may otherwise be highly urbanised.

In the context of construction projects, different categories of soil can be identified. Topsoil performs an important production value and as such forms a material asset that may be (inadvertently) removed during construction. The loss of productive topsoil can be avoided by requiring developers to stockpile topsoil during construction for reuse when the development is completed. This can be conditioned into the planning permission. For example, in planning permissions for development projects conditions may be attached to a building permit requiring the developer to retain and protect the topsoil during construction period so that it can be reused as topsoil within the green areas of the project. In some projects, soils are required to be stored for re-use to protect the native genetic seedbanks of the site prior to construction.

Case Study 20.2: The Emergence of Green Infrastructure

The term 'green infrastructure' is used to reflect the newly recognised high degree of interdependence between man and nature as illustrated by the climate change challenge. Green infrastructure is defined by the European Commission as enhancing Europe's natural capital (EC 2013) specifically:

> …a strategically planned network of natural and semi-natural areas with other environmental features designed and managed to deliver a wide range of ecosystem services such as water purification, air quality, space for recreation and climate mitigation and adaptation. This network of green (land) and blue (water) spaces can improve environmental conditions and therefore citizens' health and quality of life. It also supports a green economy, creates job opportunities and enhances biodiversity. The Natura 2000 network constitutes the backbone of the EU green infrastructure (EC 2016b).

Increasingly the importance of 'green infrastructure' is emphasised in development plans. For example, in the Greater Dublin Area the four development plans all have policies on green infrastructure as mandated by the Regional Planning Guidelines (Dublin and Mideast Regional Authorities, 2010). It is described in the Dublin City Development Plan as follows: …'*Green infrastructure is an interconnected network of green space that conserves natural ecosystem values and functions that also provides associated benefits to the human population. It is a strategically planned network of natural and semi-natural areas with other environmental features designed and managed to deliver a wide range of ecosystem services. It incorporates green spaces (or blue if aquatic ecosystems are concerned) and other physical features in terrestrial (including coastal) and marine areas*' (DCC 2016, p. 144). In the Dublin City Development Plan, reference is made to the fact that a green infrastructure network can provide for carbon and water soakage areas as well as provide quality recreational and biodiversity functions (DCC 2016, p. 49). Equally, integrating green infrastructure into the local planning process is considered essential towards meeting future challenges (DoECLG 2013b). Green infrastructure benefits are also reflected in the term 'ecosystem services' which refers to the many benefits that nature can provide to society.

The recession effects in Dublin are noticeable through paused building projects and the topic of 'temporary land use' has gained significance in urban planning terms. Here it is argued by communities affected by dereliction that not all land uses need to be permanent and that they should benefit from a site on a temporary basis—for example, for urban farming or a local park. Sometimes these activities are referred to as 'pop-up'. Given the temporary nature it may not always be feasible or necessary to plant trees in open soil, instead, it can be considered to contain the soil in containers so that removal at a later stage is easy. Also, possible contamination issues may be avoided and this can be relevant in relation to urban farming.

Urban metabolism is a concept that relates to the movement, circulation and dislocation of material resources—including soil—in the city (Lohrberg et al. 2015). The resources that a city needs and waste materials that a city produces can affect the sustainability of a city by the degree to which these resources come from outside the city boundary and the waste materials require disposal also outside the city boundary. As the emphasis on the circular economy and urban metabolism increases, soil resources will take on a higher significance and attention. There are already many examples of urban metabolism where soil plays an important role. For example, brownfield development instead of greenfield development effectively means that the soil is reused for a different purpose and that the loss of agricultural soil as a food or energy producing resource can be minimised through less conversion of greenfield land into urban use. Another example is the way in which effective composting of green waste can return waste to soil, not only reducing the waste stream but also improving the quality of the soil within the city, through the addition of a nutrient or carbon resource. Rainwater harvesting is an important component of urban metabolism because of the reduced need to cater for large amount of surface water run-off particularly during flood periods. Green roofs form an important component in achieving rainwater harvesting and reducing run-off. In order for green roofs to work effectively, the right types of soil need to be considered. The plants in green roofs capture carbon and clean the air, and these rooftops reduce energy use and help manage urban heat islands. They can also provide islands of habitat for urban wildlife, and even produce food in places where the land not so long ago, once supported farming (EC 2013).

The most common application of soil on green roofs is in the form of low-maintenance roofs with shallow soils or other growing media that are well-suited for low-growing plants like *Sedum*.

20.4 Soils and Extraction

Soil represents a valuable resource, both as a mineral resource such as for the extraction of sand or clay, or as an organic resource, such as industrial peat used as a source of fuel. Although the majority of mineral raw materials are of geological origin, mining can have major impacts on the soil resource. In Ireland, quarries are necessary as sources of construction material. The high transport cost of raw materials such as sand and gravel means that normally quarries are opened in areas where construction is taking place. In an effort to improve the industry reputation, the Irish Concrete Federation adopted an environmental code which sets out good management principles in relation to quarry operations.[1] Ireland has introduced stricter legislation for environmental assessment of quarries under the Planning Act (2010) to comply with EU Environmental Impact Assessment Directive and control the siting of quarries.

However, earlier mining, rock quarrying and sand and gravel extraction have led to scalped land, truncated at the surface. Similar effects, with shallower reach, are found during construction. In Ireland, large areas of peat removal, including complex domestic-scale turbary, and twentieth century industrial milled-peat and moss peat harvesting areas, although totally rural, could potentially be regarded as strip-mined landscapes or anthropogenically impacted soils. These soils are often included in the Technosols category, i.e. urban soils. When the surface is stable, after extraction has finished, these lands can support vegetation, making new ecosystems or will be inundated with water, making new large managed or semi-natural aquatic systems. Since 2009, Ireland has carried out restoration of some of the raised bogs which were drained in the 1980s for peat production under the Bord na Móna Raised Bog Project and post-production rehabilitation is a key objective under its Biodiversity Action Plan (Bord na Móna 2016). Since 2004, Ireland has been conducting demonstration projects with the EU to trial restoration by Coillte of formerly commercially afforested raised bog sites.[2] The first raised bog LIFE project targeted restoration of 14 sites and extension of raised bog habitat area at a cost of €2.5 million[3] with the most recent project (2011–2015) to restore 17 sites at a cost of €2.15 million.[4]

Soil associated with abandoned mine sites in Ireland can be categorised broadly as either pre- or post-mining. The former is essentially the original soil from the area surrounding the mine site that was largely undisturbed by mining. The latter includes soil that has developed naturally on the site itself, soil that has been imported, typically as part of remediation efforts, and fine-grained mine waste; all can serve as a substrate for plants where toxic elements are below critical levels.

Undisturbed soil found around abandoned mine sites may contain anomalously high concentrations of some of the elements contained in the ore. Some of this may be contamination of the soil by dust blows during and after mining, e.g. at Avoca (Gallagher and O'Connor 1997; CDM 2008), but in other cases it is an original, natural geochemical signature in the soil, reflecting the composition of the mineralised bedrock. For example, the Tellus survey has shown that topsoil in the Monaghan–Armagh area in north-east Ireland, has atypically high concentrations of lead (Pb) and zinc (Zn), reflecting widely distributed but very small-scale bedrock mineralisation, associated with brief mining activity in the nineteenth century (Young and Donald 2013; Gallagher et al. 2016). High background levels of Pb and Zn have been reported from the Silvermines area in County Tipperary, one to two orders of magnitude greater than those observed in a national survey, or when compared to the arithmetic mean of unscreened data from a global review of soils (Inter-Agency Group 2000). Discovery of several modern mines and deposits, including Navan mine (Ashton et al. 1986), has followed from the discovery of similar natural soil metal anomalies.

Modern mining is characterised by intensive mechanisation that allows for the extraction of large volumes of ore, both underground and on the surface, and the generation of very large volumes of waste rock. Now-abandoned mine sites developed or redeveloped in Ireland from the second half of the twentieth century onwards—including Avoca, Tynagh, Silvermines and Gortdrum—are large industrial sites where the surface is typically made-ground or composed of mine waste. A thin layer of soil has developed naturally on some parts of these sites, such as at Avoca where native species have gained a foothold at the margins of the site. However, the largest areas of soil are generally the products of remediation efforts, particularly on tailings ponds, where imported soil has been used to grow grass. Remediation efforts have had varying degrees of success, but, in at least some cases, the tailings ponds have been used for grazing animals.

[1]Environmental Code second edition, Irish Concrete Federation 2005.
[2]http://www.coillte.ie/coillteforest/environment/nature_conservation/life_nature_projects/.
[3]http://ec.europa.eu/environment/life/project/Projects/index.cfm?fuseaction=search.dspPage&n_proj_id=2662&docType=pdf.
[4]http://ec.europa.eu/environment/life/project/Projects/index.cfm?fuseaction=search.dspPage&n_proj_id=3838.

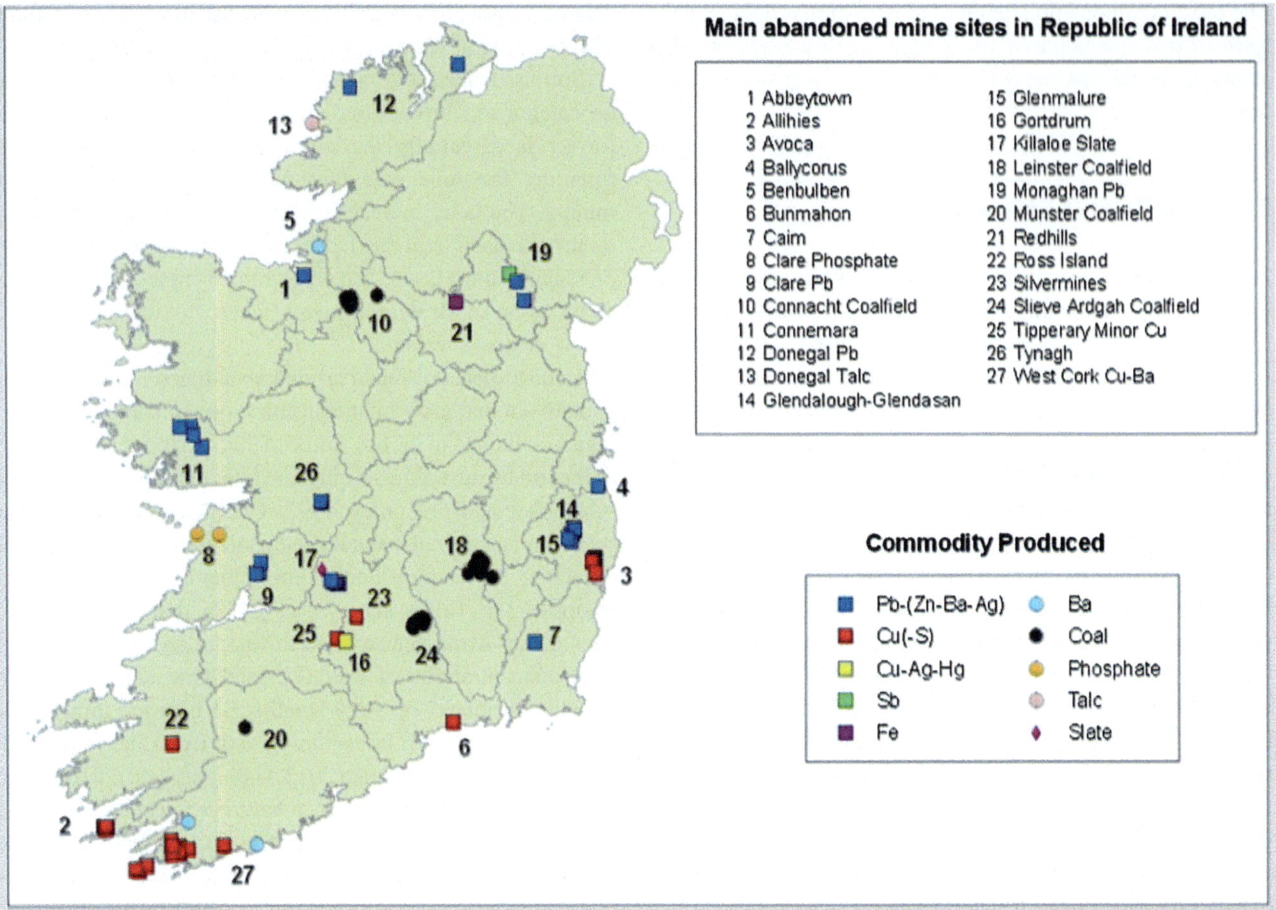

Fig. 20.2 Distribution of abandoned mine sites in Ireland. Figure originally published in Stanley et al. (2010)

Case Study 20.3: Abandoned Mine Site in Ireland

The distribution of abandoned mine sites in Ireland included in an inventory by Stanley et al. (2010) is shown in Fig. 20.2. These include very small sites (typically less than 2 ha in size) that were mined in the nineteenth century or earlier, as well as large twentieth century sites that were subjected to modern, high-volume, industrialised mining operations. The small pre-twentieth century sites were principally mined for metals such as Pb and copper (Cu), and are marked by the presence of waste heaps and dressing floors where the ore was processed by hand. Some of these sites have undergone extensive, partly natural, revegetation and are in use for agriculture, generally as pasture. Analysis of soil immediately below the root zone on such sites, particularly on or close to former dressing floors, has in numerous cases revealed significant enrichment in metals, with percentage levels of Pb and Zn and 1000s of mg/kg of Cu and arsenic (As) recorded (Stanley et al. 2010).

20.5 Conclusion

Soil represents a vital resource. Directly, soil is an important mineral resource representing a source of sand and clay as well and an important organic resource. Indirectly, the soil is a vital resource in the delivery of many soil functions already described in the preceding chapters. In this chapter, we focus more specifically on the role of soil in relation to construction. The emphasis on sustainable spatial planning contained here offers some insight into contemporary thinking around how to address broad soil related concerns, such as food production, soil sealing or water management in an increasingly urbanised context.

References

Ashton JH, Downing DT, Finlay S (1986) The geology of the Navan Zn-Pb orebody. In: Andrew CJ, Crowe RWA, Finlay S, Pennell WM, Pyne JF (eds) Geology and genesis of mineral deposits in Ireland. Irish Association for Economic Geology, Dublin

Bord na Móna (2016) Biodiversity Action Plan (2016–2021)

CDM (2008) Feasibility study for the management and remediation of the Avoca Mining Site

DCC (2009) Greater Dublin strategic drainage study, Dublin City Council

DCC (2016) Dublin City Development Plan 2016–2022, Dublin City Council

Department of Communications, Marine and Natural Resources (DCMNR) (2005) The "Top 55" deposits. Exploration and Mining Division Ireland (online). http://www.mineralsireland.ie/NR/rdonlyres/C18CED87-BE88-48E3-9149-E7AFBF075AF6/0/Top55_2005.pdf. Accessed 29 Nov 2016

DoECLG (2013a) Guidelines for planning authorities and an Bord Pleanála on carrying out Environmental Impact Assessment, Department of the Environment, Community and Local Government, 2013

DoECLG (2013b) Local Area Plans, Guidelines for planning authorities 23, Government of Ireland, (online). http://www.housing.gov.ie/sites/default/files/migrated-files/en/Publications/DevelopmentandHousing/Planning/FileDownLoad,33557,en.pdf. Accessed 12 Dec 2016

Dublin Drainage Consultancy (2005) Greater Dublin strategic drainage study (GDSDS). New Development Policy—Technical Guidance Document, vols. 1–5

EC (2013) Green infrastructure (GI)—enhancing Europe's natural capital. European Commission, Brussels

EC (2016a) Soil sealing, European Commission, Joint Research Centre European Soil Data Centre (ESDAC), (online). http://ec.europa.eu/environment/soil/sealing_guidelines.htm. Accessed 30 Nov 2016

EC (2016b) The forms and functions of green infrastructure, Environment. European Commission (online). http://extranet.novacomm-europa.eu/environment/nature/ecosystems/benefits/index_en.htm. Accessed 30 Nov 2016

EC (2016c) The EU floods directive (online). http://ec.europa.eu/environment/water/flood_risk/. Accessed 09 Dec 2016

EPA (2015) Advice notes for preparing environmental impact statements. Draft, Environmental Protection Agency

EPA (2016) Radon. Environmental Protection Agency, Johnstown Castle Estate, Co. Wexford, online. http://www.epa.ie/radiation/radon/. Accessed 29 Nov 2016

Gallagher V, O'Connor PJ (1997) Characterisation of the Avoca mine site: geology, mining features, history and soil contamination study. GSI technical report MS97/1

Gallagher V, Knights KV, Carey S, Glennon M, Scanlon R (2016) Atlas of Topsoil geochemistry of the Northern Counties of Ireland: data from the Tellus and Tellus Border Projects. Geological Survey of Ireland

Harris M, Hochstrasser T, Foley K, Cave C (2016) Urban biodiversity management in Ireland: capturing the experience of practitioners. In: Francis RA, Millington J, Chadwick MA (eds) Urban landscape ecology: science, policy and practice. Routledge Ltd, London

Hillman J (2006) City and soul (Putman: Spring Publications) as quoted in: Lohrberg et al 2015, p 116

IGI (2002) Geology in environmental impact statements. A guide. Institute of Geologists of Ireland

Inter-Agency Group (2000) Report of the investigation into the presence and influence of lead in the Silvermines area of County Tipperary. Department of Agriculture, Food and Rural Development, Dublin

Lohrberg F, Licka L, Scazzosi L, Timpe A (eds) (2015) Urban agriculture Europe. Jovis Verlag

Mitchell F, Ryan M (1998) Reading the Irish landscape. Town House and Country House, Dublin 6

Prokop G, Jobstmann H, Schönbauer A (2011) Report on best practices for limiting soil sealing and mitigating its effects. European Commission, Brussels. Technical report - 2011 – 050, pp 96–99

RPII (2006) Action plan to identify and remedy Irish Houses with Radon concentrations above the national reference level, Radiological Protection Institute of Ireland

Stanley G, Gallagher V, Ní Mhairtín F, Brogan J, Lally P, Doyle E, Farrell L (2010) Historic mine sites—inventory and risk classification, volume 1: a joint study carried out by The Environmental Protection Agency and The Geological Survey of Ireland. EPA, Wexford

Swartjes FA (ed) (2011) Dealing with contaminated sites—from theory towards practical application. Springer, Berlin

United Nations (2015) World reference base for soil resources 2014—update 2015. Food and Agriculture Organization of the United Nations, Rome

Young M, Donald A (eds) (2013) A guide to the Tellus data. Geological Survey of Northern Ireland, Belfast

Functional Land Management: A Framework for Soil Policy Formation

21

Lilian O'Sullivan, Rachel Creamer, and Rogier Schulte

Abstract

Discussions on soil policies are framed by the *Thematic Strategy for Soil Protection* (COM (2006) 231 final), launched by the European Commission (EC) in 2006. Eight main threats to soil were recognised within the EU. A review by Creamer et al. (Soil Use Manage 26:198–211, 2010) highlighted that the key threats to soil quality relevant to Ireland were loss of soil organic matter, erosion, compaction and contamination. A proposal for a Soil Framework Directive (SFD) (COM (2006) 232 final) was finally withdrawn in May 2014. The removal of the SDF represents an opportunity to refocus attention away from the threats to soil quality towards an approach that emphasises the functional capacity of soil. In Ireland, the concept of Functional Land Management (FLM) has been developed and is currently being applied to other countries across Europe. FLM is a utilitarian framework that specifically tries to optimise the delivery of five soil functions delivered through agricultural landscapes: (1) primary productivity, (2) water purification and regulation, (3) carbon storage and regulation, (4) provision of a habitat for biodiversity and (5) cycling of nutrients. In relation to matching the supply of soil functions with demand, three distinguishing pathways have been identified: (1) Implement a change in a dynamic soil property at local scale, such as nutrient concentrations. (2) Alter a static soil property to enhance one or more functions. In Ireland, this could be the implementation of subsurface land drainage systems to change the soil moisture dynamics of the soil or (3) Land use change, such as from grassland to arable. Functional Land Management does not equate to a legislative tool for the zoning of land for individual functions but rather that individual soil functions may be incentivised through existing instruments available under the EU CAP.

Keywords

Thematic strategy • Functional land management
Policy framework • Soil functions • Threats
Protection

21.1 Introduction

Discussions on soil policies are framed by the *Thematic Strategy for Soil Protection* (COM (2006) 231 final), launched by the European Commission (EC) in 2006. This strategy introduced the concept of soil functions (Chap. 13) and soil threats (EC 2006). Eight main threats to soil were recognised within the EuropeanUnion (EU); (1) erosion, (2) decline in organic matter, (3) local and diffuse contamination, (4) soil sealing, (5) compaction, (6) decline in biodiversity, (7) salinization, and (8) floods and landslides (see Chap. 13). Human activities, such as land management, urban sprawl and industrial activities are considered key drivers of increased pressure on the soil resource across the EU (EC 2012). A review by Creamer et al. (2010) based upon the agricultural management cycle for north Atlantic Europe, highlighted that the key threats to soil quality relevant to Ireland were loss of soil organic matter, erosion, compaction and contamination. A proposal for a Soil Framework Directive (SFD) (COM (2006) 232 final) was designed to provide a legislative basis for soil protection and to unite soil protection measures under one directive to respond to these threats within the EU. However, the SFD

L. O'Sullivan (✉) · R. Creamer · R. Schulte
Teagasc Agriculture and Food Development Authority, Johnstown Castle Environment Research Centre, Wexford, Ireland
e-mail: Lilian.OSullivan@teagasc.ie

L. O'Sullivan
Irish Environmental Protection Agency, Johnstown Castle, County Wexford, Ireland

R. Creamer
Wageningen University and Research Atlas—Wageningen Campus, Wageningen, The Netherlands

R. Schulte
Johnstown Castle, Wageningen University and Research Atlas—Wageningen Campus, Wageningen, The Netherlands

R. Creamer and L. O'Sullivan (eds.), *The Soils of Ireland*, World Soils Book Series,
https://doi.org/10.1007/978-3-319-71189-8_21

was never ratified by member states and was finally withdrawn in May 2014. Like water or air, the soil provides many public goods to society, however unlike water or air, soil is typically privately owned. This and the emphasis of the SFD on threats to soil quality meant that it was met with resistance and elicited a strong negative response among many stakeholders, particularly the farming community.

Despite the withdrawal of the SFD, the EC remains committed to the objective of soil protection. In 2015, the 68th United Nations (UN) General Assembly declared that year the International Year of Soils (IYS), with core objectives that included soil protection and sustainable soil management. Subsequently, the UN Development Programme officially launched the Sustainable Development Goals (SDG) in January 2016. World leaders adopted the 2030 Agenda for Sustainable Development which includes a set of 17 SDGs, four of which specifically cite soil in their title: 2.4, "land and soil quality"; 3.9, "soil pollution and contamination"; 12.4, "management of chemicals and all wastes… and significantly reduce their release to air, water and soil" and 15.3, "By 2030, combat desertification, restore degraded land and soil" (UN 2015). These initiatives indicate a growing recognition of the important role of soil in the delivery of soil functions and ecosystem services essential to sustain life on earth and represent a call to action that the soil resource must be afforded the same protection as air and water quality to guarantee the delivery of goods and services into the future. Also in 2015, the Food and Agriculture Organization published the '*Status of the World's Resources*' report, representing the first ever major global assessment of soil resources and related issues. The report concluded that the majority of the world's resources are in only fair, poor or very poor condition and also, that one-third of land is highly degraded due to erosion, salinisation, compaction, acidification and the chemical pollution of soils (FAO and ITPS 2015). Despite these conclusions, the report offers optimism that the delivery of ecosystem services can be safeguarded through sustainable soil management. In Ireland, the Environmental Protection Agency (EPA) previously determined that Irish soils are in good condition but that they are exceptionally vulnerable to land-use changes, particularly in relation to peat areas (EPA 2002). Although there is growing recognition that soil is a non-renewable resource providing many benefits for society, there is a continued lack of direct legal protection from a policy point of view.

Case Study 21.1: Soil Protection—A Fragmented Approach

As of 2016, no legislation exists at national level in Ireland that specifically targets the protection of the soil resource however there are several national policies that do indirectly offer soil protection. Agriculture is the major land based activity in Ireland with obvious implications for the soil resource. In 2010, the industry-led *Food Harvest 2020* strategy supported by government outlined the agricultural expansion plan following the removal of the EU milk quota in 2015. Targets included an increase in milk production of 50%. In acknowledgement of the increased environmental pressure associated with expansion, soil management was specifically mentioned as a necessary action for environmental sustainability (DAFF 2010). Food Harvest 2020 laid out a platform for growth driven by sustainability relying on knowledge intensification and more efficient farming. Since then, the *Food Wise 2025* strategy has superseded Food Harvest 2020 as the plan for the agricultural sector and insists that any further productivity gains are achieved through sustainable intensification. Within Food Wise 2025 soil health is recognised as a pillar of environmental sustainability with soil fertility and carbon sequestration as important areas for further research and knowledge transfer activities. Many other regulations and national policies have been developed to deliver on this strategy including the Green, Low-Carbon, Agri-Environment Scheme (GLAS), the Second National Biodiversity Plan of 2011 and Good Agricultural Practice for Protection of Waters Regulations (9/2010).

At national level no single authority has responsibility for soil protection and monitoring. This means that protection for soil tends to be fragmented with different agencies and government bodies, each with different objectives, having responsibility for different soil-related issues. Although there are no direct polices for soil at EU level, there are a number of policy statements that directly address soil as well as several policies that indirectly impact soil quality (Fig. 21.1). The Resource Efficiency Roadmap (COM (2011) 571) targets resource efficiency across many policy areas such as transport and raw materials, but in relation to soil and land, milestones include; no net land take by 2050, a reduction in soil erosion by 2050 and an increase in soil organic matter (EC 2011). The Seventh Environment Action Programme (EAP) that guides environmental policy until 2020 notes the challenge of soil degradation and includes a binding legal framework to target soil quality (EU 2013) while the preparation of "Land as a Resource" communication is expected to recognise the finite nature of land and the need for sustainable EU land management.

Several EU policies indirectly address soil and its protection. The Common Agricultural Policy (CAP) launched in 1962 is in 2016 faced with the growing challenges of global food security, climate change, sustainable management of

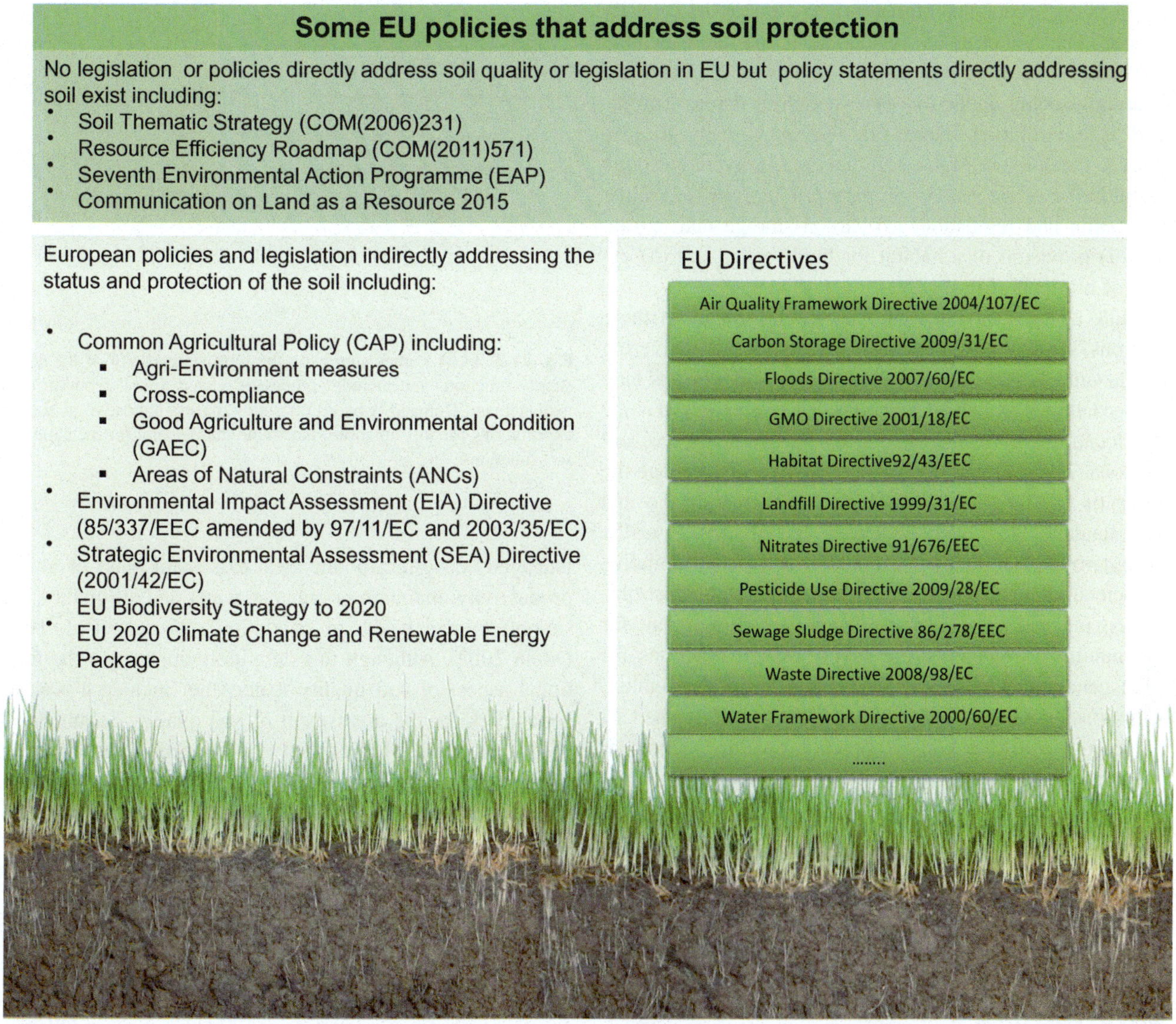

Fig. 21.1 EU policies that address soil protection directly in policy statements or indirectly through policies targeted are other aspects of the environment. An extensive cross-policy analysis of European policies has been conducted by Glæsner et al. (2014)

natural resources, maintaining the countryside and keeping the rural economy alive. Instruments emphasise the economic and environmental competitiveness of farming but increasingly incentives are focused on supporting farmers to adopt sustainable farming methods. For example, Agri-Environment Measures support the build-up of soil organic matter, the reduction of soil erosion, contamination, compaction and an increase in soil biodiversity. Cross-compliance supports standardised mechanisms for farmers to support environmental, public, animal and plant health in lieu of the single-farm payment. Also, Good Agricultural and Environmental Condition (GAEC) support certain aspects of agricultural soil protection and allow member states to specify minimum requirements around soil protection such as minimum soil cover. The Areas of Natural Constraint (ANC) aid scheme aims to conserve the countryside and support farmers where soil and land have a natural constraint. Overtime CAP supports have evolved away from production-linked subsidies towards decoupling and are increasingly targeted towards measures that can facilitate sustainable intensification.

Other policy directives are shown in Fig. 21.1 including the EU Biodiversity Strategy to 2020 that have important implications for the soil resource.

21.2 Functional Land Management

Although the removal of the SFD may be perceived as a strong lack of political will, its removal has created space to think differently about soils. It represents an opportunity to refocus attention away from the threats to soil quality towards an approach that emphasises the functional capacity of soil. The concept of Functional Land Management (FLM) was developed in Ireland by Schulte et al. (2014) and is currently being applied to other countries across Europe. FLM is a utilitarian framework that specifically tries to optimise the delivery of five soil functions delivered through agricultural landscapes: (1) primary productivity, (2) water purification and regulation, (3) carbon storage and regulation, (4) provision of a habitat for biodiversity and (5) cycling of nutrients (Fig. 21.2).

While all soils have the capacity to perform all of these functions, some parts of the landscape are better at delivering certain 'suites' of soil functions based upon soil type and land use (see Chap. 13). Not all functions can be maximised at the same location, but many synergies exist between the five soil functions. The challenge therefore is how best to manage the supply of soil functions to meet societal demands. To that end, demand for soil functions is shaped by the agri-environmental policy framework which is representative of societal demands and expectations for soil functions. Implicit in meeting societal demands is a requirement for sustainability to guarantee the supply of soil functions for future generations. This relies on maintaining soil quality. Definitions of soil quality have previously been described as, "the capacity of the soil to function, within natural or managed ecosystem boundaries, to sustain plant or animal productivity, maintain or enhance water and air quality, and support human health or habitation" (Karlen et al. 1997; Doran 2002). Although this definition emphasises the functional aspect of soil quality it does not include a scale or benchmark for the assessment of soil quality. Schulte et al. (2015) proposed "The capacity of a specific kind of soil to provide functions *to meet demands within*…" to allow a degree of elasticity to be included so that a benchmark for soil assessment can be included. This means that soil quality no longer only refers to soil properties but could also vary through a change in demand. Importantly, this allows better assessment of whether demands (framed as policy objectives) can be met by supply (soil biophysical criteria and land use). It also means that the supply of soils functions or specific functions need not be maximised but that a more balanced process of optimising soil functions can be deployed subject to the balances of supply and demand (Fig. 21.3).

Fig. 21.2 Illustrative representation of the suite of soil functions proposed by Schulte et al. (2014). The white box represents primary production; blue, water purification; black, carbon storage and regulation; green, provision of a habitat for biodiversity; purple, cycling of nutrients. Figure adapted from Fig. 1 as originally published in Schulte et al. (2014), http://dx.doi.org/10.1016/j.envsci.2013.10.002

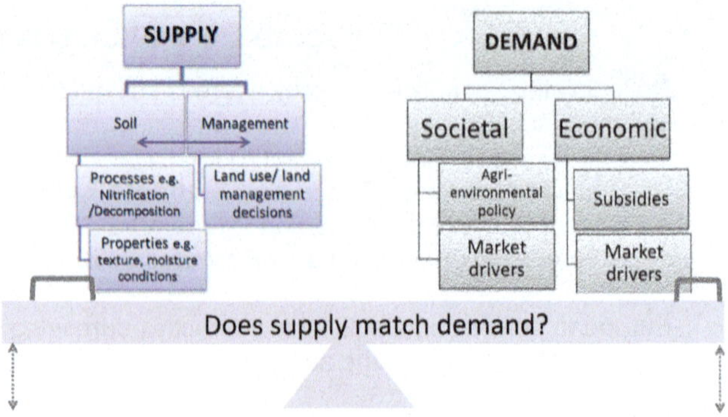

Fig. 21.3 FLM frames supply as the intrinsic capacity of the soil to supply functions and includes processes related to soil properties and management. Demand is based on policy drivers indicative of societal expectations for soil functions and frame the demand for the supply of soil functions

To further explore the FLM concept, the spatial variability and patterns of supply of each of the soil functions were mapped at Irish national level by Schulte et al. (2015). In the absence of a SFD an inventory of EU policies was defined to highlight the relevant policy drivers which were used to frame the demand for soil functions (Fig. 21.4).

When supply and demand are mapped, the results showed a high degree of granularity in relation to the supply of soil functions. While some regional variation emerged the supply of soil functions is primarily defined by local soil and land use characteristics (Fig. 21.5, top row). In contrast, demand for soil functions mapped for Ireland (Fig. 21.5, bottom row) highlighted spatial variation at small scale (primary productivity) to regional scale (nutrient cycling) and national scale (carbon sequestration).

EU Policy Driver	Function(s) of relevance[1]	Scale at which demand is framed
Common Agricultural Policy (CAP)	P	EU
Areas of Natural Constraint (ANC)	P	EU / National
Greening Measures	C H	Ubiquitous
Nitrates Directive	W N	Ubiquitous / Regional
Water Framework Directive	W H	River Basin District
Habitat & Birds Directive	H	Multiple scales
Agri-Environmental Schemes	W C H	Farm / regional
EU 2030 Climate and Energy Framework	C	EU + National targets
Sewage Sludge Directive	N	Regional

1. P = primary productivity, W = water purification, C = carbon sequestration, H = habitat provision, N = nutrient cycling

Fig. 21.4 Inventory of EU policy drivers that frame demand for the five soil functions, shown by the coloured boxes and the spatial scale to which each of these policies apply from Schulte et al. (2015). Figure as originally published in Schulte et al. (2015), http://doi.org/10.3389/fenvs.2015.00081

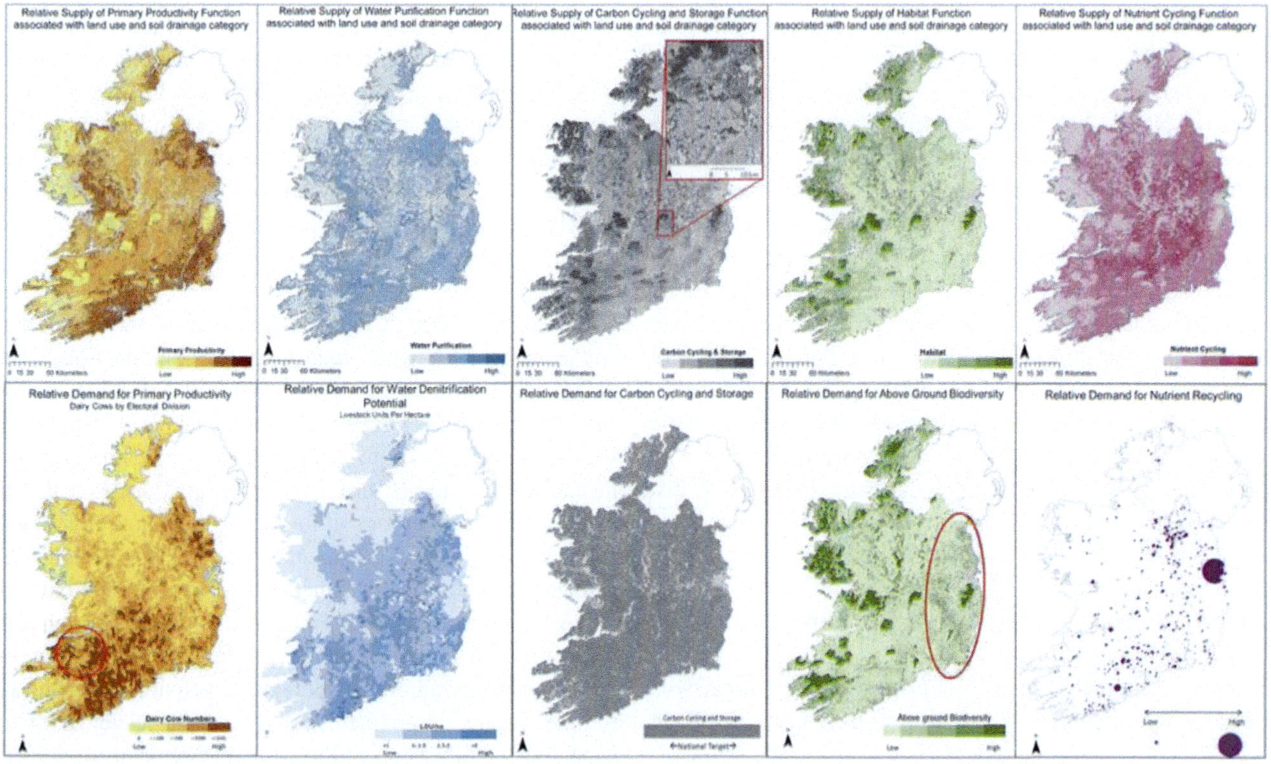

Fig. 21.5 Indicative maps of the normalised supply (top row) and demand (bottom row) for the five soil functions, from left to right: primary productivity, water purification, carbon sequestration, carbon sequestration, biodiversity and nutrient cycling from Schulte et al. (2015). Figure as originally published in Schulte et al. (2015), http://doi.org/10.3389/fenvs.2015.00081

A comparison between supply and demand shows that in relation to primary productivity, supply exceeds demand in most regions. For the most part, the demand for increased productivity aligns to those parts of the country that are associated with good agricultural soils. Exceptions to this can be found in the south-west which is characterised by poorly drained soils. To increase the supply of primary productivity in this area would require the installation of drainage systems.

To quantify the supply of the water purification function, we have defined denitrification rate as a proxy indicator. Based on this indicator the supply shows to be high in all regions, adequately meeting the demand that groundwater nitrate concentrations remain below 50 mg l^{-1} consistent which findings that a high proportion of land in Ireland is compliant with demand (Byrne and Fanning 2015).

In relation to the carbon sequestration and storage function, the supply varies greatly between soil type, land use types, and management. Demand is shown as one colour as demand for the carbon sequestration function applies only at national scale. Previously, the EU 2030 Climate and Energy Framework only specified an EU-wide target but Ireland now has a national emissions reduction target of 30% relative to the 2005 emissions.

In relation to the supply of the habitat function, designation of Natura 2000 sites has been met but the introduction of greening measures under the CAP will require an increase in this function on arable land through the allocation of Ecological Focus Areas which are traditionally associated with a low supply of this function (Schulte et al. 2015).

The demand for nutrient cycling soil function is strongly regional but low relative to the supply of this function due to the low pig and human populations in Ireland. The size of the circles represents the total area required to dispose of these nutrients. The circles represent the ideal scenario where landowners are willing to import manure but in reality accessing spread lands might require travelling long distances (Schulte et al. 2015).

21.2.1 Future Management Options

In relation to matching the supply of soil functions with demand, three distinguishing pathways have been identified:

1. Implement a change in a dynamic soil property at local scale, such as nutrient concentrations. Best management practices should allow the productivity soil function to be increased without impacting on the other soil functions. In Ireland, as few as 10% of agricultural soils have optimum pH, phosphorus and potassium concentrations (Wall et al. 2015) highlighting the opportunity to improve productivity of Irish soils. Similarly, the increased use of animal manure in arable systems will improve the carbon sequestration and nutrient cycling functions.
2. Alter a static soil property to enhance one or more functions. In Ireland, this could be the implementation of subsurface land drainage systems to change the soil moisture dynamics of the soil. This results in longer growing and grazing seasons (Tuohy et al. 2015). While this represents an increase in the primary production soil function, this may come at the expense of the water purification function (Jahangir et al. 2012) or the carbon storage functions (Kechavarzi et al. 2010). Hence, the merit of such interventions must be considered on a case by case basis.
3. Land use change, such as from grassland to arable. As with pathway two, this option typically incurs a trade-off between soil functions. In this example, ploughing for arable conversion is likely to increase the primary productivity function but is simultaneously likely to reduce the supply of the other functions. Importantly, not all trade-offs are negative and even in this case, enhancing primary production might be appropriate where there is spare capacity for the water purification function. This might represent a better alternative than expansion of the production platform into areas where the supply of other functions such as carbon storage or biodiversity is high, which could occur in areas where the land is of high nature value.

The spatial scale of the demand driver is another important consideration when managing for soil functions. The scale will determine to what extent soil functions can be offset between soils or regions. A challenge in managing for soil functions relates to their private ownership and that management, and in turn supply, is at a local scale. In contrast, the demand for soil functions may range from local, regional, national or continental scales. For example, the demand for water quality function as framed by the Nitrates Directive insists that water quality is maintained everywhere. Thus, this soil function cannot therefore be traded between fields, farms or regions, where one location could make up for the failure of another area to provide clean drinking water. At the other end of the scale, the demand for soils to sequester carbon applies ultimately at the global scale with management necessary at the national scale. The nature of this larger scale allows a degree of off-setting between soils or regions. In practice, this means that meeting national targets might not be best achieved at a local scale by every farmer off-setting their emissions through land management,

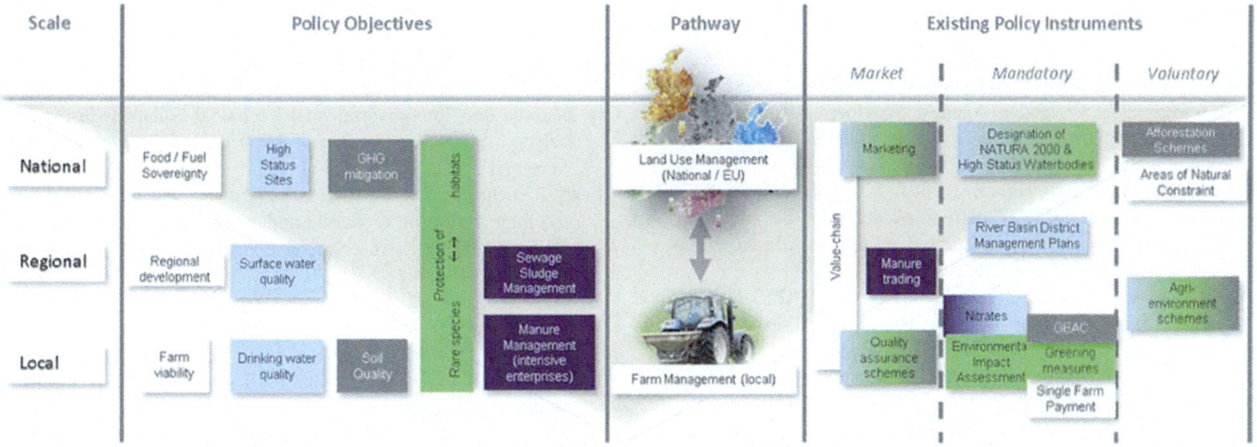

Fig. 21.6 Inventory of policy objectives and policy instruments of relevance to the management of soil functions (Schulte et al. 2015). Figure as originally published in Schulte et al. (2015), http://doi.org/10.3389/fenvs.2015.00081

such as planting forestry, but rather by incentivising forestry on those soils which are less suitable for the primary production of food. In principle, the globalised nature of food production similarly suggests some off-setting potential of the primary productivity soil function. Notably, there are advantages associated with having regional or national food access due to the reliance of global food systems on an often fragile geopolitical context.

In relation to biodiversity and nutrient cycling soil functions, these largely apply at an intermediate scale. The demand and supply for nutrient cycling relies primarily on transport considerations (Fealy and Schröder 2008) limiting the option to match supply and demand at larger scales. The demand for the biodiversity function cuts across multiple scales such as the protection of rare species at local level, entire habitats at regional scales and the protection of species throughout their natural range at national scale (Noss 1990).

These discrepancies in relation to the supply and demand for soil functions associated with scale have implications for soil and land management: some soil functions must be managed at local field and farm scale, whilst others may be offset between regions with a view to meeting national or continental targets.

Functional Land Management does not equate to a legislative tool for the zoning of land for individual functions but rather that individual soil functions may be incentivised through existing instruments available under the EU CAP. Figure 21.6 from Schulte et al. (2015) shows these instruments categorised into the type of instrument (market, mandatory, voluntary) alongside their policy objective and the scale of application.

This inventory indicates that policy makers already have a wide range of instruments to incentivise soil and land management to meet policy objectives. Some of the policy tools are 'joined up' and can readily address multiple soil functions, such as the cross-compliance, GAEC and greening requirements for the single farm payment scheme. There are however a number of relevant instruments that operate in isolation of each other, such as the re-delineation of ANC. Also, EIA, while mandatory for large-scale land management interventions for drainage systems currently only considers the habitat function, whereas a more holistic approach to soil functions such as proposed by FLM would also consider the impact of drainage on the carbon sequestration function (O'Sullivan et al. 2015) and the other soil functions. In this way, FLM offers considerable possibility as a policy integrator for sustainability with potential to utilise existing policy instruments for FLM governance.

It is important too that most instruments do not take account for differences between soils in their capacity to supply soil functions. This is likely due to this historical tendency to maximise the delivery of soil function(s) simultaneously. The availability of land is and will be an increasing limiting factor and therefore soil data is essential to account for differences between soil types and to allow for the optimisation of soil functions. While the maps presented in this chapter demonstrate the granularity in the supply of soil functions they have been developed at the 1:250,000 scale. This means that while they can aid large scale geographical patterns, they are not appropriate for farm scale. Thus, local management decisions require knowledge of local soil type which can be acquired through direct observations.

References

Byrne C, Fanning A (eds) (2015) Water quality in Ireland 2010-2012. Environmental Protection Agency, Wexford. Available at: http://www.epa.ie/pubs/reports/water/waterqua/wqr20102012/WaterQualityReport.pdf. Accessed 02 Nov 2016

Creamer RE, Brennan F, Fenton O, Healy MG, Lalor STJ, Lanigan GJ, Regan JT, Griffiths BS (2010) Implications of the proposed Soil Framework Directive on agricultural systems in Atlantic Europe—a review. Soil Use Manage 26:198–211

Department of Agriculture, Fisheries and Food (DAFF) (2010) Food harvest 2020: a vision for Irish agri-food and fisheries. DAFF, Dublin

Doran JW (2002) Soil health and global sustainability: translating science into practice. Agric Ecosyst Environ 88:119–127

European Commission (EC) (2006) Communication from the Commission to the Council, the European Parliament, the European Economic and Social Committee of the Regions, Thematic Strategy for Soil Protection. COM (2006) 231 final, EC, Brussels. Available at: http://eur-lex.europa.eu/legal-content/EN/TXT/PDF/?uri=CELEX:52006DC0231&from=EN. Accessed: 28 Oct 2016

EC (2011) Communication from the Commission to the European Parliament, the Council, the European Economic and Social Committee and the Committee of the Regions. Roadmap to a Resource Efficient Europe. COM (2011) 571 final. EC, Brussels. Available online: http://eur-lex.europa.eu/legal-content/EN/TXT/PDF/?uri=CELEX:52011DC0571&from=EN. Accessed: 01 Nov 2016

EC (2012) Report from the Commission to the European Parliament, the Council, the European Economic and Social Committee and the Committee of the Regions. The implementation of the Soil Thematic Strategy and on-going activities. COM (2012) 46 final. EC, Brussels. Available at: http://eur-lex.europa.eu/legal-content/EN/TXT/PDF/?uri=CELEX:52012DC0046&from=EN. Accessed: 28 Oct 2016

EC (2016) Proposal for a Regulation of the European Parliament and the Council on the inclusion of greenhouse gas emissions and removals from land use, land use change and forestry into the 2030 climate and energy framework and amending Regulation No 525/2013 of the European Parliament and the Council on a mechanism for monitoring and reporting greenhouse gas emissions and other information relevant to climate change COM(2016) 479 final 2016/0230 (COD). EC. Brussels. Available at: https://ec.europa.eu/transparency/regdoc/rep/1/2016/EN/1-2016-479-EN-F1-1.PDF. Accessed: 01 Nov 2016

Environmental Protection Agency (EPA) (2002) Towards setting environmental quality objectives for soil: developing a soil protection strategy for Ireland. EPA, Ireland. Available online: http://www.epa.ie/pubs/reports/land/EPA_soil_discussion.pdf. Accessed 28 Oct 2016

European Union (EU) (2013) Decision No. 1386/2013/EU of the Parliament and of the Council of 20 November 2013 on a general Union Environment Action Programme to 2020 'Living well, within the limits of our planet'. Official Journal of the European Union. Available at: http://eur-lex.europa.eu/legal-content/EN/TXT/PDF/?uri=CELEX:32013D1386&from=EN. Accessed 01 Nov 2013

FAO and ITPS (2015) Status of the World's Soil Resources (SWSR)—Main Report. Food and Agriculture Organization of the United Nations and Intergovernmental Technical Panel on Soils, Rome, Italy. Available online: http://www.fao.org/3/a-i5199e.pdf. Accessed 28 Oct 2016

Fealy R, Schröder J (2008) Assessment of manure transport distances and their impact on economic and energy costs. In: Proceedings of the international fertiliser society, vol 642. International Fertiliser Society, Colchester

Glæsner N, Helming K, de Vries W (2014) Do current european policies prevent soil threats and support soil functions? Sustainability 6:9538–9563

Jahangir MMR, Johnston P, Khalil MI, Hennessy D, Humphreys J, Fenton O, Richards KG (2012) Groundwater: a pathway for terrestrial C and N losses and indirect greenhouse gas emissions. Agric Ecosyst Environ 159:40–48. https://doi.org/10.1016/j.agee.2012.06.015

Karlen DL, Mausbach MJ, Doran JW, Cline RG, Harris RF, Schuman GE (1997) Soil quality: a concept, definition, and framework for evaluation. Soil Sci Soc Am J 61:4–10

Kechavarzi C, Dawson Q, Bartlett M, Leeds-Harrison PB (2010) The role of soil moisture, temperature and nutrient amendment on CO_2 efflux from agricultural peat soil microcosms. Geoderma 154:203–210

Noss RF (1990) Indicators for monitoring biodiversity: a hierarchical approach. Conserv Biol 4:355–364

O'Sullivan L, Creamer RE, Fealy R, Lanigan G, Simo I, Fenton O, Carfrae J, Schulte RPO (2015) Functional Land Management for managing soil functions: a case-study of the trade-off between primary productivity and carbon storage in response to the intervention of drainage systems in Ireland. Land Use Policy 47:42–54

Schulte RPO, Creamer RE, Donnellan T, Farrelly N, Fealy R, O'Donoghue C, O'hUallachain D (2014) Functional land management: a framework for managing soil-based ecosystem services for the sustainable intensification of agriculture. Environ Sci Policy 38:45–58

Schulte RPO, Bampa F, Bardy M, Coyle C, Creamer RE, Fealy R, Gardi C, Bahadur Ghaley B, Jordan P, Laudon H, O'Donoghue C, Ó'hUallacháin D, O'Sullivan L, Rutgers M, Six J, Toth GL, Vrebos D (2015) Making the most of our land: managing soil functions from local to continental scale. Front Environ Sci 3:81

Tuohy P, Fenton O, Holden NM (2015) The effects of treading by two breeds of dairy cow with different live weights on soil physical properties, poaching damage and herbage production on a poorly drained clay-loam soil. J Agric Sci 153:1424–1436

Wall DP, Plunkett M, Murphy P (2015) National soil fertility trends, in Proceedings of the Fertiliser Association of Ireland Spring Scientific Meeting 2015 Publication 50, pp 21–24. Available online at: http://www.fertilizer-assoc.ie/wp-content/uploads/2015/04/Proc-No-50-2015-FINAL.pdf. Accessed: 02 Nov 2016

United Nations (2015) Sustainable development goals. United Nations, New York. Available online: https://sustainabledevelopment.un.org/sdgs. Accessed: 28 Oct 2016

Index

R. Creamer and L. O'Sullivan (eds.), *The Soils of Ireland*, World Soils Book Series,
https://doi.org/10.1007/978-3-319-71189-8

The manufacturer's authorised representative in the EU is Springer Nature Customer Service Centre GmbH, Europaplatz 3, 69115 Heidelberg, Germany. If you have any concerns regarding our products, please contact ProductSafety@springernature.com

Printed and bound by CPI Group (UK) Ltd, Croydon, CR0 4YY
01/10/2026
02225991-0002